Sustainable Material for Biomedical Engineering Application

Wan Safwani Wan Kamarul Zaman •
Asma Abdullah Nurul
Editors

Sustainable Material for Biomedical Engineering Application

Springer

Editors
Wan Safwani Wan Kamarul Zaman
Department of Biomedical Engineering
Faculty of Engineering
Universiti Malaya
Kuala Lumpur, Malaysia

Centre for Innovation in Medical
Engineering (CIME)
Department of Biomedical Engineering
Faculty of Engineering
Universiti Malaya
Kuala Lumpur, Malaysia

Asma Abdullah Nurul
School of Health Sciences
Universiti Sains Malaysia
Kubang Kerian, Kelantan, Malaysia

ISBN 978-981-99-2266-6 ISBN 978-981-99-2267-3 (eBook)
https://doi.org/10.1007/978-981-99-2267-3

This Springer imprint is published by the registered company Springer Nature Singapore Pte Ltd.
The registered company address is: 152 Beach Road, #21-01/04 Gateway East, Singapore 189721, Singapore

Preface

Sustainability is an important aspect in biomedical engineering. In general, biomedical engineering is a multidisciplinary field which is interconnected through various networks of basic and applied sciences as well as commercialization and business. From material perspectives, sustainable materials and its derivatives are required to ensure reproducibility and give cost-effective manufacturing solutions in the biomedical engineering industry. Recent advancement in material research has shown to have great impact on biomedical and clinical applications. It is well known that materials with various characteristics and techniques are being applied to allow ease of modification but not necessarily able to provide an optimal and sustainable platform for biomedical engineering. Hence, more sustainable options are required.

Some modifications to the materials are required to enhance their characteristics and usability for various applications. However, such modification may affect its research discovery and clinical application. In addition, there are also various strategies and methodologies in materials fabrication and manufacturing that may contribute to outcome variations. This has contributed to some of the challenges in the advancement of materials technology which requires comprehensive and structured studies, more standardized and optimal methods of design and fabrication. In this regard, sustainability of materials in terms of its source and reproducibility may be able to a certain extent overcome these challenges.

This book, *Sustainable Material for Biomedical Engineering Application* discusses interdisciplinary approaches in the development of selected materials and their derivatives that are sustainable for biomedical engineering application. With potential for sustainability, the materials discussed and illustrated in this book may have the ability to increase and contribute to wider therapeutic options for patients. Some of the main aspects covered in this book are utilization of human waste, food waste, and green technology approach for materials in biomedical engineering applications such as tissue engineering, 3D printing, and biosensing. Apart from that, potential issues associated with ethics and regulatory issues are also included. This book is intended for academicians, researchers, students, and industrial players in the field of materials and biomedical engineering. To accomplish our objective, a team of experts from various disciplines have agreed to share their expertise in providing details consisting of in-depth study, analysis, and reviews with different scientific backgrounds that are mostly from Malaysia. This book also intended to showcase some of the works using locally synthesized materials that are carried out in Malaysia.

In this book, the first four chapters are dedicated to a more general and introductory perspective of sustainability materials in biomedical engineering. The next four chapters, Chaps. 5 and 6 discuss specifically on marine-based sustainable materials. While Chaps. 7 and 8 discuss selected human tissue-based sustainable materials. Conversely, Chaps. 9 and 10 discuss selected potential microbial-based sustainable materials. Chapters 11–21 are dedicated to some of the applications of sustainable materials in biomedical engineering. These applications include 3D bioprinting, biosensing, tissue engineering, stroke therapy, wound healing, and dentistry. The final two chapters, Chaps. 22 and 23, discuss some of the ethical and regulatory issues of materials usage in biomedical application.

We would like to express our sincere gratitude and appreciation to all authors who are committed and agreed to contribute and assist us in preparing this book. We hope that this book will be of benefit to everyone in the field.

Kuala Lumpur, Malaysia — Wan Safwani Wan Kamarul Zaman
Kelantan, Malaysia — Asma Abdullah Nurul

Contents

Editors and Contributors

About the Editors

Wan Safwani Wan Kamarul Zaman is currently a senior lecturer in the Department of Biomedical Engineering, Faculty of Engineering, University of Malaya, Malaysia. She is also a registered pharmacist under the Pharmacy Board of Malaysia. Her research interest is in regenerative medicine and tissue engineering with specific interest in human mesenchymal stem cells (MSCs) and its biosafety and bioefficacy, MSCs biocompatibility in bioscaffold, and the application of sustainable biomaterial in tissue engineering. She is an active reviewer for a number of national and international journals. She has more than 10 years of teaching experience in human anatomy and physiology, biochemistry and analysis, and communication and thinking skills. She has published more than 50 research articles in peer-reviewed international journal and authored or co-authored several book chapters. She is currently serving as an executive member of the Tissue Engineering and Regenerative Medicine Society of Malaysia (TESMA). She is also a member of the Malaysian Pharmaceutical Society (MPS) and a member of Society of Medical and Biological Engineering Malaysia (MSMBE). She has also served as an advisory committee member for several conferences.

Asma Abdullah Nurul is an Associate Professor and the Head of the Biomedicine Programme at the School of Health Sciences, University Science Malaysia. She has more than 10 years of academic teaching experience in molecular biology, immunology, microbiology, and stem cell and regenerative medicine. She has been engaged in multidisciplinary research which involves biomedicine, material engineering, and clinical research in the field of tissue engineering and regenerative medicine. Her main current research interests include understanding the effects of immunomodulation on the molecular mechanism of osteogenic regeneration, development of biomaterials for hard-tissue engineering (dental, bone, and cartilage), and cell-based and cell free therapy in regenerative medicine. She is also an active reviewer for a number of international journals, as well as a reviewer for university and national grants. She has published over 60 full peer-reviewed papers in the related field and authored or co-authored several book chapters. In professional bodies, she is an active member in various national and international societies. She is currently serving as an executive committee member of the Tissue Engineering and Regenerative Medicine Society Malaysia and the Malaysia Biomaterial Society.

Contributors

Muhammad Aa'zamuddin Ahmad Radzi Department of Obstetrics and Gynaecology, Sultan Ahmad Shah Medical Centre @IIUM, Kuantan, Pahang, Malaysia

Abdul Manaf Abdullah School of Mechanical Engineering, College of Engineering, Universiti Teknologi MARA, Shah Alam, Selangor, Malaysia

Hasan Zuhudi Abdullah Department of Manufacturing Engineering, Faculty of Mechanical and Manufacturing Engineering, Universiti Tun Hussein Onn Malaysia, Batu Pahat, Johor, Malaysia

Asma Abdullah Nurul School of Health Sciences, Universiti Sains Malaysia, Kubang Kerian, Kelantan, Malaysia

Suhana Ahmad Department of Immunology, School of Medical Sciences, Universiti Sains Malaysia, Kota Bharu, Malaysia

Nik Aizah Tissue Engineering Group (TEG), National Orthopaedic Center of Excellence for Research & Learning (NOCERAL), Department of Orthopaedic Surgery, Faculty of Medicine, Universiti Malaya, Kuala Lumpur, Malaysia

Ibrahim N. Amirrah Centre for Tissue Engineering and Regenerative Medicine, Faculty of Medicine, Universiti Kebangsaan Malaysia, Cheras, Kuala Lumpur, Malaysia

Anusha Wei Asohan School of Biological Science, Universiti Sains Malaysia, Gelugor, Penang, Malaysia

Mohammad Firdaus Bin Abdul Aziz Centre for Law and Ethics Law in Science and Technology (CELEST), Faculty of Law, Universiti Malaya, Kuala Lumpur, Malaysia

Yazmin Bustami School of Biological Science, Universiti Sains Malaysia, Gelugor, Penang, Malaysia

Minghooi Chai Department of Biomedical Engineering, Faculty of Engineering, Universiti Malaya, Kuala Lumpur, Malaysia

Siti Khadijah Dermawan Department of Manufacturing Engineering, Faculty of Mechanical and Manufacturing Engineering, Universiti Tun Hussein Onn Malaysia, Batu Pahat, Johor, Malaysia

Loai A. Elfawy Centre for Tissue Engineering and Regenerative Medicine, Faculty of Medicine, Universiti Kebangsaan Malaysia, Cheras, Kuala Lumpur, Malaysia

Nur Izzah Md Fadilah Centre for Tissue Engineering and Regenerative Medicine, Faculty of Medicine, Universiti Kebangsaan Malaysia, Cheras, Kuala Lumpur, Malaysia

Mh. Busra Fauzi Centre for Tissue Engineering and Regenerative Medicine, Faculty of Medicine, Universiti Kebangsaan Malaysia, Cheras, Kuala Lumpur, Malaysia

Nur Fatiha Ghazalli Basic and Medical Sciences Unit, School of Dental Sciences, Universiti Sains Malaysia, Kota Bharu, Kelantan, Malaysia

Zuratul Ain Abdul Hamid School of Materials & Mineral Resources Engineering, Engineering Campus, Universiti Sains Malaysia, Nibong Tebal, Penang, Malaysia

Tan Qing Hao Department of Biomedical Engineering, Faculty of Engineering, Universiti Malaya, Kuala Lumpur, Malaysia

Kien Hui Chua Department of Physiology, Faculty of Medicine, Universiti Kebangsaan Malaysia, Kuala Lumpur, Malaysia

Angela Ng Min Hwei Centre for Tissue Engineering and Regenerative Medicine, Faculty of Medicine, Universiti Kebangsaan Malaysia, Kuala Lumpur, Malaysia

Fatimah Ibrahim Department of Biomedical Engineering, Faculty of Engineering, Universiti Malaya, Kuala Lumpur, Malaysia
Centre for Innovation in Medical Engineering (CIME), Department of Biomedical Engineering, Faculty of Engineering, Universiti Malaya, Kuala Lumpur, Malaysia
Centre for Printable Electronics, Universiti Malaya, Kuala Lumpur, Malaysia

Suriani Binti Ibrahim Department of Mechanical Engineering, Faculty of Engineering, Universiti Malaya, Kuala Lumpur, Malaysia

Maizlinda Izwana Idris Department of Manufacturing Engineering, Faculty of Mechanical and Manufacturing Engineering, Universiti Tun Hussein Onn Malaysia, Batu Pahat, Johor, Malaysia

Ros Akmal Idris Department of Immunology, School of Medical Sciences, Universiti Sains Malaysia, Kota Bharu, Malaysia

Yanny Marliana Baba Ismail Biomaterials Niche Group, School of Materials and Mineral Resources Engineering, Engineering Campus, Universiti Sains Malaysia, Penang, Malaysia

Muhamad Zaki Jaffri Department of Manufacturing Engineering, Faculty of Mechanical and Manufacturing Engineering, Universiti Tun Hussein Onn Malaysia, Batu Pahat, Johor, Malaysia

Nurshafiqah Jasme School of Biological Science, Universiti Sains Malaysia, Gelugor, Penang, Malaysia

Jegalakshimi Jewaratnam Department of Chemical Engineering, Faculty of Engineering, Center for Separation Science & Technology (CSST), Universiti Malaya, Kuala Lumpur, Malaysia

Siti Fatimah Noor Mat Johar Reconstructive Sciences Unit, School of Medical Sciences, Universiti Sains Malaysia, Kota Bharu, Kelantan, Malaysia

Nik Nur Farisha Nik Md Noordin Kahar School of Materials & Mineral Resources Engineering, Engineering Campus, Universiti Sains Malaysia, Nibong Tebal, Penang, Malaysia

Tunku Kamarul Tissue Engineering Group (TEG), National Orthopaedic Center of Excellence for Research & Learning (NOCERAL), Department of Orthopaedic Surgery, Faculty of Medicine, University of Malaya, Kuala Lumpur, Malaysia
Advanced Medical and Dental Institute (AMDI), Universiti Sains Malaysia, Kepala Batas, Pulau Pinang, Malaysia

Salmah Binti Karman Department of Biomedical Engineering, Faculty of Engineering, Universiti Malaya, Kuala Lumpur, Malaysia

Chan Chow Khuen Department of Biomedical Engineering, Faculty of Engineering, Universiti Malaya, Kuala Lumpur, Malaysia

Zhi-Tang Wisely Koay Tissue Engineering Group (TEG), National Orthopaedic Center of Excellence for Research & Learning (NOCERAL), Department of Orthopaedic Surgery, Faculty of Medicine, Universiti Malaya, Kuala Lumpur, Malaysia

Lidawani Lambuk Present Address: Department of Immunology, School of Medical Sciences, Universiti Sains Malaysia, Kota Bharu, Malaysia

Jia Xian Law Centre of Tissue Engineering and Regenerative Medicine, Faculty of Medicine, Universiti Kebangsaan Malaysia, Kuala Lumpur, Malaysia

Yogeswaran Lokanathan Centre for Tissue Engineering and Regenerative Medicine, Faculty of Medicine, Universiti Kebangsaan Malaysia, Kuala Lumpur, Malaysia

Qi Hao Looi My Cytohealth Sdn Bhd, D353A, Lab 6, DMC Level 2, Hive 5, Taman Teknologi MRANTI, Bukit Jalil, Kuala Lumpur, Malaysia

Norhayati Luddin Prosthodontic Unit, School of Dental Sciences, Health Campus, Universiti Sains Malaysia, Kota Bharu, Kelantan, Malaysia

Manira Maarof Centre for Tissue Engineering and Regenerative Medicine, Faculty of Medicine, Universiti Kebangsaan Malaysia, Cheras, Kuala Lumpur, Malaysia

Nik Ahmad Nizam Nik Malek Department of Biosciences, Faculty of Science, Universiti Teknologi Malaysia, UTM, Johor Bahru, Malaysia

Syafira Masri Centre for Tissue Engineering and Regenerative Medicine, Faculty of Medicine, Universiti Kebangsaan Malaysia, Cheras, Kuala Lumpur, Malaysia

Juan Matmin Department of Chemistry, Faculty of Science, Universiti Teknologi Malaysia, UTM, Johor Bahru, Malaysia

Zawani Mazlan Centre for Tissue Engineering and Regenerative Medicine, Faculty of Medicine, Universiti Kebangsaan Malaysia, Cheras, Kuala Lumpur, Malaysia

Veithesswary Mogan Postgraduate Unit, School of Dental Sciences, Universiti Sains Malaysia, Kota Bharu, Kelantan, Malaysia

Dasmawati Mohamad School of Dental Sciences, Universiti Sains Malaysia, Kubang Kerian, Kelantan, Malaysia

Nur Farahiyah Binti Mohammad Faculty of Electronic Engineering & Technology Universiti Malaysia Perlis, Kampus Alam UniMAP, Pauh Putra Arau, Perlis, Malaysia
Medical Device and Life Science Cluster, Sport Engineering Research Centre (SERC), Universiti Malaysia Perlis, Kampus Alam UniMAP, Pauh Putra, Arau, Perlis, Malaysia

Rohimah Mohamud Department of Immunology, School of Medical Sciences, Universiti Sains Malaysia, Kota Bharu, Malaysia

Myat Myat-Htun Department of Physics, Bago University, Bago, Myanmar

Nashrul Fazli Mohd Nasir Faculty of Electronic Engineering & Technology Universiti Malaysia Perlis, Kampus Alam UniMAP, Pauh Putra Arau, Perlis, Malaysia
Medical Device and Life Science Cluster, Sport Engineering Research Centre (SERC), Universiti Malaysia Perlis, Kampus Alam UniMAP, Pauh Putra, Arau, Perlis, Malaysia

Sook-Luan Ng Department of Craniofacial Diagnostics and Biosciences, Faculty of Dentistry, Universiti Kebangsaan Malaysia, Kuala Lumpur, Malaysia

Marzuki Omar School of Dental Sciences, Universiti Sains Malaysia, Kubang Kerian, Kelantan, Malaysia

Siti Nurhadis Che Omar Department of Physiology, Faculty of Medicine, Universiti Kebangsaan Malaysia, Kuala Lumpur, Malaysia

Farah Amna Othman School of Health Sciences, Health Campus, Universiti Sains Malaysia, Kubang Kerian, Kelantan, Malaysia

Jing-Yit Pua Tissue Engineering Group (TEG), National Orthopaedic Center of Excellence for Research & Learning (NOCERAL), Department of Orthopaedic Surgery, Faculty of Medicine, Universiti Malaya, Kuala Lumpur, Malaysia

Nur Amalia Ra'oh Department of Ophthalmology, Faculty of Medicine, Universiti Kebangsaan Malaysia, Kuala Lumpur, Malaysia

Harishini Rajaratinam School of Health Sciences, Universiti Sains Malaysia, Kubang Kerian, Kelantan, Malaysia

Nasrul Anuar Abd Razak Department of Biomedical Engineering, Faculty of Engineering, Universiti Malaya, Kuala Lumpur, Malaysia

Siti Khatijah Abdul Razak Department of Immunology, School of Medical Sciences, Universiti Sains Malaysia, Kota Bharu, Malaysia

Mohd Riza Mohd Roslan Faculty of Electronic Engineering & Technology Universiti Malaysia Perlis, Kampus Alam UniMAP, Pauh Putra Arau, Perlis, Malaysia
Department of Engineering and Built Environment, Tunku Abdul Rahman University of Management and Technology (TAR UMT), Penang Branch Campus, Tanjong Bungah, Pulau Pinang, Malaysia

Nur Liyana Mohd Rosli Biomaterials Niche Group, School of Materials and Mineral Resources Engineering, Engineering Campus, Universiti Sains Malaysia, Penang, Malaysia

Atiqah Salleh Centre for Tissue Engineering and Regenerative Medicine, Faculty of Medicine, Universiti Kebangsaan Malaysia, Cheras, Kuala Lumpur, Malaysia

Nusaibah Sallehuddin Centre for Tissue Engineering and Regenerative Medicine, Faculty of Medicine, Universiti Kebangsaan Malaysia, Cheras, Kuala Lumpur, Malaysia

Asmaa' Mohd Satar School of Health Sciences, Health Campus, Universiti Sains Malaysia, Kubang Kerian, Kelantan, Malaysia

Munirah Sha'ban Department of Physical Rehabilitation Sciences, Kulliyyah of Allied Health Sciences, International Islamic University Malaysia (IIUM), Kuantan, Pahang, Malaysia

Nur Aina Syafiqah Suhaimi Department of Biomedical Engineering, Faculty of Engineering, Universiti Malaya, Kuala Lumpur, Malaysia

Nazmul Huda Syed School of Health Sciences, Universiti Sains Malaysia, Kubang Kerian, Kelantan, Malaysia

Safa Tahri Centre for Tissue Engineering and Regenerative Medicine, Faculty of Medicine, Universiti Kebangsaan Malaysia, Cheras, Kuala Lumpur, Malaysia

Sik-Loo Tan Tissue Engineering Group (TEG), National Orthopaedic Center of Excellence for Research & Learning (NOCERAL), Department of Orthopaedic Surgery, Faculty of Medicine, Universiti Malaya, Kuala Lumpur, Malaysia

Suat Cheng Tan School of Health Sciences, Health Campus, Universiti Sains Malaysia, Kubang Kerian, Kelantan, Malaysia

Lee Te Chuan Department of Production and Operation Management, Faculty of Technology Management and Business, Universiti Tun Hussein Onn Malaysia, Batu Pahat, Johor, Malaysia

Maheswary Thambirajoo Centre for Tissue Engineering and Regenerative Medicine, Faculty of Medicine, Universiti Kebangsaan Malaysia, Cheras, Kuala Lumpur, Malaysia

Paveethra Thegarathah Department of Chemical Engineering, Faculty of Engineering, Center for Separation Science & Technology (CSST), Universiti Malaya, Kuala Lumpur, Malaysia

Vieralynda Vitus Department of Biomedical Engineering, Faculty of Engineering, Universiti Malaya, Kuala Lumpur, Malaysia
Centre for Innovation in Medical Engineering (CIME), Department of Biomedical Engineering, Faculty of Engineering, Universiti Malaya, Kuala Lumpur, Malaysia

Roswanira Abdul Wahab Department of Chemistry, Faculty of Science, Universiti Teknologi Malaysia, UTM, Johor Bahru, Malaysia

Wan Safwani Wan Kamarul Zaman Department of Biomedical Engineering, Faculty of Engineering, Universiti Malaya, Kuala Lumpur, Malaysia
Centre for Innovation in Medical Engineering (CIME), Department of Biomedical Engineering, Faculty of Engineering, Universiti Malaya, Kuala Lumpur, Malaysia

Wan Syamimi Wan Kamarul Zaman Institute of Biological Sciences, Faculty of Science, Universiti Malaya, Kuala Lumpur, Malaysia

Farinawati Yazid Department of Family Oral Health, Faculty of Dentistry, Universiti Kebangsaan Malaysia, Kuala Lumpur, Malaysia

Norhayati Yusop Basic and Medical Sciences Unit, School of Dental Sciences, Universiti Sains Malaysia, Kota Bharu, Kelantan, Malaysia

Mohamad Hamdi Zainal-Abidin Department of Chemistry, Faculty of Science, Universiti Teknologi Malaysia, UTM, Johor Bahru, Malaysia

Norita Mohd Zain Department of Biomedical Engineering, Faculty of Engineering, Universiti Malaya, Kuala Lumpur, Malaysia

Ehsan Zeimaran Department of Biomedical Engineering, Faculty of Engineering, Universiti Malaya, Kuala Lumpur, Malaysia

Izzat Zulkiflee Centre for Tissue Engineering and Regenerative Medicine, Faculty of Medicine, Universiti Kebangsaan Malaysia, Cheras, Kuala Lumpur, Malaysia

Utilisation of Human "Wastes" as Materials in Biomedical Engineering Application

1

Vieralynda Vitus, Fatimah Ibrahim, and Wan Safwani Wan Kamarul Zaman

Abstract

The human body produces various types of waste every day as part of a normal physiological mechanism. Single individual body waste is manageable, but a population of human wastes is challenging. The accumulation of body wastes may negatively affect the environment and human health. Therefore, it requires proper waste management. In light of this, the valorisation of human waste is proposed. Thus, research on various human wastes was explored, especially as biomaterials for biomedical applications. Furthermore, the use of human wastes offers a sustainable process that complies with the Sustainable Development

V. Vitus
Department of Biomedical Engineering, Faculty of Engineering, Universiti Malaya, Kuala Lumpur, Malaysia

Centre for Innovation in Medical Engineering (CIME), Department of Biomedical Engineering, Faculty of Engineering, Universiti Malaya, Kuala Lumpur, Malaysia

F. Ibrahim
Department of Biomedical Engineering, Faculty of Engineering, Universiti Malaya, Kuala Lumpur, Malaysia

Centre for Innovation in Medical Engineering (CIME), Department of Biomedical Engineering, Faculty of Engineering, Universiti Malaya, Kuala Lumpur, Malaysia

Centre for Printable Electronics, Universiti Malaya, Kuala Lumpur, Malaysia
e-mail: fatimah@um.edu.my

W. S. Wan Kamarul Zaman (✉)
Department of Biomedical Engineering, Faculty of Engineering, Universiti Malaya, Kuala Lumpur, Malaysia

Centre for Innovation in Medical Engineering (CIME), Department of Biomedical Engineering, Faculty of Engineering, Universiti Malaya, Kuala Lumpur, Malaysia
e-mail: wansafwani@um.edu.my

W. S. Wan Kamarul Zaman, A. A. Nurul (eds.), *Sustainable Material for Biomedical Engineering Application*, https://doi.org/10.1007/978-981-99-2267-3_1

Goals of the United Nations, particularly Goals 3 and 12. Hence, several potential human wastes including hair, nails, urine, adipose tissue and post-birth waste are reviewed in this chapter.

Graphical Abstract

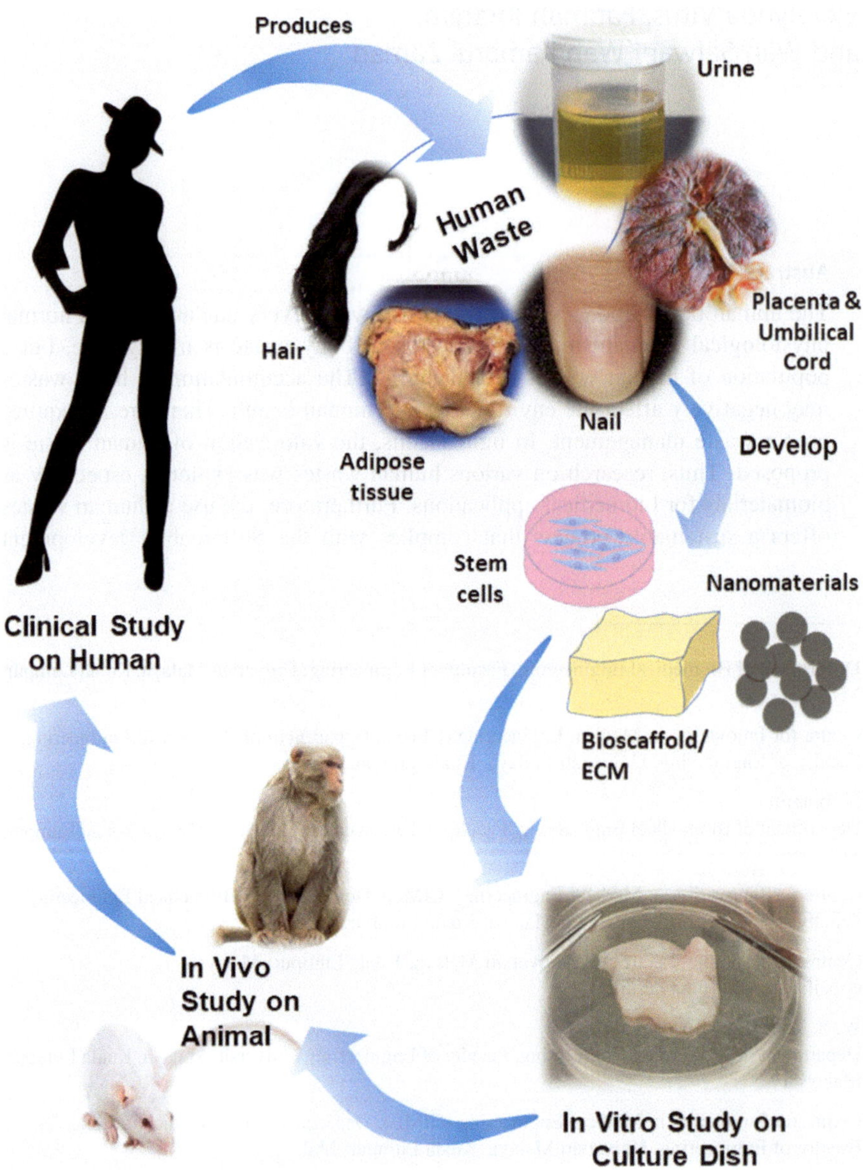

An illustration of the valorisation pathway of human wastes

Keywords

Human wastes · Biomaterials · Biomedical applications · Sustainability

Abbreviations

3D	3-dimensional
β-TCP	Beta-tricalcium phosphate
ADSCs	Adipose-derived stem cells
APDs	Asparagus diet-derived pee dots
ATMPs	Advanced therapy medicinal products
BAT	Brown adipose tissue
bFGF	Basic fibroblast growth factor
BMP4	Bone morphogenetic protein-4
BMSCs	Bone marrow-derived mesenchymal stem cells
CAGR	Compound annual growth rate
CFR	Code of federal regulations
CGT	Cell and gene therapy
CPDs	Vitamin C supplemented diet-derived pee dots
DAPI	4′,6-diamidino-2-phenylindole
DGRG	Drug registration guidance document
EC	European commission
ECM	Extracellular Matrix
EDS	Glutamic acid-aspartic acid-serine
EGF	Epidermal growth factor
EMA	European Medicine Agency
FCNDs	Fingernail-derived fluorescent carbon dots
FDA	Food and drug administration
FITC	Fluorescein isothiocyanate
GVDH	Graft-versus-host-disease
H&E	Haematoxylin and Eosin
hAEC	Human amniotic epithelial cells
hAMSCs	Human amniotic mesenchymal stem cells
hCMSC	Human chorionic mesenchymal stem cells
hCSC	C-kit+ human cardiac stem cells
hCTC	Human chorionic trophoblastic cells
HGF	Hepatocyte growth factor
HLA	Human leukocyte antigen
IDE	Investigational device exemption
IGF1	Insulin growth factors-1
IND	Investigational New Drug Application
ISO	International Organization for Standardization
LDV	Leucine-aspartic acid-valine
MDS/AML	Myelodysplastic syndrome/acute myeloid leukemia

MEF	Mouse embryonic fibroblast
MSCs	Mesenchymal stem cells
NGF	Nerve growth factor
NPRA	National pharmaceutical regulatory agency
NSCDs	Nitrogen and sulphur co-doped carbon dots-derived fingernails
PAL	Japanese Pharmaceutical Affairs Law
PDGF	Platelet-derived growth factor
PDs	Pee dots
SC-CO_2	Supercritical carbon dioxide
SDGs	Sustainable developments goals
SEM	Scanning electron microscopy
T0	Time before application
T1	Sixth month
T2	Twelfth month
TGF-b1	Transforming growth factor beta-1
TRITC	Tetramethylrhodamine
UCDs	Urine-derived carbon dots
UCMSCs	Umbilical cord-derived mesenchymal stem cells
UN	United Nations
UPDs	Unmodified diet-derived pee dots
US	United States
USCs	Urine-derived stem cells
USD	United states dollar
VEGF	Vascular endothelial growth factor
VSMC	Vascular smooth muscle cells
WAT	White adipose tissue
WJ	Wharton's jelly
WJMSCs	Wharton's jelly-derived mesenchymal stem cells

1.1 Introduction

In 2015, the United Nations (UN) members developed the 2030 Agenda for Sustainable Development, which focuses on sustainability. In its agenda, the UN established and expects that the 17 Sustainable Development Goals (SDGs) will be achieved by 2030 (United Nations (NA) 2021). Therefore, sustainability has become a crucial factor in research. Human waste is an excellent natural source of materials for various applications. Nevertheless, the accumulation of human wastes has the potential for environmental and health hazards. Therefore, attention given to human waste management is essential. Thus, developing functional products that are value-added from human wastes may become a sustainable process in compliance with the SDGs, which are Goals 3 and 12 in particular (United Nations (NA) 2021).

SDG 12 emphasises "sustainable consumption and production that improves the socioeconomics benefits of resources used while minimising the impacts." Exclusively, the use of human wastes is in line with the SDG targets of 12.2, 12.4 and 12.5, which focus on (1) "sustainable management and efficient use of natural resources"; (2) "promoting the environmentally sound management of chemicals and all wastes throughout their life cycle, following agreed international frameworks, and significantly reducing their release into the air, water and soil to minimise their adverse impacts on human health and the environment"; and (3) "stressing on the effort to reduce waste generation through prevention, reduction, recycling and reuse" (United Nations (NA) 2021).

Furthermore, using human wastes as materials to develop functional products promotes a cost-effective and sustainable approach to improving scientific research and development. Regarding this approach, industry players, mainly in the biomedical sector, can foster this effort to cultivate more sustainable industrialisation. In addition, the development of innovative technologies from wastes to value-added medical technology for biomedical applications will improve treatment strategies for addressing the current growing trend of diseases. Besides that, the intervention of cost-effective materials with simple fabrication procedures will reduce the expenditure burden in biomedical research. Thus, the development of economic medical technology will be accessible and affordable to low-income communities. This will minimise the burden of diseases and improve the general well-being and quality of community life. In this regard, the development of functional products for biomedical applications of human wastes is consistent with SDG 3, which focuses on "ensuring healthy lives and promoting well-being for all at all ages" (United Nations (NA) 2021). Hence, this chapter reviews several human wastes as sustainable materials and discusses their commercialisation potentials.

1.2 Types of Human Wastes and Their Potential Applications

1.2.1 Human Hair

1.2.1.1 Introduction to Human Hair

Human hair is an essential part of our body. Other than its aesthetic value, it also offers protection. Typically, human hair is mainly made up of α-keratin materials similar to nails, horns and wool in mammals (Yu et al. 2017a). There are three types of human hair. First, lanugo hair, the hair that shed soon after birth. Second is vellus hair, which is soft and non-pigmented. Lastly, longer, coarser and more pigmented hair is called terminal hair. Nevertheless, the vellus hair can develop into terminal hair under the influence of androgens, especially on the pubic area, beard, trunk and axillae (Hinds and Thomas 2008). Morphologically, human hair fibres have three layers: the cuticle, cortex and medulla (Fig. 1.1) (Kaneta et al. 2021; Fellows et al. 2022). The outermost layer, known as the cuticle, comprises several flattened overlapping layers of cuticle cells, which are approximately 0.5–1.0 μm thick (Fellows et al. 2022). The cuticle protects the inner cell structure, i.e., the cortex,

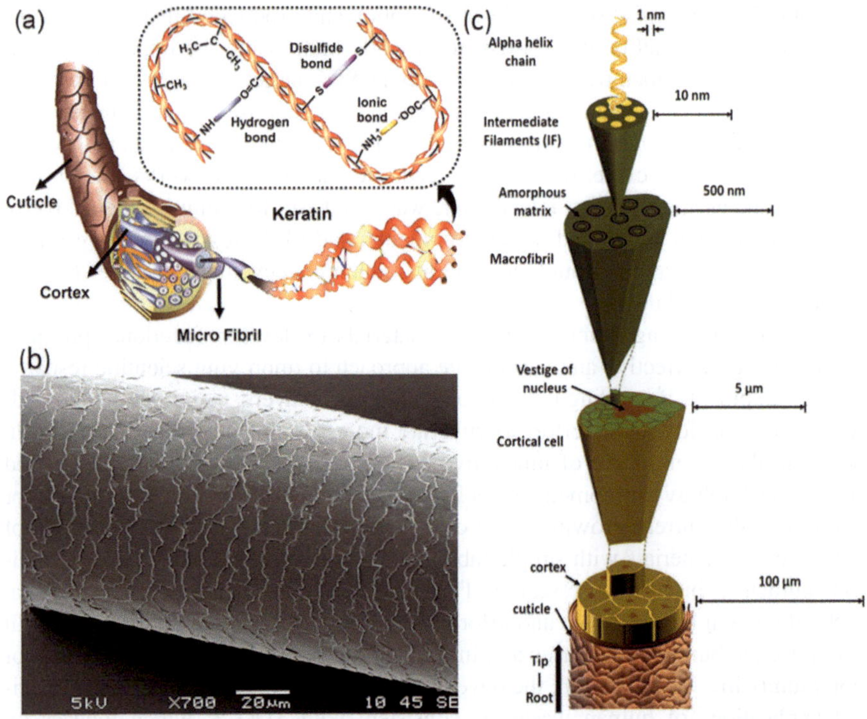

Fig. 1.1 (**a**) Schematic of keratin from human scalp hair. (Reproduced with Ko et al. (2017) with permission from the American Chemical Society). (**b**) Scanning electron microscopy (SEM) image of the hair shaft. The outer surface of the hair shaft is covered by cuticle layers. (Reproduced with Adav et al. (2018) with permission from Scientific Reports). (**c**) Schematic hierarchical structure in the human hair fibre. (Reproduced from Yu et al. (2017b) with permission from Acta Biomaterialia)

from damage caused by environmental factors such as shampoo chemicals or UV light caused by exposure to sunlight (Rogers 2019). The cortex, the main component of human hair mass, contains the most pigments that provide hair colour (Cortellini et al. 2019). The pigment is known as melanin. Further, the complex form of the cortex, which is made up of a filamentous structure, provides mechanical strength to hair (Park et al. 2018). Lastly, the innermost layer of human hair, known as the medulla, comprises air-filled vacuoles and a fibrillar structure surrounded by a cell membrane complex (Fellows et al. 2021).

1.2.1.2 Biomedical Application of Human Hair

Human hair is a biomass waste that is abundant and readily available. Human hair and its derivatives have shown positive outcomes in multiple areas of tissue engineering. A study showed that the high mechanical strength and elasticity of human hair fibres provided an excellent biomaterial that could support the continuous contraction of cardiac cells in a long-term (100 days) culture (Xiao et al. 2020). Human hair-derived keratin possesses cell-binding motifs, including leucine-aspartic acid-valine (LDV) and glutamic acid-aspartic acid-serine (EDS) binding residues, which allow for excellent cell-matrix adhesion and attachment (Agarwal et al. 2019; Gao et al. 2019a). Meanwhile, a specific protein called keratin in human hair has piqued researchers' interest in using it for the fabrication of scaffolds in tissue engineering (Fan et al. 2018; Kim et al. 2017; Veerasubramanian et al. 2018).

Meanwhile, the human hair-derived keratose scaffold promotes the differentiation of c-kit+ human cardiac stem cells (hCSC) to vascular smooth muscle cells (VSMC) in vitro by increasing the expression of VSMC genes and appears morphologically in the form of "endothelial cell tube-like." In vivo study of the keratose scaffold promotes angiogenesis and recovery of vascular tissue in a critical limb ischemia mouse model (Ledford et al. 2017). Besides, human hair-derived keratin promotes cell adhesion, migration and proliferation. Also, it induces neurotrophic factors secretion on Schwann cells and reduces four pro-inflammatory gene expressions in RAW264.7 cells (Gao et al. 2019a). In addition, the developed keratin sponges positively promote axon regeneration in the sciatic nerve crush injury model in vivo (Gao et al. 2019a). Moreover, recombinant keratin-based wound dressings could promote the proliferation and migration of HaCat keratinocyte cells in vitro. Meanwhile, the in vivo study of the dressing promotes wound healing and does not show significant organ inflammation and systemic toxicity (Gao et al. 2019b). These show the potential of keratin as a wound-dressing strategy.

Furthermore, hair-derived keratin is an excellent biomaterial for developing drug carriers because of its biocompatibility, biodegradability, and non-toxicity. For example, rod-shaped keratin nanoparticles loaded with berberine were shown to provide a sustained release of the drug in in vitro studies (Silva et al. 2022). Meanwhile, hair-derived keratin nanoparticles encapsulated with Ru-Qr promote normal liver cell viability and proliferation while inducing cell death in HeLa cells. These indicate keratin's ability to provide nutrients to normal cells while releasing a drug that induces cancer cell death (Kunjiappan et al. 2018). In addition, hair-derived keratin films infected with Trichophyton rubrum were reported to be suitable models for onychomycosis studies (Valkov et al. 2020).

Interestingly, human hair has long been used in traditional Chinese herbs as charcoal drugs for treating astringents and halting bleeding (Chen et al. 2019). Furthermore, human hair was reported as a good carbon source after being subjected to pyrolysis. During pyrolysis, all non-carbon elements were removed, leaving only a high yield of carbon structure (Pramanick et al. 2016). By subjecting human hair to the pyrolysis method, a good electrically conductive material can be obtained that contains the desired properties in bioscaffold development for tissue engineering and biomedical applications (Vitus et al. 2021). Therefore, using alternative carbon precursor materials such as human hair has attracted the attention of various researchers. Unfortunately, there are no recorded clinical trials that use products produced from human hair waste. However, both in vitro and in vivo studies have shown the feasibility of human hair products in medical applications.

1.2.1.3 Advantages and Limitations of Human Hair Waste

Human hair possesses good mechanical strength and elasticity that can uphold the pressure of contracting cells, such as cardiac cells. Plus, it can be sterilised by autoclave and used as a suture material. Meanwhile, hair-derived keratin possesses a cell-binding motif similar to proteins in the extracellular matrix and promotes cell development, including cardiac, neural and skin cells. Thus, keratin is used to construct a bioscaffold that acts as a promising drug carrier and delivery mechanism. Moreover, human hair is a source of inexpensive, easily accessible and readily available carbon precursors. However, the utilisation of human hair waste imposes several disadvantages. For instance, the effluent discharge from the human hair process results in groundwater pollution (Gupta 2014). Besides that, different individuals may be exposed to various pollutants that affect the composition of their hair, including heavy metals, pesticides, and non-biological molecules (Iglesias-Gonzalez et al. 2020). Thus, this could pollute the development of functional products from human hair, particularly bioscaffolds as cells or tissue growth platforms. On the other hand, this can also cause variation in functional hair development, leading to differences in the outcomes of cell or tissue growth and bioscaffold.

1.2.2 Nail

1.2.2.1 Introduction to Human Nail

The nail is one of several human appendages that are continuously generated throughout life and located in the dorsal region of the fingertips and toes. Nails protect the distal phalanx of the hands and feet. They also serve as an attack-defence mechanism, as well as for scratching with aesthetic value. The nail unit comprises a nail plate with a surrounding structure that includes a nail bed, nail matrix, hyponychium and proximal nail fold (Fig. 1.2) (Munteanu et al. 2021). The nail plate is a keratinised structure of compact onychocytes with a smooth longitudinal ridge surface. Meanwhile, the nail bed, which resides under the nail plate and is part

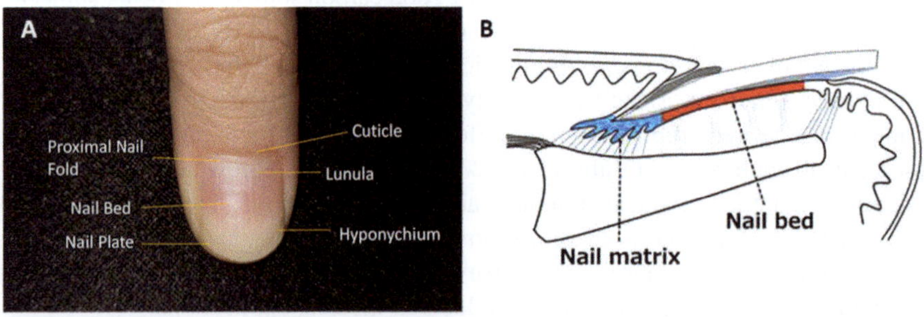

Fig. 1.2 Nail anatomy. (**A**) Gross anatomy of the human fingernail unit. (**B**) Schematic illustration of the side view image of the fingernail. (Reproduced with Munteanu et al. (2021) with permission from Elsevier)

of the nail matrix, provides attachment to the nail plate through longitudinal epidermal ridges. The complementary ridges between the nail plate and the nail bed support the nail's free edge (Johnson et al. 2021). Cutting the nail's free edge is part of body care to remove excess nails. It is a necessary routine as the free edge of the nail could potentially cause self-harm or harm to others through scratching, as the tip of the nail is quite sharp. Additionally, the free edge of nails is a potential carrier of diseases such as worm infections (Pane and Andayani 2020). Averagely, people cut their nails once a week as a routine for nail hygiene (Küçük and Günay 2016). Thus, this provides a continuous source of potential materials for biomedical applications.

1.2.2.2 Biomedical Application of Human Nail Waste

Currently, the application of human nail waste is understudied, but studies have shown the potential of the human nail in a few biomedical areas. For example, fingernail-derived fluorescent carbon dots (FCNDs) pose non-significant toxicity towards four cell lines (HEK-293, HeLa, A549 and MDA-MD-231). Additionally, FCNDs promoted significantly higher cell proliferation towards HEK-293 cells. Furthermore, imaging studies show that the FCNDs were internalised by the cells, found to be localised in the perinuclear region and were capable of emitting fluorescence images of the cells. Furthermore, FCNDs did not induce significant haemolysis and blood coagulation (Chatzimitakos et al. 2018).

A similar study showed that nitrogen and sulphur-co-doped carbon dots derived from fingernails (NSCDs) possessed good photostability and excellent biocompatibility with different cell lines. Additionally, cells could uptake NSCDs and reside in the cell nucleus, cell membrane, and cytoplasmic area. Furthermore, fluorescent images of NSCDs displayed red (330–380 nm), blue (465–495 nm), and green (515–555 nm) emissions under ultraviolet light illumination (Fig. 1.3). Thus, NSCDs could be applied for bioimaging and localisation purposes (Ni et al. 2019). Moreover, the decellularised nail bed scaffold promoted osteogenesis of bone marrow-derived mesenchymal stem cells (BMSCs) in vitro and in vivo bone regeneration in rat models with critical calvarial defects (Yu et al. 2020). Overall, the present study involving human nail waste is still in its early stages. In fact, no clinical trials using human nail waste products have been recorded on ClinicalTrials.gov. However, the present study has shown the potential of fluorescence carbon dots, derived from the human nail in vivo, for bioimaging applications.

1.2.2.3 Advantages and Limitations of Human Nail Waste

Like human hair, the human nails are made up of alpha keratin, thus providing a source of keratin similar to human hair. However, the difficulty in obtaining sufficient human nails for keratin extraction and analysis is one of the limitations of using nails in biomedical applications (Valkov et al. 2020).

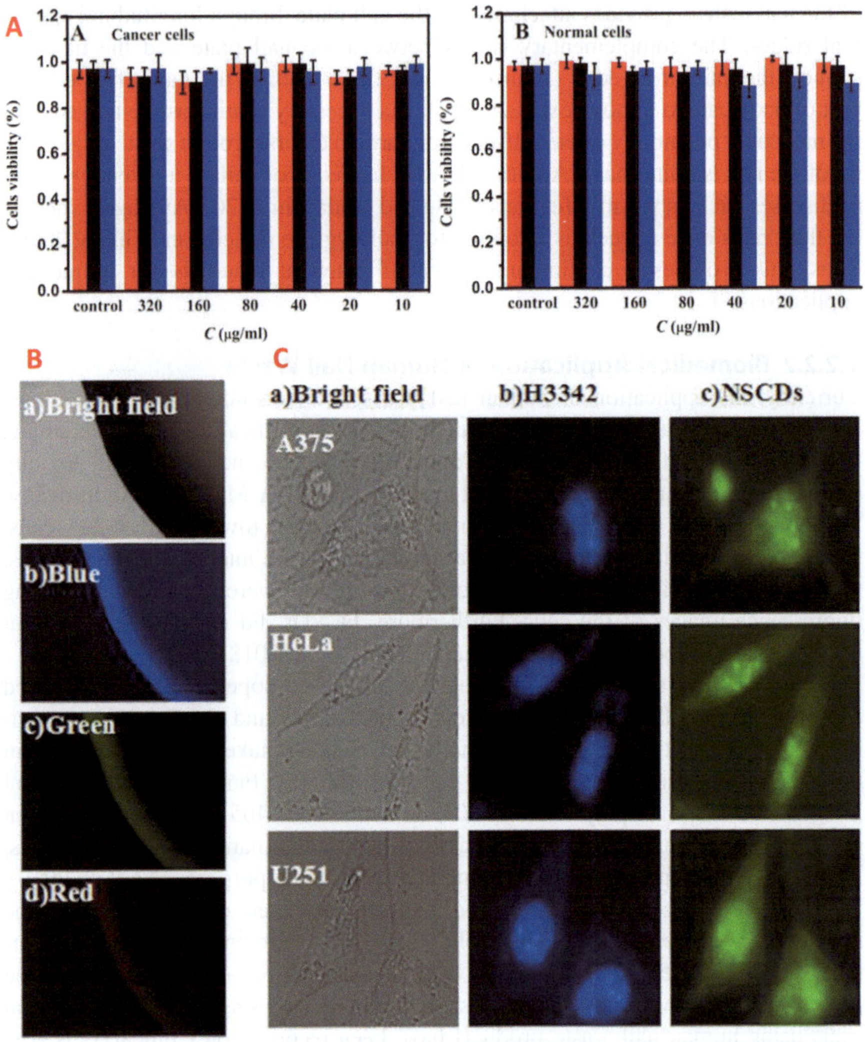

Fig. 1.3 (**A**) Cells viability percentage treated with NSCDs after 72 h. (A) Cancer cells: Red is A375 cells, Black is MCF-7 cells and Blue is A549 cells. (B) Normal cells: Red is CHEM-5 cells, Black is LO2 cells, and Blue is HUVEC. (**B**) Fluorescent image of NSCDs. (**C**) Fluorescent image of cancer cells. (Reproduced from Ni et al. (2019) with permission from Elsevier)

1.2.3 Urine

1.2.3.1 Introduction and Biomedical Application of Human Urine

Naturally, human urine is sterile without the presence of any microorganisms. However, human urine contains inorganic salts, organic compounds, and metabolic waste (Masrura et al. 2021). Although human urine is considered a disgusting by-product of humans, it has been proven that urine can be turned into a value-added product. In a study, human urine was used to produce fluorescent carbon dots called "Pee-Dots" (PDs) through the carbonisation process (Essner et al. 2016). The resulting PDs were explored for bioimaging using mouse embryonic fibroblast (MEF) mice and BT-474 human mammary gland breast/duct carcinoma cells. PDs were found to be capable of fluorescence staining on both cell lines, plus cell viability remained high even with the exposure of 1.5 mg/mL of PDs concentration (Fig. 1.4) (Essner et al. 2016).

Zhang et al. (2019), on the contrary, revealed that urine-derived carbon dots (UCDs) could be used as fluorescence stains on different types of cells and tissues. On HeLa cells treated with UCDs, the cells appeared blue (405 nm) and green fluorescence (488 nm), indicating its usage as in vitro cell imaging. Furthermore, the UCDs could be taken up by a roundworm (*Caenorhabditis elegans*) and were concentrated on the pharynx, intestine and anus. Moreover, the blue fluorescence of UCDs was observed in the cytoplasm of onion epidermal cells. Meanwhile, an in vivo study on bean sprouts revealed that the distribution of UCDs strongly appeared on the stem of bean sprouts under an excitation wavelength of 365 nm. Additionally, cytotoxicity analysis of UCDs on normal rat kidney cells appeared with no significant cytotoxicity indicating the biocompatibility of the UCDs (Zhang et al. 2019). Hence, this finding displayed the versatility of UCDs as tools in bioimaging.

Interestingly, studies have revealed that human urine is a source of stem cells. Urine-derived stem cells (USCs) have been reported to possess stem characteristics similar to those of other stem cells (Guan et al. 2015). Furthermore, Guan et al. reported that a combination of human USCs and beta-tricalcium phosphate (β-TCP) scaffold facilitated bone regeneration in a segmental femoral defect in the rat model (Fig. 1.5) (Guan et al. 2015). Meanwhile, Zhao et al. (2019) successfully induced urothelium and smooth muscle cell differentiation in human USCs. The smooth muscle cells were then further transfected with miR-199a-5p plasmids for the contractile phenotype. The cells seeded in the vessel extracellular matrix successfully developed into multilayered urothelium and organised smooth muscle cells after 2 months of in vivo ureteral reconstructions (Zhao et al. 2019). This indicates the potential of USCs in urethral regeneration.

Recently, a study compared human USCs and human (BMSCs) from the same individual. They reported the presence of USCs with higher proliferation capacity compared to that of BMSCs. Interestingly, BMSCs showed better adipogenic, chondrogenic and osteogenic differentiation potentials compared to USCs in the same passage. However, both cells promoted similar cartilage repair ability in in vivo (Sun et al. 2021). This study presented the potential of USCs as an alternative source of stem cells for cartilage tissue engineering.

Based on search terms of "urine-derived stem cells" on ClinicalTrials.gov, an ongoing clinical trial study (Identifier No. NCT04874909) utilises urine-derived renal epithelial cells as human models to develop a diagnostic and prognostic tool for patients suffering from ciliopathies leading to renal failure. The results of this study are not available yet. However, this shows that urine products can potentially be used as a model at the clinical stage.

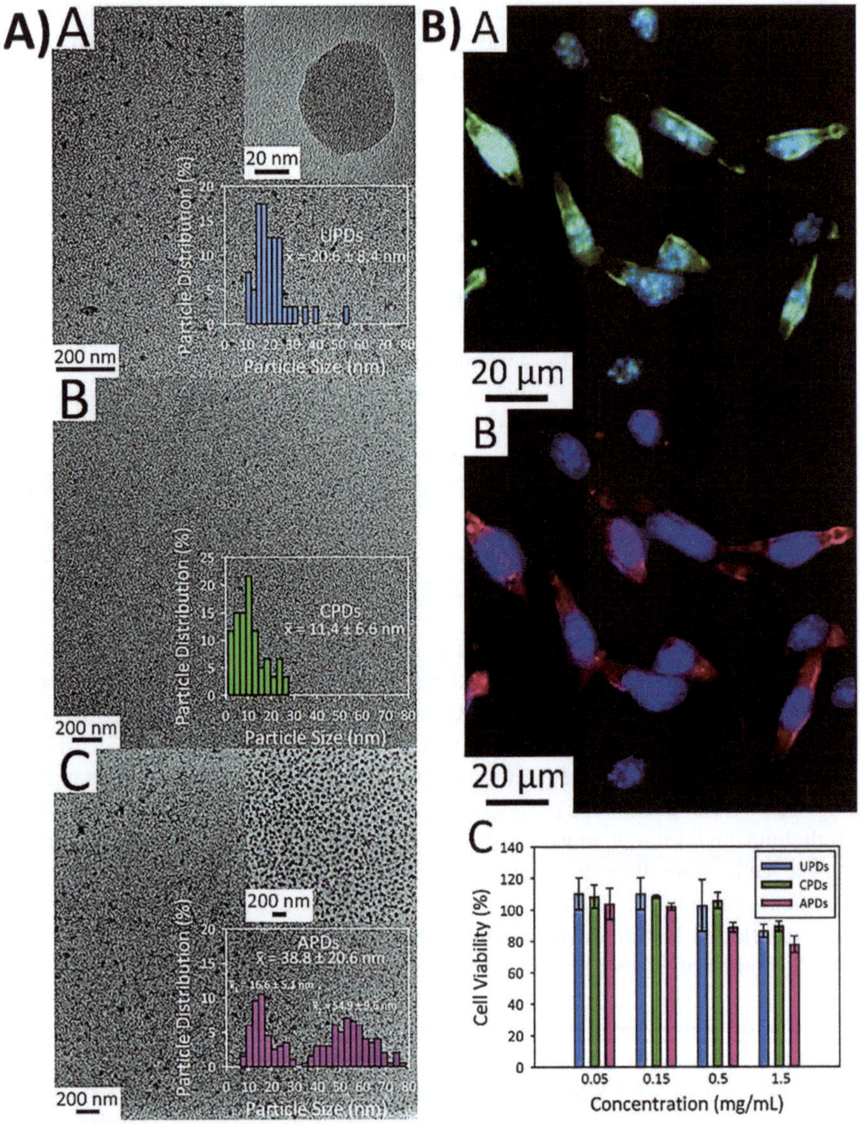

Fig. 1.4 Three images on the left show: (**A**) Transmission Electron Microscopes (TEM) image and histogram of the average particle size distribution of unmodified diet-derived pee dots (UPDs, A), Vitamin C supplemented diet-derived pee dots (CPDs, B), and Asparagus diet-derived pee dots (APDs, C), respectively. The inset of image (C) shows the bimodal particle size distribution of the APDs. Meanwhile, three images in the right diagram (**B**) show: a fluorescence image of APD-incubated MEF cells illustrating APD incorporation into the cellular cytoplasm and DAPI-stained nucleus through the FITC (A) and TRITC (B) filter cube. (C) The cell viability percentage was treated with different concentrations of three kinds of PDs. (Reproduced from Essner et al. (2016) with permission from the Royal Society of Chemistry)

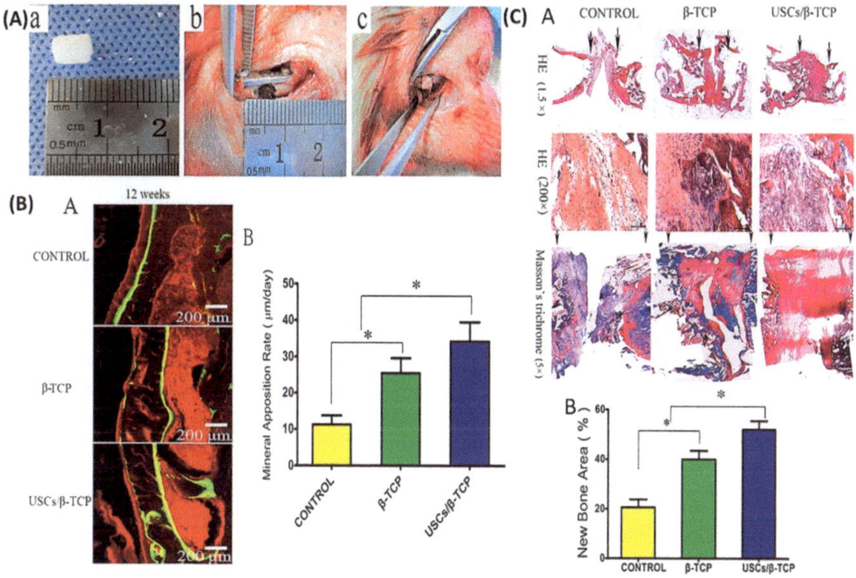

Fig. 1.5 (**a**) Surgical procedure and in vivo application: (a) β-TCP scaffold, (b) 6 mm design of femoral defect, and (c) implantation of a scaffold containing USCs. (**b**) Mineral apposition rate: (A) Alizarin red and calcein staining showed mineralisation, and (B) Mineral apposition rate between control, β-TCP, and USCs/β-TCP. Scale bar = 200 μm. (**c**) Histological analysis exhibited bone formation at 12 weeks: (A) Haematoxylin and eosin (H&E) and Masson's trichrome staining revealed bone formation in the model of induced femoral defect model (Arrows indicate the edges of host bone), and (B) The percentage of the new bone area from Masson's trichrome section image ($*P < 0.05$). Scale bar = 100 μm (H&E 200×). (Reproduced from Guan et al. (2015) with permission from Plos One. Image is under Attribution 4.0 International (CC BY 4.0))

1.2.3.2 Advantages and Limitations of Human Urine

The use of urine as biomaterials to produce functional products such as carbon dots does not pose many problems (Gao et al. 2016). However, the usage of urine-derived stem cells poses several limitations, which include obtaining sufficient cell population for treatment because some patients may be affected by diseases such as gallstones, infection or trauma, which may disturb the isolation and expansion of cells. Additionally, the effect of urine on the expansion capacity of urine-derived stem cells has not been studied (Abbas et al. 2020).

Furthermore, human urine is a rich source of nutrients and metabolites, thus this may cause secondary pollution during the development of functional products from urine, such as biochar. The chemical generated during the production of functional products and waste of excess urine needs to be treated properly. Moreover, some urine contains different pathogenic microorganisms, which is a potential occupational health hazard. Therefore, proper management of urine processing in urine treatment and field application should be looked into to avoid the release of unintentional contaminants to the environment and personnel (Masrura et al. 2021).

1.2.4 Adipose Tissue

1.2.4.1 Introduction and Biomedical Application of Human Adipose Tissue Waste

Adipose tissue, also known as fat, is a part of human appendages that most people strive to remove from their bodies. Even so, adipose tissue plays an essential role in the pathophysiology of the human body, including food intake, energy depots and distribution, insulin sensitivity, immune response and regulation of body temperature. Adipose tissue is a dynamic organ that is broadly distributed in the human body. The human body possesses two major types of adipose tissues: brown and white (Koenen et al. 2021; Cypess 2022; Sakers et al. 2022). Brown adipose tissue (BAT) mainly resides in the abdominal, axillary, supraclavicular, mediastinal, paraspinal and cervical depots. BAT regulates human body temperature to prevent hypothermia by using the stored energy as heat, via a process called non-shivering thermogenesis (Koenen et al. 2021).

In contrast, white adipose tissue (WAT) plays a role in regulating systemic energy homeostasis. Other than acting as an energy reservoir in the form of triglycerides, it also discerns energy demand and induces paracrine factors secretion to modulate metabolic tissues (Chrysafi et al. 2020). Furthermore, in an event of excessive amassing of WAT, uncontrolled stimulation of local inflammatory response in WAT will occur and may trigger insulin resistivity (Longo et al. 2019). Additionally, it can lead to adipose tissue dysfunction, increasing the risk of health complications, including hypertension, cardiometabolic disorders, cardiovascular disease and dyslipidaemia (Koenen et al. 2021; Longo et al. 2019). Therefore, reducing excess adipose tissue, particularly WAT may reduce the risk of health complications and thus, improve the quality of human health.

In this regard, various strategies are implemented to help reduce excess fat, including a healthy diet, unsweetened beverages, adequate sleep, and exercise. Moreover, a surgical procedure called liposuction offers efficient and fast removal of stubborn excess adipose tissue (Beidas and Gusenoff 2021). In a massive weight loss through liposuction, depending on the target surgical site, the removed adipose tissue volume may reach over 22 L (Cardenas-Camarena 2021). Clinically, this volume is considered high pathological waste.

Due to that, adipose tissue is used as a non-controversial source of stem cells. For example, adipose-derived stem cells (ADSCs) could potentially differentiate into osteocytes and reach maturation more quickly than stem cells derived from umbilical cords after 21 days of osteogenic differentiation (Zajdel et al. 2017). In addition, adipose tissue possesses unlimited expansion capability (Koenen et al. 2021). Apart from that, decellularised and delipidated adipose tissue provides a large amount of natural extracellular matrix (ECM) material, which could be applied in many areas of tissue engineering. Additionally, supercritical carbon dioxide (SC-CO_2) treatment on adipose tissue offers the use of a less harsh chemical with a shorter processing time while preserving important proteins, including biological factors and growth factors (Fig. 1.6) (Wang et al. 2017).

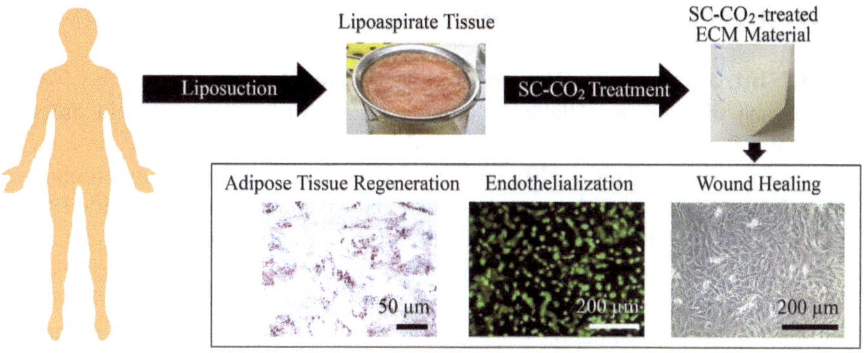

Fig. 1.6 Schematic illustration of the delipidation and decellularisation of adipose tissue by $SC\text{-}CO_2$ to obtain ECM and its subsequent application in the different fields of tissue engineering. (Reproduced from Wang et al. (2017) with permission from Elsevier)

A similar study reported the feasibility of using decellularised ECM from adipose tissue and adipose-derived stem cells in cartilage tissue engineering (Ibsirlioglu et al. 2020). In addition, adipose tissue contained a vast amount of growth factors, including basic fibroblast growth factor (bFGF), insulin growth factors-1 (IGF1), transforming growth factor beta-1 (TGF-β1), vascular endothelial growth factor (VEGF), platelet-derived growth factor (PDGF), bone morphogenetic protein-4 (BMP4), hepatocyte growth factor (HGF), nerve growth factor (NGF) and epidermal growth factor (EGF) (Chun et al. 2019). In conclusion, adipose tissue provides an indispensable source of ECM, stem cells and growth factors.

Lastly, based on the search term "adipose derived stem cells" on ClinicalTrials. gov, approximately 334 clinical studies were identified to have used adipose derived stem cells for various applications. Most studies were conducted in North America (119 studies), followed by Europe (72 studies) and East Asia (64 studies). In addition to that, there were also studies, which used adipose tissue called nano fat in several clinical trials that aimed to treat different tissues and diseases. Multiple benefits of using adipose tissue extraction have driven its application in pre clinical and clinical stages.

1.2.4.2 Advantages and Limitations of Human Adipose Tissue Waste

A high volume of adipose tissue provides a plethora of stem cell yield, ease of collection via needle biopsy or liposuction aspirates, and immune-privileged, which induce low graft-versus-host disease (GVHD) risks. Also, containing various growth factors, anti-inflammatory, anti-apoptotic, pro-angiogenic and anti-scarring effects (Azzopardi and Blundell 2018; Zheng et al. 2019; Yu et al. 2018). However, the viability and quantity of adipose stem cells can be negatively affected by local anaesthetic agents during sample collection. Furthermore, the route of adipose-derived stem cell administration could affect its efficacy and viability (Mazini et al. 2020).

1.2.5 Human Post-birth Waste

1.2.5.1 Introduction of Human Post-birth Waste

The human placenta is a flattened circular organ (Fig. 1.7) present in pregnant mammals. It acts as a barrier and provides nutrients to the foetus through the umbilical cord (Hemalatha 2020). Furthermore, it consists of basalis and trophoblast cells of the maternal decidua and mesenchymal cell line of the foetal side (Sibiak et al. 2020). Typically, the umbilical cord ranges from 40 to 60 cm in length and is surrounded by the outer amniotic epithelium. A single vein and two arteries reside in the mucoid connective tissue known as Wharton's Jelly (WJ) of the umbilical cord. WJ is avascular without the presence of neural tissue, but shows contractile capability due to the existence of mesenchymal stem cell-derived myofibroblast (Lyons and Mattei 2019). Besides that, WJ functions as a protector of the blood vessel while providing cord flexibility.

1.2.5.2 Biomedical Application of Human Post-birth Waste

Post-birth waste, including amnion fluid, placenta, and umbilical cord, is known as a medical waste that is discarded after childbirth. Nowadays, researchers, particularly in tissue engineering and regenerative medicine fields, have revealed various potential uses of post-birth waste. For example, a study group successfully hepatised a decellularised placenta while maintaining albumin and urea synthesis in vitro. Whereas, in an in vivo study, the hepatised placenta was revealed to be able to perform self-assembly and perfused well while retaining hepatic cord structure (Kakabadze et al. 2018). Additionally, the human placenta is useful as a vascular microsurgical training model due to its vasculature structure, low cost, and being ethically accepted (Del Maestro et al. 2021; Gomar-Alba et al. 2021).

Furthermore, there are several sub-groups of placenta-derived stem cells, whereby the amniotic membrane provides human amniotic epithelial cells (hAEC) and human amniotic mesenchymal stromal cells (hAMSC). Meanwhile, the placenta provides human chorionic mesenchymal stem cells (hCMSC) and human chorionic trophoblastic cells (hCTC) (Sibiak et al. 2020). Plus, the umbilical cord is a source of stem cells, known as the umbilical cord-derived mesenchymal stem cell (UCMSC)

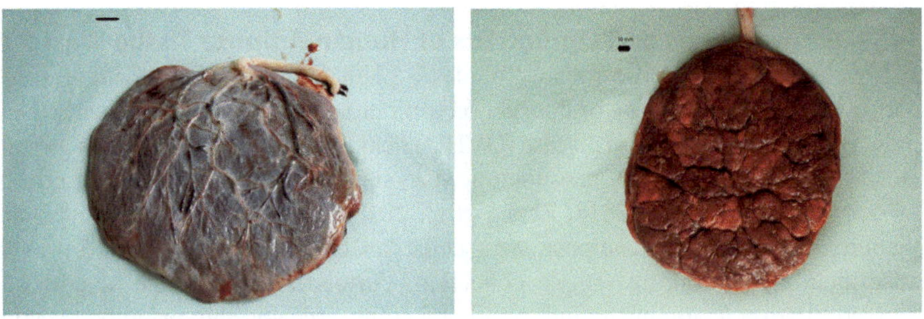

Fig. 1.7 Human placenta: foetal side (left image) and mother side (right image). (Reproduced from Domagała et al. (2020) with permission from Elsevier)

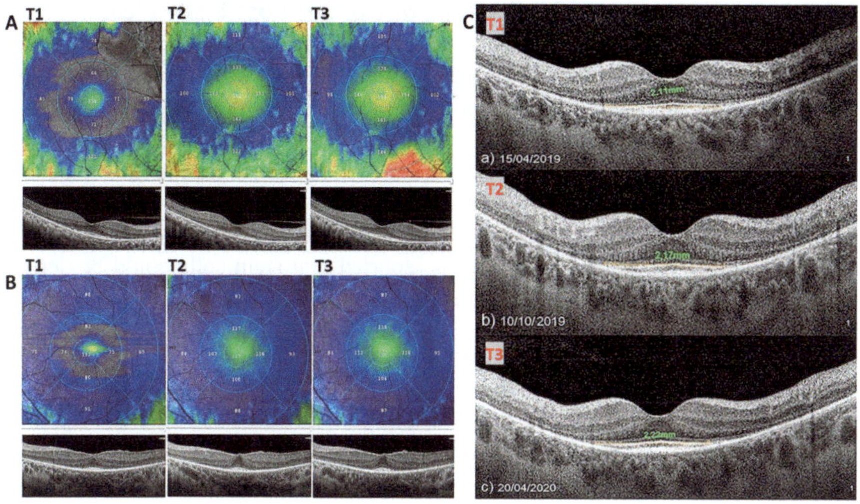

Fig. 1.8 Patient 1's right eye (**A**) and patient 15's left eye (**B**) showing the thickening of the outer retinal of the eye treated with WJMSCs according to the study time points (T0, T1, T2). (**A**) Before application, T0: 85.0 μm (a); T1: 152.20 μm (b); and T2: 154.1 μm (c). (**B**) Before application, T0: 84.8 μm (a); T1: 120.0 μm (b); and T2: 122.8 μm (c). (**C**) Patient 26's left eye, treated with WJMSC, showed an increase in the width of the horizontal ellipsoid zone according to the study time points (T0, T1, T2). Before application, T0: 2.11 mm (a); T1: 2.17 mm (b); and T2: 2.22 mm (c). (Reproduced from Özmert and Arslan (2020) with permission from Springer Nature)

(Lyons and Mattei 2019). The isolation of cells from Wharton's Jelly of the umbilical cord provides cells known as Wharton's Jelly-derived mesenchymal stem cells (WJMSCs) (Lanci et al. 2019). The effectiveness of UCMSC for various diseases has been reviewed by Azzopardi and Blundell (2018). These include patients with chronic immune thrombocytopenic purpura, idiopathic pulmonary fibrosis, atopic dermatitis, gingival recession defects, primary systemic lupus erythematosus, secondary Sjogren's syndrome and paediatric patients with MDS/AML associated with underlying GATA2 mutations (Azzopardi and Blundell 2018). In addition, UCMSC promotes chondrocyte generation in osteoarthritis chondrocytes (Wang et al. 2018).

Currently, numerous clinical trials on different phases have been completed and are ongoing involving the use of MSCs derived from post-birth wastes, including 29 trials on WJMSCs, 362 trials on UCMSCs, 25 trials on hAEC, 7 trials on amniotic membrane and hAMSC and 30 trials on placenta-derived mesenchymal stem cells, based on ClinicalTrials.gov (NIH 2020). An example of a phase 3 clinical trial using WJMSCs to treat retinitis pigmentosa has shown that there were no ocular or systemic adverse events during the 1-year follow-up period. Instead, there is an improvement in the thickness of the outer retina and the width of the horizontal ellipsoid zone (Fig. 1.8) (Özmert and Arslan 2020).

1.2.5.3 Advantages and Limitations of Human Post-birth Waste

Among the advantages of human post-birth waste is that the umbilical cord contains a higher volume of stem cells compared to bone marrow. Mesenchymal stem cells (MSCs) derived from the umbilical cord provide a low graft-versus-host disease rate due to their higher tolerance with the human leukocyte antigen (HLA) compared to BMSCs. Not only that their ease of collection does not cause pain to the mother and the baby. In fact, UCMSCs can be stored in the bank compared to BMSCs. They also reduce the transmission of infectious diseases, such as cytomegalovirus (Azzopardi and Blundell 2018). Additionally, it was reported that stem cells could be derived from amniotic fluid. However, the extraction process of amniotic fluid is difficult and may exhibit high immunogenicity, which could trigger an immune response (Yi et al. 2020).

1.3 Bioethical, Regulatory, and Translation Issues of Human "Wastes"

Collecting human wastes such as human hair, urine and nails involves a simple, non-invasive procedure without a painful surgical procedure (Gao et al. 2016). Meanwhile, human adipose tissue and post-birth waste collection involve minimally invasive surgical procedures through liposuctions and natural birth, respectively (Fujita et al. 2019; Wu et al. 2018). Furthermore, human wastes provide an inexpensive source of materials that are readily available and easily accessible for biomedical applications. Plus, autologous human wastes such as urine, adipose tissue and post-birth waste do not induce an immune response. Thus, it raised low-level ethical issues (Chen et al. 2018). Therefore, human wastes provide a better acquisition process with several excellent sources of materials. However, the use of human waste in research is relatively new. Therefore, the establishment of specific rules and regulations concerning the safe handling and environmental management of human wastes and the application of human waste products should be considered in research, especially if the technology is developed into a medical device.

The medical device industry is a healthcare industry driven by new technologies and innovations to develop products to improve healthcare therapies and services. In Malaysia, the classification of medical devices or products should be referred to the National Pharmaceutical Regulatory Agency (NPRA), Malaysia, according to Appendix II, Drug Registration Guidance Document (DRGD). The NPRA will aid the determination of the products that are not clearly defined as medical devices, drugs or cosmetics based on their intended purposes/indications and modes of action. After classification has been made, the devices or products will be regulated by relevant agencies, whereby the registration or the notification under Drugs and Cosmetics classification is regulated by the NPRA by following the requirements outlined in the Poisons Act, 1952 and its Regulations, the Sale of Drugs Act, 1952 and the Control of Drugs and Cosmetics Regulations, 1984. Whereas, the registration for products under medical device classification is regulated by the Medical

Device Authority following the requirements outlined in the Medical Device Act 2012 (Act 737).

Additionally, the safe handling of human "waste" materials can be referred to the International Organisation for Standardisation (ISO), ISO 45001, which focuses on reducing occupational injuries and diseases (ISO 2021). Meanwhile, the management of wastes produced during or after the processing of human "waste" can be referred to as ISO 14000, which emphasises on environmental management by measuring and improving any potential environmental impact (ISO 2020). Also, the biocompatibility evaluation of the human waste products can be referred to as the ISO 10993, which concentrates on the biological evaluation of the medical devices.

Studies have shown the ability of human "waste" from a different source that provides cells that are claimed as MSCs. To prove that the cells extracted from human "waste" are MSCs, specific standard criteria must be complied. Regarding that, for cells extracted from human "waste" to be defined as MSCs, they should adhere to the minimum criterion established by the Mesenchymal and Tissue Stem Cell Committee of the International Society for Cellular Therapy. First, "in a standard culture condition, the extracted cells should possess a plastic-adherent character." Next, "extracted cells should positively express CD73, CD105 and CD90 markers and negatively express surface markers CD14, CD34, CD45, CD11b, CD19 and HLA-DR." Lastly, the "extracted cells could be differentiated into adipocytes, osteoblasts and chondroblasts in vitro" (Baghaei et al. 2017; Dominici et al. 2006). In addition, MSC-like cells have shown positive outcomes in various in vitro, in vivo and early clinical studies, but a controversial debate on stem cell effectiveness in clinical usage is still ongoing.

Furthermore, under the section 351, Public Health Service Act (42 USC 262), United States (US), the isolated MSCs are classified as a minimal manipulation cellular and gene therapy (CGT) products. In the US, a CGT product is obliged to apply as an Investigational New Drug Application (IND) from the Food and Drug Administration (FDA) prior to clinical trials. Moreover, other products, such as medical devices, require Investigational Device Exemption (IDE), 21 CFR Part 812. Approval from the FDA must be obtained for a device to be used in a clinical study in order to collect safe and effective data. Additionally, the clinical trial findings should comply with regulations set by the FDA based on the Code of Federal Regulations, Title 21 (21 CFR Part 820) (Food and Drug Administration 2020). Similarly, the European Medicines Agency (EMA) requires the submission of application through The European Regulation (European Commission, EC.) No.1394/2007 for Advanced Therapy Medicinal Products (ATMPs) for conducting clinical trials. The European Regulation defines ATMPs as cell therapy medicines and tissue-engineered products, as well as combined ATMPs such as cells embedded in biodegradable scaffolds or matrices. Likewise, any study conducted to produce medical devices should comply with the country's regulations, such as the Japanese Pharmaceutical Affairs Law (PAL) in Japan and the Canada Medical Device Regulations (SOR/98-282) in Canada.

In the medical field, technology transfer or research translation is crucial for communicating and expressing knowledge and the latest discoveries. Through

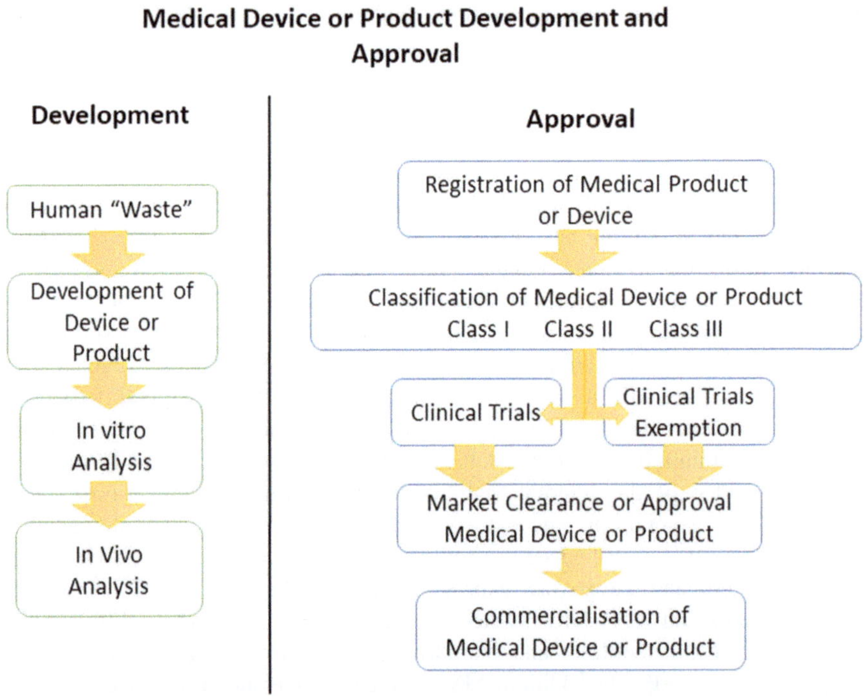

Fig. 1.9 A schematic diagram of a general medical device or product development and approval pathway

technology transfer, the efficiency of a product under development, the quality of the product, and the standardisation of cost-effective methods can be achieved (Patil et al. 2013). Commonly, the translation of medical products is the key challenge due to the increasing regulatory requirements. The ever-changing regulations pose quite a challenge to every researcher and the medical device industry to respond quickly to opportunities, including ensuring that products quickly reach the market. Furthermore, increased regulatory scrutiny has made it necessary for medical device manufacturers and companies to move into cost-effective technology, safety compliance, effectiveness and marketability (Medical Device Authority 2012). Therefore, the complex regulatory and environmental sustainability concerns will greatly impact the new technology transfer and continuously impact economic growth.

Furthermore, translation is vital for the approval of a medical device for commercialisation purposes. After a medical device has successfully undergone clinical studies with promising efficacy and safety data while complying with the regulations set by the respective regulatory body, this will provide chances for marketing approval and registration of the product. A schematic diagram shows the general steps in the development of medical devices and the approval pathway (Fig. 1.9). Thus, this is where medical translation begins to come into play. Translation of medical devices provides a medium for companies, manufacturers, and researchers to launch devices or products successfully in local and foreign markets. Strict approval processes and compliance measures are imposed to distribute and import medical devices in different countries. For example, the products should adhere to Malaysia's medical device regulations in order to launch medical devices in Malaysia.

Similarly, there is a need to adhere to the European Medical Device Regulations in order to launch a product in Europe, the Japanese Pharmaceutical Affairs Law (PAL) in Japan, and the Canadian Medical Device Regulations (SOR/98-282) in Canada. Importantly, translation of the medical documents into the country's official language is required. For example, medical device applications in Brazil should be translated into Portuguese. Generally, all records that regulators will use to evaluate a device and decide whether it can be imported and supplied to the target market must be translated, including clinical study records, administrative records, quality management system certificates and medical device instructions. Due to these prerequisites, medical translation plays a crucial role in the clearance of medical devices.

1.4 Commercialisation Potential of Human "Wastes" Products

Several human wastes were discovered to provide valuable materials that are potentially alternatives to the current sources of materials, including a non-invasive source of MSCs, a cost-effective carbon precursor, and keratin that can be used to develop a medical device. Utilisation of human wastes in medical devices for human use covers environmental sustainability concerns according to the SDGs. The present study provides a potential medical device that can be developed from human wastes and is marketable, corresponding to the worldwide demand, including biological scaffolds, nanomaterials, and stem cells.

From the marketing perspective, the current size of the worldwide market for scaffold technology worldwide was valued at USD 1.1 billion in 2020, and the compound annual growth rate (CAGR) is expected to grow by approximately 8.4% between 2021 and 2028. The increasing demand for 3-dimensional (3D) cellular models in biological studies and translational research has driven the market's growth. These include the application of 3D cell modelling, which mimics the in vivo pathophysiology state for efficient representation of disease-causing micro-environmental factors, even cancer research. Plus, bioscaffolds have also been extensively designed as drug delivery systems. Furthermore, the advancement of 3D technology has opened opportunities to develop a personalised bioscaffold for tissue engineering (GrandViewResearch 2021a). In terms of this, human wastes can be converted into functional materials to produce a 3D bioscaffold. For example, human hair derived from keratin is used to develop a bioscaffold to repair injured nerves, while adipose tissue can provide a large volume of natural ECM for biomedical applications (Gao et al. 2019a; Wang et al. 2017).

Furthermore, human wastes can be converted into nanomaterials, such as carbon dots (Zhang et al. 2019, 2021). Currently, the market size of nanomaterials worldwide is reported to be approximately USD 8.0 billion in 2020. Around 14.1% of the CAGR is expected to expand from 2021 to 2028. The growth of nanomaterials is attributed to their attractive physiochemical properties and increasing application in different industries, including healthcare, particularly biomedical (GrandViewResearch 2021b). In light of this, human wastes can be considered a material source in producing sustainable nanomaterial products.

Additionally, the market size of stem cells worldwide was estimated at USD 9.38 billion in 2020, which is comparatively higher than the market size of scaffold technology, and the CAGR is expected to increase by about 8.8% from 2021 to 2028. The growth of the stem cell market is affected by the increasing exigency for biological and technological advancements in stem cell therapeutics, focusing on precision medicines (GrandViewResearch 2021c). However, current stem cell research requires higher capital sources due to the complex process and demands longer research duration, contributing to the expensive cost of stem cell therapy. Considering this, human wastes provide an alternative source for a cheaper and non-invasive source of stem cells compared to bone marrow-derived stem cells (Lyons and Mattei 2019). Thus, using stem cells derived from human waste will help reduce research and development costs in current and future research. Overall, the commercialisation potential of products developed from human "waste" is promising.

1.5 Conclusion and Future Perspective

This chapter reviews several human wastes as potential sustainable material sources for biomedical applications, including hair, nails, urine, adipose tissue and post-birth wastes. Utilising human wastes allows for green and economic processes in research. Additionally, the utilisation of human waste offers various benefits in the biomedical field, specifically a reduction in material cost and availability. Plus, their usage presents few ethical issues. Although not discussed in depth here, it should be recognised that human wastes provide several advantages compared to other materials, including reducing the immune response with the use of autologous tissue and avoiding zoonotic infections. However, studies on human waste so far have been somewhat limited. Therefore, further research on the biomedical applications of human waste is necessary. Furthermore, the potentially harmful effects of the products derived from human wastes such as nanomaterial remain a concern in biomedical applications. Thus, further efficacy and safety evaluation of functional products derived from human wastes should be prioritised to translate these products onto the market. Additionally, understanding the immune privilege offered by human wastes in different areas of biomedical applications may revolutionise human wastes as a source of materials to produce functional products in the near future. Last but not least, the user acceptance of human waste-based biomedical devices or products needs to be further investigated as this will determine the success and sustainability of their applications.

Acknowledgements This research was supported by the Universiti Malaya Impact-Oriented Interdisciplinary Research Grant Program (IIRG007A-19HWB and IIRG007B-19HWB).

Conflicts of Interest The authors declare that they have no conflict of interest.

References

Abbas TO, Ali TA, Uddin S (2020) Urine as a main effector in urological tissue engineering—a double-edged sword. Cells 9(3):538

Adav SS, Subbaiaih RS, Kerk SK et al (2018) Studies on the proteome of human hair-identification of histones and deamidated keratins. Sci Rep 8(1):1–11

Agarwal V, Panicker AG, Indrakumar S et al (2019) Comparative study of keratin extraction from human hair. Int J Biol Macromol 133:382–390

Azzopardi JI, Blundell R (2018) Review: umbilical cord stem cells. Stem Cell Discov 08:1–11

Baghaei K, Hashemi SM, Tokhanbigli S et al (2017) Isolation, differentiation, and characterization of mesenchymal stem cells from human bone marrow. Gastroenterol Hepatol Bed Bench 10(3): 208–213

Beidas OE, Gusenoff JA (2021) Update on liposuction: what all plastic surgeons should know. Plast Reconstr Surg 147(4):658e–668e

Cardenas-Camarena L (2021) Commentary on: safety of large volume liposuction in aesthetic surgery: a systematic review and meta-analysis. Aesthet Surg J 41(9):1054–1056

Chatzimitakos TG, Kasouni AI, Troganis AN et al (2018) Carbonization of human fingernails: toward the sustainable production of multifunctional nitrogen and sulfur codoped carbon nanodots with highly luminescent probing and cell proliferative/migration properties. ACS Appl Mater Interfaces 10(18):16024–16032

Chen L, Li L, Xing F et al (2018) Human urine-derived stem cells: potential for cell-based therapy of cartilage defects. Stem Cells Int 2018:4686259

Chen Z, Ye S-Y, Yang Y et al (2019) A review on charred traditional Chinese herbs: carbonization to yield a haemostatic effect. Pharm Biol 57(1):498–506

Chrysafi P, Perakakis N, Farr OM et al (2020) Leptin alters energy intake and fat mass but not energy expenditure in lean subjects. Nat Commun 11(1):5145

Chun SY, Lim JO, Lee EH et al (2019) Preparation and characterization of human adipose tissue-derived extracellular matrix, growth factors, and stem cells: a concise review. J Tissue Eng Regen Med 16(4):385–393

Cortellini V, Carobbio A, Brescia G et al (2019) A comparative study of human and animal hairs: microscopic hair comparison and cytochrome c oxidase I species identification. Forensic Sci Med Pathol 5(1):20

Cypess AM (2022) Reassessing human adipose tissue. N Engl J Med 386(8):768–779

Del Maestro M, Rampini AD, Mauramati S et al (2021) Dye-perfused human placenta for vascular microneurosurgery training: preparation protocol and validation testing. World Neurosurg 146: e854–e864

Domagała Z, Domański J, Zimmer A et al (2020) Methodology of preparation of corrosive specimens from human placenta—a technical note. Ann Anat 228:151436

Dominici M, Le Blanc K, Mueller I et al (2006) Minimal criteria for defining multipotent mesenchymal stromal cells. The International Society for Cellular Therapy position statement. Cytotherapy 8(4):315–317

Essner JB, Laber CH, Ravula S et al (2016) Pee-dots: biocompatible fluorescent carbon dots derived from the upcycling of urine. Green Chem 18:243–250

Fan J, Yu M-Y, Lei T-d et al (2018) In vivo biocompatibility and improved compression strength of reinforced keratin/hydroxyapatite scaffold. Tissue Eng Regen Med 15(2):145–154

Fellows AP, Casford MT, Davies PB (2021) Using hybrid atomic force microscopy and infrared spectroscopy (AFM-IR) to identify chemical components of the hair medulla on the nanoscale. J Microsc 284(3):189–202

Fellows A, Casford M, Davies P (2022) Chemically characterizing the cortical cell nano-structure of human hair using atomic force microscopy integrated with infrared spectroscopy (AFM-IR). Int J Cosmet Sci 44(1):42–55

Food & Drug Administration (2020) Overview of device regulation. https://www.fda.gov/medical-devices/device-advice-comprehensive-regulatory-assistance/overview-device-regulation. Cited 24 Aug 2022

Fujita Y, Taniguchi M, Tsuzuki T et al (2019) Application of a minimally invasive liposuction technique for harvesting fat during transsphenoidal surgery: a technical note. Neurol Med-Chir 59(5):184

Gao P, Jiang D, Liu W et al (2016) Urine-derived stem cells, a new source of seed cells for tissue engineering. Curr Stem Cell Res Ther 11(7):547–553

Gao J, Zhang L, Wei Y et al (2019a) Human hair keratins promote the regeneration of peripheral nerves in a rat sciatic nerve crush model. J Mater Sci Mater Med 30(7):1–13

Gao F, Li W, Deng J et al (2019b) Recombinant human hair keratin nanoparticles accelerate dermal wound healing. ACS Appl Mater Interfaces 11(20):18681–18690

Gomar-Alba M, Parrón-Carreño T, Narro-Donate JM et al (2021) Microsurgical training: vascular control and intraoperative vessel rupture in the human placenta infusion model. Acta Neurochir 163(9):2525–2532

GrandViewResearch (2021a) Scaffold technology market size, share & trends analysis report by type, by application (stem cell, regenerative medicine, drug discovery), by disease type (cancer, dental, neurology), by end-use, and segment forecasts, 2021–2028. https://www.grandviewresearch.com/industry-analysis/scaffold-technology-market. Cited 6 Apr 2022

GrandViewResearch (2021b) Nanomaterials market size, share & trends analysis report by product (gold, silver, iron, copper), by application (aerospace, automotive, medical), by region, and segment forecasts, 2021–2028. https://www.grandviewresearch.com/industry-analysis/nanotechnology-and-nanomaterials-market. Cited 21 June 2022

GrandViewResearch (2021c) Stem cells market size, share & trends analysis report by product (mesenchymal, adult, iPSCs), by application, by technology (cell acquisition, cryopreservation), by therapy, by region, and segment forecasts, 2021–2028. https://www.grandviewresearch.com/industry-analysis/stem-cells-market. Cited 17 June 2022

Guan J, Zhang J, Li H et al (2015) Human urine derived stem cells in combination with β-TCP can be applied for bone regeneration. PLoS One 10(5):e0125253

Gupta A (2014) Human hair "waste" and its utilization: gaps and possibilities. J Waste Manag 2014: 498018

Hemalatha M (2020) Macro and microscopic study of human placenta in normal and complicated pregnancies. Kilpauk Medical College, Chennai

Hinds G, Thomas VD (2008) Malignancy and cancer treatment-related hair and nail changes. Dermatol Clin 26(1):59–68

Ibsirlioglu T, Elçin AE, Elçin YM (2020) Decellularized biological scaffold and stem cells from autologous human adipose tissue for cartilage tissue engineering. Methods 171:97–107

Iglesias-Gonzalez A, Hardy EM, Appenzeller BM (2020) Cumulative exposure to organic pollutants of French children assessed by hair analysis. Environ Int 134:105332

ISO (2020) ISO 14000 family—environmental management. https://www.iso.org/iso-14001-environmental-management.html. Cited 23 Aug 2022

ISO (2021) ISO 45000 family—occupational health and safety. https://www.iso.org/iso-45001-occupational-health-and-safety.html. Cited 23 Aug 2022

Johnson C, Sinkler MA, Schmieder GJ (2021) Anatomy, shoulder and upper limb, nails. In: StatPearls. StatPearls publishing

Kakabadze Z, Kakabadze A, Chakhunashvili D et al (2018) Decellularized human placenta supports hepatic tissue and allows rescue in acute liver failure. Hepatology 67(5):1956–1969

Kaneta D, Goto M, Hagihara M et al (2021) Visualizing intra-medulla lipids in human hair using ultra-multiplex CARS, SHG, and THG microscopy. Analyst 146(4):1163–1168

Kim JW, Kim MJ, Ki CS et al (2017) Fabrication of bi-layer scaffold of keratin nanofiber and gelatin-methacrylate hydrogel: implications for skin graft. Int J Biol Macromol 105:541–548

Ko J, Nguyen LT, Surendran A et al (2017) Human hair keratin for biocompatible flexible and transient electronic devices. ACS Appl Mater Interfaces 9(49):43004–43012

Koenen M, Hill MA, Cohen P et al (2021) Obesity, adipose tissue and vascular dysfunction. Circ Res 128(7):951–968

Küçük EE, Günay O (2016) Health-related behaviors and nutritional status of adolescents who study as boarders and day-students. Turk J Med Sci 46(4):960–966

Kunjiappan S, Panneerselvam T, Somasundaram B et al (2018) Design, in silico modeling, biodistribution study of rutin and quercetin loaded stable human hair keratin nanoparticles intended for anticancer drug delivery. Biomed Phys Eng Express 4(2):025019

Lanci A, Merlo B, Mariella J et al (2019) Heterologous Wharton's jelly derived mesenchymal stem cells application on a large chronic skin wound in a 6-month-old filly. Front Vet Sci 6:9

Ledford BT, Simmons J, Chen M et al (2017) Keratose hydrogels promote vascular smooth muscle differentiation from C-kit-positive human cardiac stem cells. Stem Cells Dev 26(12):888–900

Longo M, Zatterale F, Naderi J et al (2019) Adipose tissue dysfunction as determinant of obesity-associated metabolic complications. Int J Mol Sci 20(9):2358

Lyons FG, Mattei TA (2019) Sources, identification, and clinical implications of heterogeneity in human umbilical cord stem cells. In: Stem cells heterogeneity in different organs, pp 243–256

Masrura SU, Dissanayake P, Sun Y et al (2021) Sustainable use of biochar for resource recovery and pharmaceutical removal from human urine: a critical review. Crit Rev Environ Sci Technol 51(24):3016–3048

Mazini L, Rochette L, Admou B et al (2020) Hopes and limits of adipose-derived stem cells (ADSCs) and mesenchymal stem cells (MSCs) in wound healing. Int J Mol Sci 21(4):1306

Medical Device Authority (2012) Medical Device Act 2012. Malaysia MoH, Editor Percetakan Nasional Malaysia Berhad, Kuala Lumpur, p 49

Munteanu O, Filipoiu F-M, Cirstoiu MM et al (2021) A comprehensive study regarding the intrauterine development of nails. Organogenesis 17(1–2):14–19

Ni P, Li Q, Xu C et al (2019) Optical properties of nitrogen and sulfur co-doped carbon dots and their applicability as fluorescent probes for living cell imaging. Appl Surf Sci 494:377–383

NIH (2020) Wharton's+jelly+derived+mesenchymal+stem+cells - list results. ClinicalTrials.gov

Özmert E, Arslan U (2020) Management of retinitis pigmentosa by Wharton's jelly-derived mesenchymal stem cells: prospective analysis of 1-year results. Stem Cell Res Ther 11(1):1–17

Pane R, Andayani LS (2020) Relationship between the cleanliness of nails and the usage of footwear with the incidence of helminths infections on elementary student in Sibolga of 2019. BioEx-J 2(1):45–52

Park AM, Khan S, Rawnsley J (2018) Hair biology: growth and pigmentation. Facial Plast Surg Clin North Am 26(4):415–424

Patil J, Shete M, Chipade V et al (2013) Translation and commercialization of pharmaceutical scientific innovations: an overview. J Adv Pharm Educ Res 3(3)

Pramanick B, Cadenas LB, Kim D-M et al (2016) Human hair-derived hollow carbon microfibers for electrochemical sensing. Carbon 107:872–877

Rogers GE (2019) Known and unknown features of hair cuticle structure: a brief review. Cosmetics 6(2):32

Sakers A, De Siqueira MK, Seale P et al (2022) Adipose-tissue plasticity in health and disease. Cell 185(3):419–446

Sibiak R, Jaworski M, Dorna Z et al (2020) Human placenta-derived stem cells-recent findings based on the molecular science. Med J Cell Biol 8(4):164–169

Silva OA, Pellá MG, Popat KC et al (2022) Rod-shaped keratin nanoparticles extracted from human hair by acid hydrolysis as photothermally triggered berberine delivery system. Adv Powder Technol 33(1):103353

Sun J, Xing F, Zou M et al (2021) Comparison of chondrogenesis-related biological behaviors between human urine-derived stem cells and human bone marrow mesenchymal stem cells from the same individual. Stem Cell Res Ther 12(1):1–19

United Nations (NA) (2021) Department of Economic and Social Affairs, sustainable development. https://sdgs.un.org/goals. Cited 30 Aug 2021

Valkov A, Zinigrad M, Sobolev A et al (2020) Keratin biomembranes as a model for studying onychomycosis. Int J Mol Sci 21(10):3512

Veerasubramanian PK, Thangavel P, Kannan R et al (2018) An investigation of konjac glucomannan-keratin hydrogel scaffold loaded with Avena sativa extracts for diabetic wound healing. Colloids Surf B Biointerfaces 165:92–102

Vitus V, Ibrahim F, Wan Kamarul Zaman WS (2021) Modelling of stem cells microenvironment using carbon-based scaffold for tissue engineering application—a review. Polymers 13(23): 4058

Wang JK, Luo B, Guneta V et al (2017) Supercritical carbon dioxide extracted extracellular matrix material from adipose tissue. Mater Sci Eng C 75:349–358

Wang H, Yan X, Jiang Y et al (2018) The human umbilical cord stem cells improve the viability of OA degenerated chondrocytes. Mol Med Rep 17(3):4474–4482

Wu M, Zhang R, Zou Q et al (2018) Comparison of the biological characteristics of mesenchymal stem cells derived from the human placenta and umbilical cord. Sci Rep 8(1):1–9

Xiao Q-R, Sun S, Wu K-H et al (2020) A human hair-based platform for long-term maintenance of 3D engineered cardiac tissues in vitro. Colloids Surf A Physicochem Eng 605:125398

Yi X, Chen F, Liu F et al (2020) Comparative separation methods and biological characteristics of human placental and umbilical cord mesenchymal stem cells in serum-free culture conditions. Stem Cell Res Ther 11(1):183

Yu Y, Yang W, Wang B et al (2017a) Structure and mechanical behavior of human hair. Mater Sci Eng C 73:152–163

Yu Y, Yang W, Meyers MA (2017b) Viscoelastic properties of α-keratin fibers in hair. Acta Biomater 64:15–28

Yu J, Wang M-Y, Tai H-C et al (2018) Cell sheet composed of adipose-derived stem cells demonstrates enhanced skin wound healing with reduced scar formation. Acta Biomater 77: 191–200

Yu Y, Cui H, Zhang C et al (2020) Human nail bed extracellular matrix facilitates bone regeneration via macrophage polarization mediated by the JAK2/STAT3 pathway. J Mater Chem B 8(18): 4067–4079

Zajdel A, Kałucka M, Kokoszka-Mikołaj E et al (2017) Osteogenic differentiation of human mesenchymal stem cells from adipose tissue and Wharton's jelly of the umbilical cord. Acta Biochim Pol 64(2):365–369

Zhang X-D, Li J, Niu J-N et al (2019) Fluorescent carbon dots derived from urine and their application for bio-imaging. Methods 168:84–93

Zhang Y, Wang S, Lu F et al (2021) The neuroprotective effect of pretreatment with carbon dots from Crinis Carbonisatus (carbonized human hair) against cerebral ischemia reperfusion injury. J Nanobiotechnology 19(1):1–16

Zhao Z, Liu D, Chen Y et al (2019) Ureter tissue engineering with vessel extracellular matrix and differentiated urine-derived stem cells. Acta Biomater 88:266–279

Zheng H, Yu Z, Deng M et al (2019) Fat extract improves fat graft survival via proangiogenic, anti-apoptotic and pro-proliferative activities. Stem Cell Res Ther 10(1):1–13

The Green Approach-Based Biomaterials for Tissue Engineering Application

2

Atiqah Salleh, Manira Maarof, Zawani Mazlan, Syafira Masri, Nur Izzah Md Fadilah, Safa Tahri, Izzat Zulkiflee, Nusaibah Sallehuddin, Maheswary Thambirajoo, Loai A. Elfawy, Ibrahim N. Amirrah, Nur Amalia Ra'oh, and Mh Busra Fauzi

Abstract

The tissue engineering and regenerative medicine fields have been practising a green approach for decades by fabricating sustainable biomaterials using natural sources or by-products. These materials are not only able to significantly reduce environmental pollution, but they also have a significant impact on different applications of tissue engineering. There are various reports on utilising biological templates and applying green materials in tissue engineering. This chapter discusses the varieties of sustainable biomaterials reported in the tissue engineering field. The chapter also mentions the application of sustainable materials in tissue engineering that covers musculoskeletal to ligament tissue engineering. The challenges and safety of sustainable materials are discussed in this chapter. In conclusion, this part of the chapter discusses the potential application and future direction of sustainable materials in the tissue engineering field.

Keywords

Biomaterial · Sustainable · Tissue engineering · Natural · Green materials

A. Salleh · M. Maarof · Z. Mazlan · S. Masri · N. I. M. Fadilah · S. Tahri · I. Zulkiflee · N. Sallehuddin · M. Thambirajoo · L. A. Elfawy · I. N. Amirrah · M. B. Fauzi (✉)
Centre for Tissue Engineering and Regenerative Medicine, Faculty of Medicine, Universiti Kebangsaan Malaysia, Cheras, Kuala Lumpur, Malaysia
e-mail: fauzibusra@ukm.edu.my

N. A. Ra'oh
Department of Ophthalmology, Faculty of Medicine, Universiti Kebangsaan Malaysia, Cheras, Kuala Lumpur, Malaysia

W. S. Wan Kamarul Zaman, A. A. Nurul (eds.), *Sustainable Material for Biomedical Engineering Application*, https://doi.org/10.1007/978-981-99-2267-3_2

Abbreviations

cGMP	Current good manufacturing practice
COVID-19	Coronavirus disease
ECM	Extracellular matrix
EDS	Glutamic acid-serine
EGCG	Epigallocatechin gallate
FDA	Food and Drug Administration
H_2O_2	Hydrogen peroxide
HA	Hyaluronic acid
hADSCs	Human adipose-derived stem cells
HAp	Hydroxyapatite
LDS	Leucine-aspartate-serine
LDV	Leucine-aspartate-valine
Mg	Magnesium
MgO	Magnesium oxide
NCC	Nanocrystalline cellulose
NIR	Near-infrared spectroscopy
PCL	Polycaprolactone
PEG	Polyethylene glycol
PHAs	Polyhydroxyalkanoates
PLA	Polylactic acid
PLGA	Poly(lactic-co-glycolic) acid
POEGMA	Poly(oligo(ethylene glycol) methyl ether methacrylate)
RGD	Arginine-glycine-aspartic acid
ROS	Reactive oxygen species
SiO_2	Silicon dioxide
TE	Tissue engineering
TERM	Tissue engineering and regenerative medicine
TiO_2	Titanium oxide
USA	United State of America
ZrO_2	Zirconium dioxide

2.1 Introduction

As tissue engineering develops in the production of useful biomedical goods and services, it has become essential to apply green technologies. The term "green technology" refers to any scientific or technological advance that lowers environmental impact. For a sustainable future, scientific research specifically leads to the manufacture of green, durable, eco-friendly, and biodegradable goods (Bradu et al. 2022). Environmentally friendly materials, or "green biomaterials", are quickly developing as great substitutes for polymeric or synthetic materials, leading to

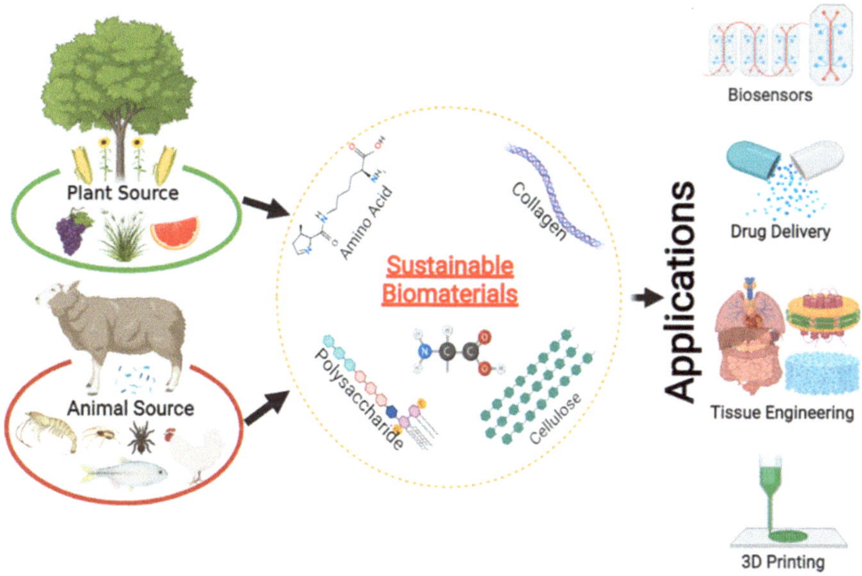

Fig. 2.1 The primary sources and applications of sustainable biomaterials in the field of tissue engineering

several possibilities for their usage especially drug delivery systems, bio-fabrication and overall tissue engineering field. The condition for good biomaterials is programmed to help in interaction with a living entity. Hence, it is important to exhibit good biocompatibility, biodegradability and non-toxic for tissue engineering application (Adetunji et al. 2021).

Majority of eco-friendly biomaterials found in nature are saccharide repeat units (Baranwal et al. 2022), which contribute to the enhancement of mechanical properties by providing crystalline regions. In addition, the incorporation of amino acid repeat units is capable of enhancing green biomaterials ability to degrade and cell compatibility for medical applications (Li 2019). There are several green biomaterials with characteristics that make them appropriate as biomaterials. Lignin and its derivatives may interact with the cell signalling cascade that affects stem cell differentiation by binding to membrane receptors (Osorio et al. 2020). Biomaterials derived from natural sources may be extracted directly from algae, plants and animals (Brovold et al. 2018), or by enzymatic or microbial fermentation (Al-Maqtari et al. 2019). The two chemical classes of green biomaterials are proteins (e.g., collagen, keratin and silk) found in nature and polysaccharides such as cellulose, alginate and hyaluronic acid (Jin et al. 2021). Figure 2.1 shows the overall ideas of green biomaterials and their application in tissue engineering fields.

As for the cell growth curve and behaviour during tissue regeneration, it is crucial to design properly bioscaffolds that resemble the native extracellular matrix (ECM). For instance, scaffolds designed for wound healing should mimic the micro-environment for human dermal fibroblasts, just as bone templates will try to achieve similar porosity and density suitable for growth of osteoblasts. Green tissue engineering requires the use of biodegradable and sustainably available natural materials and substrates to produce these scaffolds (Hussain et al. 2017). These polysaccharides generate sophisticated three-dimensional (3D) structures that have been created by chemical, physical or joint changes to have adequate physical and biocompatibilities and intensively investigated for tissue regeneration (Wu et al. 2021). Natural ingredients, such as honey or vitamins, are applied as coating for nanotechnology or integrated into 3D scaffolds or nanocosmetics (Bonsignore et al. 2021; Ajazzuddin et al. 2015). Bioceramics are used in bone tissue engineering to repair damaged bone tissues, including dental restorations, facial and cranial bone reconstruction, bony defect filling and middle ear reconstruction (Ige et al. 2012). Bioceramics consist of hydroxyapatite, coral, shell and calcium phosphate-containing clay. The applications of green technology in the tissue engineering and biomedical industries are elaborated below.

2.2 Types of Sustainable Materials Used in Tissue Engineering and Biomaterials

In recent research fields, many scientists opt for sustainable materials in their studies to fabricate bioscaffolds that come from flora and fauna, as these materials display good physicochemical characteristics such as biodegradability, compatibility, eco-friendly and biologically active in vivo and in vitro applications (Malik et al. 2021). According to Biswal and co-workers, sustainable materials or natural biopolymers render specific criteria as potential candidates for biomaterials to be used in tissue engineering (TE). These criteria are good interporosity, the high number of pore sizes, increased tensile strength, and the ideal top surface to allow the dissemination of growth factors for cell variance, cell diffusion, cell attachment, cell proliferation and regeneration of new tissue by decreasing the degradation rate (Biswal 2019). Examples of naturally derived materials are alginate, chitosan, collagen, gelatine and many more. Figure 2.2 shows the sustainable materials for tissue engineering.

2.2.1 Alginate

Alginate is a biopolymer polysaccharide naturally found in seaweeds and commercially produced from various algae species, including *Lessonia, Eclonia, Durvillea, Macrocystis, Sargassum, Laminaria and Ascophyllum.* The main component of alginate is the cell wall of *Phaeophyceae*, a brown micro-alga containing divalent alginic acid salt that moulds the intercellular matrix gel (Sahoo and Biswal 2021). As it possesses good wettability, absorbs many exudates and prevents microbial infections, alginate is an ideal wound dressing material to be incorporated into various wound dressing materials such as hydrogel, films, nanofibres, foams, wafers and topical application onto the wounded region. In addition, alginate has good

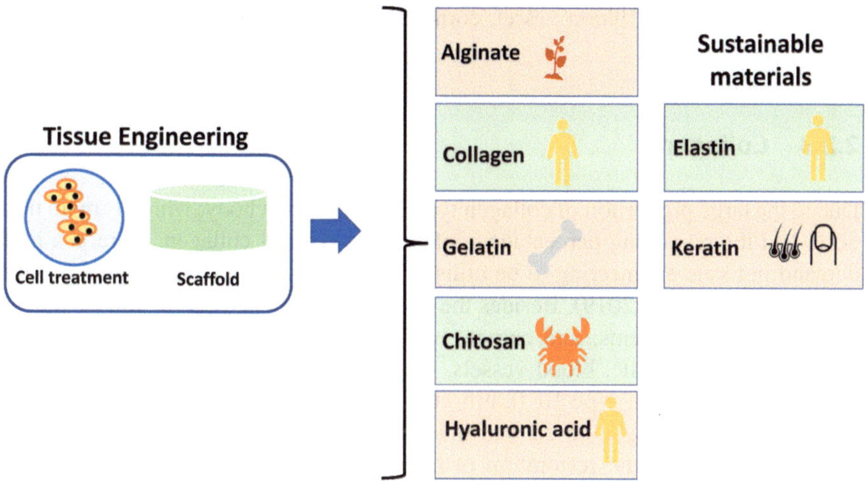

Fig. 2.2 Example of sustainable materials for tissue engineering application

biocompatibility and does not exert any immunogenic reaction; hence, it is widely combined with other polymers, such as hyaluronic acid, to increase the therapeutic effects and minimise the drawbacks of single component materials (Aderibigbe and Buyana 2018). Besides wound healing and wound dressing, alginates also been used in other application of tissue engineering such as expediting skin regeneration (Yang et al. 2019), orthopaedic field as scaffolds for cartilage reconstruction (Farokhi et al. 2020) and bone restructure (Lee et al. 2020), as vehicles to transport drugs or proteins (Hariyadi and Islam 2020), and in dentistry for tissue and bone repairing (Sancilio et al. 2018).

2.2.2 Chitosan

Chitosan is a deacetylated form of chitin acquired from the exoskeletons of crustaceans and occupies a recognised position among other biomaterials due to its unique qualities such as biodegradability, compatibility, less toxicity, water-attractants, haemostatic properties and ability to inhibit microbial infections and use as genes and drugs carrier. In tissue engineering, chitosan is a promising candidate for augmenting cell adhesion, proliferation, differentiation and cell growth by incorporating the chitosan into the scaffold to regenerate the skin or tissue. On top of that, chitosan-based scaffolds can be easily modified into desirable effects by mixing them with natural and synthetic polymers to make them more durable based on the tensile strength and stability that suits the type of application. Modified chitosan has been achieved by combining the scaffolds with biopolymers that include silk, PCL, PLA, HA, chitin and nanoceramics such as TiO_2, SiO_2, ZrO_2 and many more (Ahmed et al. 2018; Ahsan et al. 2018; Islam et al. 2020). Apart from this, chitosan is widely applicable in other medical applications such as restoration of bone tissue, cartilage, blood vessel, corneal, intervertebral disc and gingiva (Islam et al. 2020).

2.2.3 Collagen

Due to the large proportion of collagen type I in the human body, which is more than 80%, and the remaining percentage conferring to type III collagen, there is a high demand in tissue engineering to be utilised as biomaterials for skin wound healing (Davison-kotler et al. 2019). Besides the skin, type I collagen can be found in bone tissue, tendon, ligaments, and cornea, while type II is in cartilage and basement membranes. Meanwhile, blood vessels, lungs, bladder, uterus and vocal folds are composed of collagen type III (Latifi et al. 2018; Marques et al. 2019). Besides wound healing, collagen-based scaffolds, also known as biopolymers, become a viable candidate for the restoration of bone and cartilage (Marques et al. 2019). Collagen type I-based bioscaffolds gained more popularity because of their properties, biocompatibility and biodegradability. They can accelerate cell attachment, cell proliferation, cell differentiation, non-toxicity and display haemostatic characterisation (Busra et al. 2017; Terzi et al. 2018). Furthermore, this polymer hosts keratinocytes and fibroblasts that help develop the cytokines and growth factors essential in the wound repair mechanism (Muthukumar et al. 2018). Additionally, these bioscaffolds can be fabricated into different forms such as sponges, gels, and films to ev.aluate the interaction between cells and bioscaffolds and in vivo studies (Terzi et al. 2018). Though synthetic collagens are available, these polymers can be obtained naturally from human and animal sources, namely ovine, porcine, bovine, fish scales and fish bones (Rezvani Ghomi et al. 2021).

2.2.4 Gelatin

Gelatin, known as biopolymer proteins, results naturally through the partial hydrolysis of collagen derived from animal by-products such as animal skins, bones, hides, fish and insects. Gelatin can also be extracted chemically by dissolving the collagen to break down its structure in an acidic or alkaline medium under heated conditions (Valcarcel et al. 2021). Besides animal sources, gelatin can be extricated from plants such as celluloid, pectin, xanthan gum, agar and modified corn starch. Figure 2.3 shows the process of denaturation of collagen into gelatin. Gelatin serves in 3D cell culture, biomedical and tissue engineering fields as modulators for cell viability, cell growth, cell differentiation, biomaterials fabrication and drug carriers (Arun et al. 2021).

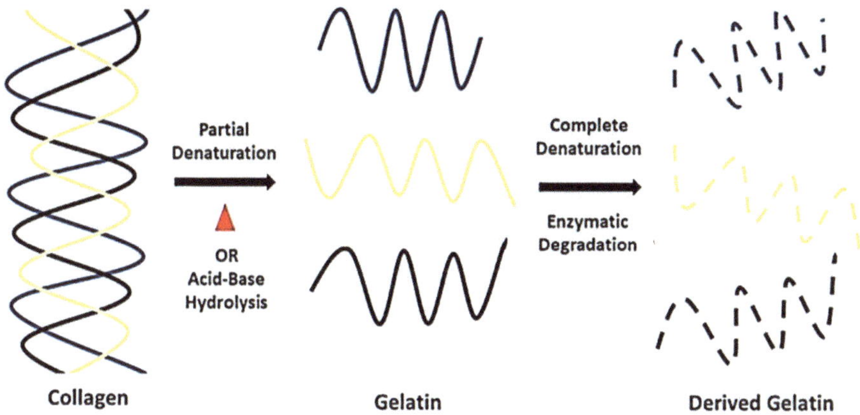

Fig. 2.3 The process of denaturation of collagen into gelatin. (The figure was adapted from Zulkiflee and Fauzi (2021) and was licensed under Creative Commons CC by 4.0)

2.2.5 Hyaluronic Acid

The naturally occurring polysaccharide hyaluronic acid (HA) which play important role in the ECM (Rezaeeyazdi et al. 2018). HA has been utilised to create hydrogel scaffolds that mimic various tissues. Due to its excellent biocompatibility and non-toxic breakdown products, HA has become a promising biomaterial and polymer for tissue engineering (Zhu et al. 2017; Hemshekhar et al. 2016). Several medical uses involve HA, including visco-supplementation, medicine delivery, eye surgery, tissue regeneration, and embryo protection (Rachmiel et al. 2021). HA highly mimics water content in human tissue, is biocompatible, and plays a vital function in promoting cell proliferation and differentiation activity. Despite its good biocompatibility, HA has poor mechanical strength and as a polyelectrolyte, contains a high concentration of free charge carriers (Tsanaktsidou et al. 2022). Therefore, molecular self-assembly can also be used to reinforce HA scaffolds. HA has evolved into the most versatile biopolymer, with applications in tissue regeneration and medicine such as ophthalmology, dermatology, and orthopaedics. Figure 2.4 shows the chemical structure of HA.

2.2.6 Keratin

Keratin is the structural protein most frequent in animal hair, nails, feathers, horns and claws (Ranjit et al. 2022). It is the most significant biopolymer encountered in animals, along with collagen. Keratin can self-assemble and form polymers by itself. It is biocompatible and capable of supporting cellular proliferation, a suitable candidate especially for specific drug delivery systems. Keratin comprises numerous cells specific binding sites that enable adhesion of cell, such as leucine-aspartate-valine (LDV) and glutamic acid-serine (EDS) integrin domains that imitate the ECM

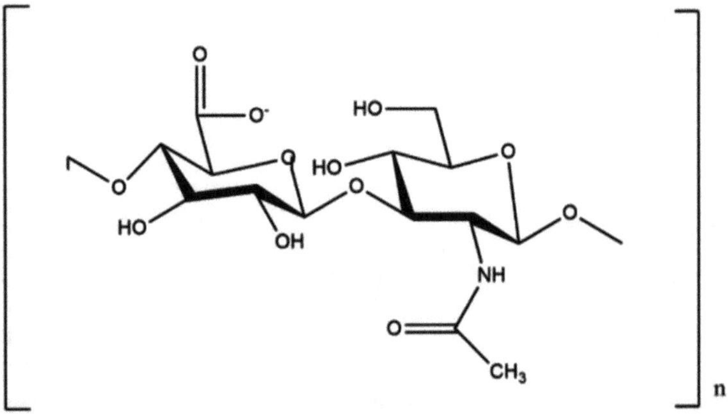

Fig. 2.4 Chemical structure of hyaluronic acid (Tsanaktsidou et al. 2022). (The figure was licensed under Creative Commons CC by 4.0)

(Stojic et al. 2021). In TE applications, keratin can be sculpted into various biomaterial models such as sponge and hydrogel for tissue regeneration. Keratin is distinguished from many conventional biomaterials by its physicochemical and inert properties, including its chemical composition, inherent self-assemble ability, existence of cell-binding domains and bacteriostatic properties.

2.2.7 Elastin

Elastin, an insoluble tropoelastin polymer of which is the primary substance in the elastic fibres, provides mechanical strength as well as elasticity to nearly all connective tissues. Elastin is a critical protein component of the ECM found in many mammalian tissues that require elasticity to function (Mora et al. 2018). Elastin can be found chiefly in tissues in the form of fibres as it provides resilience and elasticity needed in skin, arteries, ligaments and lungs (Yeo et al. 2018). Elastin allows these tissues to deform repeatedly and reversibly during regular operation, which explains its prevalence in the ECM of various extensible tissues such as arteries, skin, lungs, and cartilage. Besides, it has a well-known purpose as a supporting and protecting scaffold. Elastin's mechanical nature influences cell properties (Gilbert 2020). The architecture of the elastin matrix defines cell shape, distribution, alignment and organisation within tissues.

2.3 How Sustainable Materials Applied in Biomedical Field

Sustainable biomaterials were widely employed worldwide in the previous decades due to the vast development of these technologies. Engineered biomaterials are applied in various areas for the utilisation of musculoskeletal, bone, cartilage, ligament, and skin tissue engineering, as well as for controlled drug delivery, protein delivery, and enzyme delivery for their biomimetic and functionaliation properties.

2.3.1 Musculoskeletal Tissue Engineering

Recently, tissue engineering has been the alternative strategy to regenerate skeletal muscle via biodegradable scaffolds (Rossi et al. 2010). In skeletal muscle tissue engineering, the selection of biomaterials is crucial as it affects the myogenic process since myogenic stem cells reside and differentiate within this 3D scaffold. A variety of sustainable biomaterials of 3D scaffolds, including collagen, hyaluronan (HAp), alginate, polyethylene glycol (PEG) and hydroxyapatite (HAp) are currently being explored by researchers to be applied to skeletal tissue engineering (Rossi et al. 2011).

Allan et al. (2021) decellularised the amenity grass while retaining its natural striated topography with a The parallel vascular system in grass blades serves as a scaffold that promotes the adhesion, proliferation, alignment and differentiation of murine C2C12 myoblasts via contact guidance. This decellularised amenity grass is an inexpensive, sustainable, and passive scaffold (without further modification or functionalisation to create cellular alignment) (Allan et al. 2021). Chau et al. (2016) prepared freeze-casted hydrazone hydrogels and aerogels crosslinked with poly (oligoethylene glycol methacrylate) (POEGMA) and NCC (Chau et al. 2016). This hydrogel has tunable morphologies and anisotropies, is mechanically stable and has good biocompatibility, which promotes the differentiation of myoblast into myotubes and aligned NCC was capable of ameliorating the myotube orientation. Hence, it exhibits a potential as a scaffold for muscular tissue engineering (De France et al. 2017).

Silk fibroins are another sustainable biomaterial of choice in musculoskeletal tissue engineering applications as they have high mechanical strength, a low degradation rate and excellent biocompatibility (Zhang et al. 2021). Electrospun silk fibroin scaffolds imitate the structural organisation of natural skeletal muscles. It causes muscle tissue morphogenesis by activating myogenic cells to create more potent regenerative effects, by giving signals for aligned myotube creation, and by boosting the production of myogenic proteins (Dugan et al. 2013). Manchineella et al. (2016) also introduced antioxidant and electroactive properties into the scaffold construction, enhancing skeletal muscle regeneration (Manchineella et al. 2016). Electrospun PCL/collagen nanofibers promote cellular organisation by guiding morphogenesis. Unidirectional orientation of nanofibers in the scaffold influenced the muscle cells and myotube formed alignment (Choi et al. 2008). In another study, electrospun biodegradable block copolymer (DegraPol®) micro fibrous membrane also has good mechanical properties and supports cell adhesion, proliferation and viability, which are suitable for their application in skeletal muscle tissue engineering (Riboldi et al. 2005).

2.3.2 Bone Tissue Engineering

The objective of bone tissue engineering is to produce an ideal bone scaffold that has suitable mechanical properties, is biocompatible with the cell and promotes bone regeneration (Font Tellado et al. 2017). Hence, selecting the biomaterial and scaffold design is crucial to ensure it promotes cell proliferation, migration, differentiation and osteointegration (Orciani et al. 2017). The common sustainable biomaterials used for bone tissue engineering include calcium phosphates in the form of HAp, metals such as titanium (titanium alloys), magnesium (Mg), and polyhydroxyalkanoates (PHAs) as well as polylactic acid (PLA) (Gregor et al. 2017; Dwivedi et al. 2020; Wubneh et al. 2018).

Floreon™ is a sustainable 3D filament of blended polyesters consisting of PLA, polycaprolactone (PCL), and ceramic materials with superior mechanical properties (Delgado-Aguilar et al. 2020). The modification of the Floreon™ with a bioceramic (wollastonite) promotes the differentiation of osteoblastic cells, which is suitable for bone regeneration scaffolds (Ramos-Rodriguez et al. 2022). Sooriyaarachchi et al. (2022) also developed electrospun-aligned PCL nanofibers that can enhance the fibroblast cells' proliferation, organisation and attachments (Sooriyaarachchi et al. 2022). Electrospun PLGA/PCL scaffold coated with collagen and impregnated with silver has osteogenic and antimicrobial properties for orofacial tissue regeneration (Font Tellado et al. 2017; Qian et al. 2019). Moreover, aligned nanofibers PCL and PEG scaffold loaded with plant extract *Elaeagnus angustifolia* (EA) have antibacterial, anti-toxic and anti-inflammatory properties suitable for bone tissue engineering (Hokmabad et al. 2019; Abizov et al. 2008; Panahi et al. 2016).

Moreover, Mg is another sustainable biomaterial that has attracted a lot of attention from researchers for bone tissue engineering. The novel nanofibrous PCL scaffold loaded with Mg phosphate and Mg oxide (MgO) nanoparticles was a potential scaffold for bone engineering as it supports cell viability and spread with mesenchymal stem cells (Xing et al. 2020; Perumal et al. 2020). After 10 and 26 weeks post-implantation, the Mg alloy implant also regenerates bone tissue with just a slight alteration in blood composition (Zhang et al. 2009). Additionally, 3D-printed PLGA/TCP porous scaffolds with Mg have a higher osteogenic ability for bone defect healing (a 12 mm segment of the ulna in rabbits) (Lai et al. 2022).

2.3.3 Cartilage Tissue Engineering

Cartilage is a connective tissue that protects the bone and allows mobility. Cartilage is composed of 80% water, is alymphatic, aneural and avascular, and contains just one type of self-healing cell, chondrocytes. However, only 1–5% chondrocytes are present in total cartilage tissue, limiting the ability for spontaneous repair (Karuppal 2017). Articular cartilage defects due to traumatic injury, disease or ageing commonly occur and can lead to osteoarthritis, functional impairment and degenerative arthritis, affecting daily activities (Karuppal 2017; Lories and Luyten 2022; Chinta et al. 2021). Current clinical treatments, including microfracture and autologous chondrocyte implantation, cannot produce hyaline repair tissue consistently. Hence, creating chondrocyte cell-based scaffolds is a new strategy to regenerate cartilage tissue (Lories and Luyten 2022; Kerkhoffs et al. 2002). Fabricating scaffolds with good mechanical and biocompatible properties is challenging in ligament tissue engineering applications (Lim et al. 2019; Williams 2019).

Biocart™ II is one of the examples of commercialised bone tissue engineering product that is synthesised by using donor chondrocytes in freeze-dried fibrin and hyaluronan. Another bone template is Cartipatch composed of an alginate and agarose hydrogel. NeoCart® is bone template that is synthesised from type I collagen from a bovine source. In general, preclinical results demonstrated that natural biopolymer scaffold techniques are superior tto synthetic polymer scaffolds. The majority abovementioned product already reached Phase II or III clinical trials. NeoCart® for example have done clinical safety trial with duration of 2 years that involves eight patients diagnosed with full-thickness cartilage injury where 75% of the patients able to achieve good to complete (67–100%) healing within 24 months (Crawford et al. 2009). However, reimbursement is delayed because of the delay in research initiation and the variable clinical use of sustainable biomaterials (Williams 2019).

2.3.4 Skin Tissue Engineering

Skin is one of the biggest organs in the human body. When an injury occurs inside the skin, connective tissues or mucous membrane, structural and/or functional abnormalities of the organs may ensue (Zawani and Fauzi 2021a). The skin has an intuitive ability to regenerate and heal, however, scar tissues are deemed as cosmetically undesirable as they lose elasticity and flexibility hence tissue-engineered biomaterials are utilised. Several state-of-the-art commercialised hydrogels for tissue engineering with FDA certification, such as Apligraf® and Leucopatch®, which aid in accelerating wound healing and skin regeneration (Zawani and Fauzi 2021a). Other than that, antioxidant-embedded biomaterials are also widely studied to combat ROS and modulate redox-mediated cell homeostasis (Zawani and Fauzi 2021b). Lee et al. examined the antioxidative and ROS scavenging characteristics of EGCG-coated biomaterials intended for tissue engineering in recent research. EGCG-coated polycaprolactone (PCL) film accelerated human adipose-derived stem cells (hADSCs) attachment and proliferation in response to H_2O_2 exposure, while modulating cell signalling to reduce apoptotic genes and increasing the production of an anti-oxidative enzyme (Lee et al. 2021).

The commercialised skin templates have also shown improvement in wound conditions in clinical trials. Apligraf® is a bilayered collagen skin substitute consisting of both layers of fibroblasts and keratinocytes bioengineered to cater hard-to-heal wounds such as diabetic foot ulcers and pressure ulcers. Meanwhile, EpiFix consists of both dehydrated human amnion and chorion membrane allograft intended to improve both acute and chronic wounds. A total of 20 patients have been included in clinical study to compare these commercialised skin templates with standard care for the treatment of chronic lower diabetic ulcers. The study recorded that the commercialised skin template achieved faster wound healing with 45% of patients treated with Apligraf are completely healed by 6 weeks compared to standard treatment where only 35% of patients achieve full recovery. Meanwhile, Epifix showed the highest healing rates compared to both treatments where more than 90% of patients were able to achieve full recovery by 6 weeks (Zelen et al. 2015). The presence of bioengineered skin templates has been proven to help fasten the recovery process for the hard-to-heal wounds.

2.3.5 Drug, Enzyme and Protein Delivery

This system has accelerated drug delivery and efficiency to the immediate site. In this era, various methods are available to deliver enzymes, drugs and proteins according to the desired rate required by our body's physiological environment. Plasmid and viral-based vehicles are the most commonly used for this system as they aid in tissue repairs and delivery. However, with the advancement of technologies and scientific discoveries, various biomaterials were successfully created for gene delivery systems. In a recent study, poly(ethylene glycol) (PEG), PLA-PEG and PLGA in different nanocarriers, microcarriers, fibres and hydrogels utilised for the systems. These protein capsules loaded using one of three distinct methods, which are immersion, diffusion or immobilisation. They could be released by three significant mechanisms, which are diffusion, natural erosion caused by local stimuli (such as pH, ROS, redox, glucose or protease), or triggered degradation caused by external stimuli (visible and NIR light, magnetic field, mechanical stress) (Bizeau and Mertz 2021).

Genexol® is a commercially available polymeric micelle that has passed pharmacokinetic and clinical studies in various treatment options that include breast cancer and lung cancer. In the phase I clinical trial of Genexol®, a total of 20 patients diagnosed with cancer have administered Genexol-PM and observed for 3 weeks to determine the maximum dosage and toxicity of substances. It is determined that Genexol® does not induce any acute hypersensitivity and maximum dosage of Genexol® is at 390 mg/m^2 (Kim et al. 2004). In a phase II clinical trial, a combination of Genexol® and carboplatin was administered to 28 patients with advanced non-small cell lung cancer. Combination administration of both therapies does not induce grade 3–4 hypersensitivity in patients. However, peripheral neuropathy was observed in 42.9% of patients, which results in reduction of dose. The combination chemotherapy showed positive tumour reduction of 35.7% within 3.2 months (Kim et al. 2011). Based on the studies conducted, biomaterials have the potential to act as drug capsules and used in combined treatments for complicated diseases. However, researchers need to know the challenges of applying materials in health industry.

2.4 The Challenges and Safety of Using Materials

The primary feature of all materials is their compatibility with biological systems; hence, all materials must be biodegradable and biocompatible. Biocompatibility is the behaviour of a substance in a living tissue that does not have an unfavourable consequence, such as a poisonous, physiological or immunological reaction. Sustainable production is an important aspect of companies' survival. Regarding this, environmental issues need to be addressed to preserve the livings' well-being and green environment. In this section, we will focus on the number of challenges and safety limitations of using sustainable materials.

2.4.1 Disease Transmission from Animal Derived Materials

Animal-derived materials play an established role in tissue engineering. However, employing animal components such as collagen, which is predominantly derived from animal sources, increases the danger of disease transmission from animals to humans. Up to 2–4% of the global population may be allergic to collagen generated from pigs and cows. The risks of zoonotic diseases are one of the world's greatest worries, especially in light of the recent COVID-19 pandemic, while incidences of bovine spongiform encephalopathy have been linked to the use of prion-contaminated bovine scaffolds (Gauza-Włodarczyk et al. 2017). The 1998 publication and 2019 revision of the guideline paper entitled "Medical Devices Containing Materials Derived from Animal Sources Except for In Vitro Diagnostic Devices" have altered our knowledge of transmissible illnesses and their influence on the medical device market. The final advice gives recommendations for mitigating these risks and aims to assist producers in identifying the potential danger associated with animal tissues utilised in medical devices.

2.4.2 Control of Infection

Biomaterials have reformed modern medicine; however, they increase the risk of infection. Medical devices account for almost 25.6% of infections in the USA, and the consequences are detrimental (Magill et al. 2014). Bacterial contamination of the medical device differs from the tissue contamination (Rimondini et al. 2005). The triad of infection consists of pathogens, biomaterials and the host immune response. The most common bacteria causing biomaterial infection are those within the normal flora of the skin, including *Staphylococcus epidermidis* (polymeric material) and *Staphylococcus aureus* (metallic material) (Leake et al. 1982; Christensen 1989). Typically, the tissue contamination caused by the opportunistic infection is spontaneously eliminated by the immunological response of the host. In contrast, the presence of biomaterials induced an inflammatory response, connective tissue development, and encapsulation of fibre (Anderson 2016). This condition increases the susceptibility to bacterial adhesion and colonisation (Schierholz and Beuth

2001). Attachment of bacteria can be divided into two categories, first, primary unspecific, reversible; and second, specific, irreversible attachment. Biomaterial factors promoting bacterial adhesion include size, shape, surface, porosity, stability, chemistry and roughness of the biomaterial (Rimondini et al. 2005). Roughness and porosity augmented bacteria adhesion by encouraging specific and stable binds between the cells and the surfaces (Arciola et al. 2018). In addition, the condition of the site of implantation, time of application, and host immune response also influences the susceptibility of the medical device to bacterial adhesion.

2.4.3 Scaling Up

Concerning the translatable and producible nature of biomaterials for clinical applications is the product's scalability to a current good manufacturing standard (cGMP) procedure (Ungerleider and Christman 2014). Due to the fact that natural biomaterials are produced from human or animal origins, various batches will have varying age, time, and health condition variances, and hence different efficacy. Thus, it was determined that younger tissue supplies are more mechanically stable than older tissue sources (Sicari et al. 2012; Tottey et al. 2011). According to cGMP guidelines, scaling up to industrial quantities might take a considerable period of time. In addition, scaling up cGMP is expensive and should not be disregarded. Additionally, the efficacy of delicate processes such as cell encapsulation may be compromised. Some batches struggle to achieve encapsulation efficiencies greater than 20% even at the bench level (Mayfield et al. 2014). Preserving and transporting large quantities of biomaterials is an additional difficulty associated with scaling up (Xu et al. 2019). Moreover, as cells age, their potential for proliferation and differentiation decreases upon implantation (Mekhileri et al. 2018). Manufacturing with more durable components might be easier to scale up than unstable components such as cells or growth factors. One of the ways to tackle this problem is via biofabrication. Biofabrication aims to achieve an effective biomaterials-cells product that is able to differentiate and mature after implantation. Appropriate biofabrication techniques involving live cells are bioprinting and photolithography. The bioprinting of human endothelial and mesenchymal stem cells onto a urethane urea cardiac patch restored vascular function in the infarcted region of mice (Gaebel et al. 2011).

2.5 The Future Perspective on the Uses of Sustainable Materials

This chapter reviewed the features of biomaterials and their sustainable significance in the tissue engineering and regenerative medicine (TERM) field from a green technology viewpoint. Developing novel biomaterials is one of the fastest-growing areas in the biomedical arena. Green biomaterials and sustainable biomass such as those derived from renewable resources, including raw materials from ocean plants, agricultural processes, animal residues, and marine origin, can be used as alternatives to petrochemical polymers and can offer a new range of properties for use not only in therapeutic applications but also in diagnosis and imaging in medical practice. For instance, in recent years, materials derived from natural renewable resources such as chitosan, alginate, starch, cellulose, collagen and gelatin have attracted much attention in a wide range of biomedical implementations, that include medical implants, drug delivery and scaffolds for tissue engineering as well (Trache 2018). Along with these developments, the identification and use of renewable resources as sustainable biomass for innovative biomaterial processing is expected to contribute to increasing the sustainability needs of our world. In the main, sustainable-derived products have been broadly applied in packaging, energy storage and even biomedicine. They contribute considerably to economic viability and provide a cleaner environment. Green biomaterials are necessary for the sustainable development of contemporary global concerns in order to save the environment and lower product prices.

Recently, the TERM field has focused on finding a novel, sustainable method for developing, validating, sterilising and large-scale production of environmentally safe or green biomaterials for tissue engineering applications (Han et al. 2020). However, there are lots of challenges in producing green biomaterials that can meet all the safety regulations requirements and maintain the products' high-quality standards. The production of natural or biological materials in a controlled environment under safe manufacturing technology will be essential to prevent contamination and produce a safe, standard product. Besides, the selection and combinations of the biomaterials with specific criteria or designs need to be validated. The synthesis of green biomaterials with a unique technique that can reduce toxicity levels or harmful chemicals or compounds would be great for health and the environment.

The developed sustainable and green biomaterials (e.g., polymers, ceramics, metals, composites, nanofibers, collagen and gelatin) as a 3D scaffold would be beneficial for tissue regeneration for tissue engineering applications or in vitro models (Utami Nike et al. 2022; Fadilah et al. 2022; Sallehuddin et al. 2022; Maarof et al. 2021). The outstanding properties and efficacy, biocompatibility, biodegradability and various shapes and forms of the biomaterials provide the suitability and promising future needed to engineer tissues or organs such as skin, bone, cartilage, etc. The improvement in the technology for tissue and organ development and replacement would provide a major relief for many patients suffering from tissue or organ injury or failure. Besides that, the era of multi-dimensional digital printing technology, including 3D, four-dimensional (4D) and five-dimensional (5D), is a promising technology for multi-scale engineering towards injectable bio-ink of high complexity structures (Bishop et al. 2017). It is a relatively innovative tissue engineering technology employing natural biomaterials (safe and eco-friendly) with beneficial green technologies due to their biocompatibility and greater potential for supporting cell survival and proliferation than synthetic bio-inks. With growing understanding, green biomaterials exhibit a good potential in TERM.

Acknowledgement We are appreciative to the Faculty of Medicine at UKM for funding the completion of this chapter.

Conflict of Interest The authors declare no conflict of interests.

References

Abizov EA, Tolkachev ON, Mal'Tsev SD, Abizova EV (2008) Composition of biologically active substances isolated from the fruits of Russian olive (Elaeagnus angustifolia) introduced in the European part of Russia. Pharm Chem J 42:696–698

Aderibigbe BA, Buyana B (2018) Alginate in wound dressings. Pharmaceutics 10:42. https://doi.org/10.3390/pharmaceutics10020042

Adetunji CO, Olaniyan OT, Anani OA et al (2021) Bionanomaterials for green bionanotechnology. In: Bionanomaterials, pp 10–24

Ahmed S, Annu AA, Sheikh J (2018) A review on chitosan centred scaffolds and their applications in tissue engineering. Int J Biol Macromol 116:849–862. https://doi.org/10.1016/j.ijbiomac.2018.04.176

Ahsan SM, Thomas M, Reddy KK et al (2018) Chitosan as biomaterial in drug delivery and tissue engineering. Int J Biol Macromol 110:97–109. https://doi.org/10.1016/j.ijbiomac.2017.08.140

Ajazzuddin M, Jeswani G, Jha A (2015) Nanocosmetics: past, present and future trends. Recent Pat Nanomed 5:3–11. https://doi.org/10.2174/1877912305666150417232826

Allan SJ, Ellis MJ, De Bank PA (2021) Decellularized grass as a sustainable scaffold for skeletal muscle tissue engineering. J Biomed Mater Res A 109:2471–2482. https://doi.org/10.1002/jbm.a.37241

Al-Maqtari QA, Waleed AA, Mahdi AA (2019) Microbial enzymes produced by fermentation and their applications in the food industry—a review. Int J Agric Innov Res 8:62–82

Anderson JM (2016) Future challenges in the in vitro and in vivo evaluation of biomaterial biocompatibility. Regen Biomater 3:73–77. https://doi.org/10.1093/RB/RBW001

Arciola CR, Campoccia D, Montanaro L (2018) Implant infections: adhesion, biofilm formation and immune evasion. Nat Rev Microbiol 16:397–409. https://doi.org/10.1038/s41579-018-0019-y

Arun A, Malrautu P, Laha A, Ramakrishna S (2021) Gelatin nanofibers in drug delivery systems and tissue engineering. Eng Sci 16:71–81. https://doi.org/10.30919/es8d527

Baranwal J, Barse B, Fais A et al (2022) Biopolymer: a sustainable material for food and medical applications. Polymers (Basel) 14:1–22. https://doi.org/10.3390/polym14050983

Bishop ES, Mostafa S, Pakvasa M et al (2017) 3-D bioprinting technologies in tissue engineering and regenerative medicine: current and future trends. Genes Dis 4:185–195. https://doi.org/10.1016/j.gendis.2017.10.002

Biswal T (2019) Biopolymers for tissue engineering applications: a review. Mater Today Proc 41: 397–402. https://doi.org/10.1016/j.matpr.2020.09.628

Bizeau J, Mertz D (2021) Design and applications of protein delivery systems in nanomedicine and tissue engineering. Adv Colloid Interf Sci 287:102334. https://doi.org/10.1016/j.cis.2020.102334

Bonsignore G, Patrone M, Martinotti S, Ranzato E (2021) "Green" biomaterials: the promising role of honey. J Funct Biomater 12:72. https://doi.org/10.3390/jfb12040072

Bradu P, Biswas A, Nair C et al (2022) Recent advances in green technology and Industrial Revolution 4.0 for a sustainable future. Springer, Berlin

Brovold M, Almeida JI, Pla-Palacín I et al (2018) Naturally-derived biomaterials for tissue engineering applications. In: Chun HJ, Park K, Kim C-H, Khang G (eds) Advances in experimental medicine and biology. Springer, Singapore, pp 421–449

Busra FM, Lokanathan Y, Nadzir MM et al (2017) Attachment, proliferation, and morphological properties of human dermal fibroblasts on ovine tendon collagen scaffolds: a comparative study. Malays J Med Sci 24:33–43. https://doi.org/10.21315/mjms2017.24.2.5

Chau M, De France KJ, Kopera B et al (2016) Composite hydrogels with tunable anisotropic morphologies and mechanical properties. Chem Mater 28:3406–3415. https://doi.org/10.1021/acs.chemmater.6b00792

Chinta ML, Velidandi A, Npp P et al (2021) Assessment of properties, applications and limitations of scaffolds based on cellulose and its derivatives for cartilage tissue engineering: a review. Int J Biol Macromol 1:495–515

Choi JS, Lee SJ, Christ GJ et al (2008) The influence of electrospun aligned poly(ε-caprolactone)/collagen nanofiber meshes on the formation of self-aligned skeletal muscle myotubes. Biomaterials 29:2899–2906. https://doi.org/10.1016/j.biomaterials.2008.03.031

Christensen BE (1989) The role of extracellular polysaccharides in biofilms. J Biotechnol 10:181–202. https://doi.org/10.1016/0168-1656(89)90064-3

Crawford DC, Heveran CM, Cannon WD et al (2009) An autologous cartilage tissue implant neocart for treatment of grade III chondral injury to the distal femur: prospective clinical safety trial at 2 years. Am J Sports Med 37:1334–1343. https://doi.org/10.1177/0363546509333011

Davison-kotler E, Marshall WS, Garc E (2019) Sources of collagen for biomaterials in skin wound healing. Bioengineering 6:56. https://doi.org/10.3390/bioengineering6030056

De France KJ, Yager KG, Chan KJW et al (2017) Injectable anisotropic nanocomposite hydrogels direct in situ growth and alignment of myotubes. Nano Lett 17:6487–6495. https://doi.org/10.1021/acs.nanolett.7b03600

Delgado-Aguilar M, Puig R, Sazdovski I (2020) Polylactic acid/polycaprolactone blends F-I-PPP On the path to circular economy, substituting single-use commodity plastic products. mdpi.com. 13(11). [cited 26 Jun 2022]

Dugan JM, Collins RF, Gough JE, Eichhorn SJ (2013) Oriented surfaces of adsorbed cellulose nanowhiskers promote skeletal muscle myogenesis. Acta Biomater 9:4707–4715. https://doi.org/10.1016/j.actbio.2012.08.050

Dwivedi R, Pandey R, Kumar S, Mehrotra D (2020) Poly hydroxyalkanoates (PHA): role in bone scaffolds. J Oral Biol Craniofacial Res 10:389–392. https://doi.org/10.1016/j.jobcr.2019.10.004

Fadilah NIM, Isa IL M, Zaman WSWK et al (2022) The effect of nanoparticle-incorporated natural-based biomaterials towards cells on activated pathways: a systematic review. Polymers (Basel) 14:476. https://doi.org/10.3390/polym14030476

Farokhi M, Jonidi Shariatzadeh F, Solouk A, Mirzadeh H (2020) Alginate based scaffolds for cartilage tissue engineering: a review. Int J Polym Mater Polym Biomater 69:230–247. https://doi.org/10.1080/00914037.2018.1562924

Font Tellado S, Bonani W, Balmayor ER et al (2017) Fabrication and characterization of biphasic silk fibroin scaffolds for tendon/ligament-to-bone tissue engineering. Tissue Eng A 23:859–872. https://doi.org/10.1089/ten.tea.2016.0460

Gaebel R, Ma N, Liu J et al (2011) Patterning human stem cells and endothelial cells with laser printing for cardiac regeneration. Biomaterials 32:9218–9230. https://doi.org/10.1016/j.biomaterials.2011.08.071

Gauza-Włodarczyk M, Kubisz L, Mielcarek S, Włodarczyk D (2017) Comparison of thermal properties of fish collagen and bovine collagen in the temperature range 298–670 K. Mater Sci Eng C 80:468–471. https://doi.org/10.1016/j.msec.2017.06.012

Gilbert JL (2020) Metallic degradation and the biological environment. In: Wagner WR, Sakiyama-Elbert SE, Zhang G, Yaszemski M (eds) Biomaterials science, 4th edn. Elsevier, pp 941–954

Gregor A, Filová E, Novák M et al (2017) Designing of PLA scaffolds for bone tissue replacement fabricated by ordinary commercial 3D printer. J Biol Eng 11:1. https://doi.org/10.1186/s13036-017-0074-3

Han F, Wang J, Ding L et al (2020) Tissue engineering and regenerative medicine: achievements, future, and sustainability in Asia. Front Bioeng Biotechnol 8:83. https://doi.org/10.3389/fbioe.2020.00083

Hariyadi DM, Islam N (2020) Current status of alginate in drug delivery. Adv Pharmacol Pharm Sci 2020:8886095. https://doi.org/10.1155/2020/8886095

Hemshekhar M, Thushara RM, Chandranayaka S et al (2016) Emerging roles of hyaluronic acid bioscaffolds in tissue engineering and regenerative medicine. Elsevier B.V.

Hokmabad VR, Davaran S, Aghazadeh M et al (2019) Effect of incorporating Elaeagnus angustifolia extract in PCL-PEG-PCL nanofibers for bone tissue engineering. Front Chem Sci Eng 1:1

Hussain Z, Thu HE, Shuid AN et al (2017) Recent advances in polymer-based wound dressings for the treatment of diabetic foot ulcer: an overview of state-of-the-art. Curr Drug Targets 19:527–550. https://doi.org/10.2174/1389450118666170704132523

Ige OO, Umoru LE, Aribo S (2012) Natural products: a minefield of biomaterials. ISRN Mater Sci 2012:983062. https://doi.org/10.5402/2012/983062

Islam MM, Shahruzzaman M, Biswas S et al (2020) Chitosan based bioactive materials in tissue engineering applications—a review. Bioact Mater 5:164–183. https://doi.org/10.1016/j.bioactmat.2020.01.012

Jin M, Shi J, Zhu W et al (2021) Polysaccharide-based biomaterials in tissue engineering: a review. Tissue Eng B Rev 27:604–626. https://doi.org/10.1089/ten.teb.2020.0208

Karuppal R (2017) Current concepts in the articular cartilage repair and regeneration. J Orthop 1:2

Kerkhoffs GM, Rowe BH, Assendelft WJ et al (2002) Immobilisation and functional treatment for acute lateral ankle ligament injuries in adults. In: Kerkhoffs GM (ed) Cochrane database of systematic reviews. Wiley, Chichester

Kim TY, Kim DW, Chung JY et al (2004) Phase I and pharmacokinetic study of Genexol-PM, a Cremophor-free, polymeric micelle-formulated paclitaxel, in patients with advanced malignancies. Clin Cancer Res 10:3708–3716. https://doi.org/10.1158/1078-0432.CCR-03-0655

Kim HJ, Kim KH, Yun J et al (2011) Phase II clinical trial of Genexol® (paclitaxel) and carboplatin for patients with advanced non-small cell lung cancer. Cancer Res Treat 43:19–23. https://doi.org/10.4143/crt.2011.43.1.19

Lai Y, Li Y, Cao H, et al (2022) Osteogenic magnesium incorporated into PLGA/TCP porous scaffold by 3D printing for repairing challenging bone defect. Elsevier. [cited 26 Jun]

Latifi N, Asgari M, Vali H, Mongeau L (2018) A tissue-mimetic nano-fibrillar hybrid injectable hydrogel for potential soft tissue engineering applications. Sci Rep 8:1–18. https://doi.org/10.1038/s41598-017-18523-3

Leake ES, Gristina AG, Wright MJ (1982) Use of chemotaxis chambers for studying in vitro bacterial colonization of biomaterials. J Clin Microbiol 15:320–323. https://doi.org/10.1128/jcm.15.2.320-323.1982

Lee J, Hong J, Kim WJ, Kim GH (2020) Bone-derived dECM/alginate bioink for fabricating a 3D cell-laden mesh structure for bone tissue engineering. Carbohydr Polym 250:116914. https://doi.org/10.1016/j.carbpol.2020.116914

Lee S, Lee J, Byun H et al (2021) Evaluation of the anti-oxidative and ROS scavenging properties of biomaterials coated with epigallocatechin gallate for tissue engineering. Acta Biomater 124:166–178

Li YCE (2019) Sustainable biomass materials for biomedical applications. ACS Biomater Sci Eng 5:2079–2092. https://doi.org/10.1021/acsbiomaterials.8b01634

Lim WL, Liau LL, Ng MH et al (2019) Current progress in tendon and ligament tissue engineering. Tissue Eng Regen Med 16:549–571. https://doi.org/10.1007/s13770-019-00196-w

Lories RJ, Luyten FP (2022) The bone-cartilage unit in osteoarthritis. Nat Rev Rheumatol 2010:43–49

Maarof M, Nadzir MM, Mun LS et al (2021) Hybrid collagen hydrogel/chondroitin-4-sulphate fortified with dermal fibroblast conditioned medium for skin therapeutic application. Polymers (Basel) 13:1–14. https://doi.org/10.3390/polym13040508

Magill SS, Edwards JR, Bamberg W et al (2014) Multistate point-prevalence survey of healthcare-associated infections. N Engl J Med 370:1198–1208. https://doi.org/10.1056/NEJMoa1306801

Malik S, Sundarrajan S, Hussain T et al (2021) Sustainable nanofibers in tissue engineering and biomedical applications. Mater Des Process Commun 3:164–183. https://doi.org/10.1002/mdp2.202

Manchineella S, Thrivikraman G, Khanum KK et al (2016) Pigmented silk nanofibrous composite for skeletal muscle tissue engineering. Adv Healthc Mater 5:1222–1232. https://doi.org/10.1002/adhm.201501066

Marques CF, Diogo GS, Pina S et al (2019) Collagen-based bioinks for hard tissue engineering applications: a comprehensive review. J Mater Sci Mater Med 30:32. https://doi.org/10.1007/s10856-019-6234-x

Mayfield AE, Tilokee EL, Latham N et al (2014) The effect of encapsulation of cardiac stem cells within matrix-enriched hydrogel capsules on cell survival, post-ischemic cell retention and cardiac function. Biomaterials 35:133–142. https://doi.org/10.1016/j.biomaterials.2013.09.085

Mekhileri NV, Lim KS, Brown GCJ et al (2018) Automated 3D bioassembly of micro-tissues for biofabrication of hybrid tissue engineered constructs. Biofabrication 10:24103. https://doi.org/10.1088/1758-5090/aa9ef1

Mora AC, Schmelzer CEH, Luise C et al (2018) Biochimie degradation of tropoelastin and skin elastin by neprilysin. Biochimie 146:73–78. https://doi.org/10.1016/j.biochi.2017.11.018

Muthukumar T, Sreekumar G, Sastry TP, Chamundeeswari M (2018) Collagen as a potential biomaterial in biomedical applications. Rev Adv Mater Sci 53:29–39. https://doi.org/10.1515/rams-2018-0002

Orciani M, Fini M, Di Primio R, Mattioli-Belmonte M (2017) Biofabrication and bone tissue regeneration: cell source, approaches, and challenges. Front Bioeng Biotechnol 5:5. https://doi.org/10.3389/fbioe.2017.00017

Osorio M, Cañas A, Sanchez D et al (2020) Lignocellulosic materials for biomedical applications. In: Lignocellulosics, pp 209–248

Panahi Y, Alishiri GH, Bayat N et al (2016) Efficacy of Elaeagnus Angustifolia extract in the treatment of knee osteoarthritis: a randomized controlled trial. EXCLI J 15:203–210. https://doi.org/10.17179/excli2015-639

Perumal G, Sivakumar PM, Nandkumar AM, Doble M (2020) Synthesis of magnesium phosphate nanoflakes and its PCL composite electrospun nanofiber scaffolds for bone tissue regeneration. Mater Sci Eng C 109:110527. https://doi.org/10.1016/j.msec.2019.110527

Qian Y, Zhou X, Zhang F et al (2019) Triple PLGA/PCL scaffold modification including silver impregnation, collagen coating, and electrospinning significantly improve biocompatibility, antimicrobial, and osteogenic properties for orofacial tissue regeneration. ACS Appl Mater Interfaces 16:41

Rachmiel D, Anconina I, Rudnick-Glick S et al (2021) Hyaluronic acid and a short peptide improve the performance of a PCL electrospun fibrous scaffold designed for bone tissue engineering applications. Int J Mol Sci 22:2425. https://doi.org/10.3390/ijms22052425

Ramos-Rodriguez DH, Pashneh-Tala S, Bains AK et al (2022) Demonstrating the potential of using bio-based sustainable polyester blends for bone tissue engineering applications. Bioengineering 2022:6

Ranjit E, Hamlet S, George R et al (2022) Biofunctional approaches of wool-based keratin for tissue engineering. J Sci Adv Mater Devices 7:100398. https://doi.org/10.1016/j.jsamd.2021.10.001

Rezaeeyazdi M, Colombani T, Memic A, Bencherif SA (2018) Injectable hyaluronic acid-co-gelatin cryogels for tissue-engineering applications. Materials (Basel) 11:23–25. https://doi.org/10.3390/ma11081374

Rezvani Ghomi E, Nourbakhsh N, Akbari Kenari M et al (2021) Collagen-based biomaterials for biomedical applications. J Biomed Mater Res B Appl Biomater 109:1986–1999. https://doi.org/10.1002/jbm.b.34881

Riboldi SA, Sampaolesi M, Neuenschwander P et al (2005) Electrospun degradable polyesterurethane membranes. Biomaterials 26:4606–4615

Rimondini L, Fini M, Giardino R (2005) The microbial infection of biomaterials: a challenge for clinicians and researchers. A short review. J Appl Biomater Biomech 3:1–10

Rossi CA, Pozzobon M, De Coppi P (2010) Advances in musculoskeletal tissue engineering: moving towards therapy. Organogenesis 6:167–172. https://doi.org/10.4161/org.6.3.12419

Rossi CA, Flaibani M, Blaauw B et al (2011) In vivo tissue engineering of functional skeletal muscle by freshly isolated satellite cells embedded in a photopolymerizable hydrogel. FASEB J 25:2296–2304. https://doi.org/10.1096/fj.10-174755

Sahoo DR, Biswal T (2021) Alginate and its application to tissue engineering. SN Appl Sci 3:30. https://doi.org/10.1007/s42452-020-04096-w

Sallehuddin N, Md Fadilah NI, Hwei NM et al (2022) Characterization and cytocompatibility of collagen–gelatin–elastin (CollaGee) acellular skin substitute towards human dermal fibroblasts: in vitro assessment. Biomedicines 10:1327. https://doi.org/10.3390/biomedicines10061327

Sancilio S, Gallorini M, Di Nisio C et al (2018) Alginate/hydroxyapatite-based nanocomposite scaffolds for bone tissue engineering improve dental pulp biomineralization and differentiation. Stem Cells Int 2018:9643721. https://doi.org/10.1155/2018/9643721

Schierholz JM, Beuth J (2001) Implant infections: a haven for opportunistic bacteria. J Hosp Infect 49:87–93. https://doi.org/10.1053/jhin.2001.1052

Sicari BM, Johnson SA, Siu BF et al (2012) The effect of source animal age upon the in vivo remodeling characteristics of an extracellular matrix scaffold. Biomaterials 33:5524–5533. https://doi.org/10.1016/j.biomaterials.2012.04.017

Sooriyaarachchi D, Minière HJ, Maharubin S, Tan GZ (2022) Hybrid additive microfabrication scaffold incorporated with highly aligned nanofibers for musculoskeletal tissues. Tissue Eng Regen Med 2019:29–38

Stojic M, Ródenas-Rochina J, López-Donaire ML et al (2021) Elastin-plasma hybrid hydrogels for skin tissue engineering. Polymers (Basel) 13:7–9. https://doi.org/10.3390/polym13132114

Terzi A, Storelli E, Bettini S et al (2018) Effects of processing on structural, mechanical and biological properties of collagen-based substrates for regenerative medicine. Sci Rep 8:1–13. https://doi.org/10.1038/s41598-018-19786-0

Tottey S, Johnson SA, Crapo PM et al (2011) The effect of source animal age upon extracellular matrix scaffold properties. Biomaterials 32:128–136. https://doi.org/10.1016/j.biomaterials.2010.09.006

Trache D (2018) Nanocellulose as a promising sustainable material for biomedical applications. AIMS Mater Sci 5:201–205. https://doi.org/10.3934/matersci.2018.2.201

Tsanaktsidou E, Kammona O, Kiparissides C (2022) Recent developments in hyaluronic acid-based hydrogels for. Polymers (Basel) 14:1–34

Ungerleider JL, Christman KL (2014) Concise review: injectable biomaterials for the treatment of myocardial infarction and peripheral artery disease: translational challenges and progress. Stem Cells Transl Med 3:1090–1099. https://doi.org/10.5966/sctm.2014-0049

Utami Nike D, Md Fadilah NI, Sallehuddin N et al (2022) Genipin-crosslinking effects on biomatrix development for cutaneous wound healing: a concise review. Front Bioeng Biotechnol 10: 865014. https://doi.org/10.3389/fbioe.2022.865014

Valcarcel J, Fraguas J, Hermida-Merino C et al (2021) Production and physicochemical characterization of gelatin and collagen hydrolysates from turbot skin waste generated by aquaculture activities. Mar Drugs 19:491. https://doi.org/10.3390/md19090491

Williams DF (2019) Challenges with the development of biomaterials for sustainable tissue engineering. Front Bioeng Biotechnol 7:7. https://doi.org/10.3389/fbioe.2019.00127

Wu P, Xi X, Li R, Sun G (2021) Engineering polysaccharides for tissue repair and regeneration. Macromol Biosci 21:e2100141. https://doi.org/10.1002/mabi.202100141

Wubneh A, Tsekoura EK, Ayranci C, Uludağ H (2018) Current state of fabrication technologies and materials for bone tissue engineering. Acta Biomater 15:1–30

Xing X, Cheng G, Yin C et al (2020) Magnesium-containing silk fibroin/polycaprolactone electrospun nanofibrous scaffolds for accelerating bone regeneration. Arab J Chem 13:5526–5538. https://doi.org/10.1016/j.arabjc.2020.03.031

Xu Y, Chen C, Hellwarth PB, Bao X (2019) Biomaterials for stem cell engineering and biomanufacturing. Bioact Mater 4:366–379. https://doi.org/10.1016/j.bioactmat.2019.11.002

Yang W, Xu H, Lan Y et al (2019) Preparation and characterisation of a novel silk fibroin/hyaluronic acid/sodium alginate scaffold for skin repair. Int J Biol Macromol 130:58–67. https://doi.org/10.1016/j.ijbiomac.2019.02.120
Yeo GC, Mithieux SM, Weiss AS (2018) The elastin matrix in tissue engineering and regeneration. Curr Opin Biomed Eng 6:27–32. https://doi.org/10.1016/j.cobme.2018.02.007
Zawani M, Fauzi MB (2021a) Injectable hydrogels for chronic skin wound management: a concise review. Biomedicine 9:527. https://doi.org/10.3390/biomedicines9050527
Zawani M, Fauzi MB (2021b) Epigallocatechin gallate: the emerging wound healing potential of multifunctional biomaterials for future precision medicine treatment strategies. Polymers (Basel) 13(21):3656
Zelen CM, Gould L, Serena TE et al (2015) A prospective, randomised, controlled, multi-centre comparative effectiveness study of healing using dehydrated human amnion/chorion membrane allograft, bioengineered skin substitute or standard of care for treatment of chronic lower extremity diabetic ulcers. Int Wound J 12:724–732. https://doi.org/10.1111/iwj.12395
Zhang E, Xu L, Yu G et al (2009) In vivo evaluation of biodegradable magnesium alloy bone implant in the first 6 months implantation. J Biomed Mater Res A 90:882–893
Zhang L, Zhang W, Hu Y et al (2021) Systematic review of silk scaffolds in musculoskeletal tissue engineering applications in the recent decade. ACS Biomater Sci Eng 7:817–840. https://doi.org/10.1021/acsbiomaterials.0c01716
Zhu Z, Wang Y-M, Yang J, Luo X-S (2017) Hyaluronic acid: a versatile biomaterial in tissue engineering. Plast Aesthet Res 4:219. https://doi.org/10.20517/2347-9264.2017.71
Zulkiflee I, Fauzi MB (2021) Gelatin-polyvinyl alcohol film for tissue engineering: a concise review. Biomedicine 9:979. https://doi.org/10.3390/BIOMEDICINES9080979

3 Food Waste-Derived Sources: Synthesis, Properties and Applications in Biomedical Engineering

Norita Mohd Zain and Minghooi Chai

Abstract

The great potential of low-cost waste-derived materials that are abundant worldwide has received growing attention in recent years. Developing these waste materials into new alternatives for various applications will lead to sustainable development globally, with the potential for fully utilising biomass sources. This chapter focuses on materials derived from food wastes specifically for biomedical applications. Initially, we present the current data on the world's food and food waste production by analysing each stage of the value chain. We then describe the synthesis parameters involved in converting food wastes into biomaterials such as bacterial cellulose (BC) and hydroxyapatite (HA) and discuss the basic requirements and properties for various biomedical applications. A list of commercial brands of these biomedical/healthcare products and their manufacturing companies is also included for reference.

Keywords

Biomass · Food waste · Biomedical applications · FAO data · Food-waste recovery

N. Mohd Zain (✉)
Department of Biomedical Engineering, Faculty of Engineering, Universiti Malaya, Kuala Lumpur, Malaysia

Centre of Advanced Materials, Faculty of Engineering, Universiti Malaya, Kuala Lumpur, Malaysia
e-mail: nmz1969@um.edu.my

M. Chai
Department of Biomedical Engineering, Faculty of Engineering, Universiti Malaya, Kuala Lumpur, Malaysia

W. S. Wan Kamarul Zaman, A. A. Nurul (eds.), *Sustainable Material for Biomedical Engineering Application*, https://doi.org/10.1007/978-981-99-2267-3_3

Abbreviations

β-TCP	β-Tricalcium phosphate
BC	Bacterial cellulose
FAO	Food and Agriculture Organisation of the United Nations
HA	Hydroxyapatite
HS	Hestrin and Schramm
TCP	Tricalcium phosphate

3.1 Introduction

Management of the food system and its wastes requires a complex approach involving many parties along the value chain. Therefore, a sustainable food system is needed to secure food and nutrition supplies for future generations. In 2020, world production was 11.2 billion tonnes for livestock and crops and 214 million tonnes for fisheries, enough to feed a human population of 7.8 billion people worldwide (Food and Agriculture Organization of the United Nations 2022a, b). Table 3.1 shows world production for 13 selected food types and their major producer countries. Different regions of the world are blessed with different resources and environmental conditions to produce certain types of food. Through innovation, the amount of food supply must progressively increase to support the continual growth of the human population.

The value chain in food supply describes the roles of all parties involved in cultivating, harvesting, storing, transporting, processing, distributing and selling food to end consumers. A value chain can be broken down into different stages, where food quantity may be lost at each stage, or the quality may degrade. These are known as food loss and waste. Between 2012 and 2021, total food loss and waste worldwide is estimated at 5.0% (Food and Agriculture Organization of the United Nations 2022c), as depicted in Fig. 3.1. The degree of food loss and waste depends on the type of food and stage of the value chain. Generally, they tend to be higher towards the downstream stages. For the whole value chain from start to end, loss and waste are 6.0–12.0% for fruits and vegetables, 4.0–12.7% for potatoes, and 1.5–22.9% for meat and animal products, respectively. For example, the loss and waste for every stage of chicken eggs throughout the value chain from harvest to retail are below 4.0%. However, for households, the loss and waste are 8.0% due to the discarding of eggshells. Similarly, for cattle meat, loss and waste at every stage are below 5.0% except for households, which increase to 11.0%.

To protect our environment, food scraps discarded from the above sources must be appropriately managed to protect our environment as they impart a negative value when sent to landfill or incineration. First, there is a waste of resources such as land and nutrients to produce food in the first place. Second, more resources, such as land and energy, are used to process waste. Third, landfilling and incineration processes generate greenhouse gases. Fourth, processed waste does not contribute to society in

Table 3.1 World production in 2020 for selected food across five continents

Food	World production 2020	Production percentage by continent					Top 3 producer countries, quantity	[a] Other selected producer countries, quantity
		Africa	Americas	Asia	Europe	Oceania		
Banana	119.9	17.7%	26.3%	54.0%	0.5%	1.4%	1. India, 31.5 2. China, 11.9 3. Indonesia, 8.2	5. Ecuador, 6.0 6. Philippines, 6.0 14. Vietnam, 2.2 17. Thailand, 1.4 39. Malaysia, 0.3
Cattle meat	67.9	8.8%	48.6%	22.6%	15.5%	4.5%	1. USA, 12.4 2. Brazil, 10.1 3. China, 6.0	4. Argentina, 3.2 5. Australia, 2.4 6. Mexico, 2.1
Cheese[b]	23.3	4.4%	35.0%	7.8%	49.6%	3.2%	1. USA, 6.2 2. Germany, 2.6 3. France, 1.6	8. Egypt, 0.5 12. Argentina, 0.4 13. Australia, 0.4
Chicken eggs[c]	1642.9	4.6%	18.8%	64.5%	11.7%	0.4%	1. China, 604.7 2. India, 114.4 3. Indonesia, 112.1	6. Mexico, 56.9 8. Japan, 43.9 12. Colombia, 16.4 18. Malaysia, 13.4 38. Romania, 5.2
Corn	1162.4	7.8%	50.1%	31.4%	10.7%	0.04%	1. USA, 360.3 2. China, 260.9 3. Brazil, 104.0	6. India, 30.2 8. Indonesia, 22.5 9. South Africa, 15.3 14. Romania, 10.9 23. Poland, 6.7
Olive	23.6	23.7%	3.4%	13.2%	59.6%	0.2%	1. Spain, 8.1 2. Italy, 2.2 3. Tunisia, 2.0	5. Turkey, 1.3 7. Egypt, 0.9 11. S. Arabia, 0.4
Orange	75.5	12.9%	40.5%	37.6%	8.5%	0.5%	1. Brazil, 16.7 2. India, 9.9 3. China, 7.6	5. Mexico, 4.6 8. Indonesia, 2.7 15. Vietnam, 1.1

(continued)

Table 3.1 (continued)

Food	World production 2020	Production percentage by continent					Top 3 producer countries, quantity	[a] Other selected producer countries, quantity
		Africa	Americas	Asia	Europe	Oceania		
Pear	23.1	3.1%	6.7%	77.5%	12.2%	0.4%	1. China, 16.1 2. Italy, 0.62 3. USA, 0.61	4. Argentina, 0.6 5. Turkey, 0.5 6. South Africa, 0.4
Pork	109.8	1.5%	22.0%	48.5%	27.6%	0.5%	1. China, 42.1 2. USA, 12.8 3. Germany, 5.1	4. Spain, 5.0 7. Vietnam, 3.6 8. Canada, 2.3 14. Philippines, 1.5 30. Romania, 0.4
Potato[d]	448.6	12.3%	10.9%	52.3%	24.0%	0.6%	1. China, 127.4 2. India, 52.5 3. Ukraine, 20.8	9. Poland, 7.8 17. Egypt, 5.7 18. Turkey, 5.2
Seafood[e]	214.0	5.7%	10.5%	74.9%	8.1%	0.8%	1. China, 83.9 2. Indonesia, 21.8 3. India, 14.2	4. Vietnam, 8.0 10. Japan, 4.2 11. Norway, 4.1 15. Thailand, 2.6 16. Egypt, 2.0 17. Malaysia, 1.8
Sorghum	58.7	46.8%	34.6%	15.7%	2.2%	0.7%	1. USA, 9.5 2. Nigeria, 6.4 3. Ethiopia, 5.1	5. Mexico, 4.7 6. China, 3.5 37. S. Arabia, 0.1
Sugar cane	1869.7	5.1%	54.0%	39.2%	0.0%	1.7%	1. Brazil, 757.1 2. India, 370.5 3. China, 108.7	5. Thailand, 75.0 9. Indonesia, 28.9 12. Philippines, 24.4

Unless otherwise stated, all data is in a million tonnes compiled from the FAO database (Food and Agriculture Organization of the United Nations 2022a, b)

[a] Prefix number at producer countries represents world rank in terms of production quantity

[b] Production data for cheese is in the year 2019

[c] Unit for chicken eggs is in a million nos

[d] Potato includes sweet potato

[e] Seafood includes captured and aquaculture

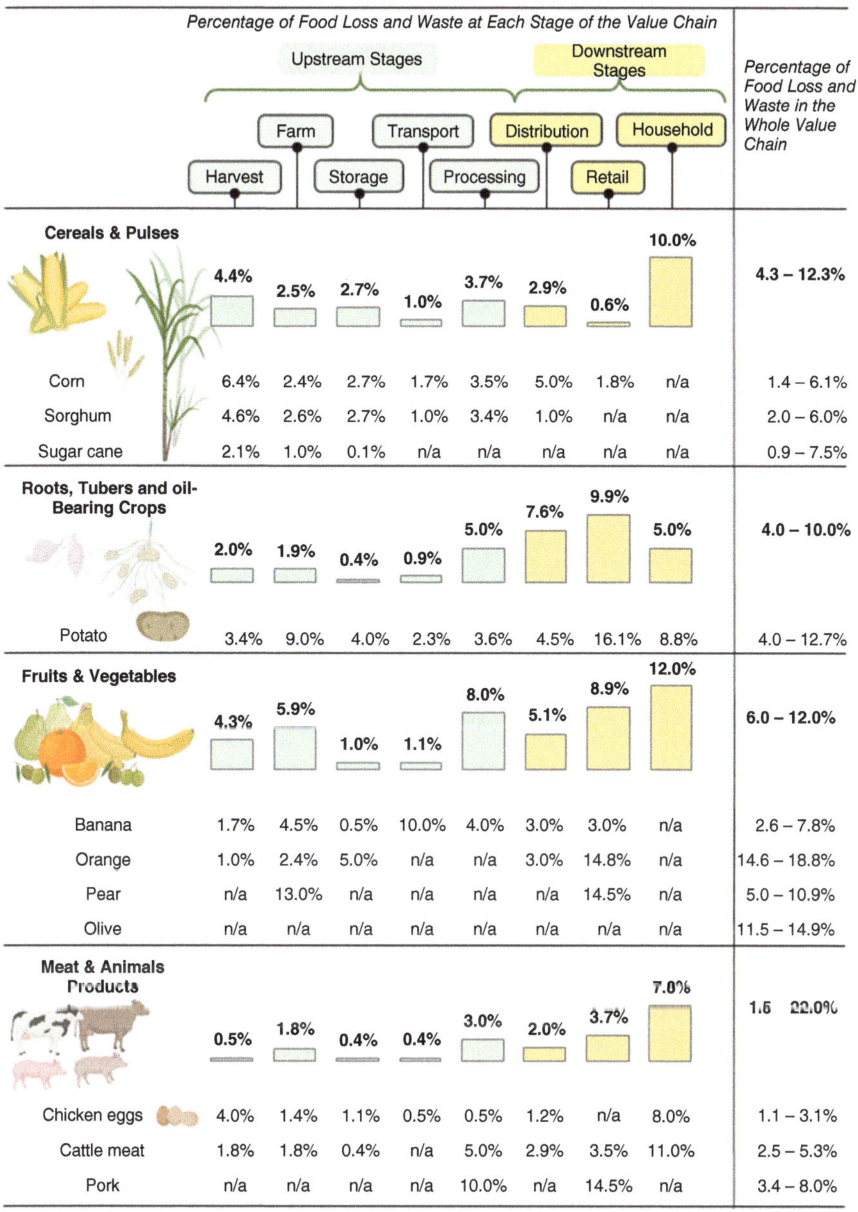

Fig. 3.1 Worldwide food loss and waste at each stage of the value chain for different types of food. (Data compiled from FAO database from 2012 to 2021 (Food and Agriculture Organization of the United Nations 2022c))

a meaningful way. These wastes contain chemical and energy values that can be recovered through valorisation to enhance food products valued higher than their input cost. Therefore, scientists look for opportunities in other industries to convert food waste into valuable products. In biomedical engineering, food wastes are valorised by converting them into products used in the healthcare industry.

3.2 From Food Waste to Bacterial Cellulose

The most common strain of bacteria used in the production of bacterial cellulose (BC) is the gram-negative *Komagataeibacter xylinus* (previously named *Gluconoacetorbacter xylinus*). At the cellular level, glucose is converted to uridine diphosphoglucose in the cytoplasm, followed by conversion to cellulose at the plasma membrane, where cellulose fibrils are secreted out of the cell into the culture medium (Glaser 1958; Ross et al. 1991). Thus, the cellulose fibrils produced by bacteria, known as BC, can be harvested, and used in biomedical applications. Bacteria may use a wide variety of carbon sources during their synthesis processes, such as glucose, fructose, molasses, glycerol, and mannitol (Mishra et al. 2022). However, a widely accepted cultural medium is the one presented in Hestrin and Schramm (HS) (Hestrin and Schramm 1954), which is now commercially available.

3.2.1 Types of Food Waste Sources

Various types of food waste have been used as carbon sources to grow BC. Most of them are by-products from the food and agriculture industries and are sourced from local producers where the research is carried out. Such an approach is important to ensure that the food waste resource is sustainable, its processing is economically viable, and the environmental impact is reduced. Table 3.2 shows various types of food waste used in BC research and the locations where they were sourced.

Table 3.2 Types of food waste and their sourcing locations in recent research on bacterial cellulose production

The carbon source for bacterial cellulose	Location	Reference
Banana leaf	Ecuador Malaysia	Fiallos-Cárdenas et al. (2021) Joo Shun et al. (2019)
Cheese whey	Greece	Lappa et al. (2021)
Corncob	Nigeria	Moyinoluwa et al. (2022)
Olive mill wastewater	Turkey	Sar and Akbas (2022)
Orange peel	Italy	Kunjalukkal et al. (2022)
Pear peel	China	Ma et al. (2021)
Potato juice	Poland	Ciecholewska-Juśko et al. (2021)
Potato peel	Egypt	Abdelraof et al. (2019)
Sorghum	China	Wang et al. (2021)
Sugarcane bagasse	Nigeria	Moyinoluwa et al. (2022)

3.2.2 Process Parameters and Properties

Commercially available Hestrin-Schramm (HS) culture medium to grow BC typically contains 20.00 g/L D-glucose, 5.00 g/L peptone, 5.00 g/L yeast extract, 7.00 g/L N sodium phosphate dodecahydrate and 1.00 g/L citric acid. Using food waste as a carbon source to substitute in part or entirely for the cultural medium leads to reduced costs and lessened environmental impact from the valorisation of the food waste. In most cases, the production yield of BC using food waste-derived carbon sources is at least comparable to that of using a commercial HS culture medium (Lappa et al. 2021; Moyinoluwa et al. 2022; Sar and Akbas 2022; Ma et al. 2021; Ciecholewska-Juśko et al. 2021; Abdelraof et al. 2019; Wang et al. 2021).

Food wastes usually undergo pre-treatment before being fed into a bioreactor to grow BC. The primary purpose of the pre-treatment process is to extract polysaccharides from the food waste and further break them down into monosaccharides such as glucose and fructose. Table 3.3 shows various pre-treatment processes depending on the types of food waste and their resultant final glucose content.

During the synthesis process of BC, the most common incubation temperature is 30 °C (Hestrin and Schramm 1954; Lappa et al. 2021; Moyinoluwa et al. 2022; Sar and Akbas 2022; Ma et al. 2021; Wang et al. 2021). However, incubation temperatures as low as 28 and 35 °C have also been reported (Ciecholewska-Juśko et al. 2021; Abdelraof et al. 2019). The initial pH for the synthesis process is often set at near neutral, between 5 and 7 (Hestrin and Schramm 1954; Lappa et al. 2021; Moyinoluwa et al. 2022; Sar and Akbas 2022; Ma et al. 2021; Ciecholewska-Juśko et al. 2021), and the inoculum size at between 8% and 10% (Lappa et al. 2021; Ma et al. 2021; Abdelraof et al. 2019). The incubation period may range from 6 to 10 days depending on the production yield, where an agitated culture will typically result in a higher yield, thus requiring a shorter incubation period (Lappa et al. 2021; Moyinoluwa et al. 2022; Sar and Akbas 2022; Ma et al. 2021; Ciecholewska-Juśko et al. 2021; Abdelraof et al. 2019; Wang et al. 2021). Generally, the production yield for BC is relatively low, between 4 and 11 g/L (Sar and Akbas 2022; Ma et al. 2021; Ciecholewska-Juśko et al. 2021; Abdelraof et al. 2019), which is the main reason for its high production cost.

The synthesised BC is then collected from the culture, often in pellicle form, and immersed into a hot solution of sodium hydroxide at 80 °C, where the concentration is below 5%w/w (Lappa et al. 2021; Moyinoluwa et al. 2022; Sar and Akbas 2022; Ma et al. 2021; Ciecholewska-Juśko et al. 2021; Wang et al. 2021). A concentration of sodium hydroxide as low as 0.1% has been used (Lappa et al. 2021) but more commonly at between 0.3% and 1% (Sar and Akbas 2022; Ma et al. 2021; Ciecholewska-Juśko et al. 2021; Wang et al. 2021). After about 1 h of immersion, the BC is washed with water until a neutral pH is obtained and then dried to a constant weight. The drying process may occur between 60 and 105 °C, whereas 105 °C requires only 6 h, but 60 °C will take 24 h (Abdelraof et al. 2019; Wang et al. 2021).

Table 3.3 Various pre-treatment processes based on types of food waste and the resultant final glucose content

Carbon source	Pre-treatment process and final glucose content	Reference
Banana leaf	1. Wash with water and sun-dry for 2 h 2. Press with a sugarcane juice machine for three passes 3. Centrifuge the extract at 3500 rpm for 35 min at 4 °C 4. Collect the supernatant layer and filter at 11 μm Final glucose content: **16 g/L** 5. Use the filtrate as feedstock to produce bacterial cellulose	Fiallos-Cárdenas et al. (2021), Joo Shun et al. (2019)
Cheese whey	1. Hydrolyse using β-galactosidase at loading 15 U/mL at the temperature of 60–65 °C for 36 h Final glucose content: **23 g/L** 2. Use the hydrolysate as feedstock to produce bacterial cellulose	Lappa et al. (2021)
Corncob	1. Sun-dry the corncob followed by milling to particle size 0.5–2.0 mm 2. Add water to 10%w/v and heat to 160 °C in an oil bath for 10 min 3. Cool down the mixture, collect the solid fraction, and oven-dry at 70 °C overnight 4. Hydrolyse using cellulase enzyme at loading 15 FPU/g dry weight of substrate for 48 h. Control the process at pH 5.5 and a temperature of 35 °C Final glucose content: **16 g/L** 5. Use the hydrolysate as feedstock to produce bacterial cellulose	Moyinoluwa et al. (2022)
Olive mill wastewater	1. Sterilise at 121 °C for 15 min 2. Centrifuge at 4000 rpm for 15 min 3. Collect the supernatant layer Final glucose content: **7 g/L** 4. Use as feedstock to produce bacterial cellulose	Sar and Akbas (2022)
Orange peel	1. Mix orange peel and pulp with water at a weight ratio of 1:2 2. Heat to 90 °C for 30–60 min 3. Filter with filter paper and cool to room temperature Final glucose content: **25 g/L** 4. Use the filtrate as feedstock to produce bacterial cellulose	Kunjalukkal et al. (2022)
Pear peel	1. Mix the pear residue with water at a 1:6 w/v ratio 2. Store at 20 °C with a relative humidity of 80–85% Final glucose content: **35 g/L** 3. Use as feedstock to produce bacterial cellulose	Ma et al. (2021)

(continued)

Table 3.3 (continued)

Carbon source	Pre-treatment process and final glucose content	Reference
Potato juice	1. Dilute potato tuber juice with water at a 1:1 ratio Final glucose content: **3.5 g/L** 2. Use as feedstock to produce bacterial cellulose	Ciecholewska-Juśko et al. (2021)
Potato peel	1. Dilute the potato peel waste with water to 30% w/v 2. Acidify with nitric acid 2.0 M. 3. Heat to 100 °C for 3 h 4. Neutralise with sodium hydroxide to pH 6.0. 5. Centrifuge at 6000 g for 15 min 6. Collect the supernatant layer and sterilise at 121 °C for 15 min Final glucose content: **20 g/L** 7. Use as feedstock to produce bacterial cellulose	Abdelraof et al. (2019)
Sorghum (root, stalk and leaf)	1. Hydrolyse the sorghum parts with a solution containing 70.2% phosphoric acid and 5.2% hydrogen peroxide, mixed at a ratio of 1:10 w/w 2. Keep at 40 °C for 2 h with mild agitation 3. Dilute the liquor with water at a ratio of 1:3 w/w 4. Filter and collect the solid residue 5. Wash with distilled water to a neutral pH 6. Storage at −20 °C Final glucose content: **43 (root), 67 (stalk) and 48 (leaf) g/L** 7. Use as feedstock to produce bacterial cellulose	Wang et al. (2021)
Sugarcane bagasse	1. Sun-dry the sugarcane bagasse followed by milling to particle size 0.125–2.0 mm 2. Add water to 10%w/v and heat to 160 °C in an oil bath for 10 min 3. Cool down the mixture; collect the solid fraction and oven-dry at 70 °C overnight 4. Hydrolyse using cellulase enzyme at loading 15 FPU/g dry weight of substrate for 48 h Control the process at pH 5.5 and a temperature of 35 °C Final glucose content: **16 g/L** 5. Use as feedstock to produce bacterial cellulose	Moyinoluwa et al. (2022)

Depending on its production process, the resultant BC may vary in density, tensile strength, stiffness, and crystallinity index. These are shown in Table 3.4. Their decomposition temperatures are generally above 300 °C.

Table 3.4 Properties of bacterial cellulose made from food waste resources

Properties	Values	Reference
Density	0.19–1.55 g/cm^3	Ciecholewska-Juśko et al. (2021), Wang et al. (2021)
Tensile strength	1.35–76.2 MPa	Moyinoluwa et al. (2022), Ma et al. (2021), Ciecholewska-Juśko et al. (2021), Wang et al. (2021)
Stiffness	30–129 MPa	Wang et al. (2021)
Crystallinity index	57–83%	Moyinoluwa et al. (2022), Sar and Akbas (2022), Abdelraof et al. (2019)
Decomposition temperature	>300 °C	Moyinoluwa et al. (2022), Sar and Akbas (2022)

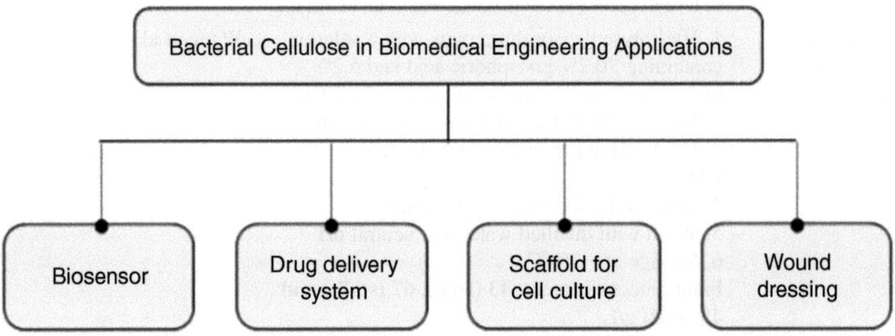

Fig. 3.2 Types of biomedical engineering applications that use bacterial cellulose

3.2.3 Biomedical Engineering Applications of Bacterial Cellulose

Bacterial cellulose is used for various Biomedical Engineering applications, as in Fig. 3.2, in manufacturing biosensors, drug delivery systems, cell culture scaffolds and wound dressing.

There are three types of cellulose materials that are used in biosensors: cellulose nanofibers, cellulose nanocrystals and bacterial nanocellulose (Dai et al. 2020). BC is produced from renewable resources, and it is used to produce disposable biosensors that are flexible, cost-effective and eco-friendly. In a cell-based biosensor, BC is integrated with a carbon nanotube to form a structure to encapsulate laccase. When the enzyme laccase reacts with organic matter, it generates electricity, thus making the design a self-powered biosensor (Li et al. 2019). When BC is embedded with curcumin, it is used as a test kit to detect drugs and albumin present in bodily fluid (Faham et al. 2019; Naghdi et al. 2019), while embedding it with fluorescent elements makes a biosensor that detects a level of biothiols in the body (Abbasi-Moayed et al. 2018). Biothiol is a biomarker for the health condition of a person and an abnormally high level of biothiol is indicative of certain diseases (Sahoo 2020). Sputtering the surface with the gold makes the BC optically sensitive for its use in biosensors based on light transmission (Cennamo et al. 2019). Very often, BC is combined with another material to adjust to the desired properties for its use as a biosensor. When impregnated with an ionic liquid compound, it develops electroconductivity so that it can function as an electrode. When the electrode is subjected to an external force, it gives an electric signal through piezoelectricity. Composite of BC with poly(3,4-ethylene dioxythiophene) polystyrene sulphonate (Pasquale et al. 2019) and lamellas of titanium carbide bonded with BC (Chen et al. 2019) are examples of electrodes for piezoelectric biosensors.

Bacterial cellulose is used in drug delivery systems because its structure contains many hydroxyl groups. These hydroxyl groups enable the formation of temporary bonds with functional groups present in the specific chemical structure of drugs. Due to its high purity and non-immunogenic properties, BC is an excellent material to carry drug substances to the injured site of the body. First, a drug is immobilised on the surface of BC. Then the immobilised BC is applied to the injury site. By designing release kinetics between the drug and BC, the drug can slowly release to the tissue at a level below toxicity and for a prolonged period. Various drugs have been studied for their release kinetics from a BC delivery system, such as diclofenac (Adepu et al. 2021; Beekmann et al. 2020), doxorubicin (Abbasian et al. 2019), famotidine (Badshah et al. 2018), hydroxychloroquine (Zambuzi et al. 2021), ketorolac (Orasugh et al. 2019), piroxicam (Plappert et al. 2019) and tizanidine (Badshah et al. 2018). The release kinetics of the system is adjusted by modifying the porosity of the BC during its manufacture (Adepu et al. 2021) or by chemical modification of the BC surface (Badshah et al. 2018). More recently, immobilising a BC surface with a human IgG antibody has been proposed for use in cancer treatment (Chung et al. 2022), where the aim was to deliver an antibody to a local region where a tumour is present without reaching other parts of the body to prevent systemic toxicity.

Bacterial cellulose is used in tissue engineering as a scaffold in cell culture due to its biocompatibility, non-immunogenicity, and non-toxicity. The highly porous structure of the scaffold provides anchoring sites for the cells to migrate, grow and proliferate. Generally, a static culture during BC production can result in a product in pellicle form, making it suitable for cutting or layering into the desired size and thickness and therefore used as a tissue engineering scaffold. Post-processing fabrication methods can also prepare a modified structure of BC to adjust its porosity and shape. Each method has its advantages and limitations. For example, freeze-drying will produce an interconnected nanostructure of the BC (Andree et al. 2021), porogen-leaching will achieve a controlled porosity (Khan et al. 2018), and 3D printing can produce complex shapes for the scaffold (Gutierrez et al. 2019).

Bacterial cellulose is used in wound dressing due to its high water-holding capacity (up to 99%), (Anton-Sales et al. 2018), thus enabling a moist condition at the injury site and absorption of exudates. Besides, BC is a barrier against infection during the healing process. The therapeutic effects are relieving pain and accelerating healing time. Table 3.5 lists the commercial products of BC for medical applications.

Table 3.5 Commercial products of bacterial cellulose used in medical-pharmaceutical grade

Company	Product
Axcelon Biopolymers Inc., Canada	Nanoderm™
	Nanoderm™ Ag
Bionext Produtos Biotecnológicos Ltda., Brazil	Bionext®
Bowil Biotech Sp., Poland	CelMat®
Coreleader Biotech Co. Ltd., Taiwan	Bio-skinG®
DMC Importação e Exportação de Equipamentos Ltda., Brazil & USA	BioCel®
Innovatec Ltda., Brazil	Nanoskin®
Jenpolymer Materials UG & Co. KG, Germany	Basyc®
Kusano Sakko Inc., Japan	Fibnano®
Lohmann & Rauscher GmbH & Co. KG, Germany & Austria	Suprasorb®
S2Medical AB, Sweden	EpiProtect®
Seven Indústria de Produtos Biotecnológicos Ltda., Brazil	CutiCell®
	Dermafill® (previously known as BioFill)
	NexFill®
Vuelo Pharma Ltda., Brazil	MembraCel®
Xylos Corporation, USA	XCell®

Data in this table is adapted from Choi et al. (2022), Diego Marestoni et al. (2020), Ho et al. (2022), and Zhong (2020)

3.3 From Food Waste to Hydroxyapatite

Hydroxyapatite (HA) is an inorganic compound made up of calcium and phosphate. In the biological system, it forms the mineral phase of the bone. Producing HA from natural resources means extracting the naturally occurring phosphate and/or calcium and crystallising the mineral into HA which requires high energy input. HA is represented by the molecular formula $Ca_{10}(PO_4)_6(OH)_2$. As such, 10 moles of calcium are chemically bonded with 6 moles of phosphate, giving its stoichiometric molar ratio of Ca/P at $10/6 = 1.67$.

3.3.1 Types of Food Waste Sources

Different types of sources contain slightly different compositions of minerals. As a result, when they are subjected to elemental analysis, the produced HA from these sources deviates from the stoichiometric molar ratio. The presence of cations such as sodium ion (Na^+) and magnesium ion (Mg^{2+}) will substitute for calcium ion (Ca^{2+}), thus lowering the Ca/P ratio. Likewise, an anion such as carbonate ion (CO_3^{2-}) will substitute phosphate ion (PO_4^{3-}), thus increasing the Ca/P ratio. When selecting the source of HA, it is important to consider a consistent feedstock of food waste.

Therefore, research should consider the continuous supply or availability, of such food waste at the local condition for future scale-up purposes. Table 3.6 shows various types of food waste that have been used in HA research and the locations where they were sourced.

3.3.2 Process Parameters and Properties

Derivation of HA from a mammalian bone is relatively straightforward. It begins with boiling water for 2–8 h, followed by oven-drying (Luthfiyah et al. 2022; Mohd Pu'ad et al. 2020a; Tran and Bui 2021; Bui and Linh 2019). The oven-dried bone is sintered at temperature above 750 °C, as shown in Table 3.7. The exact process has also been adopted for fishbone.

Another method for producing HA from food waste is described in Table 3.8. Calcium carbonate naturally occurring in an abalone shell, chicken eggshell, clamshell, cuttlefish bone, mussel shell and snail shell can be converted to calcium oxide by heating to an elevated temperature. This process releases carbon dioxide gas.

$$\text{Calcination}: \text{CaCO}_3(\text{s})\ \text{heat} \rightarrow \text{CaO}(\text{s}) + \text{CO}_2(\text{g})$$

Alternatively, calcium source in fishbone, mussel shell and snail shell in the form of calcium carbonate is directly ground to powder size. The solution of calcium oxide forms an aqueous alkaline solution of calcium hydroxide.

$$\text{Dissolution}: \text{CaO}(\text{s}) + \text{H}_2\text{O}(\text{L}) \rightarrow \text{Ca}(\text{OH})_2(\text{aq})$$

A phosphate source is added to the solution to form a precipitate of HA. The following reaction shows the neutralisation of calcium hydroxide with phosphoric acid to form a precipitate of HA:

$$\begin{aligned}\text{Precipitation}: 10\text{Ca}(\text{OH})_2(\text{aq}) + 6\text{H}_3\text{PO}_4(\text{aq}) &\rightarrow \text{Ca}_{10}(\text{PO}_4)_6(\text{OH})_2(\text{s}) \\ &+ 18\text{H}_2\text{O}(\text{L})\end{aligned}$$

The oven-dried precipitate of HA from the above reaction is then sintered at 700–1000 °C to re-crystallise the mineral. Studies have shown that sintering temperature affects the final Ca/P of the derived hydroxyapatite, as demonstrated in Fig. 3.3.

3.3.3 Biomedical Engineering Applications of Hydroxyapatite

Figure 3.4 depicts the most well-known application of HA in bone graft and dental restoration, besides other uses, including drug delivery systems and metallic coating for implant devices.

Table 3.6 Types of food waste and their sourcing locations in recent research on hydroxyapatite production

Food waste sources to produce hydroxyapatite	Location	Reference
Livestock		
Bovine bone	Indonesia	Luthfiyah et al. (2022)
	Malaysia	Mohd Pu'ad et al. (2020a)
Bovine bone (femur)	Mexico	Hincapie et al. (2022)
	Vietnam	Tran and Bui (2021)
Bovine cortical	Colombia	Forero et al. (2021a)
Cattle bone waste	Colombia	Hernández-Barreto et al. (2022)
Cattle horn core	Cameroon	Ngueagni et al. (2020)
Porcine bone	Columbia	Forero et al. (2021b)
	Vietnam	Bui and Linh (2019)
Porcine bone (femur)	Colombia	Forero et al. (2021a)
Poultry		
Chicken eggshell	Indonesia	Amanda et al. (2022)
	Malaysia	Mohd Pu'ad et al. (2020b)
	Romania	Dumitrescu et al. (2021)
Ostrich bone	Nepal	Malla et al. (2020)
Seafood		
Cuttlefish bone	Croatia	Rogina et al. (2020)
Fishbone	India	Irfan et al. (2020)
Fishbone (catfish, kingfish, mackerel, grouper, snapper, tilapia and tuna)	Indonesia	Kusumawati et al. (2022)
Fishbone (Nile tilapia)	Mexico	Truite et al. (2022)
Fishbone (skipjack tuna)	Thailand	Aenglong et al. (2022)
Fish discards from the aquaculture process (red scorpion fish, Atlantic horse mackerel and salmon)	Spain	Fernández Arias et al. (2021)
Shell (abalone)	Indonesia	Permatasari et al. (2021)
Shell (clam)	Malaysia	Ng et al. (2022)
Shell (mussel)	India	Irfan et al. (2020)
	Italy	Scialla et al. (2020)

(continued)

Table 3.6 (continued)

Food waste sources to produce hydroxyapatite	Location	Reference
Shell (snail)	Egypt	Youssef et al. (2022)
	Ghana	Asimeng et al. (2020)
	Indonesia	Zuliantoni et al. (2022)

Table 3.7 Sintering temperature during the production of hydroxyapatite by direct sintering of the food waste source

Source	Sintering temperature (°C)	Reference
Bovine bone	850 and 900	Luthfiyah et al. (2022)
Bovine bone	900	Tran and Bui (2021)
Porcine bone	750	Bui and Linh (2019)
Fishbone (skipjack tuna)	900	Aenglong et al. (2022)
Fishbone (red scorpionfish, Atlantic horse mackerel, and salmon)	750 and 950	Fernández Arias et al. (2021)

Due to trauma or disease, bone defects that cannot be corrected by autograft bone transplantation need the implantation of a substitute grafting material. This material delivers growth factors to promote bone ingrowth and as a matrix for bone cells to adhere and grow (Finkemeier 2002). HA is a suitable material because it is osteoconductive, bioactive and resembles the natural mineral in bone. Optimum strength and degradability for the implant can be designed by varying the ratio of HA with β-tricalcium phosphate (β-TCP), where increasing the proportion of β-TCP reduces its strength but increases the degradation rate (Yamada et al. 1997). During healing, when new bone tissue advances, the implant resorbs away for the tissue to take over the cavity. HA used in bone graft comes in the form of powder, injectable, or post-fabricated into porous granules, beads or scaffolds using spray drying, porogen-leaching or 3D printing.

For dental restoration, HA is preferred over tricalcium phosphate (TCP) due to its slow resorption. Thus its ratio is increased to adjust the Ca/P to above 1.5 so that the apatite structure is preserved (Khan and Syed 2018). Table 3.9 lists the commercial products of HA for bone grafts, ocular implants, and dental restoration. HA has also been incorporated into oral products such as toothpaste to remineralise enamel and dentine, reduce hypersensitivity of dentine, and prevent bacteria from binding to the enamel surface (Chen et al. 2021).

As a carrier in the drug delivery system, HA has been loaded with drugs to treat osteosarcoma, osteomyelitis and osteoporosis (Jahil et al. 2022). Various drugs have been studied for their controlled release from a HA delivery system, such as acetazolamide (Kaneko et al. 2022), ciprofloxacin (Agalya et al. 2022; Ferreira-

Table 3.8 Calcination temperature, source of phosphate and sintering temperature during the production of hydroxyapatite via derivation of calcium oxide from the food waste source

Steps:	(a) Derive calcium oxide (CaO) from the food source. Then dissolve the calcium oxide in water to form an alkaline aqueous solution of calcium hydroxide (Ca $(OH)_2$)			
	(b) Add a phosphate source (PO_4^{3-}) into the calcium hydroxide solution to form a precipitate of hydroxyapatite, then filter and oven-dry the precipitate			
	(c) Heat the precipitate to a sintering temperature to re-crystallise the mineral			
Source	Method to derive CaO (in step A)	Source of PO_4^{3-} (in step B)	Sintering temperature or microwave power (in step °C)	Reference
Abalone shell	Calcination at 1000 °C	$(NH_4)_2HPO_4$	650 °C	Permatasari et al. (2021)
Chicken eggshell	Calcination at 800 °C	$(NH_4)_2HPO_4$	Microwave, <1000 W	Dumitrescu et al. (2021)
	Calcination at 900 °C	H_3PO_4	700 °C	Mohd Pu'ad et al. (2020b)
Clamshell	Calcination at 1000 °C	H_3PO_4	1200–1300 °C	Ng et al. (2022)
Cuttlefish bone	Calcination at 200 °C, 5 bar	$NH_4H_2PO_4$	800 °C	Rogina et al. (2020)
Fishbone	Crush with mortar and pestle	Na_2HPO_4	Microwave, 750 W	Irfan et al. (2020)
Mussel shell	Calcination at 700 °C	H_3PO_4/ $(NH_4)_2HPO_4$	1000 °C	Scialla et al. (2020)
	Calcination at 1000 °C	H_3PO_4/ $(NH_4)_2HPO_4$	1000 °C	Scialla et al. (2020)
	Crush with mortar and pestle	Na_2HPO_4	Microwave, 750 W	Irfan et al. (2020)
Snail shell	Calcination at 900 °C	H_3PO_4	1000 °C	Youssef et al. (2022)
	Pound with a wooded hammer	$(NH_4)_2HPO_4$	700 °C	Zuliantoni et al. (2022)

Ermita et al. 2020), cisplatin (Andronescu et al. 2013), doxorubicin (Sumathra et al. 2020; Zhao et al. 2019), fampridine (Marincas et al. 2021), gentamicin (Ionescu et al. 2019; Suchy et al. 2019), strontium nitrate (Wei et al. 2021), tinidazole (Zhao et al. 2019) and vancomycin (Suchy et al. 2019; Lv et al. 2022).

Finally, HA is used to coat metallic components such as the stainless steel of an implant device to resist corrosion within the physiological environment, reduce the immunogenic effect, and promote bone ingrowth where applicable (Ahmed et al. 2020). Commercial products of HA for metallic coating components of implants are Biotal (Plasma Biotal, UK), Medipure (Medicoat, France) and Metco (Oerlikon, Switzerland) (Ganvir et al. 2021).

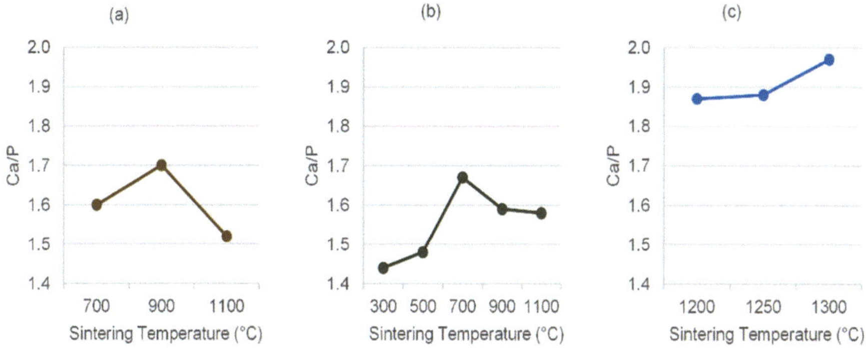

Fig. 3.3 Effect of sintering temperature on Ca/P ratio of hydroxyapatite made from (**a**) bovine bone, (**b**) eggshell and (**c**) clamshell. (Data in this figure is adapted from Mohd Pu'ad et al. (2020a, b), Ng et al. (2022))

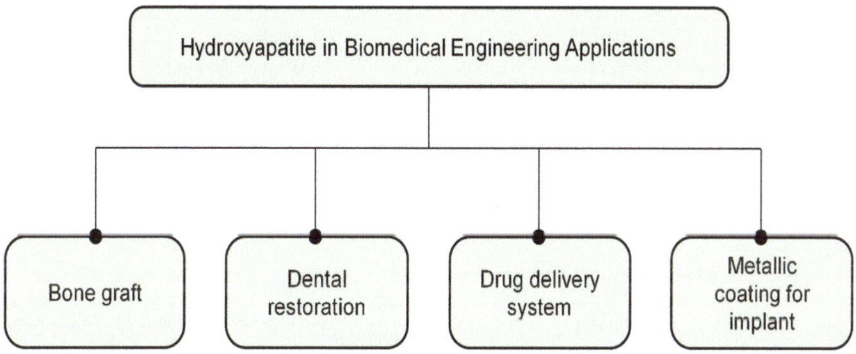

Fig. 3.4 Types of biomedical engineering applications that use hydroxyapatite

3.4 Conclusion and Future Perspective

A continuous increase in the world population simultaneously results in the accumulation of food waste globally. Food scraps from households and retail are among the highest percentages of food waste, often discarded as domestic waste. If these food wastes are not controlled and well managed in the dumping land, it can constitute severe environmental pollution and societal health issues. Since food waste consists of nutrients and valuable, sustainable bioproducts, it is critical to identify the most appropriate methods and options for transforming this waste into various potential applications. This reality is in line with Goal 12 of the United Nations 2015 Sustainable Development Goals (SDG), which documented that food waste generation needs to be reduced through prevention, recycling and reuse to ensure sustainable consumption and production patterns. Since then, different approaches have been extensively investigated to uphold the goal of recycling

Table 3.9 Commercial products of hydroxyapatite used in bone, ocular, and dental restoration

Country	Company	Product(s)
Czech Republic	Lasak	OssaBase-HA
France	Biometlante	MBCP
	Ceraver	Cerapatite
	Depuy-Bioland	Bioroc
	SBM	Synatite
	Teknimed	Ceraform
Germany	Curasan	Cerasorb
	Haraeus Kulzer	Ostim
	Karl Storz Recon	Synthacer
	Osartis	PerOssal
India	G.Surgiwear	G-Bone, G-Graft, QOssHeal
	IFGL Bio Ceramics	BioGraft
Ireland	Premier Biomaterials	PurAtite
Japan	Covalent Materials	Neobone
	Hoya Corp.	Apaceram
	Pentax	Bonefil, Bonetite
	Sumitomo Osaka Cement	Boneceram
Latvia	PrevDent Baltic	DeSensiDent
Poland	Chema-Elektromet	HA Biocer
Portugal	Fluidinova	nanoXim
Russia	VladMiVA	Bioplast-Dent, EoDent
South Korea	Dentis	Ovis Bone HA
	DIO	Genesis BCP
	Genoss	New Bone, Osspol Dental, Osteon
	Inobone	Frabone Dental
	MegaGen	Mega-Oss Bovine
	OssGen	DualPor
	SK Bioland	Collaoss
Spain	Ziacom Medical	Osseos BCP
Sweden	Promimic	HA nano Surface
Switzerland	Geistlich Pharma	Bio-Oss
	Mathys	PermaOS
UK	ApaTech	Actifuse, ApaPore
USA	Ceramed	Ostegraf
	Cook-Waite Laboratories	Alveograf, Periograf
	Integrated Orbital Implants	Bio-Eye
	SigmaGraft	BoneSigma BCP
	Stryker Orthopaedics	BoneSource
	Zimmer	Calcitite

Data in this table is adapted from Coelho et al. (2019), Dorozhkin (2013), Fiume et al. (2021), Hernández-Ruiz et al. (2022), Krishnamurithy et al. (2015), Ku et al. (2019), and Wang et al. (2020)

food waste. It has been established that the most efficient and cost-effective method to recover food waste with high production yield is using modern biotechnological approaches. However, these approaches require support from local policies and large-scale investment to improve the infrastructure and technological solutions in harvesting, storage, transport and distribution.

In this chapter, we briefly discussed the properties of food-waste-derived BC and HA from different sources and preparation methods for various biomedical applications. Food waste such as cheese whey, orange peel, pear peel, and sorghum contain glucose that can be extracted, making them valuable feedstock to grow BC. A renewable, biocompatible and biodegradable BC with different morphologies and physiochemical properties can be fabricated by adding biomaterials or other components to the culture medium. This approach has made BC possible for extensive applications in many medical devices as substrates for detecting and measuring a body's biological and chemical states, a vehicle to deliver drugs to targeted tissue in the body and artificial skin on the injured skin to protect from infection and promote healing. However, several issues need revision for the industrial production and application development of BC. For example, labour and static fermentation times are crucial factors in production capability. The production efficiency can be improved in several ways, including developing new culture media and fermentation reactors, utilising automated equipment and exploring new potential applications of BC.

On the other hand, naturally occurring calcium and phosphate found in food waste such as bone, eggshells and seashells supply an essential precursor in the formation of bioceramic HA. It has earned many distinctions for dental and orthopaedic applications due to its fascinating biological properties, such as bioactivity, biocompatibility, non-toxicity, osteoconductivity and ability to provide immediate chemical bonding with natural bone tissue. Different methods for HA production from food wastes have been researched, such as the hydrothermal conversion of calcified materials with phosphorus precursors, enzymatic hydrolysis, extraction with ionic liquids and the most common thermal treatment at temperatures of 600–1200 °C. The thermal decomposition method is the most simple and cost-effective and has established a commercial status. However, collagen proteins and other valuable compounds that might be present in the raw material are eliminated by this method. The transformation of food waste into HA has not only addressed the environmental issues but also created an economic opportunity to manufacture products required for biomedical-related industries. Besides, the cost of using expensive and high-purity chemicals for HA production can be reduced. Currently, food-wastes-derived HA is an alternative to chemical-based HA in biomedical applications due to its resemblance with the mineral phase of our bone and structure.

In conclusion, the properties of food-waste-derived BC and HA intended for biomedical applications should be carefully investigated. As supporting evidence, their potential bioavailability and metabolism should be evaluated using a toxicological database compiled from in vitro and in vivo experiments.

Acknowledgements The authors gratefully acknowledge the financial support of the University Malaya Research Grant (RU grant number: RF019A-2018).

Conflict of Interest The authors declare no conflict of interest.

References

Abbasian M et al (2019) Chemotherapy of breast cancer cells using novel pH-responsive cellulose-based nanocomposites. Adv Pharm Bull 9:122–131

Abbasi-Moayed S et al (2018) A rainbow ratiometric fluorescence sensor array on bacterial nanocellulose for visual discrimination of biothiols. Analyst 143(14):3415–3424

Abdelraof M, Hasanin M, Saied H (2019) Ecofriendly green conversion of potato peel wastes to high productivity bacterial cellulose. Carbohydr Polym 211:75–83. https://doi.org/10.1016/j.carbpol.2019.01.095

Adepu S, Kalyani P, Khandelwal M (2021) Bacterial cellulose-based drug delivery system for dual mode drug release. Trans Indian Natl Acad Eng 6:265–271

Aenglong C et al (2022) Synthesis of soluble calcium compound from skipjack tuna bones using edible weak acids. LWT 162:113460. https://doi.org/10.1016/j.lwt.2022.113460

Agalya P et al (2022) Ultrasonic assisted in situ mineralisation of hydroxyapatite nanoparticles in the presence of drug molecule: an insight on biowaste derived materials for the local drug delivery. Inorg Chem Commun 143:109788. https://doi.org/10.1016/j.inoche.2022.109788

Ahmed Y, Yasir M, Ur Rehman MA (2020) Fabrication and characterisation of zein/hydroxyapatite composite coatings for biomedical applications. Surfaces 3(2):237–250

Amanda H, Elline E, Fibryanto E (2022) Synthesis and physical characterization of nano-hydroxyapatite-collagen-epigallocatechin-3-gallate hydrogel composite. J Indones Dent Assoc 5:7–13

Andree V et al (2021) Influence of drying methods on the physical properties of bacterial nanocellulose. Mater Res Express 8:025402. https://doi.org/10.1088/2053-1591/abe016

Andronescu E et al (2013) Collagen-hydroxyapatite/cisplatin drug delivery systems for locoregional treatment of bone cancer. Technol Cancer Res Treat 12:275–284

Anton-Sales I et al (2018) Opportunities of bacterial cellulose to treat epithelial tissues. Curr Drug Targets 20(8):808–822

Asimeng B, Afeke D, Tiburu E (2020) Biomaterial for bone and dental implants: synthesis of B-type carbonated hydroxyapatite from biogenic source. In: Vizureanu P, Ferreira Botelho C (eds) Biomaterials. IntechOpen, London, pp 121–130

Badshah M et al (2018) Surface modification and evaluation of bacterial cellulose for drug delivery. Int J Biol Macromol 113:526–533

Beekmann U et al (2020) Process control and scale-up of modified bacterial cellulose production for tailor-made anti-inflammatory drug delivery systems. Carbohydr Polym 236:116062. https://doi.org/10.1016/j.carbpol.2020.116062

Bui X, Linh T (2019) The extraction of pure hydroxyapatite from porcine bone by thermal process. Metall Mater Eng 25(1):47–58

Cennamo N et al (2019) An eco-friendly disposable plasmonic sensor based on bacterial cellulose and gold. Sensors 19:4894. https://doi.org/10.3390/s19224894

Chen Z et al (2019) Compressible, elastic, and pressure-sensitive carbon aerogels derived from 2D titanium carbide nanosheets and bacterial cellulose for wearable sensors. Chem Mater 31. https://doi.org/10.1021/acs.chemmater.9b00259

Chen L et al (2021) Hydroxyapatite in oral care products—a review. Materials 14:4865. https://doi.org/10.3390/ma14174865

Choi S et al (2022) Bacterial cellulose and its applications. Polymers 14:1080. https://doi.org/10.3390/polym14061080

Chung CK et al (2022) Bacterial cellulose as drug delivery system for optimizing release of immune checkpoint blocking antibodies. Pharmaceutics 14:1351. https://doi.org/10.3390/pharmaceutics14071351

Ciecholewska-Juśko D et al (2021) Potato juice, a starch industry waste, as a cost-effective medium for the biosynthesis of bacterial cellulose. Int J Mol Sci 22:10807. https://doi.org/10.3390/ijms221910807

Coelho CC et al (2019) Nano-hydroxyapatite in oral care cosmetics: characterisation and cytotoxicity assessment. Sci Rep 9:11050. https://doi.org/10.1038/s41598-019-47491-z

Dai L et al (2020) Ultrasensitive physical, bio, and chemical sensors derived from 1-, 2-, and 3-D nanocellulosic materials. Small 16:1906567. https://doi.org/10.1002/smll.201906567

Diego Marestoni L et al (2020) Commercial and potential applications of bacterial cellulose in Brazil: ten years review. Polímeros 30:e2020047. https://doi.org/10.1590/0104-1428.09420

Dorozhkin S (2013) Calcium orthophosphate-based bioceramics. Materials 6:3840–3942

Dumitrescu C et al (2021) Nano-hydroxyapatite vs. xenografts: synthesis, characterization, and in vitro behavior. Nanomaterials 11:2289. https://doi.org/10.3390/nano11092289

Faham S et al (2019) A nanocellulose-based colorimetric assay kit for smartphone sensing of iron and iron-chelating deferoxamine drug in biofluids. Anal Chim Acta 1087:104–112. https://doi.org/10.1016/j.aca.2019.08.056

Fernández Arias M et al (2021) Biogenic calcium phosphate from fish discards and by-products. Appl Sci 11:3387. https://doi.org/10.3390/app11083387

Ferreira-Ermita D et al (2020) Characterisation and in vivo biocompatibility analysis of synthetic hydroxyapatite compounds associated with magnetite nanoparticles for a drug delivery system in osteomyelitis treatment. Results Mater 5:100063. https://doi.org/10.1016/j.rinma.2020.100063

Fiallos-Cárdenas M et al (2021) Bacterial nanocellulose derived from banana leaf extract: yield and variation factors. Resources 10:121. https://doi.org/10.3390/resources10120121

Finkemeier C (2002) Bone-grafting and bone-graft substitutes. J Bone Joint Surg Am 84:454–464

Fiume E et al (2021) Hydroxyapatite for biomedical applications: a short overview. Ceramics 4(4):542–563

Food and Agriculture Organization of the United Nations (2022a) FAOSTAT

Food and Agriculture Organization of the United Nations (2022b) Fisheries and aquaculture

Food and Agriculture Organization of the United Nations (2022c) Food loss and waste database

Forero P et al (2021a) Biogenic hydroxyapatite powders: effects of source and processing methodologies on physicochemical properties and bioactive response. Mater Charact 173:110950. https://doi.org/10.1016/j.matchar.2021.110950

Forero P et al (2021b) Temperature effect in physicochemical and bioactive behavior of biogenic hydroxyapatite obtained from porcine bones. Sci Rep 11:11069. https://doi.org/10.1038/s41598-021-89776-2

Ganvir A et al (2021) Deposition of hydroxyapatite coatings by axial plasma spraying: influence of feedstock characteristics on coating microstructure, phase content and mechanical properties. J Eur Ceram Soc 41(8):4637–4649

Glaser L (1958) The synthesis of cellulose in cell-free extracts of Acetobacter xylinum. J Biol Chem 232:627–636

Gutierrez E et al (2019) 3D printing of antimicrobial alginate/bacterial-cellulose composite hydrogels by incorporating copper nanostructures. ACS Biomater Sci Eng 5(11):6290–6299

Hernández-Barreto D, Hernández-Cocoletzi H, Moreno-Piraján J (2022) Biogenic hydroxyapatite obtained from bone wastes using CO_2-assisted pyrolysis and its interaction with glyphosate: a computational and experimental study. ACS Omega 7:23265–23275

Hernández-Ruiz K et al (2022) Hydroxyapatite recovery from fish by-products for biomedical applications. Sustain Chem Pharm 28:100726. https://doi.org/10.1016/j.scp.2022.100726

Hestrin S, Schramm M (1954) Synthesis of cellulose by Acetobacter xylinum. II. Preparation of freeze-dried cells capable of polymerising glucose to cellulose. Biochem J 58:345–352

Hincapie J et al (2022) Effect of the modification of spray drying parameters on the fabrication of bovine-derived hydroxyapatite microspheres for biomedical applications. Mater Today Commun 31:103838. https://doi.org/10.1016/j.mtcomm.2022.103838

Ho Y-S, Halim AFM, Islam M (2022) The trend of bacterial nanocellulose research published in the science citation index expanded from 2005 to 2020: a bibliometric analysis. Front Bioeng Biotechnol 9:795341. https://doi.org/10.3389/fbioe.2021.795341

Ionescu O et al (2019) Bone-graft delivery systems of type PLGA-gentamicin and collagen-hydroxyapatite-gentamicine. Mater Plast 56(3):534–537

Irfan M et al (2020) Microwave-assisted one-step synthesis of nanohydroxyapetite from fish bones and mussel shells. Mater Lett 282:128685. https://doi.org/10.1016/j.matlet.2020.128685

Jahil M et al (2022) The natural and commerical sources of hydroxyapatite/collagen composites for biomedical applications: a review study. Int J Appl Pharm 14:77–87

Joo Shun T et al (2019) Banana frond juice as novel fermentation substrate for bioethanol production by Saccharomyces cerevisiae. Biocatal Agric Biotechnol 21:101293. https://doi.org/10.1016/j.bcab.2019.101293

Kaneko K, Miyasaka R, R. (2022) Hayman Nano-hydroxyapatite improves intestinal absorption of acetazolamide (BCS Class IV drug)–but how? PLoS One 17:e0268067. https://doi.org/10.1371/journal.pone.0268067

Khan A, Syed M (2018) A review of bioceramics-based dental restorative materials. Dent Mater J 38(2):163–176

Khan S et al (2018) Preparation and structural characterisation of surface modified microporous bacterial cellulose scaffolds: a potential material for skin regeneration applications in vitro and in vivo. Int J Biol Macromol 117:1200–1210

Krishnamurithy G et al (2015) Proliferation and osteogenic differentiation of mesenchymal stromal cells in a novel porous hydroxyapatite scaffold. Regen Med 10(5):579–590

Ku J-K et al (2019) Dental alloplastic bone substitutes currently available in Korea. J Korean Assoc Oral Maxillofac Surg 45:51–67

Kunjalukkal S et al (2022) Sustainable production of stiff and crystalline bacterial cellulose from orange peel extract. Sustainability 14:2247. https://doi.org/10.3390/su14042247

Kusumawati P et al (2022) Nano-calcium powder properties from six commercial fish bone waste in Indonesia. Squalen Bull Mar Fish Postharvest Biotechnol 17:1–12

Lappa I et al (2021) A comprehensive bioprocessing approach to foster cheese whey valorisation: on-site β-galactosidase secretion for lactose hydrolysis and sequential bacterial cellulose production. Fermentation 7:184. https://doi.org/10.3390/fermentation7030184

Li X et al (2019) Encapsulation of enzyme by metal-organic framework for single-enzymatic biofuel cell-based self-powered biosensor. Nano Energy 68:104308. https://doi.org/10.1016/j.nanoen.2019.104308

Luthfiyah S et al (2022) Comparing properties of bovine bone derived hydroxyapatite and synthetic hydroxyapatite. J Appl Sci Eng 25:1045–1051

Lv X-F et al (2022) Nano sized hydroxyapatite-polylactic acid-vancomycin in alleviation of chronic osteomyelitis. Drug Des Devel Ther 16:1983–1993

Ma X et al (2021) Coproduction of bacterial cellulose and pear vinegar by fermentation of pear peel and pomace. Bioprocess Biosyst Eng 44:1–14. https://doi.org/10.1007/s00449-021-02599-3

Malla K et al (2020) Extraction and characterisation of novel natural hydroxyapatite bioceramic by thermal decomposition of waste ostrich bone. Int J Biomater 2020:1690178. https://doi.org/10.1155/2020/1690178

Marincas L et al (2021) Hydroxyapatite and silicon-modified hydroxyapatite as drug carriers for 4-aminopyridine. Crystals 11:1124. https://doi.org/10.3390/cryst11091124

Mishra S et al (2022) Biochemistry, synthesis and applications of bacterial cellulose: a review. Front Bioeng Biotechnol 10:780409. https://doi.org/10.3389/fbioe.2022.780409

Mohd Pu'ad N et al (2020a) Extraction of biological hydroxyapatite from bovine bone for biomedical applications. Mater Sci Forum 1010:579–583

Mohd Pu'ad N et al (2020b) Synthesis of eggshell derived hydroxyapatite via chemical precipitation and calcination method. Mater Today Proc 42(1):172–177

Moyinoluwa A et al (2022) Bacterial cellulose production from agricultural residues by two Komagataeibacter sp. strains. Bioengineered 13:10010–10025. https://doi.org/10.1080/21655979.2022.2062970

Naghdi T et al (2019) Lab-on-nanopaper: an optical sensing bioplatform based on curcumin embedded in bacterial nanocellulose as an albumin assay kit. Anal Chim Acta 1070:1–24. https://doi.org/10.1016/j.aca.2019.04.037

Ng C et al (2022) Characterisation and sintering properties of hydroxyapatite bioceramics synthesised from clamshell biowaste. IIUM Eng J 23:228–236

Ngueagni P et al (2020) Adsorption of Pb(II) and Cd(II) ions onto modified biogenic slaughterhouse waste: equilibrium and kinetic analysis. Int J Environ Anal Chem 102:4344–4363. https://doi.org/10.1080/03067319.2020.1784409

Orasugh J et al (2019) Sustained release of ketorolac tromethamine from poloxamer 407/cellulose nanofibrils graft nanocollagen based ophthalmic formulations. Int J Biol Macromol 140:441–453

Pasquale G et al (2019) A bacterial cellulose based mass sensor. In: IEEE International Symposium on Measurements & Networking (M&N), 2019, pp 1–4

Permatasari H et al (2021) Nano-carbonated hydroxyapatite precipitation from abalone shell (Haliotis asinina) waste as the bioceramics candidate for bone tissue engineering. Nanomater Nanotechnol 11:1–9

Plappert S et al (2019) Anisotropic nanocellulose gel–membranes for drug delivery: tailoring structure and interface by sequential periodate–chlorite oxidation. Carbohydr Polym 226:115306. https://doi.org/10.1016/j.carbpol.2019.115306

Rogina A et al (2020) The bioactivity of titanium-cuttlefish bone-derived hydroxyapatite composites sintered at low temperature. Powder Metall 63(4):300–310

Ross P, Mayer R, Benziman M (1991) Cellulose biosynthesis and function in bacteria. Microbiol Rev 55:35–58

Sahoo P (2020) Detection of biothiols using some novel chemosensors: an overview. Mini Rev Org Chem 18(7):867–884

Sar T, Akbas M (2022) Potential use of olive oil mill wastewater for bacterial cellulose production. Bioengineered 13:7659–7669. https://doi.org/10.1080/21655979.2022.2050492

Scialla S et al (2020) Mussel shell-derived macroporous 3D scaffold: characterization and optimization study of a bioceramic from the circular economy. Mar Drugs 18:309. https://doi.org/10.3390/md18060309

Suchy T et al (2019) Evaluation of collagen/hydroxyapatite electrospun layers loaded with vancomycin, gentamicin and their combination: comparison of release kinetics, antimicrobial activity and cytocompatibility. Eur J Pharm Biopharm 140:50–59

Sumathra M et al (2020) In vivo assessment of a hydroxyapatite/κ-carrageenan–maleic anhydride–casein/doxorubicin composite-coated titanium bone implant. ACS Biomater Sci Eng 6(3):1650–1662

Tran H, Bui T (2021) A new thermoshock-based method for rapid preparation of ultra-pure hydroxyapatite nanodebris from bovine bone. Adv Nat Sci Nanosci Nanotechnol 12:035014. https://doi.org/10.1088/2043-6262/ac2952

Truite C et al (2022) Bioperformance studies of biphasic calcium phosphate scaffolds extracted from fish bones impregnated with free curcumin and complexed with Β-cyclodextrin in bone regeneration. Biomol Ther 12:383. https://doi.org/10.3390/biom12030383

Wang T-Y et al (2020) Mega-oss and Mega-TCP versus Bio-Oss granules fixed by alginate gel for bone regeneration. BDJ Open 6:14. https://doi.org/10.1038/s41405-020-0042-8

Wang Q et al (2021) Full utilisation of sweet sorghum for bacterial cellulose production: a concept of material crop. Ind Crop Prod 162:113256. https://doi.org/10.1016/j.indcrop.2021.113256

Wei X et al (2021) Osteoinductive potential and antibacterial characteristics of collagen coated iron oxide nanosphere containing strontium and hydroxyapatite in long term bone fractures. Arab J Chem 14:102984. https://doi.org/10.1016/j.arabjc.2020.102984

Yamada S et al (1997) Osteoclastic resorption of calcium phosphate ceramics with different hydroxyapatite/beta-tricalcium phosphate ratios. Biomaterials 18:1037–1041

Youssef H et al (2022) Synthesis of natural nano-hydroxyapatite from snail shells and its biological activity: antimicrobial, antibiofilm, and biocompatibility. Membranes 12:408. https://doi.org/10.3390/membranes12040408

Zambuzi G et al (2021) Modulating the controlled release of hydroxychloroquine mobilised on pectin films through film-forming pH and incorporation of nanocellulose. Carbohydr Polym Technol Appl 2:100140. https://doi.org/10.1016/j.carpta.2021.100140

Zhao Q et al (2019) Adsorption behavior of drugs on hydroxyapatite with different morphologies: a combined experimental and molecular dynamics simulation study. Ceram Int 45(15): 19522–19527

Zhong C (2020) Industrial-scale production and applications of bacterial cellulose. Front Bioeng Biotechnol 8:605374. https://doi.org/10.3389/fbioe.2020.605374

Zuliantoni Z et al (2022) Extraction and characterisation of snail shell waste hydroxyapatite. Results Eng 14:100390. https://doi.org/10.1016/j.rineng.2022.100390

Malaysian Seashells Based Hydroxyapatite for Biomedical Application

4

Nur Farahiyah Mohammad, Mohd Riza Mohd Roslan, and Nashrul Fazli Mohd Nasir

Abstract

Hydroxyapatite (HA, $Ca_{10}(PO_4)_6(OH)_2$) is widely used in dentistry and orthopaedics as a bone substitute or replacement material. The material is chemically and crystallographically identical to the mineral component of human biological bone. Natural materials such as animal bones, clay, eggshells, corals, and seashells can be used as a main source of calcium precursor to synthesise HA. Using Malaysian shell waste materials is sustainable and economical. Thus, these advantages have attract researchers to explore potential products from shell trash. This chapter presents the synthesis method and characterisation of hydroxyapatite synthesised from several Malaysian local seashells such as Paphia textile (lala), Corbiculacea (etok), and Polymesoda expansa (lokan) shells.

N. F. Mohammad (✉) · N. F. M. Nasir
Faculty of Electronic Engineering & Technology Universiti Malaysia Perlis, Kampus Alam UniMAP, Pauh Putra Arau, Perlis, Malaysia

Medical Device and Life Science Cluster, Sport Engineering Research Centre (SERC), Universiti Malaysia Perlis, Kampus Alam UniMAP, Pauh Putra, Arau, Perlis, Malaysia
e-mail: farahiyah@unimap.edu.my; nashrul@unimap.edu.my

M. R. M. Roslan
Faculty of Electronic Engineering & Technology Universiti Malaysia Perlis, Kampus Alam UniMAP, Pauh Putra Arau, Perlis, Malaysia

Department of Engineering and Built Environment, Tunku Abdul Rahman University of Management and Technology (TAR UMT), Penang Branch Campus, Tanjong Bungah, Pulau Pinang, Malaysia
e-mail: mohdriza@tarc.edu.my

W. S. Wan Kamarul Zaman, A. A. Nurul (eds.), *Sustainable Material for Biomedical Engineering Application*, https://doi.org/10.1007/978-981-99-2267-3_4

Keywords

Hydroxyapatite · Paphia textile · Anadara granosa · Corbiculacea · Polymesoda expansa · Precipitation

Abbreviations

EDS Energy dispersive X-ray spectroscopy
FTIR Fourier transform infrared
HA Hydroxypatite
XRD X-ray diffraction

4.1 Introduction

Bioceramics is defined as any ceramic materials that are specially developed for interfacing with the biological system to determine, treat, enhance, or replace any tissue, organ, or body function. Bioceramics is widely used in medical applications due to their high affinity for living bone tissue and bioactivity. In addition, it is resistant to corrosion, non-toxic, and non-carcinogenic, and it does not facilitate foreign body reactions (Punj et al. 2021). Examples of bioceramics are calcium phosphates, alumina, zirconia, and bioactive glasses. Out of the different types of calcium phosphates, hydroxyapatite (HA, $Ca_{10}(PO_4)_6(OH)_2$) is widely used as a material for bone graft, orthopaedic implant coating, and drug delivery (Singh et al. 2020). A synthetic HA with a stoichiometric composition of Ca/P = 1.67 remains the primary choice for biomedical applications due to its similarity with the mineral constituents of human bone and teeth (Mohammad et al. 2013). Hydroxyapatite has remarkable biological properties such as biocompatible, non-toxic, bioactive, osteoconductive, osteointegrative, and osteoinductive (Priyadarshini and Vijayalakshmi 2018). Hydroxyapatite can be synthesised from synthetic or natural sources (Singh et al. 2020). To synthesise HA from the synthetic sources, various combinations of calcium and phosphorous precursors for precipitation and hydrothermal synthesis of HA nanoparticles have been employed. The following are the commonly used calcium precursors: calcium nitrate tetrahydrate ($Ca(NO_3)_2 \cdot 4H_2O$) (Mohammad et al. 2015a, b), calcium chloride ($CaCl_2$) (Shiba et al. 2016), and calcium D-pantothenate monohydrate ($C_{18}H_{32}CaN_2O_{10}$) (Ng et al. 2010). The phosphorus precursors include the following: diammonium hydrogen phosphate ($(NH_4)_2HPO_4$) (Mohammad et al. 2015a, b), dipotassium hydrogen phosphate trihydrate (K_2HPO_4) (Shiba et al. 2016), ammonium dihydrogen phosphate ($NH_4H_2PO_4$) (Yang et al. 2012), phosphoric acid (H_3PO_4), sodium monophosphate dihydrate ($NaH_2PO_4 \cdot 2H_2O$), and potassium dihydrogen phosphate (KH_2PO_4) (Singh et al. 2020).

Natural sources such as those from minerals, mammals, marine life, and shells are preferred for the synthesis of HA due to their accessibility, ease of acquisition, increased calcium content, and ease of processing. Mineral resources such as limestone (Sirait et al. 2020; Mohd Pu'ad et al. 2019) and dolomite (Amirthalingam et al. 2019) are the main resources of inorganic minerals and contain a high percentage of calcium carbonate ($CaCO_3$) that becomes the calcium precursor for the synthesis of HA. The mammalian sources include porcine bone, bovine bone, chicken bone, and dromedary bone (Arokiasamy et al. 2022). Fish bone (Asadollahzadeh et al. 2019; Shi et al. 2018; Bahraminasab et al. 2021), fish scale (Pon-On et al. 2016), cuttlefish bone (Venkatesan et al. 2018), and corals (Naredla et al. 2022; Siswanto et al. 2020) are examples of sources from marine life. Shell sources such as eggshells (Wu et al. 2015, 2016), snail shells (Bambaeero and Bazargan-Lari 2021), and seashells (Roslan et al. 2019, 2020, 2021; Mohd Roslan et al. 2021a) are some of the examples of calcium-rich biological wastes that can be used to synthesise HA.

As previously stated, HA is a material that has been applied in biomedical applications as well as bone tissue scaffolds to facilitate new bone tissue (Sirait et al. 2020). As a result, there is a rising need for HA materials, but the price of chemicals makes them expensive to synthesise. Consequently, over time, researchers have increasingly used biological natural sources such as fish bone, seashell, egg-shell, and bovine bone as main sources of calcium precursor of HA. Normally, due to the expensive cost of disposal options such as incineration, these natural sources are frequently treated as waste materials that could be thrown away after the flesh is digested (Saharudin et al. 2018).

The shells of *Paphia textile* (*lala*) (Saharudin et al. 2018; Mohammad et al. 2012a), *Anadara granosa* (*blood cockle*) (Saharudin et al. 2018; Mohammad et al. 2010, 2012b), *Corbiculacea* (*etok*) (Roslan et al. 2019, 2021; Mohd Roslan et al. 2021a), *Polymesoda expansa* (*lokan*) (Roslan et al. 2020), and *Pholas orientalis* (*mentarang*) (Mohd Roslan et al. 2021b), which are marine life that can be found in Malaysian seas, have been proven to show tremendous potential to serve as a calcium precursor in the production of HA (Saharudin et al. 2018) due to their high concentration of calcium carbonate ($CaCO_3$), which is more than 95% by weight (Mohd Roslan et al. 2021a; Barua et al. 2018). Additionally, seashells continue to be underutilised and disposed of in landfills without any further processing (Saharudin et al. 2018). Seashell powder can be easily converted to calcium hydroxide ($Ca(OH)_2$) or calcium oxide (CaO) as both can be utilised as calcium sources (Roslan et al. 2019, 2020, 2021; Mohd Roslan et al. 2021a).

4.2 *Paphia textile* (*lala*)-Based Hydroxyapatite

Paphia textile (*P. textile*), which is locally known as *lala* or *kepah retak seribu*, is a popular seafood in Malaysia. It is well-liked because of its mouthwatering texture and sweet, juicy, and distinctive flavour. This marine life can be found along mudflats and other freshwater habitats. However, the shells of *P. textile* are not

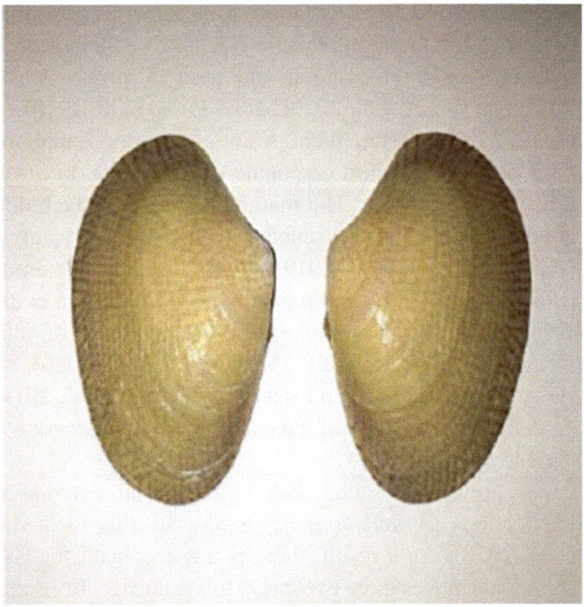

Fig. 4.1 *Paphia textile* (*lala*)

known to be used commercially and are usually discarded to waste. The shells contain a high percentage of calcium carbonate ($CaCO_3$) (94–99%) or also known as calcite (George et al. 2019). Therefore, *P. textile* shell can act as a calcium precursor to synthesise hydroxyapatite. Based on its mineral composition, *P. textile* shell has been proposed as a possible alternative for a bone substitute material. Moreover, *P. textile* is usually found in the intertidal and sublittoral zones of the coastal ecosystem in the sandy-muddy bottoms. This bivalve species has a strong, comparable, inequilateral shell with greyish brown zigzag lines banded with beige to pale yellow external colour (Argente and Estacion 2014), as shown in Fig. 4.1.

4.2.1 Synthesis of HA from *P. textile* (*lala*)

P. textile (*lala*) was obtained from the wet market at Kuala Perlis, Perlis, Malaysia. The dirt on the shells' surface was removed using water, and then they were repeatedly rinsed with distilled water. Then, the shells were dried in the oven at 200 °C for 2 h. The shells were then ground into a fine calcium carbonate ($CaCO_3$) powder with an agate mortar and pestle. The sample was calcined in a furnace at 1000 °C for 3 hours at a heating rate of 10 °C/min. The sample that consisted of calcium carbonate ($CaCO_3$) turned into an amorphous calcium oxide (CaO) after the calcination process by releasing carbon dioxide (CO_2). Equation 4.1 demonstrates the chemical reaction that occurs during the calcination process:

$$CaCO_3 \rightarrow CaO + CO_2 \tag{4.1}$$

Hydroxyapatite was synthesized using precipitation method. The calcium hydroxide ($Ca(OH)_2$) and phosphoric acid (H_3PO_4) were used as the calcium and phosphate precursors, respectively. Calcium oxide (CaO) was converted into calcium hydroxide ($Ca(OH)_2$) by adding 300 mL of distilled water to a beaker holding 5 g of calcium oxide (CaO). This produced a 0.25 M calcium hydroxide ($Ca(OH)_2$) solution. Equation 4.2 shows the chemical reaction of the conversion:

$$CaO + H_2O \rightarrow Ca(OH)_2 \tag{4.2}$$

The mixture was stirred for 2 hours by using a magnetic stirrer at 45 °C at 1500 rpm until a uniform mixture of calcium hydroxide ($Ca(OH)_2$) solution was achieved. Then, 2.4 mL of 0.15 M phosphoric acid (H_3PO_4) was slowly dripped into the solution at a rate of 0.1 mL/min at room temperature. The pH of the mixture was maintained at pH 9 by the addition of ammonia (NH_4OH) solution. The mixture turned into a milky solution and aged at room temperature for 24 hours. Then, the white precipitate was collected and washed by centrifuging at a rotation speed of 3000 rpm for 5 minutes with distilled water. These processes were repeated three times to eliminate the remaining ammonia (NH_4OH) in the sample. Then, the sample was dried for 6 hours in the oven at 110 °C. The procedure was repeated at different pH values of 10 and 11. The anhydrous white precipitates were calcined in a furnace at 800 °C for 2 hours and then ground using a mortar and pestle to produce a very fine powder.

4.2.2 Effect of the Synthesis pH on the Phase and Crystallographic Structures *P. textile*-Based HA

The synthesised HA products obtained from the reaction of calcium hydroxide ($Ca(OH)_2$) with phosphoric acid (H_3PO_4) via the precipitation method at pH values of 9, 10, and 11 were characterised using X-ray diffraction (XRD). Sharp peaks showed that HA crystals had formed with a high degree of crystallinity.

The XRD patterns of HA derived from *P. textile* synthesised at different pH values are shown in Fig. 4.2. The XRD patterns of HA-pH 11 were attributed to HA reflections, and no characteristic peaks of the secondary phase were observed, a finding which was consistent with JCPDS file no. 01-074-0566 (HA). However, non-pure HA was obtained at pH 9 and pH 10 as several peaks of calcium hydroxide ($Ca(OH)_2$) (18.2°, 34.2°, and 47.3°) matched with JCPDS 01-07600570 ($Ca(OH)_2$). The presence of calcium hydroxide ($Ca(OH)_2$) in the secondary phase in HA indicated the incomplete reaction of calcium precursor in which calcium oxide (CaO) absorbed water from the air to form calcium hydroxide ($Ca(OH)_2$) (Bahraminasab et al. 2021). This phenomenon also occurred due to the insufficient time for the reaction between calcium hydroxide ($Ca(OH)_2$) and phosphoric acid (H_3PO_4) during the synthesis process (Sy Mohamad et al. 2016).

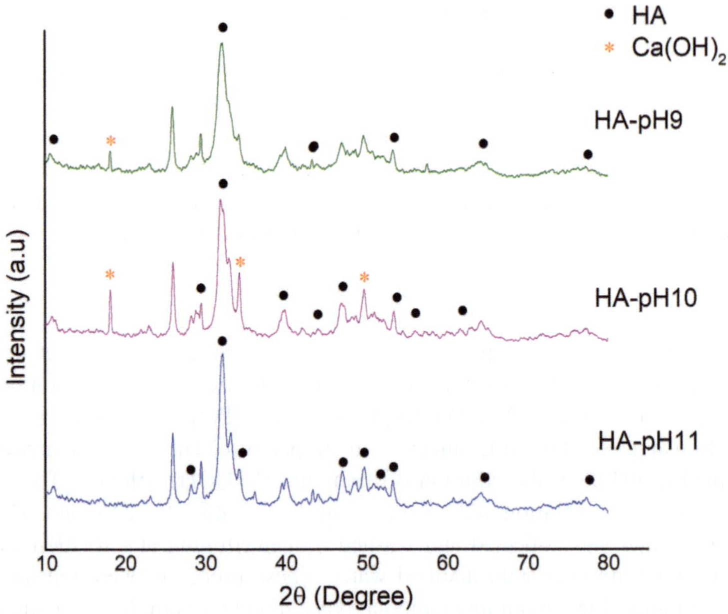

Fig. 4.2 XRD of *P. textile*-based HA synthesis at different pH values

In comparison to HA-pH 9 and HA-pH 10, HA-pH 11 showed the sharpest and most intense peaks indicating that HA-pH 11 was more crystalline compared to other samples. The crystallite size of the HA powder was determined using Debye Scherrer's equation, as shown in Eq. 4.3 where k is the shape constant (0.9), λ is the wavelength of the radiation, β_{hkl} is the full width at half maximum, and θ is the peak position (Rabiei et al. 2020).

$$D = \frac{k\lambda}{\beta_{hkl}\cos\theta} \tag{4.3}$$

The sharpest diffraction peaks at $2\theta = 32.1°$ were chosen to calculate the crystalline size of the sample. The average crystallite sizes of the HA powder were found to be 63.63 nm, 47.73 nm, and 23.87 nm for pH values of 9, 10, and 11, respectively, showing that the crystallite size reduced with rising pH. Thus, it can be concluded that the pH of the synthesis process affected the purity and crystallinity of HA.

4.3 *Corbiculacea* (*etok*)-Based Hydroxyapatite

Corbiculacea (*etok*) is a freshwater clam (see Fig. 4.3) that can be found at paddy fields, brackish rivers and isolated rivers all over Malaysia except in Langkawi Island and Penang Island (Rak et al. 2020). The shell of *Corbiculacea* is yellow-

Fig. 4.3 *Corbiculacea* (*etok*). (Reproduced with permission from Draheim (2012))

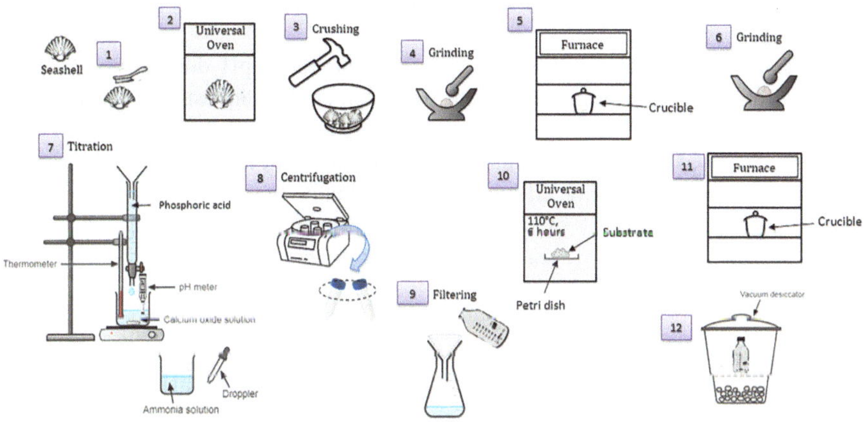

Fig. 4.4 Synthesis process of HA from *Corbiculacea* (*etok*) shells (Mohd Roslan et al. 2021a) (Reproduced with permission from Springer Nature Customer Service Centre GmbH: Springer, Intelligent Manufacturing and Mechatronics. Lecture Notes in Mechanical Engineering, Synthesizing and Optimization the Hydroxyapatite Based on Corbiculacea Seashells, Mohd Riza Mohd Roslan, Nashrul Fazli Mohd Nasir, Nur Farahiyah Mohammad et. al., 2021)

green in colour, with concentric growth rings of the prismatic layer, which can be seen on the outer layer of the shells.

4.3.1 Synthesis of HA from *Corbiculacea* (*etok*)

The synthesis process of HA from *Corbiculacea* (*etok*) is shown in Fig. 4.4. The shells were washed, cleaned, and dried in an oven at 100 °C for 1 hour. Then, the shells were crushed and ground into a micrometre size. Next, the *Corbiculacea* powder was calcined in a furnace at 1100 °C for 3 hours to turn calcium carbonate ($CaCO_3$) into calcium oxide (CaO). The calcium oxide (CaO) powder was then grounded into fine powder.

Next, 5 g of calcium oxide (CaO) powder was added into 300 mL of distilled water and stirred at 500 rpm. The calcium hydroxide ($Ca(OH)_2$) solution was then aged at 45 °C for 2 hours. After that, 2.8 mL of phosphoric acid (H_3PO_4) was titrated into the calcium hydroxide ($Ca(OH)_2$) solution at a rate of 5.5 mL/min. By adding 28% ammonia (NH_4OH) solution, the pH of the solution was altered to pH 10 or pH 11. The milky white precipitate was then aged in a room temperature for 24 hours. The precipitate was then filtered, washed with distilled water, and dried in the oven for 6 hours at 110 °C to completely remove the water. The white dried HA paste was then crushed using a mortar and pestle. The powder was finally sintered in a 500 °C furnace.

4.3.2 Characterisation of *Corbiculacea*-Based HA

Various synthesis parameters such as Ca/P molar ratio, pH during the synthesis process, and sintering temperature can be considered to investigate the physicochemical properties of *Corbiculacea*-based HA. A fourier transform infrared (FTIR) analysis was used to identify the functional groups present in the HA powders and to validate the HA. The x-ray diffraction (XRD) analysis was used to evaluate the phase and the crystallinity of the HA powder. Meanwhile, the energy dispersive X-ray (EDS) spectroscopy analysis was performed to determine the elemental composition of HA.

4.3.3 Effect of Phosphate Amount on Ca/P Molar Ratio

Roslan et al. (2019) manipulated the Ca/P molar ratio by using different amounts of phosphoric acid (H_3PO_4) during the synthesis of HA, namely, 1.80 mL, 1.64 mL, and 1.35 mL, which were respective to the 1.50, 1.67, and 2.00 of the Ca/P molar ratio (theoretically calculated). Table 4.1 shows the actual Ca/P molar ratios calculated from the EDS analysis that were slightly different from the theoretically calculated ratios. These results were most probably due to the increase in oxygen, O element, which was possibly contributed by the formation of oxide during calcination.

Table 4.1 Ca/P molar ratio of the HA synthesised using different amounts of phosphate precursor (Roslan et al. 2019)

Amount of phosphoric acid (H_3PO_4) (mL)	HA Ca/P molar ratio (theoretically calculated)	HA Ca/P molar ratio (calculated from EDS)
1.8	1.5	1.58
1.64	1.67	1.64
1.35	2.00	2.05

4.3.4 Effect of pH and Calcination Temperature

Roslan et al. (2019) demonstrated that pH value and calcination temperature during the synthesis process might affect the physicochemical properties of HA. In this experiment, four batches of HA were synthesised (Mohd Roslan et al. 2021a), as shown in Table 4.2.

The FTIR results (Fig. 4.5) show that pure HA was successfully synthesised at pH 10 and pH 11, with or without the calcination process. The main significant finding that can be observed from the spectrum is the CO_3^{2-} peak for HA-10 (880.25 cm^{-1}) and HA-11 (873.68 cm^{-1}) disappeared after calcination at 500 °C.

Table 4.2 pH value and calcination temperature of HA

Sample name	pH	Calcination temperature (°C)
HA-10	10	Without sintering
HA-10-500	10	500
HA-11	11	Without sintering
HA-11-500	11	500

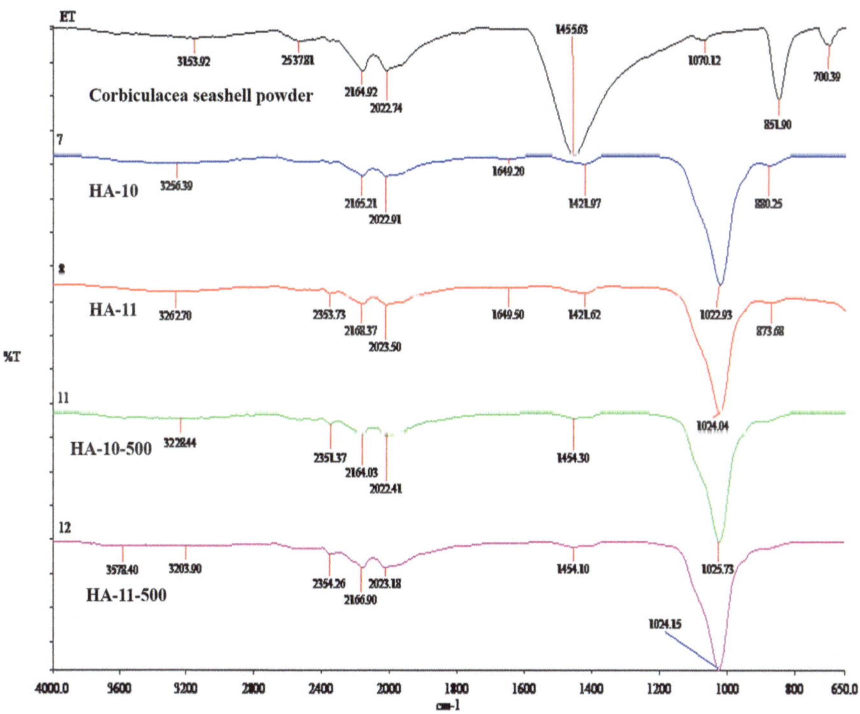

Fig. 4.5 FTIR results of HA synthesised at different pH values and calcination temperatures (Mohd Roslan et al. 2021a). (Reproduced with permission from Springer Nature Customer Service Centre GmbH: Springer, Intelligent Manufacturing and Mechatronics. Lecture Notes in Mechanical Engineering, Synthesizing and Optimization the Hydroxyapatite Based on Corbiculacea Seashells, Mohd Riza Mohd Roslan, Nashrul Fazli Mohd Nasir, Nur Farahiyah Mohammad et. al., 2021)

Thus, it can be concluded that the calcination process may reduce the amount of carbonate ion in the HA samples.

4.4 *Polymesoda expansa* (*lokan*)-Based Hydroxyapatite

Polymesoda expansa (*P. expansa*) is a heavy bivalve marine life that can be found buried in the stiff mud of the landward fringe of mangroves (see Fig. 4.6). In Malaysia, *P. expansa* can be found in the mangrove forests of Peninsular Malaysia, Sabah, and Sarawak. The chemical compositions of *P. expansa* shell are 67.7% calcium oxide (CaO), 3.06% silicon dioxide (SiO_2), 1.19% magnesium oxide (MgO), 0.02% iron(III) oxide (Fe_2O_3), and 2.71% others (Febrianto 2021).

4.4.1 Synthesis of HA from *P. expansa* (*lokan*)

P. expansa (*lokan*) was acquired from a local market in Sabah, Malaysia. The fleshes were removed from the shells. Then, the shells were cleaned using distilled water and ground into fine powder using a mortar and pestle. Next, the fine *P. expansa* was calcined in a furnace at 900 °C for 4 h at a heating rate of 10 °C/min. Next, calcium hydroxide (CaOH) solution was prepared by adding 2.4 g of calcium oxide (CaO) powder gained from the calcination of *P. expansa* into 100 mL of distilled water under constant stirring. Then, phosphoric acid (H_3PO_4) was titrated at a rate of 5.5 mL/min into the calcium hydroxide (CaOH) solution under continuous stirring. The ammonia (NH_4OH) solution was dropped into the mixture to set the pH of the solution to 9 and 13 during the titration process. The solution was then aged for 24 h in room temperature. Later, the white precipitate was further filtered and rinsed using distilled water. The precipitates were then dried in an oven at 110 °C for 6 h. Finally, using a mortar and pestle, the calcined HA was pulverised into a powder form.

Fig. 4.6 *Polymesoda expansa* (*lokan*). (Reproduced with permission from Pinay06 (2004))

4.4.2 Effect of the Synthesis pH on the Ca/P Molar Ratio of *P. expansa*-Based HA

The synthesised HA products obtained from the reaction of calcium hydroxide (Ca $(OH)_2$) (converted from *P. expansa* shells) with phosphoric acid (H_3PO_4) via the precipitation method at pH values of 9 and 13 were characterised using EDS. The Ca/P molar ratio of the HA synthesised at pH 9 was 1.78, while at pH 13 was 1.97. These results contradicted the ideal pH of HA, which is 1.67, obtained using the calcium and phosphate precursor brought to a pH 11–12 using ammonia (NH_4OH) solution (Gokcekaya et al. 2021). Similarly, these findings again proved that pH of the synthesis process might affect the purity of HA.

4.5 Conclusion

Local Malaysian seashells such as *P. textile* (*lala*), *Corbiculacea* (*etok*), *and P. expansa* (*lokan*) have been utilised as natural sources of calcium precursors of hydroxyapatite. Several synthesis parameters such as pH, calcination temperature, and amount of phosphate precursor have been considered and manipulated to investigate their effects on the physicochemical properties and Ca/P molar ratio of HA. The results demonstrated that altering the pH value during the HA synthesis process may affect the phase, crystallography structure, and Ca/P molar ratio of HA. Single-phase HA was obtained at pH 11. The crystallite size decreased while crystallinity increased with the increase in pH. In conclusion, the Malaysian seashells have been proven to be the potential natural sources for the synthesis of HA. This may lead to the commercialisation of Malaysian seashells as new sources of biomaterials.

Acknowledgements The authors would like to acknowledge the support from the Fundamental Research Grant Scheme (FRGS) under a grant number of FRGS/1/2021/TK0/UNIMAP/02/59 from the Ministry of Higher Education Malaysia and from Research Materials Fund (RESMATE) under a grant number of 9001-00626 from the Universiti Malaysia Perlis.

Conflict of Interest The authors declare that they have no conflict of interest.

References

Amirthalingam N, Deivarajan T, Paramasivam M (2019) Mechano chemical synthesis of hydroxy-apatite using dolomite. Mater Lett 254:379–382

Argente FA, Estacion JS (2014) Effect of different harvesting practices on the dynamics of Paphia textile (Gmelin 1792) (Bivalvia: Veneridae) populations at two sites in Zamboanga del Norte, Southern Philippines. Environ Exp Biol Med 12:113–120

Arokiasamy P, Al Bakri Abdullah M, Abd Rahim SZ, Luhar S, Sandu AV, Jamil NH, Nabiałek M (2022) Synthesis methods of hydroxyapatite from natural sources: a review. Ceram Int 48(11): 14959–14979

Asadollahzadeh M, Rabiee SM, Salimi-Kenari H (2019) In vitro apatite formation of calcium phosphate composite synthesized from fish bone. Int J Appl Ceram Technol 16(5):1969–1978

Bahraminasab M, Doostmohammadi N, Alizadeh A (2021) Low-cost synthesis of nano-hydroxyapatite from carp bone waste: effect of calcination time and temperature. Int J Appl Ceram Technol 18(3):573–582

Bambaeero A, Bazargan-Lari R (2021) Simultaneous removal of copper and zinc ions by low cost natural snail shell/hydroxyapatite/chitosan composite. Chin J Chem Eng 33:221–230

Barua E, Deoghare AB, Deb P, Lala SD (2018) Naturally derived biomaterials for development of composite bone scaffold: a review. IOP Conf Ser Mater Sci Eng 377:012013

Draheim R (2012) Corbicula. https://www.flickr.com/photos/52133016@N08/8030428537. Accessed 5 Nov

Febrianto A (2021) Penggunaan Cangkang Kerang Lokan (Polymesoda Expansa) Kompilasi Dengan Bentonit Untuk Memperbaiki Kualitas Ph, Logam Besi (Fe) Dan Warna Pada Air Gambut. Teknik Lingkungan, Indonesia

George P, Hamid ZA, Bakar@Zakaria MZA, Perimal EK, Bharatham H (2019) A short review on cockle shells as biomaterials in the context of bone scaffold fabrication. Sains Malays 48(7): 1539–1545

Gokcekaya O, Ergun C, Webster TJ, Bahadir A, Ueda K, Narushima T, Nakano T (2021) Effect of precursor deficiency induced Ca/P ratio on antibacterial and osteoblast adhesion properties of Ag-incorporated hydroxyapatite: reducing Ag toxicity. Materials (Basel) 14(12):3158

Mohammad NF, Zahid MA, Awang SA, Zakaria Z, Abdullah AA (2010) Synthesis and characterization of bioceramic from Malaysian cockle shell. In: Industrial Electronics & Applications (ISIEA), 2010 IEEE symposium on, pp 413–416

Mohammad NF, Muhammed MH, Zakaria Z, Abdullah AA, Mohammad IS (2012a) Characterization of calcium phosphate bioceramic from Paphia undulata shells. In: 2012 international conference on biomedical engineering (ICoBE), pp 114–116

Mohammad NF, Kanesvaran G, Othman SM, Mohammad IS (2012b) Effect of particulate size on mechanical strength of Anadara granosa bioceramic. In: 2012 international conference on biomedical engineering (ICoBE), pp 104–108

Mohammad NF, Fei Yee Y, Othman R (2013) Synthesis of nanoporous carbonated hydroxyapatite using non-ionic pluronics surfactant. Adv Mater Res 686:33–34

Mohammad NF, Othman R, Yeoh FY (2015a) Pore characteristics of mesoporous carbonated hydroxyapatite synthesised with different nonionic surfactant and carbonate concentration. Mater Sci Forum 819:353–360

Mohammad NF, Othman R, Yeoh FY (2015b) Controlling the pore characteristics of mesoporous apatite materials: hydroxyapatite and carbonate apatite. Ceram Int 41(9, Part A):10624–10633

Mohd Pu'ad NAS, Koshy P, Abdullah HZ, Idris MI, Lee TC (2019) Syntheses of hydroxyapatite from natural sources. Heliyon 5(5):e01588

Mohd Roslan MR, Mohd Nasir NF, Mohammad NF, Meng CE, Mohd Amin NA, Abdul Khalid MF, Zakaria MZ, Azizan MM, Jusoh M (2021a) In: Bahari MS et al (eds) Intelligent manufacturing and mechatronics. Springer, Singapore, pp 975–981

Mohd Roslan MR, Mohd Nasir NF, Mohamad NF, Nur H (2021b) Hydroxyapatite (HA) from Mentarang (Pholas Orientalis) shells. Multidiscipl Appl Res Innov 2(3):194–198

Naredla M, Osmani RA, Meenakshi S, Gupta MS, Gowda DV (2022) Potential applications of coral sand in bone healing and drug delivery. J Drug Deliv Sci Technol 69:103150

Ng S, Guo J, Ma J, Loo SCJ (2010) Synthesis of high surface area mesostructured calcium phosphate particles. Acta Biomater 6(9):3772–3781

Pinay06 (2004) Shellfish. https://commons.wikimedia.org/wiki/File:Shellfish4.jpg. Accessed 5 Nov

Pon-On W, Suntornsaratoon P, Charoenphandhu N, Thongbunchoo J, Krishnamra N, Tang IM (2016) Hydroxyapatite from fish scale for potential use as bone scaffold or regenerative material. Mater Sci Eng C 62:183–189

Priyadarshini B, Vijayalakshmi U (2018) Development of cerium and silicon co-doped hydroxyapatite nanopowder and its in vitro biological studies for bone regeneration applications. Adv Powder Technol 29(11):2792–2803

Punj S, Singh J, Singh K (2021) Ceramic biomaterials: properties, state of the art and future prospectives. Ceram Int 47(20):28059–28074

Rabiei M, Palevicius A, Monshi A, Nasiri S, Vilkauskas A, Janusas G (2020) Comparing methods for calculating nano crystal size of natural hydroxyapatite using X-ray diffraction. Nanomaterials 10(9):1627

Rak AE, Azizan AT, Yaacob MR, Hamzah Z, Syed Omar SA, Zakaria MN, Ismail M, Wan Ibrahim WK, Mohamad Rani WSF, Zaki MZ, Mohammed A (2020) Traditional processing method of smoked Corbicula fluminea (Etak): case of Etak vendor in Kelantan, Malaysia. IOP Conf Ser Earth Environ Sci 596(1):012057

Roslan MR, Nasir NFM, Khalid MFA, Mohammad NF, Meng CE, Hashim NNN, You BC, Majid MSA, Amin NAM (2019) The optimization of the hydroxyapatite (HA) material characteristics produced from Corbiculacea (Etok) shells. J Phys Conf Ser 1372(1):012077

Roslan MR, Nasir NFM, Gilani MA, Mohammad NF, Cheng EM, Khalid MFA, Zoolfakar AS, Amin NAM, Khan SF (2020) Preliminary study of the polymesoda expansa based hydroxyapatite for medical devices coating application. AIP Conf Proc 2306(1):020033

Roslan MRM, Nasir NFM, Mohammad NF, Meng CE, Mohamed MS, Abdullah MN (2021) The influences of sintering process on the characteristics of Corbiculacea (Etok) shells based hydroxyapatite powder. J Phys Conf Ser 2071(1):012007

Saharudin SH, Shariffuddin JH, Ismail A, Mah JH (2018) Recovering value from waste: biomaterials production from marine shell waste. Bull Mater Sci 41(6):162

Shi P, Liu M, Fan F, Yu C, Lu W, Du M (2018) Characterization of natural hydroxyapatite originated from fish bone and its biocompatibility with osteoblasts. Mater Sci Eng C 90:706–712

Shiba K, Motozuka S, Yamaguchi T, Ogawa N, Otsuka Y, Ohnuma K, Kataoka T, Tagaya M (2016) Effect of cationic surfactant micelles on hydroxyapatite nanocrystal formation: an investigation into the inorganic–organic interfacial interactions. Cryst Growth Des 16(3): 1463–1471

Singh G, Singh RP, Jolly S (2020) Customized hydroxyapatites for bone-tissue engineering and drug delivery applications: a review. J Sol-Gel Sci Technol 94(3):505–530

Sirait M, Sinulingga K, Siregar N, Siregar RSD (2020) Synthesis of hydroxyapatite from limestone by using precipitation method. J Phys Conf Ser 1462(1):012058

Siswanto S, Hikmawati D, Kulsum U, Rudyardjo DI, Apsari R, Aminatun A (2020) Biocompatibility and osteoconductivity of scaffold porous composite collagen–hydroxyapatite based coral for bone regeneration. Open Chem 18(1):584–590

Sy Mohamad SF, Mohamad S, Jemaat Z (2016) Study of calcination condition on decomposition of calcium carbonate in waste cockle shell to calcium oxide using thermal gravimetric analysis. J Eng Appl Sci 11:9917–9921

Venkatesan J, Rekha PD, Anil S, Bhatnagar I, Sudha PN, Dechsakulwatana C, Kim S-K, Shim MS (2018) Hydroxyapatite from cuttlefish bone: isolation, characterizations, and applications. Biotechnol Bioprocess Eng 23(4):383–393

Wu S-C, Hsu H-C, Hsu S-K, Chang Y-C, Ho W-F (2015) Effects of heat treatment on the synthesis of hydroxyapatite from eggshell powders. Ceram Int 41(9, Part A):10718–10724

Wu S-C, Hsu H-C, Hsu S-K, Chang Y-C, Ho W-F (2016) Synthesis of hydroxyapatite from eggshell powders through ball milling and heat treatment. J Asian Ceramic Soc 4(1):85–90

Yang L-X, Yin J-J, Wang L-L, Xing G-X, Yin P, Liu Q-W (2012) Hydrothermal synthesis of hierarchical hydroxyapatite: preparation, growth mechanism and drug release property. Ceram Int 38(1):495–502

Chitosan from Marine Biowaste: Current and Future Applications in Tissue Engineering

5

Nazmul Huda Syed, Harishini Rajaratinam, and Asma Abdullah Nurul

Abstract

Chitosan is a natural polysaccharide-based polymer that is synthesised from chitin through a deacetylation process. Chitin occurs naturally as organised crystalline microfibrils that form structural components in arthropod exoskeletons, fungus or yeast cell walls. Commercial chitosan is primarily derived from marine biowaste, such as crab and shrimp shells. Chitosan possesses excellent biological properties such as biocompatibility, antibacterial activity, wound healing, coagulating activity, bioadhesivity, and regenerative capabilities. This chapter will provide an overview of the various marine biowaste sources used to produce chitosan and its derivatives, as well as its applicability in combination with other polymers, emphasising current and prospective applications in tissue engineering.

Keywords

Biowaste · Chitosan · Marine · Tissue engineering · Polymers

Abbreviations

CNT	Carbon nanotube
ECM	Extracellular matrix
GAG	Glycosaminoglycans
MSC	Mesenchymal stem cells
PCL	Poly ε-caprolactone

N. H. Syed · H. Rajaratinam · A. A. Nurul (✉)
School of Health Sciences, Universiti Sains Malaysia, Kubang Kerian, Kelantan, Malaysia
e-mail: nurulasma@usm.my

W. S. Wan Kamarul Zaman, A. A. Nurul (eds.), *Sustainable Material for Biomedical Engineering Application*, https://doi.org/10.1007/978-981-99-2267-3_5

PEG	Poly(ethylene glycol)
PEM	Polyelectrolyte multilayer
PHB	Polyhydroxybutyrate
PLGA	Poly(lactic-co-glycolic acid)
PVA	Polyvinyl alcohol

5.1 Introduction

Seafood commerce generates approximately 106 tons of waste each year, most of it is directly discarded into the marine environment without prior treatment (Schmitz et al. 2019). However, this marine biowaste is an excellent source of raw materials to produce biocompounds (Mohanty et al. 2018). On the other hand, approximately 140,106 tons of synthetic polymers are manufactured every year (Abo Elsoud and El Kady 2019). These polymers have limited stability and biodegradability. This necessitates the use of environmentally friendly biodegradable polymers. Chitin and chitosan, biopolymers that are derived from residues of marine biowaste, have recently attracted the interest of researchers.

Chitin and chitosan, common biopolymers in nature, are crucial elements in the outer shells of many different organisms and as seafood industry by-products. They are found in around 40% of the seafood waste generated. Following cellulose, chitin is the major prevalent biopolymer in nature. Chitosan is an amino polysaccharide that is derived from alkali chitin deacetylation (Arasukumar et al. 2019). Chitosan is widely utilised as a biomaterial due to its biological properties, including biocompatibility, biodegradability, and nontoxicity. Moreover, it has antibacterial and antifungal properties, deeming it useful as a therapeutic agent for application in medicine, agriculture, food, the environment, and textile industries (Santos et al. 2020).

Natural polymers have received a lot of attention over the last few decades. Chitosan is known to have excellent biocompatibility due to the structural similarity to glycosaminoglycans (GAGs) of the extracellular matrix (ECM). Chitosan's hydrophilic structure has been shown to promote cellular processes like adhesion, differentiation, and proliferation. Furthermore, the presence of free amine groups inside its backbone chain enables for additional chemical modifications to be made to improve biomedical functionality. Chitosan's abundant availability and unique biological properties make it a potential candidate for successful tissue regeneration (Sultankulov et al. 2019). This chapter provides an insight into the structure and preparation of chitosan from marine biowaste, as well as the chitosan's present and future applications in tissue engineering.

5.2 Chitosan: Structure

Chitosan is produced from chitin, whose molecular weight varies depending on its source, ranging from 300 to 1000 kDa. It is frequently confused with cellulose due to their nearly identical chemical backbones; however, chitin has acetamide groups (CH_3CONH_2) at the carbonyl groups rather than hydroxyl groups (Dutta et al. 2012). Chitin's chemical structure contains 1-4 linked 2-acetamido-2-deoxy-β-D-glucopyranose, whilst chitosan has β-(1-4) linked 2-amino-2-deoxy-β-D-glucopyranose. Chitosan is an N-deacetylated derivative of chitin that is produced through converting acetamide groups into key amino groups (Fig. 5.1) (Aranaz et al. 2009).

There are three types of crystalline amorphic chitin found in nature, including α-, β-, and γ-chitin. Chitinous substances obtained from invertebrates, crustaceans, and arthropods, especially those with hard exoskeleton, are normally α-shaped. The arrangement is compact due to the stronger hydrogen bonding between different α-chitin chains (Sajomsang and Gonil 2010). The purity, crystallinity, and polymer chain composition of chitin differ based on the amount present in its source (Zainol Abidin et al. 2020). In its natural state, chitin is flexible, extremely strong, transparent, and elastic. When chitin is combined with calcium carbonate ($CaCO_3$), it forms an even tougher composite, as seen in crustacean and mollusc exoskeletons (Azuma et al. 2014). Chitin deacetylation generates acetate ions and a $-NH_2$ group that is produced when the acetamide group is hydrolysed.

In chitosan, glucosamine to *N*-acetyl glucosamine ratio is used to determine deacetylation. Chitin is enriched in *N*-acetyl glucosamine when compared with glucosamine; chitosan, however, is richer in glucosamine (Ramírez et al. 2010; Benhabiles et al. 2012). Chitosan by far is more active than chitin because of amine groups on every deacetylated unit, as well as primary and secondary hydroxyl groups on each repeat unit. The amine groups within chitin and chitosan enables distinct biological functions and modification reactions (Kumar 2000). Based on the

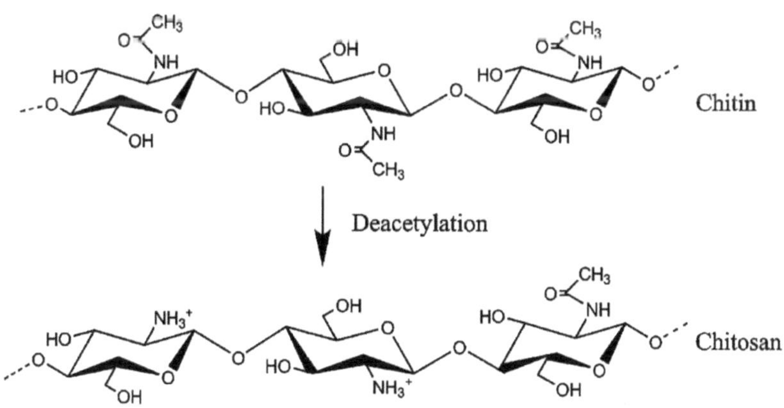

Fig. 5.1 Chemical structures and process demonstration of chitin into chitosan

experimental settings, the deacetylation degree impacts chemical properties (such as flexibility, viscosity, surface area, conductivity, tensile strength, porosity, and solubility) and biological properties (such as bioavailability, biodegradability, biocompatibility, adsorption, and antioxidant) (Park and Kim 2010; Fan et al. 2012).

5.3 Chitosan: Sources and Extraction

Figure 5.2 depicts the most common chitin and chitosan sources including crustaceans (crab and shrimp shell), insects (ants, beetle cocoon, and cuticle), fungi (*Aspergillus nidulans* and *Penicillium chrysogenum*), and molluscs (clams and squids). Despite the fact that chitin is abundant in many natural sources, chitosan is found in limited quantities. It is usually extracted from chitin for industrial and research applications using chemical or enzymatic methods (Ibrahim and Zairy 2015).

5.4 Marine Biowaste as Chitosan Source

Previous studies indicate that crustaceans are the top-priority commercial sources for chitosan production. On a large scale, marine biowaste, including crustacean exoskeletons, is the major resource of chitin and chitosan (Elieh-Ali-Komi and

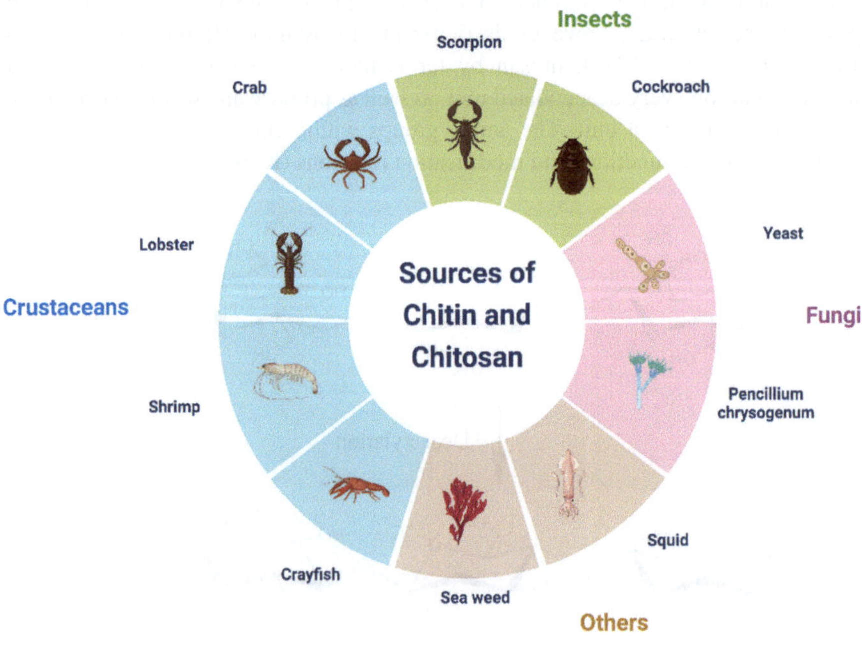

Fig. 5.2 Common sources of chitin and chitosan

Hamblin 2016). Crabs and shrimps, as well as many other crustaceans such as lobsters, prawns, and crayfish, have been extensively used as sources of chitin in the past (Yao et al. 2011; Arbia et al. 2013). Chitin percentage (wt%) varies by organism; in crustacean shell waste, chitin content ranges from 20% to 30% by weight, with some lobster genera having the highest chitin content of 60–70% by weight (Synowiecki and Al-Khateeb 2003; No and Meyers 1995). The use of crustacean by-products (animal body, claws, and thorax) containing approximately 20% chitin has yielded promising results in the production of chitin or chitosan, making it a less expensive and more readily available source of chitin production on an industrial scale (Vázquez et al. 2013). Because approximately half of the crustacean's total body mass ends up as waste and is dumped into the sea, recent studies illustrate environmental and economic applications of these sources for preparation of chitin (No and Meyers 1995).

5.5 Preparation of Chitosan

Multitudinous studies have focused on producing marine chitin as well as chitosan. Generally, chitosan extraction is classified into two types: chemical and enzymatic methods (Fig. 5.3).

5.5.1 Chemical Methods

Chitosan extraction from chitin undergoes four processing steps: demineralization, deproteination, decolorization, and deacetylation. Demineralization entails removing calcium carbonates ($CaCO_3$) and calcium and phosphates with a strong acid (HCl); deproteination entails removing protein and other organic components with an alkali solution (NaOH); decolorization entails removing pigment with an alcohol (ethanol); and deacetylation entails converting chitin to chitosan with an alkali solution (NaOH) (Ruiz and Corrales 2017). Although earlier research revealed a deproteinization step prior to demineralization, the yield and quality of the chitin generated have not been perceived to be significantly affected by this particular sequence. Recent literature suggests using the processing steps in the order of demineralization > deproteinization > decolorization > deacetylation. The demineralization step initially increases the shells' surface area by removing $CaCO_3$, thereby accelerating the following deproteinization reaction (Younes and Rinaudo 2015).

There are several parameters that have been identified to affect each processing step during chitosan production. Creating an environment with excessive acid by using both higher acid concentrations and lower shell-to-acid ratios to assist the reaction in progressing towards demineralization. Two factors enhance the efficacy of the demineralization response: creating an environment with excessive acid by using both higher acid concentrations and reduced shell-to-acid ratios, as well as increasing the reaction duration (Kou et al. 2021). When seeking to improve the

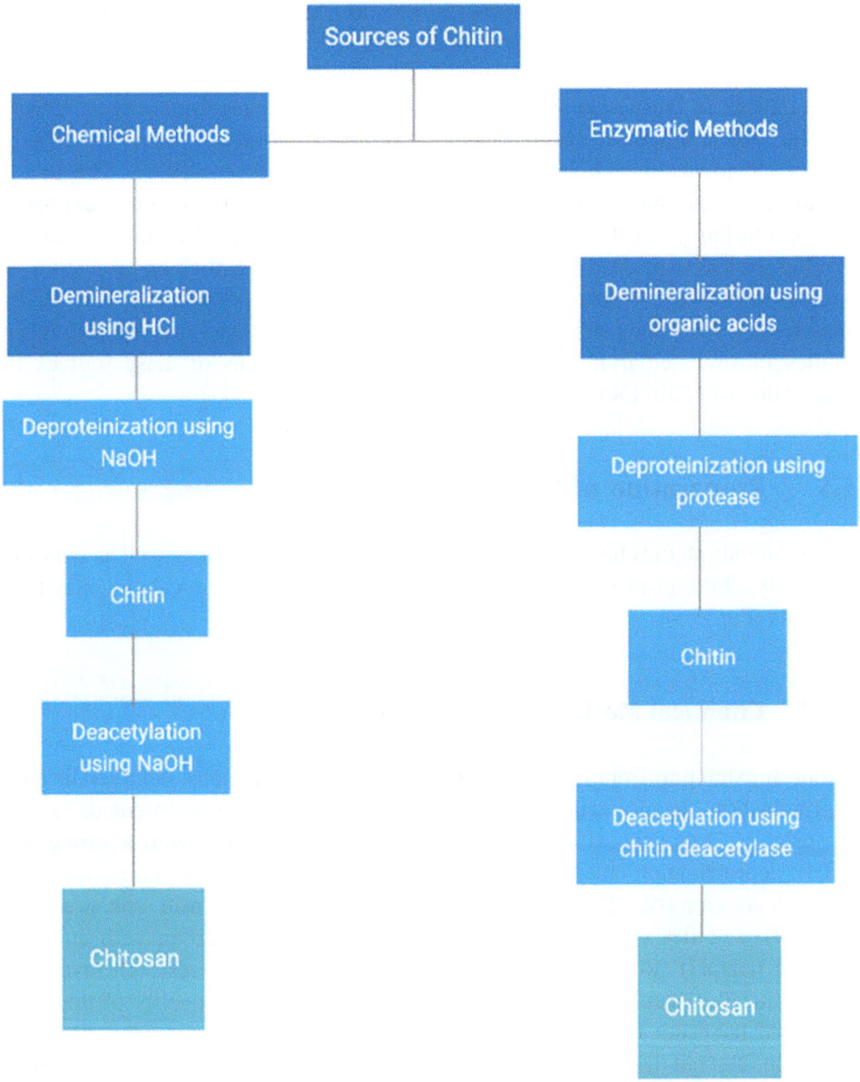

Fig. 5.3 Preparation of chitosan using chemical and enzymatic methods

deproteinization efficacy, the reaction temperature, alkali concentration, and marine species utilised are the key variables to take into account. Deacetylation increases primarily with increasing reaction temperature, base concentration, and reaction duration (Kou et al. 2021).

5.5.2 Enzymatic Methods

Like chemical methods, enzymatic methods use acids to remove $CaCO_3$ from the shell during demineralization. Enzymatic methods, on the other hand, employ enzymes at low temperatures to replace alkaline and high reaction temperatures in deproteinization and deacetylation reactions. Enzymatic deproteinization has been accomplished using a variety of proteinase enzymes (Kim 2010), developed from various sources such as fish intestines and microbes (Younes et al. 2014). Enzymatic approaches have major drawbacks when compared with chemical approaches, even with the milder reaction parameters. The most significant would be the operation's cost, particularly for large-scale production. Moreover, enzymatic methods are less effective when compared with chemical methods due to their failure to completely remove the last 10% shell proteins during deproteinization and lower deacetylation values (Kaczmarek et al. 2019; Suresh 2019). However, the limitation of reaction inefficiency can be overcome by combining enzymatic reactions with chemical reactions for effective refinement of the product (El Knidri et al. 2018).

5.6 Chitosan Applications in Tissue Engineering

Over the past few decades, chitosan has emerged as an appealing material for tissue engineering. Chitosan is well known for its non-toxic, biodegradable, and biocompatible properties, as well as its anti-microbial activity, and can be adjusted to produce multifaceted structures with morphology resembling that of the natural matrix. Furthermore, chitosan scaffolds are an attractive scaffold material for tissue engineering (Ahmed et al. 2018). A three-dimensional scaffold is identical to that of the extracellular matrix (ECM), acting as a template for the host cells to grow into and as a physical support for the cells to proliferate and differentiate into functional tissues or organs. Chitosan consists of amino, hydroxyl, and carboxyl functional groups. These functional groups easily combine chitosan with other natural and synthetic materials to produce composite systems.

5.6.1 Skin Tissue Engineering

The skin is the body's biggest organ, which serves as the first line of defence against external influences and stimuli such as viruses and mutagenic substances (Kabashima et al. 2019). Any thermal, chemical, or electrical stimulus that damages the skin can result in cutaneous complications, which in the worst-case scenario can result in wounds that are persistent and difficult to cure. Treatment alternatives and skin grafting methods demonstrate inappropriate and delayed healing as well as trigger unfavourable immunogenic reactions. Tissue engineering offers a potential solution in this area since it matches the 'shape' of natural systems and encourages efficient recovery. This method also permits the growth and multiplication of cells in three dimensions, and growth factor incorporation into biomaterials enables the preferred transformation and skin cell migration (Pereira and Bartolo 2016; Liu et al. 2018; Shpichka et al. 2019).

There are several methods implemented in the preparation of chitosan biomaterials, which are used in skin tissue engineering, including solvent casting (Madni et al. 2021; Dutta et al. 2011), freeze drying (Fernandes et al. 2011), electrospinning (Prasad et al. 2015), microwave assisted (Piątkowski et al. 2019), and dense gas foaming (Ji et al. 2011). Furthermore, the implementation of biomaterials based on chitosan in skin tissue engineering can be classified into four main categories which include chitosan-based membranes, chitosan-based hydrogels, chitosan-based sponge material, and chitosan-based fibre material.

Ma et al. proposed collagen/chitosan scaffold cross-linked by glutaraldehyde as a solid construct for skin tissue engineering. The scaffold displayed dermal equivalent characteristics of skin with enhanced biostability and beneficial biocompatibility (Ma et al. 2003). This study was amongst the earliest studies that incorporated the application of chitosan in skin tissue engineering. A study in 2006 detailed the production of porous scaffolds of biopolymer merges between collagen and chitosan using two techniques which were freeze drying and dehydrothermal crosslinking techniques (Tangsadthakun et al. 2006). The addition of chitosan markedly increased the collagen scaffold's resistance to lysozyme breakdown, and this property could be controlled by altering the chitosan's blend composition. The collagen and chitosan scaffold possessed potential mechanical strength, biodegradability, and capacity to promote cell proliferation that are essential for application in tissue engineering (Tangsadthakun et al. 2006).

Han et al. conducted a thorough study using a new gelatin-chitosan sponge scaffold, in which the chitosan and gelatin ratio was examined for its implications on mechanical properties, fluid uptake, and biodegradation characteristics. For efficient wound healing, a biocompatibility and in vivo study reported that the temperament of chitosan and gelatin sponge scaffold was favourable to regulate cell adhesion, multiplication, good support, as well as adherence to cells (Han et al. 2014). The reduction of bacterial infection is the major aspects of skin tissue engineering. Ma et al. designed chitosan membranes and added glycerol and antibacterial materials; this addition increased the chitosan membrane stability, whereas the antibacterial materials teamed well against the progression of *Escherichia coli* and *Staphylococcus aureus* (Ma et al. 2017).

A prospective automated platform called bioprinting makes it possible to create intricate three-dimensional tissue constructions by simultaneously depositing various cell types and biomaterials. Ng et al. modified the chitosan-based biomaterials for operational three-dimensional bioprinting in skin tissue engineering. The developed chitosan-based hydrogels displayed optimal printability at room temperature, displayed an appropriate shape fidelity of the printed three-dimensional builds, and portrayed promising biocompatibility with fibroblast skin cells (Ng et al. 2016).

Besides the three-dimensional scaffold, the concept of two-dimensional skin tissue engineering is often manipulated to treat wounds. Various two-dimensional scaffolds, like films and porous membranes, are potential candidates for skin tissues, expanding the range of structures accessible to solve the concerns with skin healing.

Ficai et al. developed nanostructured chitosan-based films with an additional microbial activity (Ficai et al. 2018). Recently, Pansara et al. formulated chitosan-stabilised silver nanoparticle film aimed to heal skin wounds. The healing film showed great wound healing outcomes and additional microbial inhibition properties in vitro and in vivo (Pansara et al. 2020). Chitosan films can be further categorised as chitosan free-standing films and chitosan thin films (Croisier and Jérôme 2013). Xia et al. developed a porous cellulose membrane with chitosan-based nanofibers for cutaneous wound injury. They reported that the cellulose-chitosan membrane displayed great wettability, hydrophilicity, and permeability of gases, good transmittance of light, mechanical properties, biocompatibility, and microbial inhibition (Xia et al. 2020). These features are definitely appropriate to treat cutaneous wound damage. Another example of chitosan porous nanofiber membranes was highlighted by Gao et al. where antibacterial electrospun nanofibrous bilayered wound dressings made from quaternised silicone and quaternised chitosan were developed. The bilayered nanofibrous membranes exhibited appropriate antibacterial activities, good hydrophilicity, acceptable performance of homeostasis, inhibition of scarring, and wound healing characteristics (Gao et al. 2021a). As noted by newly published innovative ideas, the extensive range of characteristics possessed by chitosan and its processed materials offers it a very bright future as a biomaterial, particularly in the areas of skin tissue engineering and wound healing.

5.6.2 Cartilage and Bone Tissue Engineering

Cartilage tissue consists of highly specialised chondroblasts as well as chondrocytes-producing proteins related to ECM (Morille et al. 2013). Articular cartilage links bone joints, performing major biomechanical functions including shock absorption, load bearing, and wear resistance (Dash et al. 2011). It is slightly complicated to administer injury to the cartilage tissue, since it is avascular, possesses an intricate structure, and has minimal cell population with increased heterogeneity levels (Balasundaram and Storey 2014). Furthermore, cartilage tissue functions in a demanding environment. Natural self-repair ability becomes unreliable with the damage diameter exceeding 4 mm. As a result, external treatments for such injuries should be recommended. Mosaicplasty, autologous chondrocyte injection, and microfracture are treatment options for cartilage tissue injury (Kock et al. 2012). However, these alternatives do not have the same configurational structure as native cartilage (Kock et al. 2012).

The ECM is a multipart, cell-regulating environment encompassing collagen fibrils and various macromolecules including adhesive and mineral-binding proteins, as well as proteoglycans (GAGs). Chitosan can be used as an ECM because it physically and chemically mimics GAG with *N*-acetyl glucosamine and glucosamine units (Chicatun et al. 2017). Numerous studies have advocated the application of chitosan as cartilage and bone tissue engineering biomaterial due to its malleability into numerous shapes, such as fibres, sponges, and hydrogels.

Garcia et al. have recently utilised electrospun chitosan-based fibre mats as a framework for chondrocyte formation to take advantage of their biocompatibility with native tissues. Chitosan substrates may provide an appropriate milieu for native cells to adhere, colonise, and regenerate damaged tissue (Garcia et al. 2021). Electrospun poly(hydroxybutyrate)/chitosan mixed fibrous scaffolds were proposed by Sadeghi et al. for cartilage tissue engineering. It was reported that the addition of chitosan could enhance the mass loss rate (and percentage) as well as hydrophilicity of electrospun PHB scaffolds by keeping the mechanical properties within an acceptable range for the planned use (Sadeghi et al. 2016). Panjapheree et al. developed a biphasic scaffold of silk fibroin film attached to silk fibroin/chitosan sponge which is perfect for osteoarthritic cartilage damage (Panjapheree et al. 2018). The biphasic scaffold of silk/chitosan was designed for osteoarthritis surgery and displayed promising physical performance and good biological functionality for the purpose of osteoarthritis surgery. Besides chitosan fibres and sponges-based scaffolds, Park et al. introduced injectable hydrogel consisting of chitosan and hyaluronic acid for uses in cartilage tissue engineering (Park et al. 2013). Furthermore, the superiority numbers of chondrocytes encapsulated in chitosan hydrogels were highlighted. The addition of hyaluronic acid in the chitosan hydrogel enhanced encapsulated chondrocyte expansion and accumulation of cartilage ECM.

When compared to cartilage, bone's dynamics can be thought of as an active body tissue with a built-in capacity for regeneration and healing after encountering minor injuries (Ranganathan et al. 2019). Bone tissue is more rigid as compared to cartilage. Bone defect can be described as the bone tissue deficiency in the parts that are affected (Smrke et al. 2013). Bone losses are divided into two major types: bone defects and structural damage (Ranganathan et al. 2019). Chitosan is particularly suitable as a bone scaffold biomaterial because it promotes osteoclast adherence and multiplication, as well as the synthesis of mineralised bone matrix (Levengood et al. 2014).

The bone composition consists of reasonable quantities of minerals and proteins. Hydroxyapatite is the secure type of calcium phosphate, occurring in bones as the key component (60–65%), together with other components such as keratin sulphate, collagen, lipids, and chondroitin sulphate (Venkatesan and Kim 2010). The chemical composite of chitosan involving hydroxyapatite is very suitable for bone tissue engineering since it is biocompatible and biodegradable with the bone type (Venkatesan and Kim 2010). Chitosan and hydroxyapatite are amongst the ideal bioactive substances in bone tissue engineering and are well known for their incredible biocompatibility with the environment of the human body (Wang 2003).

For bone tissue engineering, the relationship between the structure, characteristics, and processes of three-dimensional chitosan-nanohydroxyapatite and pure chitosan scaffolds, as well as each biological response, was compared (Thein-Han and Misra 2009). Pre-osteoblasts (MC3T3-E1) were shown to respond biologically on nanocomposite scaffolds better than the chitosan scaffolds based on enhanced cell attachment, increased multiplication, and evenly distributed morphology (Thein-Han and Misra 2009). Pallela et al. designed a biomimetic scaffold with chitosan and hydroxyapatite originating from the bone and collagen of *Thunnus obesus* and marine sponge (*Ircinia fusca*), respectively, through freeze drying and lyophilisation methods. Cell proliferation in the biomimetic scaffold was greater than in pure chitosan (Pallela et al. 2012). Demirtaş et al. reported the application of bioprintable chitosan hydrogel in bone tissue engineering. They also demonstrated that chitosan and chitosan-hydroxyapatite hydrogels had been stable under controlled physiological conditions and promoted osteogenic and mineralization activities (Demirtaş et al. 2017).

5.6.3 Cardiac Tissue Engineering

Cardiac tissue engineering mainly focuses on electrical conductance as the cardiac muscle functions on action potentials as part of the cardiac tissue engineering. Martins et al. included carbon nanofibers as a doping substance to design an extremely conductive chitosan-based material for cardiac tissue engineering. Surprisingly, the cardiac cell metabolic activity in the chitosan/carbon composite was found to be significantly higher when compared with cells grown on chitosan scaffold alone. Additionally, the addition of carbon nanofibers has increased the cardiac-specific gene expression implicated in electrical coupling and muscle contraction (Martins et al. 2014). A multi-layered three-dimensional scaffold incorporating chitosan/gelatin hydrogel and polycaprolactone was designed as a biodegradable patch that could be used for surgical reconstruction to overcome congenital heart defects. Pok et al. reported that the polycaprolactone core allows for surgical handling, makes suitability for sutures, and enhanced initial tensile strength, while the chitosan/gelation hydrogel scaffold enabled for cell adhesion, and the pore size and mechanical properties of the scaffold were suitable to cardiomyocyte function and migration (Pok et al. 2013).

A more recent study by Baei et al. directed on application of gold nanoparticle-chitosan hydrogels and electrical conductivity for cardiac tissue engineering. The study reported that addition of electro-conductive gold nanoparticles to the chitosan hydrogels stimulated the myocardial construct properties (Baei et al. 2016). Recently, a biocompatible chitosan–selenium nanoparticles film with the ability to conduct electricity was constructed for the use in cardiac tissue engineering. The study addressed the ability of chitosan–selenium nanoparticles film to be used as materials in promoting cellular behaviour through electrical stimulation, mechanical strength, and as biocompatible film to treat cardiomyopathy (Kalishwaralal et al. 2018). More recently, Mombini and colleagues designed chitosan-polyvinyl alcohol (PVA)-carbon nanotube (CNT) nanofibers as electrically conductive scaffolds. The study reported that the nanofibers containing 1% of carbon nanotube have ideal properties for cardiac cell differentiation (Mombini et al. 2019).

In 2020, Tamimi et al. designed a scaffold fabrication using solubilised cardiac ECM combined with chitosan and alginate. Interestingly, combining the cardiac ECM with chitosan and alginate significantly enhanced its tensile strength (Tamimi et al. 2020). Due to its positive charge and hydrophilicity, chitosan is capable of generating a soft tissue environment, particularly when coupled with biomolecules.

Chitosan scaffolds provide stem cells with increased mechanical support while they divide and develop. With a deeper understanding of the biological and mechanical principles of cardiac tissue regeneration, chitosan-based methods for cardiac tissue regeneration can be adopted and improved with greater efficacy (Patel et al. 2021).

5.6.4 Vascular Tissue Engineering

Because of the scarcity of donor sites and the immunological reaction to allograft and xenograft, vascular illnesses like blood vessel damage, atherosclerosis, and aneurysms continue to present a challenge for physicians. The best substitute for blood vessels has been thought to be tissue-engineered blood vessels (Zhu et al. 2009).

Du et al. developed a three-dimensional chitosan/poly ε-caprolactone (PCL) nanofibrous vessel scaffold through serial quantity assessment co-electrospinning to address thrombopoiesis, which aids the lining of the endothelial cells on the lumen of blood vessels. The chitosan/PCL scaffold offered a new method for developing small-diameter blood vessel grafts with innate anticoagulation and accelerated reendothelialization properties of mammalian vessels (Du et al. 2012). The serious vascular patches scarcity has surfaced as a big challenge to the ones who desperately need vascular surgery, as the occurrence of cardiovascular-related diseases has increased. Zhang et al. designed polyelectrolyte multilayer (PEM) vascular patches adding heparin and chitosan alternately onto decellularized scaffold coated with polyurethane. Furthermore, enhanced biocompatibility including reduced haemolysis rates, extended in vitro coagulation durations, and suitable platelet adhesion resistance was demonstrated. These patches further demonstrated improved adherence and endothelial progenitor cell proliferation (Zhang et al. 2019).

With similar properties, Wang et al. previously created an asymmetric small-diameter chitosan-based vascular graft with antithrombotic and antibacterial properties for vascular applications in tissue engineering. The study highlighted that the addition of chitosan had an antibacterial effect, while carboxymethyl chitosan addition had an antithrombotic effect, both of which are consistent with the primary goals of vascular tissue engineering. The study also highlighted the appropriate mechanical characteristics of the asymmetric vascular graft compared to those of the native vessels (Wang et al. 2020). Recently, Gao et al. constructed vascular patches that have biomechanical characteristics resembling natural blood vessels. A decellularized scaffold was coated with a poly(ethylene glycol) (PEG) film and its surface was modified with heparin-chitosan PEM to create a biomimetic vascular patch. The vascular patch created by Gao et al. showed comparable mechanical properties as those observed in natural blood vessels (Gao et al. 2022). Such assembly of vascular patches initiated by Gao et al. is almost similar to the ones developed by Zhang et al. (2019). In a previous study, electrospun PCL and chitosan blend nanofibers were used to make tissue-engineered vascular grafts, which were then implanted in mice and sheep. The vascular grafts were found to have mechanical properties that were the same as the carotid artery. After 6 months, the vascular grafts were reported to be clear without any signs of calcification, aneurysm, or thrombosis (Fukunishi et al. 2016).

5.6.5 Nerve Tissue Engineering

Neurons are different from many other cells in the body because they cannot divide as easily. Once nerve tissue has been damaged, it cannot be regenerated easily, so the function it controls will be impaired (Fields et al. 1989). Ding et al. developed a unique design that carried the purpose of tissue-engineered nerve grafts. The designed graft was a neural scaffold added to autologous MSCs based on chitosan/poly(lactic-co-glycolic acid) (PLGA), and it was employed as a substitute for linking the dog sciatic nerve long gaps (50 mm). They further reported that the autologous MSCs with neural scaffold enhanced sciatic nerve regeneration as well as its functional role recovery (Ding et al. 2010).

The application of chitosan tubes with distinct acetylation levels in bridging peripheral nerve defects is another example of how chitosan is utilised in nerve tissue engineering. Haastert-Talini et al. indicated that the chitosan tubes allowed morphological and functional regeneration of nerves as those observed in autologous nerve grafts (Haastert-Talini et al. 2013). In 2015, Gonzalez-Perez et al. evaluated the regeneration characteristics of chitosan tubes to silicone tubes and nerve autografts for bridging critical peripheral nerve gaps in the sciatic nerve of rat models. The regeneration characteristics of chitosan tubes were superior to silicone tubes but inferior to nerve autografts (Gonzalez-Perez et al. 2015), thus suggesting that chitosan guides are reliable conduits to build artificial nerve grafts bridging longer nerve gaps at the periphery. With the similar use of chitosan-based scaffold, Gu et al. designed a chitosan medium as well as silk fibroin fibres, cultivated with Schwann cells for ECM deposition, subjected to decellularization. Such a methodology paved the way for the development of a chitosan/silk fibroin-based, Schwann cell-derived ECM-modified scaffold that was utilised for sciatic nerve gap bridging in rat models (Gu et al. 2014).

In recent times, the end-to-end sutures remain the gold standard to treat nerve repairs without tension (short gap, <5 mm) whilst for larger or severe gaps (>5 mm) nerve autografts act as the standard of practice (Huang et al. 2012). According to the visions shared by Barton et al. an alternative nerve restoration method that did not require the use of sutures was photochemical tissue bonding utilising rose Bengal combined with chitosan bioadhesive. The combination of rose Bengal-chitosan adhesive treated nerves exhibited comparable results when compared to the conventional suture repair (Barton et al. 2015). In order to increase the effectiveness of axonal regeneration and enhance functional recovery across the critical defect length of 15 mm in the sciatic nerves of diabetic and healthy rats, Meyer et al. have attempted a less complex intraluminal alteration to the chitosan nerve guides. Such an attempt was a success where the chitosan films stimulated chitosan nerve guides and the regeneration outcome was comparable to those of autologous nerve grafts (Meyer et al. 2016). Baniasadi et al. studied the development and characterisation of conductive chitosan/gelatin-based scaffolds containing polyaniline/graphene nanocomposite. The study implied that the assembly exhibited appropriate electrical conductivity, mechanical characteristics, and good biocompatibility with Schwann cells (Baniasadi et al. 2019). On the other hand, Sadeghi et al. evaluated the chitosan effect on the hydrophilicity and bioactivity of conductive electrospun composite scaffold. It was confirmed that the inclusion of chitosan imposed great advancement in the hydrophilicity of the scaffold. In terms of bioactivity, the incorporation of chitosan supported cell adherence, spreading, and revealed an enhancement in the proliferation of PC12 cells (Sadeghi et al. 2019). Furthermore, an alginate/chitosan hydrogel for the purpose of sciatic nerve tissue regeneration was constructed by Salehi et al. The hydrogel contained olfactory ectomesenchymal stem cells. The outcomes of the study revealed that by applying the chitosan-based hydrogel with olfactory ectomesenchymal stem cells to the damaged sciatic nerve stimulated better healing compared to the control group and hydrogel without such cells (Salehi et al. 2019). Overall, chitosan-based scaffolds, hydrogels, tubules, and bioadhesives have emerged as excellent options for treating nerve damage and bridging nerve gaps.

5.7 Future Perspectives and Conclusion

With a good comprehension of the compositions of the human organs, biological functionality, biomechanics, disease causation, and potential injury, a collaborative effort between physicians, biologists, chemists, engineers, medical practitioners, and polymer scientists would be useful to the development of custom-made polymers for future tissue engineering applications. Chitosan is now among the effective biopolymers for the purpose of tissue engineering due to its widespread accessibility and distinctive characteristics. It is necessary to modify the scaffold's characteristics for specific tissue requirements.

A new method called three-dimensional bioprinting has emerged in which cells of interest are combined with polymeric materials to create bio-inks, which are then printed layer by layer to create three-dimensional tissue or organ analogues (Ji et al. 2011). Chitosan appears to meet nearly all of the prerequisites for the production of bio-ink. Several studies to date have reported initiatives in which chitosan is optimised with other polymers to produce bio-inks. Tsukamoto et al. evaluated the three-dimensional moulding capacity of hydroxybutyl chitosan, a thermoresponsive polymer, using a robotic dispensing three-dimensional printer. Furthermore, the study co-cultured human heart microvascular endothelial cells producing cardiac tissue mimics that were highly vascularised and structured (Tsukamoto et al. 2020). However, despite these advancements, chitosan-based three-dimensional bioprinting application in cardiac tissue engineering remains not fully explored to its best abilities (Li et al. 2019). The specific cardiac tissue engineering applications of chitosan are restricted at the moment and are anticipated to increase in the future years. Similar application using chitosan has been incorporated to create a unique bio-ink for three-dimensional printed neural structures and neural applications

(Gu et al. 2016). In addition, chitosan has also found an upcoming utility in dental tissue engineering (EzEldeen et al. 2021; Tang et al. 2020).

The chitosan-based functional materials are capable of size control, simple to prepare, and may accomplish the desired design by streamlining the production process using microfluidic technology. Additionally, the use of chitosan-based multifunctional materials with microfluidic assistance was ramped up and optimised (Ly et al. 2021; Majedi et al. 2013; Lee et al. 2010). As a result, the mass production of chitosan functional materials with complex structures using simple microchannel structures and the incorporation of other materials has great promise and merits further research (Gao et al. 2021b). There is still a need to better optimise the manufacturing process and enhance the material properties as most investigations of the microfluidic-aided chitosan-based multifunctional materials are still in their initial stages and far from being implemented in industrial production (Gao et al. 2021b).

Although the manufacturing of scaffolds using three-dimensional printing has revolutionised, there is still a long way to go before we can replicate organs. The difficulty of producing a functioning organ with numerous circulatory lumens of various sizes is one of the biggest issues with three-dimensional printing (Sahranavard et al. 2020). Future research should take into account the few restrictions in chitosan three-dimensional printing. The goal of chitosan-based biomaterials research will continue to be the modification of chitosan that produces significant derivatives with better qualities as sought for scaffolds. In this crucial scope of biomaterials, researchers are constantly investigating the potential of chitosan with the goal of overcoming the limitations of current ones and reaching the majority of the desired scaffold qualities. Investigators and medical practitioners anticipate that the promising application of chitosan-derived substances in different tissues and organs including blood vessels, skin, cardiac, cartilage, bone, and nerves implies their successful future in the science of tissue engineering.

Acknowledgement This study was supported by the Prototype Development Research Grant (PRGS/1/2022/SKK0/USM/02/1), Ministry of Higher Education, Malaysia. The funding body had no role in study design or in the collection and interpretation of data

Conflict of Interest The authors have no conflicts of interest to declare that are relevant to the content of this article.

References

Abo Elsoud MM, El Kady EM (2019) Current trends in fungal biosynthesis of chitin and chitosan. Bull Natl Res Cent 43(1):1–2

Ahmed S, Ali A, Sheikh J (2018) A review on chitosan centred scaffolds and their applications in tissue engineering. Int J Biol Macromol 116:849–862

Aranaz I, Mengíbar M, Harris R et al (2009) Functional characterization of chitin and chitosan. Curr Chem Biol 3(2):203–230

Arasukumar B, Prabakaran G, Gunalan B et al (2019) Chemical composition, structural features, surface morphology and bioactivities of chitosan derivatives from lobster (Thenus unimaculatus) shells. Int J Biol Macromol 135:1237–1245

Arbia W, Arbia L, Adour L et al (2013) Chitin extraction from crustacean shells using biological methods–a review. Food Technol Biotechnol 51(1):12–25

Azuma K, Ifuku S, Osaki T et al (2014) Preparation and biomedical applications of chitin and chitosan nanofibers. J Biomed Nanotechnol 10(10):2891–2920

Baei P, Jalili-Firoozinezhad S, Rajabi-Zeleti S et al (2016) Electrically conductive gold nanoparticle-chitosan thermosensitive hydrogels for cardiac tissue engineering. Mater Sci Eng C 63:131–141

Balasundaram G, Storey DM (2014) Novel nano-rough polymers for cartilage tissue engineering. Int J Nanomedicine 9:1845

Baniasadi H, Ahmad Ramazani AR, Mashayekhan S (2019) Fabrication and characterization of conductive chitosan/gelatin-based scaffolds for nerve tissue engineering. Int J Biol Macromol 74:360–366

Barton MJ, Morley JW, Stoodley MA et al (2015) Long term recovery of median nerve repair using laser-activated chitosan adhesive films. J Biophotonics 8(3):196–207

Benhabiles MS, Salah R, Lounici H et al (2012) Antibacterial activity of chitin, chitosan and its oligomers prepared from shrimp shell waste. Food Hydrocoll 29(1):48–56

Chicatun F, Griffanti G, McKee MD et al (2017) Collagen/chitosan composite scaffolds for bone and cartilage tissue engineering. In: Biomedical composites. Woodhead Publishing, New Delhi, pp 163–198

Croisier F, Jérôme C (2013) Chitosan-based biomaterials for tissue engineering. Eur Polym J 49(4): 780–792

Dash M, Chiellini F, Ottenbrite RM et al (2011) Chitosan—a versatile semi-synthetic polymer in biomedical applications. Prog Polym Sci 36(8):981–1014

Demirtaş TT, Irmak G, Gümüşderelioğlu M (2017) A bioprintable form of chitosan hydrogel for bone tissue engineering. Biofabrication 9(3):035003

Ding F, Wu J, Yang Y et al (2010) Use of tissue-engineered nerve grafts consisting of a chitosan/ poly (lactic-co-glycolic acid)-based scaffold included with bone marrow mesenchymal cells for bridging 50-mm dog sciatic nerve gaps. Tissue Eng Part A 16(12):3779–3790

Du F, Wang H, Zhao W et al (2012) Gradient nanofibrous chitosan/poly ε-caprolactone scaffolds as extracellular microenvironments for vascular tissue engineering. Biomaterials 33(3):762–770

Dutta PK, Rinki K, Dutta J (2011) Chitosan: a promising biomaterial for tissue engineering scaffolds. In: Jayakumar R, Prabaharan M, Muzzarelli R (eds) Chitosan for biomaterials II. Advances in polymer science, vol 244. Springer, Berlin. https://doi.org/10.1007/12_2011_112

Dutta J, Tripathi S, Dutta PK (2012) Progress in antimicrobial activities of chitin, chitosan and its oligosaccharides: a systematic study needs for food applications. Food Sci Technol 18(1):3–4

El Knidri H, Belaabed R, Addaou A et al (2018) Extraction, chemical modification and characterization of chitin and chitosan. Int J Biol Macromol 120:1181–1189

Elieh-Ali-Komi D, Hamblin MR (2016) Chitin and chitosan: production and application of versatile biomedical nanomaterials. Int J Adv Res Publ 4(3):411

EzEldeen M, Loos J, Nejad ZM et al (2021) 3D-printing-assisted fabrication of chitosan scaffolds from different sources and cross-linkers for dental tissue engineering. Eur Cells Mater 40:485–501

Fan W, Yan W, Xu Z et al (2012) Formation mechanism of monodisperse, low molecular weight chitosan nanoparticles by ionic gelation technique. Coll Surf B 90:21–27

Fernandes LL, Resende CX, Tavares DS et al (2011) Cytocompatibility of chitosan and collagen-chitosan scaffolds for tissue engineering. Polimeros 21:1–6

Ficai D, Ardelean IL, Holban AM et al (2018) Manufacturing nanostructured chitosan-based 2D sheets with prolonged antimicrobial activity. Room J Morphol Embryol 59(2):517–525

Fields RD, Le Beau JM, Longo FM (1989) Nerve regeneration through artificial tubular implants. Prog Neurobiol 33(2):87–134

Fukunishi T, Best CA, Sugiura T (2016) Tissue-engineered small diameter arterial vascular grafts from cell-free nanofiber PCL/chitosan scaffolds in a sheep model. PLoS One 11(7):e0158555

Gao Z, Su C, Wang C (2021a) Antibacterial and hemostatic bilayered electrospun nanofibrous wound dressings based on quaternized silicone and quaternized chitosan for wound healing. Eur Polym J 159:110733

Gao Y, Ma Q, Cao J et al (2021b) Recent advances in microfluidic-aided chitosan-based multifunctional materials for biomedical applications. Int J Pharm 600:120465

Gao H, Hu P, Sun G et al (2022) Decellularized scaffold-based poly (ethylene glycol) biomimetic vascular patches modified with polyelectrolyte multilayer of heparin and chitosan: preparation and vascular tissue engineering applications in a porcine model. J Mater Chem B 10(7): 1077–1084

Garcia CE, Lardy B, Bossard F et al (2021) Chitosan based biomaterials for cartilage tissue engineering: chondrocyte adhesion and proliferation. Food Hydrocoll 1:100018

Gonzalez-Perez F, Cobianchi S, Geuna S et al (2015) Tubulization with chitosan guides for the repair of long gap peripheral nerve injury in the rat. Microsurgery 35(4):300–308

Gu Y, Zhu J, Xue C et al (2014) Chitosan/silk fibroin-based, Schwann cell-derived extracellular matrix-modified scaffolds for bridging rat sciatic nerve gaps. Biomaterials 35(7):2253–2263

Gu Q, Tomaskovic-Crook E, Lozano R et al (2016) Functional 3D neural mini-tissues from printed gel-based bioink and human neural stem cells. Adv Health Mater 5(12):1429–1438

Haastert-Talini K, Geuna S, Dahlin LB et al (2013) Chitosan tubes of varying degrees of acetylation for bridging peripheral nerve defects. Biomaterials 34(38):9886–9904

Han F, Dong Y, Su Z (2014) Preparation, characteristics and assessment of a novel gelatin–chitosan sponge scaffold as skin tissue engineering material. Int J Pharm 476(1–2):124–133

Huang W, Begum R, Barber T et al (2012) Regenerative potential of silk conduits in repair of peripheral nerve injury in adult rats. Biomaterials 33(1):59–71

Ibrahim HM, Zairy EE (2015) Chitosan as a biomaterial — structure, properties, and electrospun nanofibers. In: Bobbarala V (ed) Concepts, compounds and the alternatives of antibacterials. IntechOpen, London. https://doi.org/10.5772/61300. https://www.intechopen.com/chapters/49246. Accessed 30 Oct 2022

Ji C, Annabi N, Khademhosseini A (2011) Fabrication of porous chitosan scaffolds for soft tissue engineering using dense gas CO2. Acta Biomater 7(4):1653–1664

Kabashima K, Honda T, Ginhoux F (2019) The immunological anatomy of the skin. Nat Rev Immunol 19(1):19–30

Kaczmarek MB, Struszczyk-Swita K, Li X et al (2019) Enzymatic modifications of chitin, chitosan, and chitooligosaccharides. Front Bioeng Biotechnol 7:243

Kalishwaralal K, Jeyabharathi S, Sundar K et al (2018) A novel biocompatible chitosan–Selenium nanoparticles (SeNPs) film with electrical conductivity for cardiac tissue engineering application. Mater Sci Eng C 92:151–160

Kim SK (2010) Chitin, chitosan, oligosaccharides and their derivatives: biological activities and applications, 1st edn. CRC Press, Boca Raton, FL. https://doi.org/10.1201/EBK1439816035

Kock L, van Donkelaar CC, Ito K (2012) Tissue engineering of functional articular cartilage: the current status. Cell Tissue Res 347(3):613–627

Kou SG, Peters LM, Mucalo MR (2021) Chitosan: a review of sources and preparation methods. Int J Biol Macromol 169:85–94

Kumar MN (2000) A review of chitin and chitosan applications. React Funct Polym 46(1):1–27

Lee KH, Shin SJ, Kim CB et al (2010) Microfluidic synthesis of pure chitosan microfibers for bio-artificial liver chip. Lab Chip 10(10):1328–1334

Levengood SK, Zhang M et al (2014) Chitosan-based scaffolds for bone tissue engineering. J Mater Chem 2(21):3161–3184

Li S, Tian X, Fan J et al (2019) Chitosans for tissue repair and organ three-dimensional (3D) bioprinting. Micromachines 10(11):765

Liu H, Wang C, Li C et al (2018) A functional chitosan-based hydrogel as a wound dressing and drug delivery system in the treatment of wound healing. RSC Adv 8(14):7533–7549

Ly KL, Hu P, Luo X (2021) Flow-assembled chitosan membranes in microfluidics: recent advances and applications. J Mater Chem B 9(15):3258–3283

Ma L, Gao C, Mao Z et al (2003) Collagen/chitosan porous scaffolds with improved biostability for skin tissue engineering. Biomaterials 24(26):4833–4841

Ma Y, Xin L, Tan H et al (2017) Chitosan membrane dressings toughened by glycerol to load antibacterial drugs for wound healing. Mater Sci Eng C 81:522–531

Madni A, Kousar R, Naeem N et al (2021) Recent advancements in applications of chitosan-based biomaterials for skin tissue engineering. J Bioresour Bioprod 6(1):11–25

Majedi FS, Hasani-Sadrabadi MM, Emami SH et al (2013) Microfluidic assisted self-assembly of chitosan based nanoparticles as drug delivery agents. Lab Chip 13(2):204–207

Martins AM, Eng G, Caridade SG et al (2014) Electrically conductive chitosan/carbon scaffolds for cardiac tissue engineering. Biomacromolecules 15(2):635–643

Meyer C, Stenberg L, Gonzalez-Perez F et al (2016) Chitosan-film enhanced chitosan nerve guides for long-distance regeneration of peripheral nerves. Biomaterials 76:33–51

Mohanty B, Mohanty U, Pattanaik SS et al (2018) Future prospects and trends for effective utilization of fish processing wastes in India. Innov Farm 3(1):1–5

Mombini S, Mohammadnejad J, Bakhshandeh B et al (2019) Chitosan-PVA-CNT nanofibers as electrically conductive scaffolds for cardiovascular tissue engineering. Int J Biol Macromol 140: 278–287

Morille M, Van-Thanh T, Garric X (2013) New PLGA–P188–PLGA matrix enhances TGF-β3 release from pharmacologically active microcarriers and promotes chondrogenesis of mesenchymal stem cells. J Control Release 170(1):99–110

Ng WL, Yeong WY, Naing MW (2016) Polyelectrolyte gelatin-chitosan hydrogel optimized for 3D bioprinting in skin tissue engineering. Int J Bioprint 2(1):1

No HK, Meyers SP (1995) Preparation and characterization of chitin and chitosan—a review. J Aquat Food Prod Technol 4(2):27–52

Pallela R, Venkatesan J, Janapala VR et al (2012) Biophysicochemical evaluation of chitosan-hydroxyapatite-marine sponge collagen composite for bone tissue engineering. J Biomed Mater Res A 100(2):486–495

Panjapheree K, Kamonmattayakul S, Meesane J (2018) Biphasic scaffolds of silk fibroin film affixed to silk fibroin/chitosan sponge based on surgical design for cartilage defect in osteoarthritis. Mater Des 141:323–332

Pansara C, Mishra R, Mehta T et al (2020) Formulation of chitosan stabilized silver nanoparticle-containing wound healing film: in vitro and in vivo characterization. J Pharm Sci 109(7): 2196–2205

Park BK, Kim MM (2010) Applications of chitin and its derivatives in biological medicine. Int J Mol Sci 11(12):5152–5164

Park H, Choi B, Hu J et al (2013) Injectable chitosan hyaluronic acid hydrogels for cartilage tissue engineering. Acta Biomater 9(1):4779–4786

Patel B, Manne R, Patel DB et al (2021) Chitosan as functional biomaterial for designing delivery systems in cardiac therapies. Gels 7(4):253

Pereira RF, Bartolo PJ (2016) Traditional therapies for skin wound healing. Adv Wound Care 5(5): 208–229

Piątkowski M, Radwan-Pragłowska J, Janus Ł et al (2019) Microwave-assisted synthesis and characterization of chitosan aerogels doped with Au-NPs for skin regeneration. Polym Test 73:366–376

Pok S, Myers JD, Madihally SV et al (2013) A multilayered scaffold of a chitosan and gelatin hydrogel supported by a PCL core for cardiac tissue engineering. Acta Biomater 9(3): 5630–5642

Prasad T, Shabeena EA, Vinod D et al (2015) Characterization and in vitro evaluation of electrospun chitosan/polycaprolactone blend fibrous mat for skin tissue engineering. J Mater Sci Mater Med 26(1):1–3

Ramírez MA, Rodríguez AT, Alfonso L et al (2010) Chitin and its derivatives as biopolymers with potential agricultural applications. Biotech Appl 27(4):270–276

Ranganathan S, Balagangadharan K, Selvamurugan N (2019) Chitosan and gelatin-based electrospun fibers for bone tissue engineering. Int J Biol Macromol 133:354–364

Ruiz GA, Corrales HF (2017) Chitosan, chitosan derivatives and their biomedical applications. In: Shalaby EA (ed) Biological activities and application of marine polysaccharides. IntechOpen, London. https://doi.org/10.5772/66527. https://www.intechopen.com/chapters/53455

Sadeghi D, Karbasi S, Razavi S et al (2016) Electrospun poly (hydroxybutyrate)/chitosan blend fibrous scaffolds for cartilage tissue engineering. J Appl Polym Sci 133(47):44171

Sadeghi A, Moztarzadeh F, Mohandesi JA (2019) Investigating the effect of chitosan on hydrophilicity and bioactivity of conductive electrospun composite scaffold for neural tissue engineering. Int J Biol Macromol 121:625–632

Sahranavard M, Zamanian A, Ghorbani F et al (2020) A critical review on three dimensional-printed chitosan hydrogels for development of tissue engineering. Bioprinting 17:e00063

Sajomsang W, Gonil P (2010) Preparation and characterization of α-chitin from cicada sloughs. Mater Sci Eng C 30(3):357–363

Salehi M, Bagher Z, Kamrava SK et al (2019) Alginate/chitosan hydrogel containing olfactory ectomesenchymal stem cells for sciatic nerve tissue engineering. J Cell Physiol 234(9): 15357–15368

Santos VP, Marques NS, Maia PC (2020) Seafood waste as attractive source of chitin and chitosan production and their applications. Int J Mol Sci 21(12):4290

Schmitz C, González Auza L, Koberidze D et al (2019) Conversion of chitin to defined chitosan oligomers: current status and future prospects. Mar Drugs 17(8):452

Shpichka A, Butnaru D, Bezrukov EA et al (2019) Skin tissue regeneration for burn injury. Stem Cell Res Ther 10(1):1–6

Smrke D, Rožman P, Veselko M et al (2013) Treatment of bone defects — allogenic platelet gel and autologous bone technique. In: Andrades JA (ed) Regenerative medicine and tissue engineering. IntechOpen, London. https://doi.org/10.5772/55987. https://www.intechopen.com/chapters/44620

Sultankulov B, Berillo D, Sultankulova K et al (2019) Progress in the development of chitosan-based biomaterials for tissue engineering and regenerative medicine. Biomolecules 9(9):470

Suresh PV (2019) Enzymatic technologies of chitin and chitosan. In: Enzymatic technologies for marine polysaccharides, 1st edn. CRC Press, Boca Raton, FL

Synowiecki J, Al-Khateeb NA (2003) Production, properties, and some new applications of chitin and its derivatives. Crit Rev Food Sci Nutr 43(2):145–171

Tamimi M, Rajabi S, Pezeshki-Modaress M (2020) Cardiac ECM/chitosan/alginate ternary scaffolds for cardiac tissue engineering application. Int J Biol Macromol 164:389–402

Tang G, Tan Z, Zeng W et al (2020) Recent advances of chitosan-based injectable hydrogels for bone and dental tissue regeneration. Front Bioeng Biotechnol 8:587658

Tangsadthakun C, Kanokpanont S, Sanchavanakit N et al (2006) Properties of collagen/chitosan scaffolds for skin tissue engineering. J Met Mater Miner 16(1):37

Thein-Han WW, Misra RDK (2009) Biomimetic chitosan–nanohydroxyapatite composite scaffolds for bone tissue engineering. Acta Biomater 5(4):1182–1197

Tsukamoto Y, Akagi T, Akashi M (2020) Vascularized cardiac tissue construction with orientation by layer-by-layer method and 3D printer. Sci Rep 10(1):1–1

Vázquez JA, Rodríguez-Amado I, Montemayor MI et al (2013) Chondroitin sulfate, hyaluronic acid and chitin/chitosan production using marine waste sources: characteristics, applications and eco-friendly processes: a review. Mar Drugs 11(3):747–774

Venkatesan J, Kim SK (2010) Chitosan composites for bone tissue engineering—an overview. Mar Drugs 8(8):2252–2266

Wang M (2003) Developing bioactive composite materials for tissue replacement. Biomaterials 24(13):2133–2151

Wang Y, He C, Feng Y et al (2020) A chitosan modified asymmetric small-diameter vascular graft with anti-thrombotic and anti-bacterial functions for vascular tissue engineering. J Mater Chem B 8(3):568–577

Xia J, Zhang H, Yu F (2020) Superclear, porous cellulose membranes with chitosan-coated nanofibers for visualized cutaneous wound healing dressing. ACS Appl Mater Interfaces 12(21):24370–24379

Yao K, Li J, Yao F et al (2011) Chitosan-based hydrogels: functions and applications, 1st edn. Taylor & Francis Group, CRC Press, Boca Raton, FL. https://doi.org/10.1201/b11048

Younes I, Rinaudo M (2015) Chitin and chitosan preparation from marine sources: structure, properties and applications. Mar Drugs 13(3):1133–1174

Younes I, Hajji S, Frachet V et al (2014) Chitin extraction from shrimp shell using enzymatic treatment. Antitumor, antioxidant and antimicrobial activities of chitosan. Int J Biol Macromol 69:489–498

Zainol Abidin NA, Kormin F, Zainol Abidin NA et al (2020) The potential of insects as alternative sources of chitin: an overview on the chemical method of extraction from various sources. Int J Mol Sci 21(14):4978

Zhang J, Wang D, Jiang X et al (2019) Multistructured vascular patches constructed via layer-by-layer self-assembly of heparin and chitosan for vascular tissue engineering applications. J Chem Eng 370:1057–1067

Zhu C, Fan D, Duan Z et al (2009) Initial investigation of novel human-like collagen/chitosan scaffold for vascular tissue engineering. J Biomed Res B App Mater 89(3):829–840

Natural Hydroxyapatite from Black Tilapia Fish Bones and Scales for Biomedical Applications

6

Hasan Zuhudi Abdullah, Maizlinda Izwana Idris, Lee Te Chuan, Siti Khadijah Dermawan, and Muhamad Zaki Jaffri

Abstract

Hydroxyapatite is one of the biomaterials that is widely used for bone implants due to its composition and structure, which is like human bone. The hydroxyapatite (HAp) component primarily consists of calcium (Ca) and phosphorus (P). The aim of this study was to extract HAp from natural resources. Natural hydroxyapatite can be derived from animal, shell and fish sources. In this chapter, the natural HAp addressed was produced from Black Tilapia fish bones and scales (fish by-product wastes). A simple method was used based on the heat treatment process (600–1000 °C) to extract natural hydroxyapatite. Natural hydroxyapatite powders with controlled characteristics, which included morphology, chemical composition and crystallinity for getting good or better hydroxyapatite powders were prepared. Biological testing was conducted to analyse the powder, especially its biological activity. Natural hydroxyapatite from fish bones and scales has the potential to induce the formation of apatite and exhibits good biocompatibility.

Keywords

Natural hydroxyapatite · Fish bone · Fish scale · Black Tilapia fish · Fish waste · Biocompatible

H. Z. Abdullah (✉) · M. I. Idris · S. K. Dermawan · M. Z. Jaffri
Department of Manufacturing Engineering, Faculty of Mechanical and Manufacturing Engineering, Universiti Tun Hussein Onn Malaysia, Parit Raja, Batu Pahat, Johor, Malaysia
e-mail: hasan@uthm.edu.my; izwana@uthm.edu.my

L. Te Chuan
Department of Production and Operation Management, Faculty of Technology Management and Business, Universiti Tun Hussein Onn Malaysia, Parit Raja, Batu Pahat, Johor, Malaysia
e-mail: tclee@uthm.edu.my

W. S. Wan Kamarul Zaman, A. A. Nurul (eds.), *Sustainable Material for Biomedical Engineering Application*, https://doi.org/10.1007/978-981-99-2267-3_6

Abbreviations

β-TCP	Beta tri-calcium phosphate
Ca/P	Calcium/phosphorus
$CaCl_2$	Calcium chloride
CPC	Calcium phosphate crystal
DMSO	Dimethyl sulfoxide
EDX	Energy dispersive X-Ray spectroscope
FAO	Food and Agricultural Organization
HAp	Hydroxyapatite
HCl	Hydrochloric acid
hFOB	Human fetal osteoblastic
JCPDS	Joint Committee on Powder Diffraction Standards
$K_2HPO_4 \cdot 3H_2O$	Dipotassium hydrogen phosphate trihydrate
KCl	Potassium chloride
KOH	Potassium hydroxide
$MgCl_2 \cdot 6H_2O$	Magnesium chloride hexahydrate
MTT	3-(4, 5-dimethylthiazol-2-yl)-2, 5-diphenyl tetrazolium)
Na_2SO_4	Sodium sulfate
NaCl	Sodium chloride
$NaHCO_3$	Sodium bicarbonate
NaOH	Sodium hydroxide
SBF	Simulated body fluid
SEM	Scanning electron microscope
SM	Secondary material
Tris	tris(hydroxymethyl)aminomethane
XRD	X-Ray diffraction

6.1 Introduction to Hydroxyapatite (HAp)

Hydroxyapatite (HAp) is a bioceramics in the family of calcium phosphate. It has a high similarity in chemical composition with human bone, which makes them very important for bone replacement. HAp was found to have high biocompatibility, biodegradability and osteoconductivity with bone that makes it useful in medical implants or orthopaedic surgeries to heal bone defects or fractures. Furthermore, HAp has attracted researchers as the most promising candidate for bone replacement due to its chemical and biological properties (Ahmad Fara et al. 2015). It has been used as implants in hard tissue engineering, such as for skeleton, bone and joints, and in dental applications (Mohd Pu'ad et al. 2019). The standard HAp has a chemical formula that consists of $Ca_{10}(PO_4)6(OH)_2$, and its ratio of Ca/P: 1.67 can be synthesized chemically. Normally, natural HAp has a different structure compared to synthetic HAp since it contains trace elements of, for example, magnesium (Mg), natrium (Na), silicon (Si), zinc (Zn), barium (Ba), kalium (K), fluoride (F) and carbonate (CO_3) which are preserved from the original resources (Akram et al. 2014; Milovac et al. 2014).

6.2 Natural Hydroxyapatite (HAp)

Recently, natural hydroxyapatite has been extensively extracted from natural resources due to its outstanding composition, which is close to the human bones composition compared to synthetic HAp (Akram et al. 2014; Milovac et al. 2014). This natural hydroxyapatite is mostly found in animal bones and marine by-products, such as bovine and porcine animal bones from cows and pigs, respectively, as well as marine sources from fish bones and scales. It was also reported that several researchers have stated interest in the extraction or chemical treatment of shells such as cockle shells, eggshells, oyster shells and mussel shells since they exhibit great potential as natural hydroxyapatite resources as well (Mohd Pu'ad et al. 2019).

6.2.1 Chemical Composition: Ion/Trace Elements

Natural hydroxyapatite specifically extracted from animal bones and marine by-products significantly differs from the synthetic sources due to it containing additional ions or trace elements such as magnesium, fluoride, zinc, silicon, and strontium in its chemical compositions (Fig. 6.1). These additional trace elements give the non-stoichiometric factor in the chemical compound of natural hydroxyapatite (Mohd Pu'ad et al. 2019; Hussin et al. 2022). This means that the Ca/P ratio is not equal to 1.67 or even higher. These ions or trace elements have significant biological effects, such as stimulating osteoblast activities and promoting cell adhesion and proliferation (Fig. 6.2). Table 6.1 describes the biological effects of the ions or trace elements (Hussin et al. 2022).

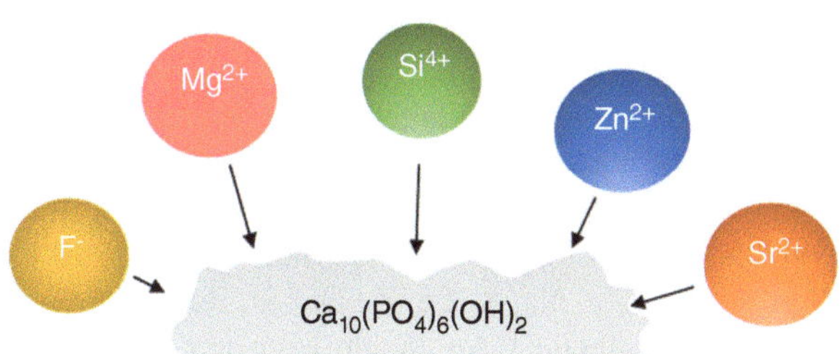

Fig. 6.1 Ions or trace elements of natural hydroxyapatites

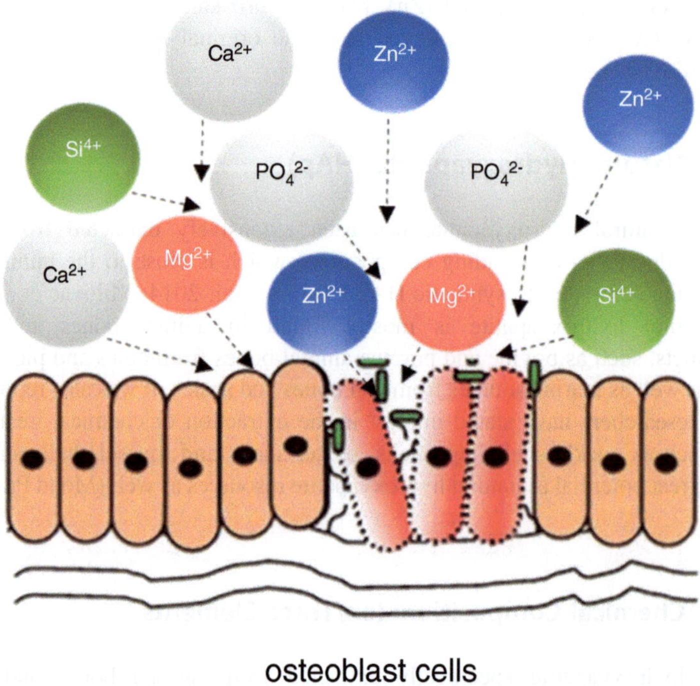

Fig. 6.2 Schematic diagram of biological effects from ions or trace elements on the osteoblast cells

6.3 Fish Bones and Scales as Hydroxyapatite Sources

Fish wastes have significantly increased due to the rising global fish consumption. It is noteworthy that fish wastes include scales, skin, bone, internal organs, tails and fins. Owing to this, fish wastes have grown to be a significant environmental issue. According to Food and Agricultural Organization (FAO) statistics, approximately 20–30 million tonnes of fish waste are discarded globally each year (Lee et al. 2022; Ideia et al. 2020).

The recovery of fish wastes into valuable materials such as biomaterials has created a new phenomenon in research nowadays. The extraction of HAp from fish wastes has attracted a great deal of attention recently because it has the potential to lessen the impact of environmental issues due to fish wastes (Lee et al. 2022). The high calcium, phosphate and carbonate content in fish bones and scales make them potential sources for HAp extraction. Furthermore, the existence of tracer elements, for example, Mg, Na, K, Fe, Zn and Sr, improves the biocompatibility of extracted HAp over synthetic HAp. Aside from tracer elements, CO_3^{2-} is also commonly found in biological HAp, particularly HAp extracted from fish bones and scales. It may promote the formation of bone-like apatite by initiating negative charge carriers

Table 6.1 Biological effects of ions or trace elements (Reproduced with permission from Hussin et al. (2022) Elsevier)

Ion/trace element	Biological effects	References
Zinc (Zn^{2+})	Inhibit osteoblast cell formation	Webster et al. (2001), Shepherd et al. (2014)
	Avoid osteoporosis	Ito et al. (2005)
	Increase angiogenesis	De Lima et al. (2010)
	Improve the differentiation of osteogenic cells	Ishikawa et al. (2002), Ikeuchi et al. (2003), Bhattacharjee et al. (2014), Yu et al. (2017), Cuozzo et al. (2020), Matsunaga (2008)
Magnesium (Mg^{2+})	Cell adhesion and enhance bioactivity	Yanli (2011), Kavitha et al. (2015), Arul and Ramya (2017)
	Stimulate cell differentiation	Da Silva et al. (2020)
Fluoride (F^{-})	Stimulate osteoblast activity	Li et al. (2015)
	Avoid osteoporosis	Aoba (1997)
	Hinder osteoclast proliferation	Liu et al. (2019)
	Enhance strength and corrosion resistance	Swadi et al. (2012)
Silicon (Si^{2+})	Enhance cell differentiation	Gao et al. (2016), Sun et al. (2018)
	Improve osteogenic differentiation	Rodrigues et al. (2017), Yao et al. (2016)
	Enhance mechanical property	Yamada et al. (2019)
Strontium (Sr^{2+})	Absence of cytotoxicity	Ehret et al. (2017)
	Enhance osteoblast activity and proliferation	Olivier et al. (2020)
	Promote cell adhesion, proliferation, and alkaline phosphatase activity	Ge et al. (2018)

and increasing the solubility of extracted HAp (Kumar and Mohanty 2022; Zakaria et al. 2022).

In most cases, the fish bones or scales will be boiled to remove any skin, raw flesh, salt, organic impurities and other unwanted materials. Following that, HAp will be extracted using various methods such as calcination, alkaline hydrothermal hydrolysis and thermal calcification extraction to eliminate the excess of other impurities such as protein, lipids and oils. Table 6.2 shows previous studies on the extraction of HAp from fish bones and scales.

In summary, most studies found that the particle shape of HAp extracted from different fish sources was irregular, spherical, oval, hexagonal, rod-like and so on. Despite the fact that the same extraction method, species, part (bone or scale) was used, the morphology showed different shapes. In comparison to calcination, alkaline hydrolysis and thermal calcification are the most common methods for producing nano-size HAp. It means that the ratio of Ca/P for the extracted HAp from fish sources was found to be not equal to 1.67, which is the stoichiometric ratio of HAp. HAp extracted from fish seems to be non-stoichiometric, either less or greater than 1.67. This was mainly due to a Ca or P deficiency. In most studies, Ca is

Table 6.2 Previous studies on the extraction of HAp from different fish bones and scales

Sources	Extraction method	Findings	References
Crucian carp scale	Deep eutectic solvent (choline chloride/1,4-butanediol)	*Yield:* 40.58% *Morphology:* Agglomerated irregular non-needles, sticks and flakes *Particle size:* 0.235–117.1 μm	Liu et al. (2021)
Redmouth grouper (Aethaloperca rogaa) bone	Calcination at 750 °C	*Ca/P ratio*: 1.72 *Tracer element:* Na, Mg, K, Sr	Dabiri et al. (2018)
Sardinella longiceps bone	Alkaline hydrolysis (treated with 5% NaOH solution)	*Morphology:* Dense nano-sized particles *Particle size:* 19.65 nm	Surya et al. (2021)
Osteoglossum bicirrhosum scale	Alkaline treatment (NaOH) followed by calcination at 600 and 800 °C	*Crystallite size:* 20.17–68.96 nm *Particles size:* 200–400 nm	Horta et al. (2021)
Black Tilapia scale	Calcination at 600–800 °C	*Morphology:* Hexagonal fragments form *Tracer element:* Na, Mg	Jaffri et al. (2022)
Labeo rohita scale	Alkaline hydrothermal treatment (0.6% KOH)	*Morphology:* Spherical and oval *Ca/P ratio:* 1.52 *Particle size*: 33–479 μm	Sarkar and Das (2021)
Black scabbard bone (BS) and Grey triggerfish skin (GT)	Calcination at 400–1000 °C	*Crystalline phases*: Presence of β-TCP after calcined at 600 °C (BS), presence of β-TCP, halite, and rhenanite after calcined at 600 °C (GT) *Tracer element:* Na, Mg, Sr *Ca/P ratio:* 1.47–1.54 (BS), 1.62–1.65 (GT) *Specific surface area:* 2.5–133.5 m^2/g (BS), 2.1–110.4 m^2/g (GT)	Esposti et al. (2020)
Lates calcarifer bone	Thermal calcination	*Particle size:* Less than 40 nm *Tracer element:* Fe, K, Mg, Na, Se, Zn	Ho et al. (2022)
Micropogonias furnieri bone	Calcination at 800 °C	*Morphology:* Porous with average pore size of ~10 μm *Tracer element:* Mg, Na, K	Prado et al. (2021)
Hypopthalmichthys molitrix (silver carp) bone	Alkaline hydrolysis (AH) and thermal calcification (TC)	*Particle size:* 58.3 nm (AH), 64.3 nm (TC) *Ca/P ratio:* 1.65 (AH), 1.45 (TC)	Acharya et al. (2022)
Tuna fish bone	Calcination at 700–1100 °C	*Crystalline phases*: Presence of β-TCP after calcined at 900 °C *Morphology:* Spherical *Ca/P ratio:* 1.60 *Tracer element:* Mg, Sr, Na, K and Zn	Latif et al. (2020)

a common vacancy in HAp, where tracer elements such as Na^+, Mg^{2+}, K^+, Sr^{2+} and others were used to replace the calcium position. Meanwhile, CaO_3^{2-} was substituted in the P vacancy. Some studies also discovered β-TCP at higher temperatures (>600 °C). At high temperatures, during calcination, the process may convert HAp to other phases of calcium phosphate such as β-TCP. Because of HAp's high biocompatibility and β-TCP's good resorbability, it is a promising material that could be used in biomaterials, particularly in bone grafts (Esposti et al. 2020; Latif et al. 2020).

It has also been demonstrated that HAp extracted from fish sources has excellent in-vitro testing and in-vivo biocompatibility. Several studies have proved that it has high cell viability, mineralization ability, proliferation and is non-toxic. Deb and Deoghare (2019) extracted HAp from Catla scales using a combination of acid and alkaline treatment, followed by 1000 °C calcination. They discovered that after 28 days of immersion in simulated body fluid, the extracted HAp induced the formation of bone-like apatite, indicating that it is compatible to be used for biomedical applications.

Further, Shi et al. (2018) employed thermal calcination to extract nano-sized HAp from the bones of other fish, e.g. salmon, cod and rainbow trout. They discovered that nano-sized HAp extracted from salmon bones and rainbow trout are more biocompatible than cod bone due to the presence of CO_3^{2-} and Mg^{2+}. After 7 days of incubation, it was discovered that nano-sized HAp extracted from salmon and rainbow trout bones increased the viability of osteoblasts cells. Furthermore, they observed that the alkaline phosphate bioactivity of nano-sized HAp was high, indicating that nano-sized HAp extracted from fish has stronger interactions.

Meanwhile, Sathiskumar et al. (2019) employed alkaline heat treatment to prepare agglomerated rod-shaped nano-sized HAp from Cirrhinus mrigala fish scales. The biocompatibility of as-synthesized natural HAp with human osteosarcoma cells (MG-63 cells) at various dosages was investigated. Compared to commercial HAp, natural HAp promotes the MG-63 cell's growth and demonstrates superior ALP activity and cell viability.

HAp extracted from fish waste is simple to prepare and has excellent biocompatibility compared to synthetic HAp. Aside from that, it has a positive impact on the environment by reducing fish wastes. This chapter describes HAp derived from fish as a promising option that should be investigated further for biomedical applications.

6.4 Extraction and Characterization of HAp from Black Tilapia Fish Bones and Scales

6.4.1 Samples Preparation

The raw materials (Black Tilapia fish bones and scales) were specifically obtained from the G3 lake inside the grounds of Universiti Tun Hussein Onn Malaysia, Johor, Malaysia. As a waste by-product, the fishes were processed and separated from the meat carefully. The Black Tilapia fishes (bones and scales) were simmered in boiling water for 1 h prior to the treatment to remove any skin, raw flesh or other unwanted materials. Both samples were then rinsed and dried to remove any remaining fish meat. The samples were consecutively dried out for 3 h at 95 °C in an oven (Termoline, China), ground into a nice powder and pelletized with a hydraulic presser.

6.4.2 Characterization of HAp Powders

The HAp samples obtained from the fish (bones and scale) were characterized by using an XRD diffractometer (Bruker Advance D8) to analyze a crystalline structure and phase composition. The parameter of XRD used was Cu Kα ($\lambda = 1.5405$ Å) radiation at 40 mA (current) and 40 kV (accelerating voltage). An energy dispersive X-ray spectroscope (EDX) and a scanning electron microscope (SEM) (SU1510, Hitachi, Japan) with a 15-kV accelerating voltage were used, after which the microstructure of the obtained HAp samples was established. Each sample had a gold coating sputtered on it before investigation. The materials' elemental composition was discovered by EDX analysis.

6.4.3 Heat Treatment Technique (Calcination)

The fish (bones and scales) powder samples were calcined at two temperatures (i.e. 600 and 1000 °C) in a furnace (Protherm). To remove all impurity layers and all undesired biological traits, the calcination process was carefully maintained for 3 h (10 °C/min heating rate and natural cooling). To prevent contamination and lessen water surface captivation, the HAp samples were put in a desiccator after the calcination treatment.

6.4.4 Outcome of Black Tilapia-Derived HAp Analysis

6.4.4.1 Characterization of HAp Powder from Fish Bone

6.4.4.1.1 Morphological Analysis

Figure 6.3 displays the SEM images (morphology) of fish bone samples (raw and calcined at 600 and 1000 °C). The raw fish bone was discovered as a solid surface owing to the organic substances (protein, collagen and lipids) and water (Ahmad Fara et al. 2015). This contrasts with the raw sample, where the morphology was porous on the calcined powders. This happened because of the raw fish bone's organic components evaporated during the heat treatment process (calcination) and left residues of inorganic elements. At 1000 °C, secondary material (SM) which were small particles with brighter colour were present. The particle size increased as the temperature increased (from 600 to 1000 °C) due to the crystals beginning to coalesce into one body, at 1.54 μm and 12.35 μm respectively.

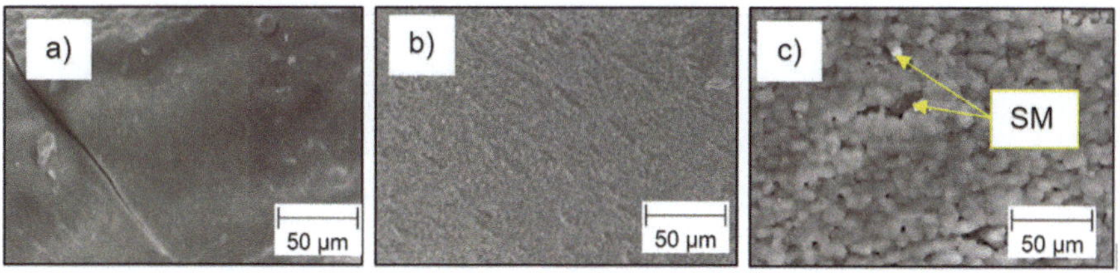

Fig. 6.3 SEM images of fish bones (**a**) Raw fish bone: compacted, (**b**) Calcined at 600 °C: porous surface and (**c**) Calcined at 1000 °C: porous surface with the presence of secondary material (SM)

Table 6.3 The Ca/P ratio for uncalcined and calcined fish bone (600 and 1000 °C)

Element (wt.%)	Raw fish bone	Calcined (600 °C)	Calcined (1000 °C)
Calcium (Ca)	55.62	61.28	56.39
Phosphorous (P)	39.65	34.38	34.62
Magnesium (Mg)	1.32	3.19	2.00
Sodium (Na)	3.41	1.15	6.98
Ca/P ratio	**1.40**	**1.78**	**1.63**

6.4.4.1.2 Chemical Composition Analysis

As demonstrated in Table 6.3, the ratio of Ca/P for fish bone is close to stoichiometric HAp, at 1.67. The closest amount of Ca/P was at 1000 °C, which is 1.63. The major elements of concentration are Ca and P with trace elements of Na and Mg present. The variation in the Ca/P ratio may be explained by the presence of trace elements from the source. Mg and Na are significant during biological tests (in-vitro and in-vivo) that simulate the natural bone development and increase the creation of new bone (Cacciotti et al. 2009).

6.4.4.1.3 Crystallinity Analysis

The XRD patterns of raw fish bone and calcined (600 and 1000 °C) are displayed in Fig. 6.4. The HAp's spectra are more intense in the calcined samples than in the raw. This demonstrates that the mineral crystallinity had increased and supports the finding in the SEM images where the collagen and organic substances were removed during the calcination process with the increasing calcination temperature. As expected, the crystallite size was increased, which relates to shrinking of the diffraction peaks. The peaks belong to the wilkeite element. Wilkeite was present in the raw fish bone and calcined sample at 1000 °C, which is also commonly found in natural HAp (Roopalakshmi et al. 2017). At 1000 °C, secondary material (SM) was observed to be β-TCP as indicated in the SEM analysis as tiny grains.

6.4.4.2 Characterization of HAp Powder from Fish Scale

6.4.4.2.1 Morphological Analysis

The SEM images were used to inspect and optimize HAp powder's microstructure. Figure 6.5a displays the surface morphology fish scale samples that have been

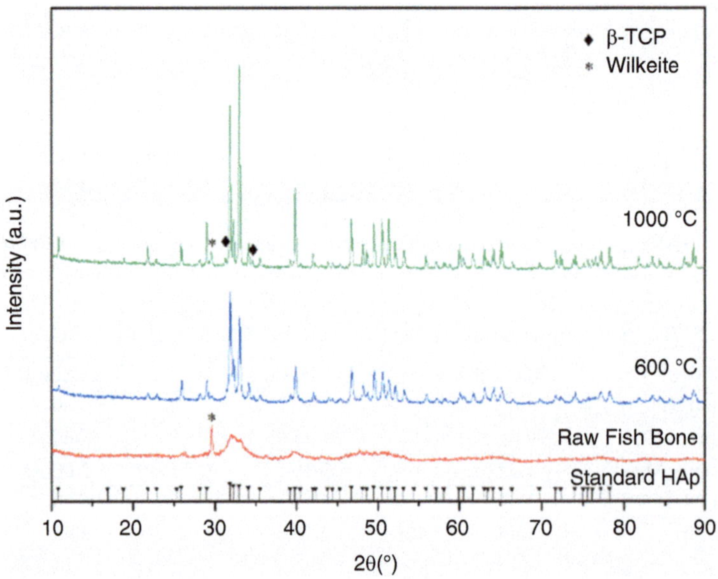

Fig. 6.4 The peaks in the XRD pattern of the raw and calcined fish bone at 600 and 1000 °C (peaks get sharper and more intense as the temperature rises)

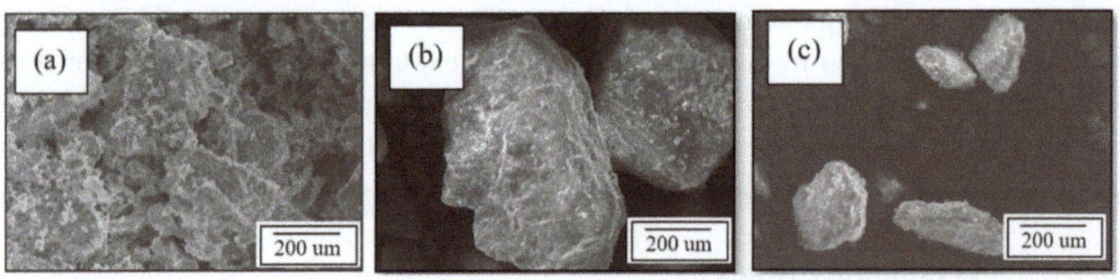

Fig. 6.5 SEM images of fish scales (**a**) Large-connected particles on the raw sample, (**b**) Large loose particles on the scales after calcination at 600 °C, and (**c**) Small loose particles after calcination at 1000 °C after removal of organic compounds

crushed. From the observation, the raw sample size was bigger than the samples that had been calcined from both temperatures. In this state, the samples still contain the organic parts such as gelatin, HAp and collagen type I, mimicking the structure of bone (Lin et al. 2019).

Figure 6.5b, c shows the Black Tilapia fish scale calcined at 600 and 1000 °C. Utilizing SEM observation, the theory from XRD was validated. With regard to particle size, shape and microstructure, SEM micrographs provided insight into the composition of the powder. HAp powder that had been calcined at 600 °C was thicker compared to powder calcined at 1000 °C. The fish scale HAp also has an enormous surface area, a clustering algorithm of similar shape, and a leaky formation, according to the result observed from the SEM analysis.

Table 6.4 Ca/P molar ratio for HAp samples from fish scale

Element (wt.%)	Raw fish scale	Calcined (600 °C)	Calcined (1000 °C)
Calcium (Ca)	47.68	58.45	62.51
Phosphorous (P)	39.40	40.31	35.92
Ca/P ratio	**1.21**	**1.45**	**1.74**

6.4.4.2.2 Chemical Composition Analysis

The chemical characteristics of the raw fish scale and calcined (600 and 1000 °C) were assessed by using EDX. Table 6.4 shows the molar ratio for raw samples and after being calcined at 600 and 1000 °C. The Ca/P for fish scale raw samples was 1.21. This phenomenon with regard to the existence of the organic compound makes the result of the Ca/P ratio smaller than the pure HAp. For samples that were calcined at 600 and 1000 °C, the results were 1.45 and 1.74 respectively. From the Ca/P result, HAp samples calcined at 1000 °C have the closest ratio compared to the standard Hap, which is 1.67 (Lu et al. 2018). The Ca/P ratio increase was due to the organic compound that was demolished during the calcination process.

6.4.4.2.3 Crystallinity Analysis

The XRD spectroscopy was used to analyze the presence and crystallinity of the HAp within the sample as part of the structural analysis of the fish scale HAp sample. The development of the ideal hexagonal crystalline composition of HAp and diffraction angle 2θ was contrasted and shown by the XRD pattern. The most pronounced or clearest peaks were discovered to link to high crystalline standard HAp materials according to the XRD result pattern as shown in Fig. 6.6. According to the materials library, from X-pert high score software, the spectrum also closely resembled the typical XRD pattern of pure HAp. This result indicates that the calcination method was the source of the high crystallinity at 1000 °C. The B type carbonate group and the standard JCPDS 00-009-0432 were identical (HAp) (Sathiskumar et al. 2019). Compared with the raw fish scale and the sample calcined at 600 °C, there were lesser peaks found based on the analysis. Results indicated that there were no impurities in the Black Tilapia fish scale HAp, and an excellent crystalline structure was discovered.

6.4.5 Biological Testing (In-Vitro)

6.4.5.1 Sample Preparation

The 1.0 g of powder samples calcined at 1000 °C were formed into pellets under five tonnes of pressure. A 13 mm diameter particle was made using Carver Hydraulic Laboratory Pellet Presses. The pellets were then sintered at 1000 °C for 3 h.

6.4.5.2 Bioactivity Assessment

The bioactivity of the HAp was assessed with the use of the simulated body fluid (SBF) solution. Kokubo's traditional method was used by mixing the reagents

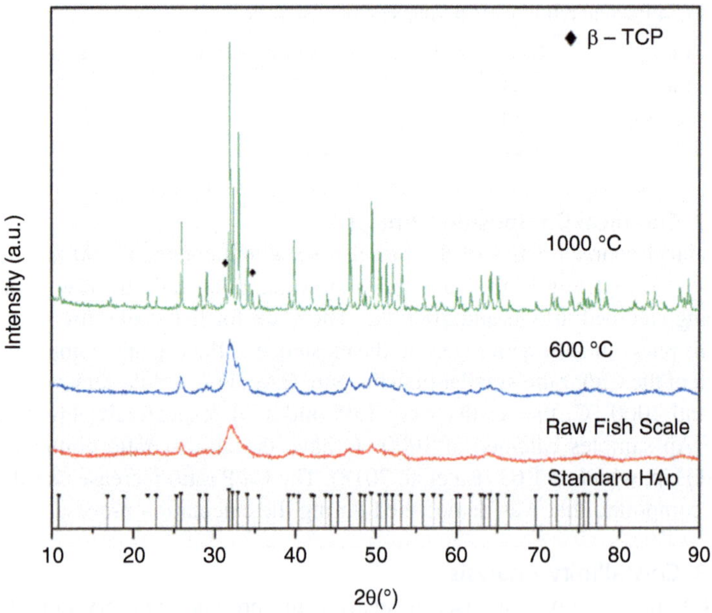

Fig. 6.6 The peaks in the XRD pattern of the raw and calcined fish scale at 600 and 1000 °C (peaks get sharper and more intense as the temperature rises)

Table 6.5 The reagents used for SBF preparation (1000 mL) (Kokubo and Takadama 2006)

No.	Reagent	Amount (g)	No.	Reagent	Amount (g)
1	NaCl	8.035	6	1.0 M HCl	39 mL
2	$NaHCO_3$	0.355	7	$CaCl_2$	0.292
3	KCl	0.225	8	Na_2SO_4	0.072
4	$K_2HPO_4 \cdot 3H_2O$	0.231	9	Tris	6.118
5	$MgCl_2 \cdot 6H_2O$	0.311	10	1.0 M HCl	0–5 mL

according to the order in Table 6.5 to create an SBF solution with a pH of 7.4 at 36.5 °C that was comparable to the ionic concentration of human blood plasma (Kokubo and Takadama 2006). The pellet samples were soaked in SBF for 3, 7, and 14 days in a humidified atmosphere before being rinsed and dried. The samples were then examined using SEM to analyze the growth of apatite on the surface.

6.4.5.3 Cell Viability Assessment

Colorimetric analysis using the 3-(4, 5-dimethylthiazol-2-yl)-2, 5-diphenyl tetrazolium) (MTT) test was used to measure the reaction of samples towards the metabolic activity of osteoblast cells (Life Technologies). The UV light (254 nm wavelength at about 76 W/cm^2) was used to sterilize the samples for an hour to reduce contamination. A 24-well plate (containing sterilized samples) was seeded (n = 3) with cultivated osteoblast cells at a concentration of 6.0 × 103 cells/mL in each well. The positive control was chosen to be the untreated cells (without the sample). Cells were cultivated for 1 or 2 days for both treated and not treated cells. After the allotted incubation time, the samples were taken out, put in a new 24-well plate, and PBS washed to remove any non-attached cells. Each well received 250 mL of medium and 50 mL of MTT solution (5 mg/mL) before being incubated for 4 h to complete the reaction. After that, 62.5 mL of each well's solution was discarded, and 125 mL of DMSO (dimethyl sulfoxide, 0.2 μ sterile filtered) was pipetted into the remaining solution to dissolve the formazan crystals. The formazan products were measured with a microplate reader (wavelength: 540 nm) after they had been dissolved (ThermoFisher Multiskan FC, USA). The absorbance measurements were shown as cell viability (percentage) using Eq. 6.1 (Mohd Daud et al. 2018):

$$\%\text{cell viability} = \frac{\text{treated cell OD}_{540}}{\text{non} - \text{treated cell OD}_{540}} \times 100 \tag{6.1}$$

6.4.5.4 Biological Analysis for HAp Extracted from Fish Bone

6.4.5.4.1 Bioactivity Analysis

Figure 6.7 reveals the SEM analysis for the creation of irregular calcium phosphate crystals (CPCs) on the pellet surface. Apatite formation shows that all the calcined pellets interact with the SBF medium (Sridevi et al. 2020). The apatite found on the pellet surface demonstrated the biomineralization process. In the processes leading to the sample's ensuing osseointegration, creating apatite layer was a significant stage. The ability of HAp to nucleate CPC on its surfaces was used to determine its bioactivity. From the results, it means that all the calcined fish bone samples were bioactive.

6.4.5.4.2 Cell Viability Analysis

Figure 6.8 illustrates the percentages of cell viability that were greater than 100% for all of the samples, whether for Day 1 or Day 2 (Rohmadi et al. 2021). After a day of incubation, the average cell viability was 120.10%, and after 2 days, it was 162.62%. This is due to the samples' low toxicity towards the hFOB 1.19 cell lines. Additionally, the medium used gave each well's cells appropriate nutrition. Therefore, cells

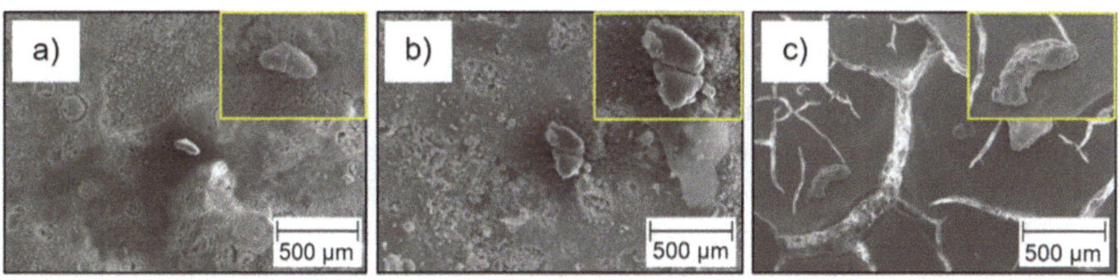

Fig. 6.7 SEM images showing the HAp precipitates on pellet bone samples calcined at 1000 °C for (**a**) 3 days, (**b**) 7 days and (**c**) 14 days of SBF immersion

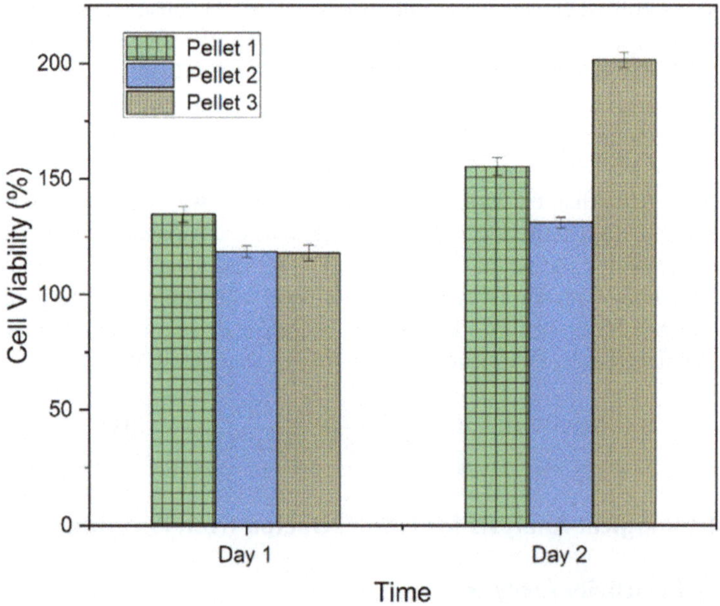

Fig. 6.8 MTT assay of hFOB 1.19 cells for fish bone incubated for 1 day and 2 days

can live and be viable under those specific conditions. These results suggested that the extracted HAp, which was previously thought to be a waste product, has possible applications in the field of bone healing.

6.4.5.5 Biological Analysis for HAp Extracted from Fish Scale

6.4.5.5.1 Bioactivity Analysis

Figure 6.9 shows the morphologies of the HAp samples after immersion in the SBF solution for 3, 7 and 14 days. The surfaces demonstrated the formation of apatite particles on the HAp pallet sample, albeit with higher density. The phosphate ion in the fluid interacted with the HAp on the surfaces of the HAp pallet sample, and as a result of this contact (which led to changes in the surface structure), apatite was formed (Baino and Yamaguchi 2020). On the other hand, as the surfaces of the HAp pallet samples settle in the SBF solutions, they appear to gradually crystallize into apatite.

6.4.5.5.2 Cell Viability Analysis

Figure 6.10 displays the percentages of fish scale samples' cell viability that are more than 100%. The average cell viability was 78.80% after a day of incubation and 117.22% after 2 days. This means that the samples were not toxic to the cells. This is due to the best environment for cell growth and enough nutrients in the medium served, which provided for higher cell viability. These findings provided proof that the biomaterial characteristics have a major impact on cell activity and viability.

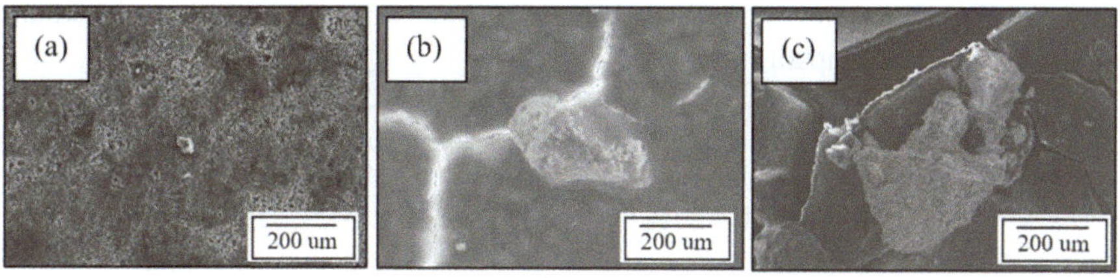

Fig. 6.9 SEM image of apatite formation after immersion in SBF solution for (**a**) 3 days, (**b**) 7 days and (**c**) 14 days

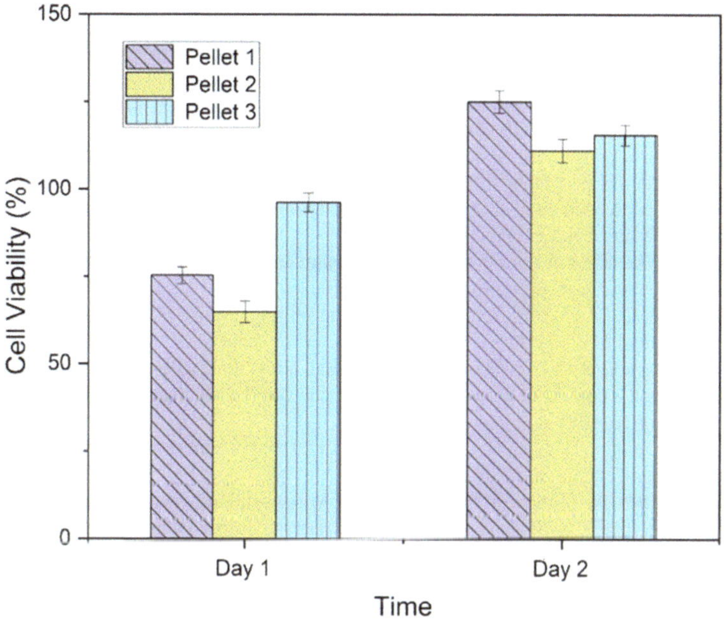

Fig. 6.10 Percentage of cells viability for fish scale

6.5 Conclusion

Tilapia fish bone and scale (waste) have great potential in the production of natural hydroxyapatite for biomedical applications. HAp powder from Black Tilapia fish bone showed good chemical properties and biological activities. It has a close Ca/P ratio to the stoichiometric of standard HAp (1.67), which are 1.63 for bone and 1.74 for scale. β-TCP phase occurred in the calcined powder for scale and bone at 1000 °C. The morphology of both HAp (bone and scale) pellet samples showed the formation of apatite after 3, 7 and 14 days of immersion in SBF. The HAp of both bone and scale were also shown to be not toxic to the cell since more than 100% of average cell viability after 2 days of incubation was obtained. Natural hydroxyapatite from fish bones and scales has the potential to induce the formation of apatite and exhibits good biocompatibility.

Acknowledgements The authors would like to express their gratitude to Universiti Tun Hussein Onn Malaysia (UTHM) and the Ministry of Higher Education (MOHE) through their provision of the Fundamental Research Grant Scheme (FRGS) FRGS/1/2018/STG07/UTHM/02/2.

Conflict of Interest The authors declare no conflict of interest.

References

Acharya P, Kupendra M, Fasim A et al (2022) Synthesis of nano hydroxyapatite from Hypopthalmichthys molitrix (silver carp) bone waste by two different methods: a comparative biophysical and in vitro evaluation on osteoblast MG63 cell lines. Biotechnol Lett 44(10): 1175–1188. https://doi.org/10.1007/s10529-022-03292-5

Ahmad Fara ANK, Yahya MA, Abdullah HZ (2015) Preparation and characterization of biological hydroxyapatite (HAp) obtained from tilapia fish bone. Adv Mater Res 1087:152–156

Akram M, Ahmed R, Shakir I et al (2014) Extracting hydroxyapatite and its precursors from natural resources. J Mater Sci 49:1461–1475

Aoba T (1997) The effect of fluoride on apatite structure and growth. Crit Rev Oral Biol Med 8(2): 136–153

Arul KT, Ramya JR (2017) Multifunctional behaviour of magnesium/calcium phosphate. Int J Eng Sci Res Technol 6(8):15–18

Baino F, Yamaguchi S (2020) The use of simulated body fluid (SBF) for assessing materials bioactivity in the context of tissue engineering: review and challenges. Biomimetics 5(4):1–19. https://doi.org/10.3390/biomimetics5040057

Bhattacharjee P, Begam H, Chanda A et al (2014) Animal trial on zinc doped hydroxyapatite: a case study. J Asian Ceramic Soc 2:44–51

Cacciotti I, Bianco A, Lombardi M et al (2009) Mg-substituted hydroxyapatite nanopowders synthesis, thermal stability and sintering behaviour. J Euro Ceram Soc 29(14):2969–2978. https://doi.org/10.1016/j.jeurceramsoc.2009.04.038

Cuozzo RC, Sartoretto SC, Resende RFB et al (2020) Biological evaluation of zinc-containing calcium alginate-hydroxyapatite composite microspheres for bone regeneration. J Biomed Mater Res B Appl Biomater 108:2610–2620

Da Silva LM, Tavares DS, dos Santos EA (2020) Isolating the effects of Mg^{2+}, Mn^{2+}, and Sr^{2+} ions on osteoblast behavior from those caused by hydroxyapatite transformation. Mater Res 23(2): e20200083

Dabiri SMH, Rezaie AA, Moghimi M et al (2018) Extraction of hydroxyapatite from fish bones and its application in nickel adsorption. BioNanoScience 8:823–834

De Lima IR, Alves GG, Fernandes GVO et al (2010) Evaluation of the in vivo biocompatibility of hydroxyapatite granules incorporated with zinc ions. Mater Res 13(4):563–568

Deb P, Deoghare AB (2019) Effect of acid, alkali and alkali–acid treatment on physicochemical and bioactive properties of hydroxyapatite derived from Catla fish scales. Arab J Sci Eng 44:7479–7490

Ehret C, Aid-Launais R, Sagardoy T et al (2017) Strontium-doped hydroxyapatite polysaccharide materials effect on ectopic bone formation. PLoS One 12(9):e0184663

Esposti PILD, Miguel CC, Admiano A et al (2020) Extraction and characterization of hydroxyapatite-based materials from grey triggerfish skin and black scabbardfish bones. Int J Appl Ceram Technol 18:235–243. https://doi.org/10.1111/ijac.13625

Gao J, Wang M, Shi C et al (2016) Synthesis of trace element Si and Sr codoping hydroxyapatite with non-cytotoxicity and enhanced cell proliferation and differentiation. Biol Trace Elem Res 174:208–217

Ge M, Ge K, Gao F et al (2018) Biomimetic mineralized strontium-doped hydroxyapatite on porous poly(l-lactic acid) scaffolds for bone defect repair. Int J Nanomedicine 13:1707–1721

Ho KHL, Dao VH, Pham XK et al (2022) Physicochemical properties, acute and subchronic toxicity of nano-hydroxyapatite obtained from Lates calcarifer fish bone. Reg Stud Mar Sci 55:102560

Horta MKDS, Moura FJ, Aguilar MS et al (2021) In vitro evaluation of natural hydroxyapatite from Osteoglossum bicirrhosum fish scales for biomedical application. Int J Appl Ceram 18(6): 1930–1937. https://doi.org/10.1111/ijac.13846

Hussin MFS, Abdullah HZ, Idris MI et al (2022) Extraction of natural hydroxyapatite for biomedical applications—a review. Heliyon 8(8):e10356

Ideia P, Pinto J, Ferreira R et al (2020) Fish processing industry residues: a review of valuable products extraction and characterization methods. Waste Biomass Valor 11:3223–3246

Ikeuchi M, Ito A, Dohi Y et al (2003) Osteogenic differentiation of cultured rat and human bone marrow cells on the surface of zinc-releasing calcium phosphate ceramics. J Biomed Mater Res 67A:1115–1122

Ishikawa K, Miyamoto Y, Yuasa T et al (2002) Fabrication of Zn containing apatite cement and its initial evaluation using human osteoblastic cells. Biomaterials 2(2):423–428

Ito A, Otsuka M, Kawamura H et al (2005) Zinc-containing tricalcium phosphate and related materials for promoting bone formation. Curr Appl Phys 5:402–406

Jaffri MZ, Abdullah HZ, Ismail ZMM et al (2022) Characterization of the halal natural hydroxyapatite extract from Black Tilapia fish scale. Key Eng Mater 908:165–170

Kavitha L, Gopi B, Ramya S (2015) Synthesis, characterization and in vitro studies of magnesium, fluoride co-substituted hydroxyapatite nanoparticles for dental application. 3rd Euro Cong Expo Dent Oral Health 14(3)

Kokubo T, Takadama H (2006) How useful is SBF in predicting in vivo bone bioactivity? Biomaterials 27(15):2907–2915. https://doi.org/10.1016/j.biomaterials.2006.01.017

Kumar R, Mohanty S (2022) Hydroxyapatite: a versatile bioceramic for tissue engineering application. J Inorg Organomet Polym 32(12):1–17. https://doi.org/10.1007/s10904-022-02454-2

Latif AFA, Mohd Pu'ad NAS, Ramli NAA et al (2020) Extraction of biological hydroxyapatite from tuna fish bone for biomedical applications. Mater Sci For 1010:584–549

Lee TC, Mohd Pu'ad NAS, Alipal J et al (2022) Tilapia wastes to valuable materials: a brief review of biomedical, wastewater treatment, and biofuel applications. Paper presented at the international symposium of reaction engineering, catalysis & sustainable energy (RECaSE 2021), Malaysia, 6 Apr 2021

Li Z, Huang B, Mai S et al (2015) Effects of fluoridation of porcine hydroxyapatite on osteoblastic activity of human MG63 cells. Sci Technol Adv Mater 15:035006

Lin K, Zhang D, Macedo MH et al (2019) Advanced collagen-based biomaterials for regenerative biomedicine. Adv Func Mater 29(3):1–16. https://doi.org/10.1002/adfm.201804943

Liu S, Zhou H, Liu H et al (2019) Fluorine-contained hydroxyapatite suppresses bone resorption through inhibiting osteoclasts differentiation and function in vitro and in vivo. Cell Prolif 52: e12613

Liu Y, Liu M, Ji S et al (2021) Preparation and application of hydroxyapatite extracted from fish scale waste using deep eutectic solvents. Ceram Int 47:9366–9372

Lu J, Yu H, Chen C (2018) Biological properties of calcium phosphate biomaterials for bone repair: a review. RSC Adv 8(4):2015–2033. https://doi.org/10.1039/c7ra11278e

Matsunaga K (2008) First-principles study of substitutional magnesium and zinc in hydroxyapatite and octacalcium phosphate. J Chem Phys 128:245101

Milovac D, Gamboa-martínez TC, Ivankovic M et al (2014) PCL coated hydroxyapatite scaffold derived from cuttle fish bone: in vitro cell culture studies. Mater Sci Eng C 42:264–272

Mohd Daud N, Hussein Al-Ashwal R, Abdul Kadir MR et al (2018) Polydopamine-assisted chlorhexidine immobilization on medical grade stainless steel 316L: apatite formation and in vitro osteoblastic evaluation. Ann Anat 220:29–37. https://doi.org/10.1016/j.aanat.2018.06.009

Mohd Pu'ad NAS, Koshy P, Abdullah HZ et al (2019) Syntheses of hydroxyapatite from natural sources. Heliyon 5(5):e01588

Olivier F, Rochet N, Delpeux-Ouldriane S et al (2020) Strontium incorporation into biomimetic carbonated calcium-deficient hydroxyapatite coated carbon cloth: biocompatibility with human primary osteoblasts. Mater Sci Eng C Mater Biol Appl 116:111192

Prado JPS, Yamamura H, Magri AMP et al (2021) In vitro and in vivo biological performance of hydroxyapatite from fish waste. J Mater Sci Mater Med 32:109

Rodrigues AI, Reus RL, Blitterswijk CA et al (2017) Calcium phosphates and silicon: exploring methods of incorporation. Biomater Res 21(6):1–11

Rohmadi R, Harwijayanti W, Ubaidillah U et al (2021) In vitro degradation and cytotoxicity of eggshell-based hydroxyapatite: a systematic review and meta-analysis. Polymers 13(19):1–17. https://doi.org/10.3390/polym13193223

Roopalakshmi S, Ravishankar R, Belaldavar S et al (2017) Investigation of structural and morphological characteristic of hydroxyapatite synthesized by sol-gel process. Mater Today Proc 4(11): 12026–12031. https://doi.org/10.1016/j.matpr.2017.09.126

Sarkar C, Das M (2021) Isolation and characterization of hydroxyapatite powder from the scale of freshwater fish Labeo rohita using green technology. Indian J Nat Prod Resour 12:359–368

Sathiskumar S, Vanaraj S, Sabarinathan D et al (2019) Green synthesis of biocompatible nanostructured hydroxyapatite from Cirrhinus mrigala fish scale—a biowaste to biomaterial. Ceram Int 45(6):7804–7810. https://doi.org/10.1016/j.ceramint.2019.01.086

Shepherd DV, Kauppinen K, Brooks RA et al (2014) An in vitro study into the effect of zinc substituted hydroxyapatite on osteoclast number and activity. J Biomed Mater Res 102:4136–4141

Shi P, Liu M, Fan F et al (2018) Characterization of natural hydroxyapatite originated from fish bone and its biocompatibility with osteoblast. Mater Sci EngC 90:706–712

Sridevi S, Sutha S, Kavitha L et al (2020) Physicochemical and biological behaviour of biogenic derived hydroxyapatite and carboxymethyl cellulose/sodium alginate biocomposite coating on Ti6Al4V alloy for biomedical applications. Mater Chem Phys 254:123455. https://doi.org/10.1016/j.matchemphys.2020.123455

Sun T, Wang M, Shao Y et al (2018) The effect and osteoblast signaling response of trace silicon doping hydroxyapatite. Biol Trace Elem Res 181:82–94

Surya P, Nithin A, Sundaramanickam A et al (2021) Synthesis and characterization of nano-hydroxyapatite from Sardinella longiceps fish bone and its effects on human osteoblast bone cells. J Mech Behav Biomed Mater 119:104501

Swadi AK, Jassim ZW, Kadum MU et al (2012) Influence of fluoride addition on hydroxyapatite prepared for medical applications. Baghdad Sci J 9(3):541–547

Webster TJ, Ergun C, Doremus RH et al (2001) Hydroxylapatite with substituted magnesium, zinc, cadmium, and yttrium. II. Mechanisms of osteoblast adhesion. J Biomed Mater Res 59:312–317

Yamada Y, Inui T, Kinoshita Y et al (2019) Silicon-containing apatite fiber scaffolds with enhanced mechanical property express osteoinductivity and high osteoconductivity. J Asian Ceram Soc 7(2):101–108

Yanli C (2011) Simultaneous incorporation of magnesium and fluorine ions in hydroxyapatite coatings on Ti6Al4V implant PhD. Nanyang Technology University, Singapore

Yao W, Xing WX, Dalin W (2016) Silicon-doped nano-hydroxyapatite: solubility, anti-fracture and compressive strengths. Chinese J Tissue Eng Res 20:435–440

Yu W, Sun TW, Qi C et al (2017) Evaluation of zinc-doped mesoporous hydroxyapatite microspheres for the construction of a novel biomimetic scaffold optimized for bone augmentation. Int J Nanomedicine 12:2293–2306

Zakaria KA, Yatim NI, Ali N et al (2022) Recycling phosphorus and calcium from aquaculture waste as a precursor for hydroxyapatite (HAp) production: a review. Environ Sci Pollut Res 29: 46471–46486

Human Amnion as a Cell Delivery Vehicle for Tissue Engineering and Regenerative Medicine Applications

7

Sik-Loo Tan, Nik Aizah, Zhi-Tang Wisely Koay, Jing-Yit Pua, Qi Hao Looi, and Tunku Kamarul

Abstract

Human amnion (HAM) is a biomaterial used in clinical settings with good safety records since it was first introduced as an effective means for managing open wounds and burns. It is a great resource for tissue engineering, a multidisciplinary field in science that involves cells, matrix, and biological factors that provide cues to the cells. Regenerative medicine is an emerging field of science that aims to replace or regenerate human cells, tissues, or organs to establish, restore, or even enhance normal tissue function after an injury or degeneration due to ageing. HAM is a promising biomaterial that can be used as a cell carrier that promotes stem cell proliferation as well as differentiation for tissue regeneration. This chapter discusses the potential applications of HAM in

S.-L. Tan (✉) · N. Aizah · Z.-T. W. Koay · J.-Y. Pua
Tissue Engineering Group (TEG), National Orthopaedic Center of Excellence for Research & Learning (NOCERAL), Department of Orthopaedic Surgery, Faculty of Medicine, Universiti Malaya, Kuala Lumpur, Malaysia
e-mail: tansikloo@um.edu.my; aizahn@um.edu.my

Q. H. Looi
My Cytohealth Sdn Bhd, D353A, Lab 6, DMC Level 2, Hive 5, Taman Teknologi MRANTI, Bukit Jalil, Kuala Lumpur, Malaysia
e-mail: dr.daniellooi@cytoholdings.com

T. Kamarul (✉)
Tissue Engineering Group (TEG), National Orthopaedic Center of Excellence for Research & Learning (NOCERAL), Department of Orthopaedic Surgery, Faculty of Medicine, Universiti Malaya, Kuala Lumpur, Malaysia

Advanced Medical and Dental Institute (AMDI), Universiti Sains Malaysia, Bertam, Kepala Batas, Pulau Pinang, Malaysia
e-mail: tkzrea@um.edu.my

W. S. Wan Kamarul Zaman, A. A. Nurul (eds.), *Sustainable Material for Biomedical Engineering Application*, https://doi.org/10.1007/978-981-99-2267-3_7

tissue engineering and regenerative medicine as an "off-the-shelf" biomaterial derived from biowaste (or clinical waste) and as a sustainable green technology-related product.

Keywords

Amnion · Tissue engineering · Orthopaedics · Regenerative medicine · Bio-waste product · Green technology

Abbreviations

AD-MSC	Adipose-derived MSC
BBCL	Biological bandage contact lens
ECM	Extracellular matrix
GCP	Good clinical practice
GLP	Good laboratory practice
GMP	Good manufacturing practice
HAM	Human amnion
haMSC	Human amniotic mesenchymal stromal cells
HBD3	Human beta-defensin
mDHACM	Micronized dehydrated human amnion
MSC	Mesenchymal stem cell
OCOP	Office of combination products
PICS	Pharmaceutical Inspection Cooperation Scheme
PMOA	Primary mode of action
QSR	Quality system requirements
RCT	Randomized control trials
SLPI	Secretory leukocyte protease inhibitor
TE	Tissue engineering

7.1 Introduction

Biomaterials are materials designed and used for medical applications to augment natural tissue function. They can be natural or synthetic and are usually a composite of multiple components that can interact and integrate with cellular components in the body system. Traditionally, biomaterials used for clinical applications are naturally derived, such as gelatine, silk, and collagen (Mandla et al. 2018). Tissue engineering (TE) enhances tissue regeneration by combining cells, scaffolds, and biological factors. To effectively deliver cells into a recipient body for tissue regeneration, the development of native or synthetic scaffolding biomaterials that mimics the extracellular matrix (ECM) plays a crucial role. With the advancement in TE research, human amnion (HAM) has emerged as an important biomaterial for TE

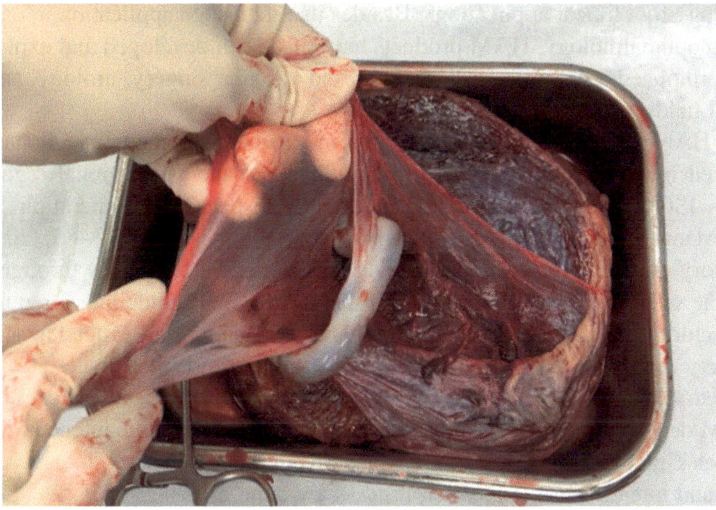

Fig. 7.1 Human amnion (HAM) is derived from the innermost layer of the placenta mass, which can be easily peeled off from its chorion layer

and regenerative medicine. HAM is derived from the placenta mass obtained after the delivery of an infant, which is usually discarded as surgical waste (Fig. 7.1). Over the years, it has had a good track record for clinical applications, especially in ophthalmology and dermatology (Davis 1909; Dua et al. 2004). It is an ideal biomaterial that is readily available, highly abundant, and cost-effective and does not raise ethical issues. HAM has been regarded as an "off-the-shelf" and ready-to-use biomaterial for TE and regenerative medicine, as it can be stored for a long period after processing (air-dried, freeze-dried, or glycerol preserved). It can promote wound healing as it is a biocompatible immune-privileged tissue that exerts anti-inflammatory, antimicrobial, antifibrotic, and antimutagenic effects (Kubo et al. 2001; Ricci et al. 2013). In TE and regenerative medicine, HAM has been used as a natural scaffold for cell seeding to promote stem cell attachment, proliferation, and differentiation (Lindenmair et al. 2014; Tan et al. 2011; Zarei et al. 2019). Various TE approaches have been developed to improve its application as a cell delivery vehicle and clinically qualified TE product.

HAM can be processed by various techniques, either by air-drying, freeze-drying (lyophilization), or glycerol preservation, to extend its shelf life for longer storage. The air-dried HAM has additional advantages compared to lyophilized HAM as it does not require cold chain delivery from the HAM scaffold production laboratory to the cell culture laboratory for cell seeding. Based on a patent search in the Lens – a global patent and scholarly knowledge search database (lens.org), at least 1500 active records related to HAM processing methods can be found, of which 20 have been patented. All these intellectual properties aimed to preserve HAM for longer storage and/or convert it into a medical device such as for wound dressing, suture, tissue grafts, etc. This suggests a high commercial value in HAM as a green technology for clinical applications. Besides its established application in dermatology and ophthalmology, HAM products have also been developed and explored for use in orthopaedic surgery, cardiothoracic and vascular surgery, urology, and other medical fields.

For HAM to be used as biomaterials in regenerative medicine, it has to be processed in a laboratory in compliance with local and international regulations, such as ISO 13485: 2016 and Pharmaceutical Inspection Co-operation Scheme Good Manufacturing Practice (PIC/S GMP) guidelines. ISO 13485: 2016 is a mandatory requirement for medical devices to manage their quality throughout the life cycle of a medical device, whereas the PIC/S GMP guideline is required for manufacturing sterile medicinal products. These regulations are crucial to safeguard the safety and health of patients. When used as an implanted material, HAM is classified as a class D, or high-risk, medical device. Nevertheless, the classification of HAM depends on the claims made by the manufacturer on its intended use. The actual classification of HAM is considered individually, taking into account its design and intended use, e.g. when used as a wound dressing for burns or chronic ulcerated wounds, it is classified as a class C, or moderate-high risk, medical device.

7.2 Human Amnion as a Biomaterial

The HAM is the innermost layer of the placenta that forms an amniotic cavity surrounding a developing foetus, providing the foetus with physical and biological protection. It consists of two membranes: the inner HAM and an outer chorionic membrane. The HAM consists of three main layers: a monolayer of human amnion epithelial cells (hAECs, Fig. 7.2), which is in direct contact with the amniotic fluid, and a mesenchymal layer composed of human amniotic mesenchymal stromal cells (hAMSCs) dispersed in the matrix (Antoniadou and David 2016). A thick basal lamina (or basement membrane) lies in between these two layers and is abundant with high amounts of collagen fibres and a variety of extracellular matrix (ECM) proteins such as laminin, fibronectin, glycoproteins, and proteoglycans (Fénelon et al. 2021). These ECM proteins are crucial in supporting cell adhesion, proliferation, and growth.

HAMs are used clinically as natural scaffolds, especially in dermatology and ophthalmology procedures due to their low immunogenicity (Bose 1979; Szekeres-Bartho 2002) and their anti-inflammatory properties (Higa et al. 2005; Rocha and Baptista 2015). HAMs also show antimicrobial properties, as HAM plays a crucial role in maintaining a sterile environment for the foetus until birth. Several papers have documented reduced bacterial counts and enhanced healing in infected wounds due to the local defence mechanism in the placenta (Stock et al. 2007; Zare-Bidaki et al. 2017). Antimicrobial compounds such as human beta-defensin (HBD3), Elafin, and secretory leukocyte protease inhibitor (SLPI) have also been reported to be found in HAMs, which show inhibitory effects against various pathogens (Stock et al. 2007; Szukiewicz et al. 2016). These properties are beneficial when used for treatment in specific cases where the immune system is defective, or levels of

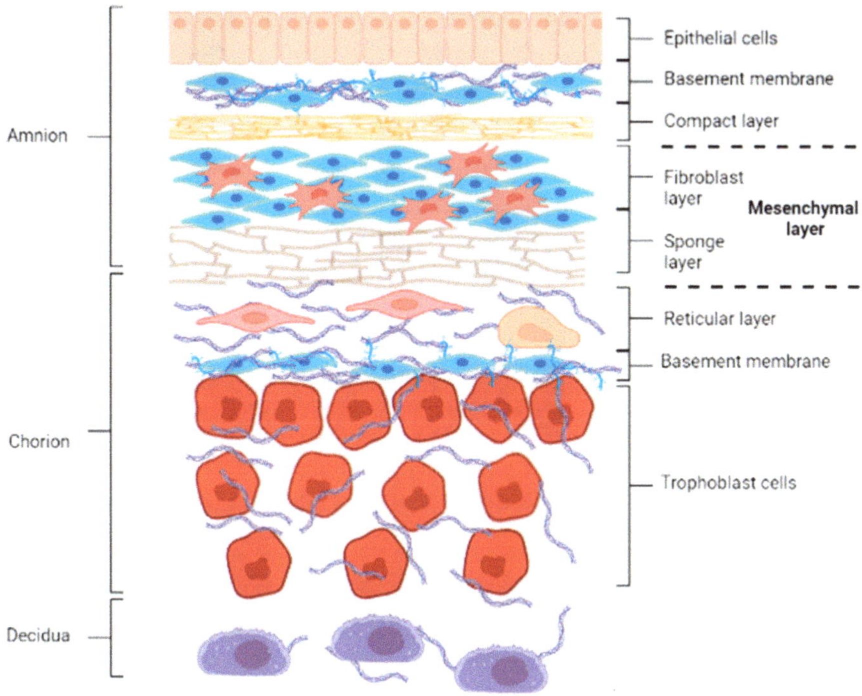

Fig. 7.2 Schematic diagram of the human foetal membrane which illustrates the amnion (HAM) and chorion layer. Adapted from (Uchide et al. 2012)

infection in the patient are elevated, which would impede the natural wound healing process.

Freshly procured HAM has been used for clinical applications without any preservation (Demirkan et al. 2002; Dogramaci and Duman 2016; Nicodemo et al. 2017; Ozboluk et al. 2010; Ozgenel 2004; Saini and Mirakhur 2014; Saini et al. 1998). HAM can also be processed to extend its shelve life through various methods, such as air-drying (Tan et al. 2011) (Fig. 7.3), glycerol-based cryopreservation (Leppanen et al. 2017; Liu et al. 2016; Prakash et al. 2020), lyophilization or freeze-drying (Hadi and Sukmawan 2014; Liu et al. 2020), and hyperdrying (Zukawa et al. 2022). The air-drying method is among the simplest method, which only requires air-drying the HAM under a class II biosafety cabinet. In glycerol-based preservation, HAM is kept in a cryopreservation media in a cryovial for long-term storage in a −80 °C deep freezer (Prakash et al. 2020). Cryopreserved HAMs must then be defrosted and rinsed with sterile phosphate-buffered saline (PBS) prior to use (Prakash et al. 2020). Lyophilization or freeze-drying involves placing the HAM in a freeze-dryer and freezing it at 0–50 °C for 1–10 h (Liu et al. 2020). The freezing chamber ensures a cold trap temperature is maintained at a range of −10 °C to −56 °C (Liu et al. 2020). The vacuum pump is opened, and the water content ranges from 1% to 20% (Liu et al. 2020). After that, cobalt 60 irradiation is used to

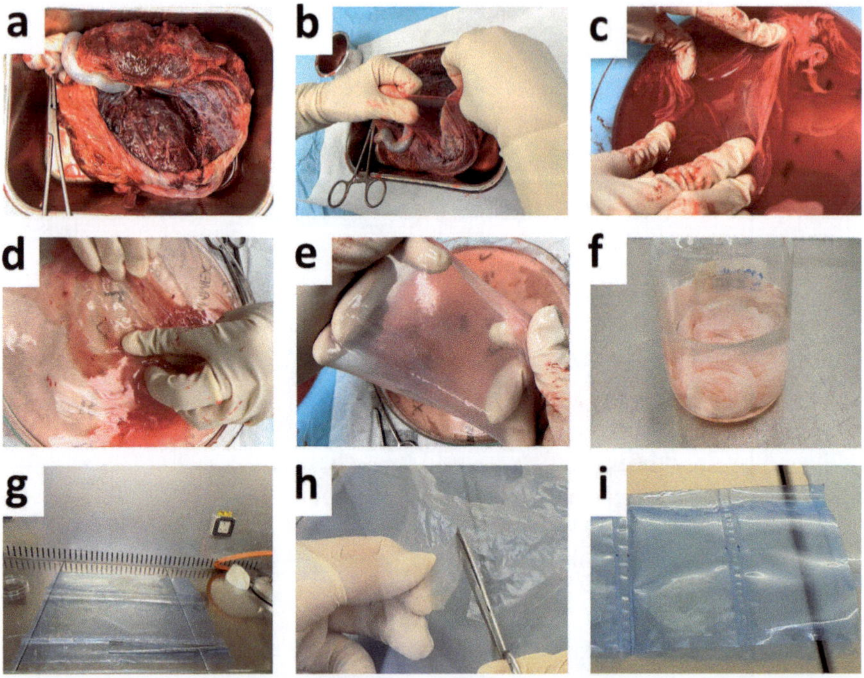

Fig. 7.3 Human amnion (HAM) as a biowaste is processed into a sustainable green technology-related product for tissue engineering and regenerative medicine applications. Figures (**a–i**) illustrate the steps in the air-drying processing method to convert HAM into a medical device (cell carrier for clinical transplantation). (**a**) Placenta mass is immediately procured after delivery in a sterile stainless steel surgical tray. (**b**) HAM is then peeled from the placenta mass. (**c**) After detaching the chorion layer, HAM is transferred into a sterile petri dish for cleaning with copious amounts of sterile normal saline. (**d**) The blood clots on the HAM are removed from the HAM gently. (**e**) HAM has a transparent appearance after washing and removal of blood clots. (**f**) HAM is stored in a sterile bottle with sterile normal saline and transported to the processing laboratory to further process it via washing steps with distilled water, sodium hypochlorite, and normal saline (with an aseptic technique in a class II biosafety cabinet). (**g**) After washing, the HAM is spread flat on a clean work surface inside a class II biosafety cabinet for overnight air-drying. (**h**) The overnight air-dried HAM is trimmed into the desired shape. (**i**) The trimmed HAM is sealed into individual packing and sent for gamma sterilization before it can be used for tissue engineering or regenerative medicine research applications

sterilize the freeze-dried HAMs (Liu et al. 2020). Hyperdrying is another method that uses far-infrared rays, depression of air, and microwaves before it is sterilized by γ-ray irradiation (Okabe et al. 2014).

Besides the processing methods to extend the storage time of HAM, the decellularization approach is used to further reduce the immunogenicity of this membrane to avoid post-implantation rejection (Jin et al. 2007; Sang et al. 2020). Decellularized HAM involves removing the epithelial layer of the amnion chemically by incubating the amnion in ethylenediaminetetraacetic acid 0.05% and mechanically by the manual scraping of the amnion surface (Liu et al. 2018). The latest advancement in HAM processing includes the use of electrospinning technology to add nanofiber layers around HAM. This approach enhances the structural integrity and mechanical strength of HAM whilst providing a three-dimensional environment that promotes growth factor permeation to support the proliferation of tenocytes and fibroblasts (Liu et al. 2020) (Fig. 7.3).

7.3 Recent Advancement in Amnion Research

The concept of using HAM as a substitute for skin was first proposed by Davis (1909), and since then, the use of HAM in surgery has been expanded into various applications (Fig. 7.4). The first clinical report of successful grafting of HAM in treatment for skin burns and ulcerations was soon reported in 1913 (Stern 1913). However, interest in HAM research diminished due to the consequent fear of virus transmission.

It was not until the late 1970s that new methods for processing and storage of HAM were established and gained scientific interest again. Rao and Chandrasekharam (1981) introduced the use of dried human and bovine amnions. Drying and sterilizing the amnion extends the life of the product to more than 9 months and also enables it to be stored at room temperature. Their method helped address issues with preserving amnions and maintaining their supply in amnion banks. Subsequently, clinical trials were carried out by Visuthikosol et al. (1992) to prepare lyophilized foetal membranes as biological dressings. The membrane was applied to a variety of wounds, and the clinical study showed that the use of lyophilized HAM as a biological dressing has promising results.

The use of HAM further extended into modern ophthalmology, where researchers Kim and Tseng (1995) reported the successful use of preserved HAM for corneal surface reconstruction in a rabbit model. The success of this study encouraged the application of this technique to patients with severe ocular surface diseases. Later, in 2004, in a search for alternative sources of mesenchymal stem cells (MSCs), In 't Anker et al. (2004) discovered that second-trimester and term third-trimester amnion contains a rich source of foetal MSCs and is a readily available resource that does not involve an invasive procedure to obtain. This serves as a good alternative source of MSCs compared to adult bone marrow MSC which has limitations in quantity and also requires an invasive procedure to obtain.

In 2007, Portmann-Lanz et al. (2007) developed a method of creating a cell-free extracellular matrix scaffold to provide support for the growth of amniotic cells. The manufactured acellular amnion matrix retains structural and functional properties required for cell outgrowth in vitro and may be a useful tool for repairing prematurely ruptured foetal membranes. Following that, Kitagawa et al. (2009) described the hyperdry method to improve the preservation of HAM. This method uses far-infrared rays and microwaves, followed by sterilization using gamma-ray irradiation, which allows it to be stored at room temperature instead of freezers that are

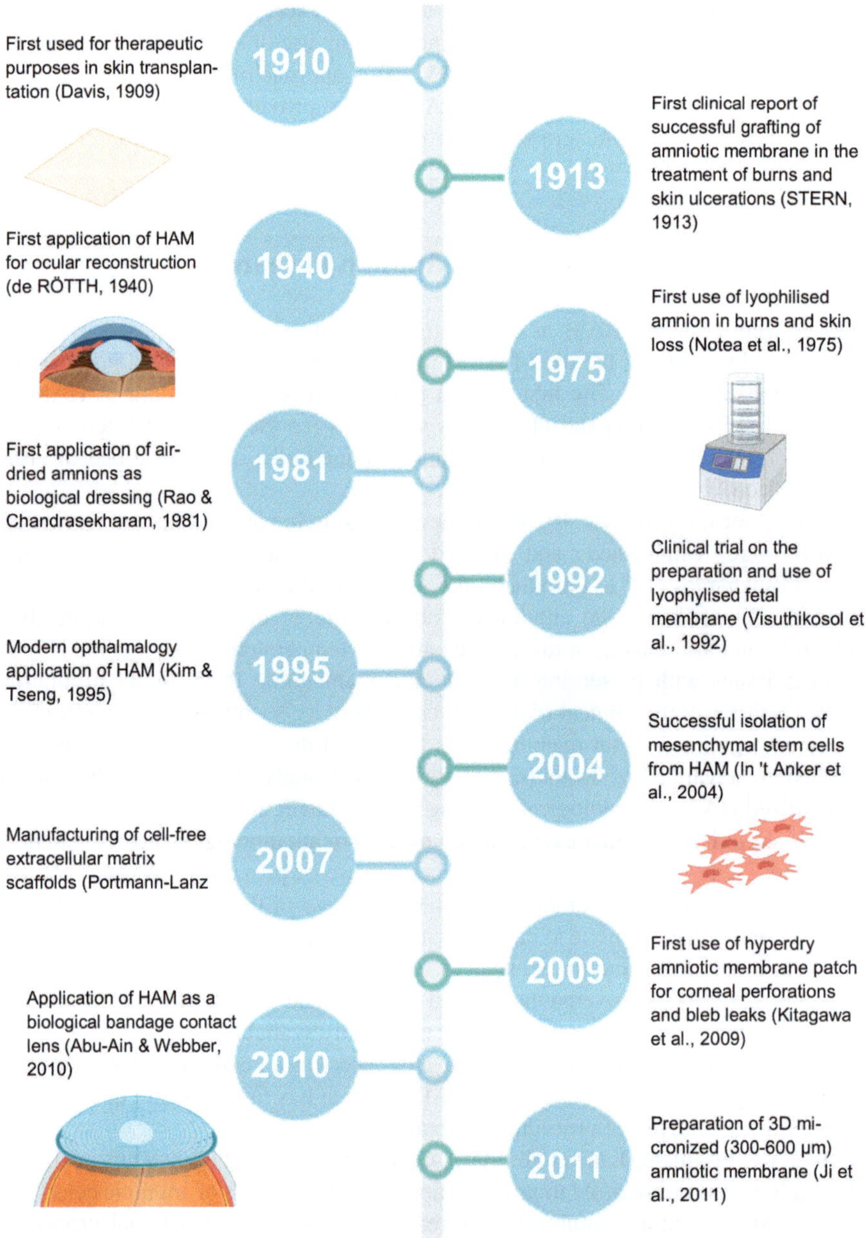

Fig. 7.4 Milestones in the history of HAM as a natural biomaterial in clinical applications and tissue engineering

space-consuming. The hyperdry HAM was used as a patch to repair bleb leaks or corneal perforations and showed promising results.

In 2010, Abu-Ain and Webber (2010) demonstrated the use of HAM as a biological bandage contact lens (BBCL) to treat persistent corneal epithelial defects. The authors described the novel method as having more advantages over conventional fixing methods such as with glue, suture, or conformers. The BBCL is cheap, easy to perform and can be easily removed or replaced whenever necessary. In another major stride in HAM processing and application, the preparation of 3D micronized (300–600 mm) amniotic membrane by repeated freeze-thawing cycles was introduced by Ji et al. (2011). By depleting cell components and homogenizing the amniotic membrane with a macrohomogenizer in liquid nitrogen, the newly prepared membrane possesses the characteristics of a microcarrier and can still completely retain the basement membrane structure and active substances in the HAM matrix.

Besides being utilized as a membranous graft, HAM has also been reported to support MSCs attachment and proliferation for use as a cell carrier in TE and regenerative medicine (Tan et al. 2011), either as intact HAM or as a decellularized membrane (Jin et al. 2007; Lindenmair et al. 2014; Tan et al. 2011). In a recent study, it is reported that HAM is more suitable for the direct culture of undifferentiated adipose-derived MSC (AD-MSC) than the culture of differentiated cells (Zarei et al. 2019). Other than in laboratorial experiments, HAM is also becoming the focus of clinical trials. A single-centre randomized control trial using HAM as a treatment for plantar fasciitis has been reported (Zelen et al. 2013). This clinical trial showed that micronized dehydrated HAM (mDHACM) is a viable option for reducing inflammation in patients. Nevertheless, more clinical studies with a high level of evidence are still required to support and validate their findings, as well as to further bring HAM application from bench to bedside.

7.4 Human Amnion as a Sustainable Biomaterial for Tissue Engineering and Regenerative Medicine (Term) Applications

7.4.1 Dermatology

One of the most common applications of HAM is in dermatology, where it can be applied as a biological dressing to treat and manage wounds. Several reports have described its use in treating acute and chronic wounds, such as ulcers and burns of varying degrees (Insausti et al. 2010; Singh et al. 2004). The regenerative properties of HAM promote wound healing and act as a physical and biological barrier against the external environment (Murphy et al. 2017).

There are similarities in the ultrastructure and functionality between HAM and normal human skin, which allows HAMs to be used as a scaffold to create a living skin-equivalent (LSE) for wound closure. It has been shown that skin epidermal morphology is improved by incorporating the use of HAMs as it retains key basement membrane components (Yang et al. 2006). Other than that, composite skin scaffolds can also be formed by combining HAMs with synthetic or natural polymers. Composite scaffolds have shown a good response for skin regeneration, likely due to the presence of growth factors, cytokines, and regulatory molecules that promote the healing of wounds (Koob et al. 2014).

7.4.2 Orthopaedic Surgery

Several studies have looked at the application of HAM in orthopaedic surgery, particularly at the osteoinductive and bone regeneration ability as well as the angiogenesis potential of HAMs. It has been shown that the use of biocompatible HAMs was able to improve stem cell proliferation and osteogenic differentiation, which promotes bone regeneration in defects (Mohr et al. 2010; Tang et al. 2017). The unique biological properties of HAMs give them the potential to replace conventional synthetic materials that were used for bone regeneration such as ceramics, metals, and polymers due to their low biodegradability and biocompatibility.

Other than bones, HAMs are also able to promote chondrogenic differentiation to help in cartilage regeneration. HAMs used as a chondrocyte carrier were shown to be able to improve chondrocyte proliferation and glycosaminoglycans (GAG) expression (Krishnamurithy et al. 2011), making HAMs a potential option as a substrate or carrier for cell-based cartilage therapy.

The application of HAM in tendon healing showed that they could prevent the formation of adhesions after injury and reconstruction (Liu et al. 2018; Ozgenel 2004). Adhesion formation is one of the most common complications during the repair of injured tendons, and it leads to poor recovery and loss of tendon function. Non-biological materials such as polyethylene films prevent this by mechanical isolation; however, it lacks permeability that might result in the possibility of tendon necrosis to occur (Liu et al. 2018). This highlights the benefits of using a semi-permeable membrane such as HAM to overcome this limitation. Moreover, HAMs show adequate mechanical properties such as elasticity, stiffness, and tensile strength due to its composition and orientation of collagen fibrils in the ECM. The ECM components such as hyaluronic acid, laminin, proteoglycans, and collagen found in HAMs also provide good cell adhesion capacity and allow more effective seeding of cells in tissue engineering (Fénelon et al. 2021). In short, HAMs could provide great potential as a naturally derived scaffold for bone, cartilage, as well as tendon healing due to their beneficial properties.

7.4.3 Ophthalmology

In ophthalmology, the use of HAMs on ocular surface reconstruction is also becoming popular. Studies have shown that transplantation of HAM is able to decrease the vascularization of the ocular surface due to its anti-angiogenic properties (Shao et al.

2004). The tissue is used to treat epithelial defects or cover the ocular surface to promote healing through surgical techniques such as inlay transplantation (graft), onlay transplantation (patch), or combinatorial technique (sandwich technique).

Inlay transplantation involves using the HAM as a graft to replace lost tissue due to the removal of defective tissues from the site. Either single or multilayer grafts may be used depending on the depth of the wound. HAM grafts can be integrated into the corneal stroma as they form hemidesmosomes and desmosomes, which mediate cell–cell adhesion and improve the structural quality of the tissue by providing stability and anchorage. The HAM is placed epithelial side up as the HAM is intended to act as a scaffold for epithelial cells to grow on (Röck et al. 2018).

Onlay transplantation, also known as patch amnion transplantation, is done by the addition of HAM over the periphery of a superficial wound. The HAM is used epithelial side down as the transplant would be able to act as a physical barrier to protect the wound from external threats (Malhotra and Jain 2014). Proteins secreted from the HAM will limit fibrosis and scarring allowing for more effective wound healing (Dua 1999).

The sandwich approach utilizes both inlay and onlay methodologies as a combined transplantation. The graft provides structural integrity, while the patch protects the eye with a physical barrier. As the graft has its epithelial side up and the patch has its epithelial side down, the epithelium would be expected to grow between these two layers. A multiple-layer approach can also be carried out with layers of HAMs stacked one on top of the other to fill an area of corneal melt or thinning (Dua et al. 2004). The graft is usually incorporated long-term into the substratum of the host epithelium, while the patch would either detach naturally over time or is manually removed.

7.4.4 Cardiovascular Surgery

Vascular endothelial cells can also be cultured using HAMs with preservation methods such as cryopreservation or lyophilization, with the latter being reported to show better attachment to the basement membrane (Niknejad et al. 2011). The ECM components of the HAMs create an almost native scaffold for cell seeding. In a study, HAM used as grafts for the construction of veins showed no evident inflammation or fibrosis even after 48 weeks. The grafts were still patent with no signs of dilation, and normal endothelial cell morphology was observed (Peirovi et al. 2012). In another study, a multilaminate rolling approach using HAM to generate a tubular construct for blood vessel regeneration was used. The decellularized and tubularized HAM was quickly repopulated with endothelial cells and vascular smooth muscle cells, becoming an ideal replacement graft similar in composition to natural vessels (Amensag and McFetridge 2012).

7.4.5 Urology

HAMs have also been used for regenerative purposes in the urinary tract, such as in the kidneys, urinary bladder, ureter, and urethra. Similar to other applications elsewhere, composites can also be created to aid in reconstructive urology. HAMs are preserved and processed to form a biocomposite that would help improve mechanical resistance and facilitate regeneration of the urothelial and smooth muscle layer (Adamowicz et al. 2016). The reason for reinforcing the HAMs is to overcome the fragility of unprocessed HAMs, which poses a limitation when applied in bladder reconstruction due to the extensive change in the volume of the organ.

7.4.6 Dental Surgery

The usage of HAM has also been extended to dental surgery, specifically in periodontal regenerative therapy. The purpose of periodontal regenerative therapy is to restore the supporting structures of the tooth that have been lost as a result of periodontitis. HAM has been used in several reports to preserve the socket after tooth extraction (Hassan et al. 2017), promote healing in periodontal bony defects (Holtzclaw and Toscano 2013), and guide bone regeneration (Bhide and Tenenbaum 2021). The various growth factors in HAMs promote the proliferation and migration of periodontal ligament-derived cells and angiogenesis, thus improving periodontal healing (Imamura et al. 2022).

7.5 Conversion of a Biowaste into a Medical Device

HAM is a naturally derived medical device, or a green technology, which is converted from clinical waste into an innovative Eco-product. It is procured from the placenta of consented mothers who underwent elective caesarean sections, which is a sustainable source and would not be short in supply. Selection criteria for the donor recruitment ensure that only donors who are seronegative for human immunodeficiency virus, human hepatitis B and C, and syphilis are included in the HAM procurement. After procurement from the operation theatre, HAM can be processed into biomaterial, which is easy to handle, durable, and can be sterilized as an "off-the-shelf" medical device for use in TE and regenerative medicine. From one single HAM procured from one caesarean case, at least 15 small pieces of HAM (with an area of 3 cm × 3 cm) can be made available for use in patients. The total cost needed for HAM processing, for instance, air-drying process, only requires the use of sterile normal saline and a small volume of sodium hypochlorite solution. The equipment needed for HAM processing is basic and is commonly available in most cell culture laboratories or in Good Manufacturing Practice (GMP) compliance laboratories (i.e. biosafety cabinet and a shaker). This simple processing procedure translates into a low manufacturing cost for HAMs to become medical devices. This eases the expansion of HAM manufacturing and the possibility for HAM to become widely available for use in clinical settings. Since the placenta is usually discarded post-partum, it is readily available, and HAM can be harvested without much concern from the consented mothers. The ethical concerns about the usage of HAMs are minimal as it does not pose any harm to the delivered infant or the mother.

HAM products are among a few of the discoveries and innovations that are at the forefront of biomedical research that offer enormous promise (Bai et al. 2018). However, the roadmap from academic biomedical research to clinical bedside application is not straightforward or clearly defined in any academic training programs or guidelines. For example, despite a majority of research initiatives indicating that HAM serves as an effective and natural scaffold for various applications, several challenges including ethics approval, in vitro and in vivo pre-clinical studies, human clinical trials, manufacturing guidelines, as well as the quality system in practices for HAM need to be established before HAM could be moved from bench to bedside (Contopoulos-Ioannidis et al. 2003; Morris et al. 2011). This is also one of the reasons why a significant number of basic science research never translated into human clinical trials (Butler 2008; Fernandez-Moure 2016). The roadmap of HAM, from proof of concept to commercialization, is illustrated in Fig. 7.5. For the registration of HAM as a medical device, it has to be completed with preclinical in vitro and in vivo studies [with compliance to OECD-Good Laboratory Practice (OECD-GLP) guidelines] for all the claims made by the manufacturer on its intended use. After the completion of the preclinical studies, then only one can move on to human clinical trials [with compliance to Good Clinical Practice (GCP) and the Pharmaceutical Inspection Convention and Pharmaceutical Inspection Co-operation Scheme (PICS/GMP) Guidelines] for evaluating its safety and efficacy before randomized control trials can be conducted. With the completion of preclinical and clinical studies, then only can a medical device be registered as a medical product for human use.

This section focuses on the description of the global GMP quality system requirements (QSR) and the quality system practices required for a biomaterial like HAM to be commercialized as a medical device. This section discusses the QSR required in general for the EU and some countries in South East Asia.

ISO 13485 is a universal standard for manufacturers that produce medical devices. It provides tried-and-true guidelines for maintaining assurance and controlling risk (Jain et al. 2018). According to ISO 13485, "*an organization needs to demonstrate its ability to provide medical devices and related services that consistently meet customer and applicable regulatory requirements*". Providing clients and end users with secure medical devices and improving patient outcomes are two issues that ISO 13485 addresses for medical device manufacturers (*ISO 13485:2016 Medical devices—Quality management systems—Requirements for regulatory purposes* 2016). An organization uses this standard for designing, developing, producing, installing, and maintaining medical equipment. Internal and external stakeholders, including certification organizations, also utilize ISO 13485: 2016 to evaluate an organization's capacity to satisfy customer and legal needs. Its main goal is to make it easier to comply with standardized medical device regulatory criteria. The standard is founded on eight core quality management principles:

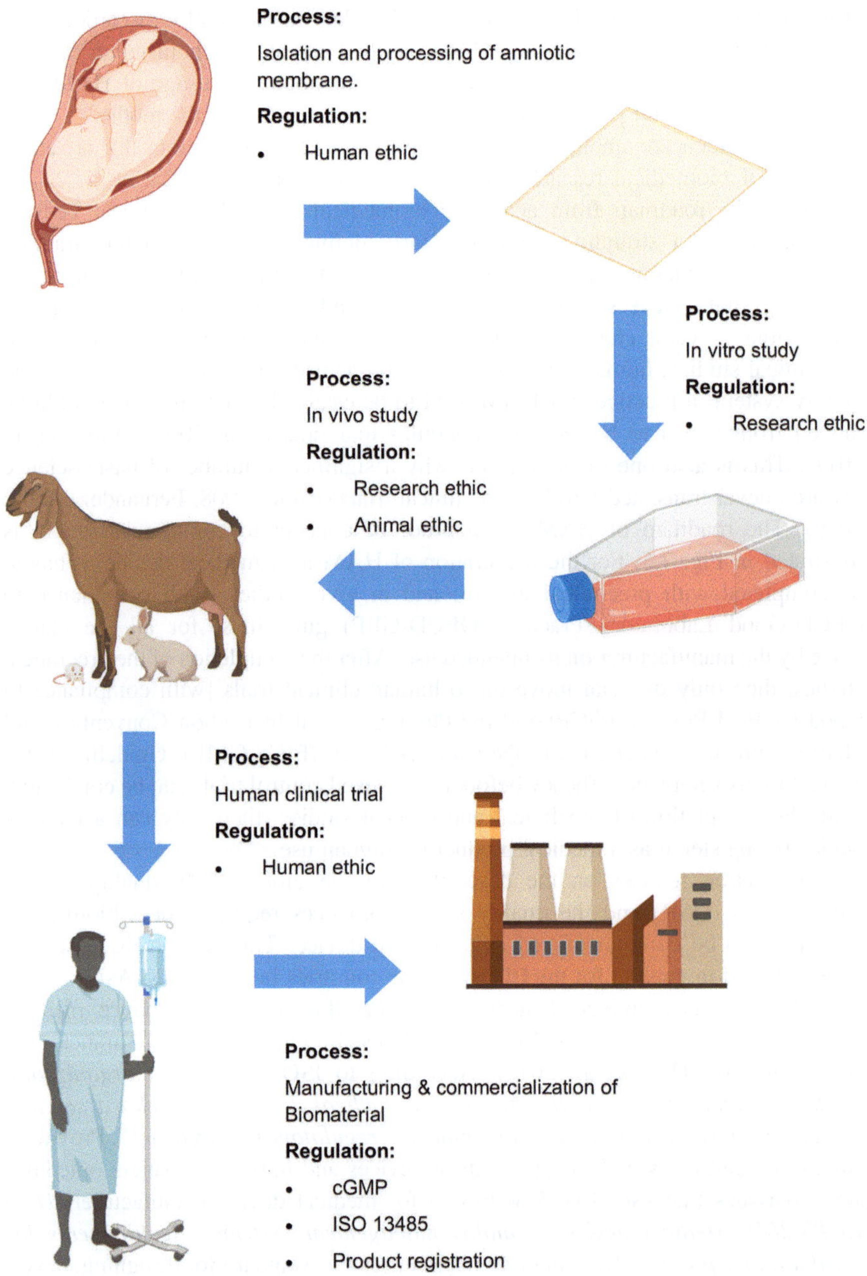

Fig. 7.5 The roadmap from basic science preclinical research to human clinical trials in HAM products. The development of HAM as a medical device starts with basic science research followed by proof of concept in vivo preclinical studies in animal models. After the proof of concept in preclinical models, human clinical trials need to be conducted to show its safety and efficacy before it can be tested in a larger cohort of patients in randomized control trials (RCTs) and finally be registered as a medical device for clinical applications

customer focus, leadership, participation of people, process- and system-based management, continuous improvement, fact-based decision-making, and advantageous supplier relationships (Tarabah 2015).

Manufacturers of finished medical devices marketed in the US, the UK, and several South East Asian nations are subject to the standards of 21 CFR 820 (*Title 21 Food and Drugs—Chapter I—Food and Drug Administration Department of Health and Human Services—Subchapter H—Medical Devices—Part 820 Quality System Regulation* 2016); but, depending on the countries in which they operate, some items or manufacturers may be subject to only a subset of those criteria (Jain et al. 2018). In contrast, ISO standards are accepted in most countries across the world.

In most circumstances, a biomaterial like HAM, which is a medical device, is intended to be used with either autologous or allogeneic-derived stem cells (e.g. MSCs). When HAM is intended to be used as MSCs seeded HAM for implantation purposes, it is a medical product comprising of two or more regulated components, i.e. biologics and device, that are physically or chemically combined and produced as a single entity. When a medical device is derived from a combination of two or more single-entities, it will not fit exclusively into either drug or medical device categories. A "drug" has a different definition from a medical device. A "drug" includes any substance, product, or article intended to be used or capable, or purported to be capable, of being used on humans or any animal, whether internally or externally, for a medicinal purpose.

According to the primary mode of action (PMOA), the FDA established the Office of Combination Products (OCPs) in the US in 2002. This office is in charge of designating each combination product as a drug, medical device, biological, or combination product (Lauritsen and Nguyen 2009). However, guideline and Good Manufacturing Practice (GMP) requirements for combination products are less clearly defined compared to single-entity products, as combination products usually involve a complex mode of action and manufacturing process. In 2009, the FDA issued proposed rules to (a) codify the current GMP requirements applicable to combination products and (b) clarify the post-marketing safety reporting requirements that apply to combination products. In some countries, combination products of biologics/medical device, such as MSCs seeded HAM, must comply with ISO13485 standard for medical devices, and GCP as well as GMP guidelines (as required for Cell and Gene Therapy Products) ("Title 21 Food and Drugs—Chapter I—Food and Drug Administration Department of Health and Human Services—Part 4 Current good manufacturing practice requirements for combination products: proposed rule" 2016).

7.6 Conclusion

HAM is a naturally derived biomaterial that has a huge potential for TE and regenerative medicine applications, either directly as a tissue graft or in combination with other biologics like stem cells, to promote tissue regeneration. Various HAM processing methods have been established to prolong storage and ease transportation for use at the bedside. Over the years, the use of HAM has evolved from being used as a skin dressing for burns and ocular reconstructions to being used as a cell carrier for stem cells. The roadmap for developing HAM as a medical device for use in clinical applications requires the complete research flow from in vitro and in vivo preclinical studies based on OECD-Good Laboratory Practice (GLP), to clinical trials with compliance to GMP and GCP for safety and efficacy studies, before it can be registered as a medical device.

Acknowledgement This chapter was prepared as part of the collaborative work under the Prototype Research Grant Scheme (PRGS) project funded by the Ministry of Higher Education (reference number: PR001-2021 and PRGS/1/2021/SKK01/UM/02/1).

Conflict of Interest Authors declare no conflict of interest.

References

Abu-Ain MS, Webber SK (2010) The biological bandage contact lens: a novel technique for using the amniotic membrane in the treatment of persistent corneal epithelial defects. Eye 24(7): 1306–1307. https://doi.org/10.1038/eye.2010.1

Adamowicz J, Pokrywczyńska M, Tworkiewicz J, Kowalczyk T, van Breda SV, Tyloch D et al (2016) New amniotic membrane based biocomposite for future application in reconstructive urology. PLoS One 11(1):e0146012

Amensag S, McFetridge PS (2012) Rolling the human amnion to engineer laminated vascular tissues. Tissue Eng Part C Methods 18(11):903–912. https://doi.org/10.1089/ten.TEC.2012.0119

Antoniadou E, David AL (2016) Placental stem cells. Best Pract Res Clin Obstet Gynaecol 31:13–29. https://doi.org/10.1016/j.bpobgyn.2015.08.014

Bai X, Gao M, Syed S, Zhuang J, Xu X, Zhang XQ (2018) Bioactive hydrogels for bone regeneration. Bioact Mater 3(4):401–417. https://doi.org/10.1016/j.bioactmat.2018.05.006

Bhide VM, Tenenbaum HC (2021) Horizontal guided bone regeneration using a dehydrated amnion/chorion membrane: a case report. Int J Periodontics Restorative Dent 41(3):375–381. https://doi.org/10.11607/prd.4634

Bose B (1979) Burn wound dressing with human amniotic membrane. Ann R Coll Surg Engl 61(6): 444–447. https://pubmed.ncbi.nlm.nih.gov/496235

Butler D (2008) Translational research: crossing the valley of death. Nature 453(7197):840–842. https://doi.org/10.1038/453840a

Contopoulos-Ioannidis DG, Ntzani E, Ioannidis JP (2003) Translation of highly promising basic science research into clinical applications. Am J Med 114(6):477–484. https://doi.org/10.1016/s0002-9343(03)00013-5

Davis JS (1909) II. Skin grafting at the Johns Hopkins Hospital. Ann Surg 50(3):542–549. https://doi.org/10.1097/00000658-190909000-00002

Demirkan F, Colakoglu N, Herek O, Erkula G (2002) The use of amniotic membrane in flexor tendon repair: an experimental model. Arch Orthop Trauma Surg 122(7):396–399. https://doi.org/10.1007/s00402-002-0418-3

Dogramaci Y, Duman IG (2016) Reinforcement of the flexor tendon repair using human amniotic membrane(a biomechanical evaluation using the modified Kessler method of tendon repair). J Am Podiatr Med Assoc 106(5):319–322. https://doi.org/10.7547/15-036

Dua HS (1999) Amniotic membrane transplantation. Br J Ophthalmol 83(6):748. https://doi.org/10.1136/bjo.83.6.748

Dua HS, Gomes JA, King AJ, Maharajan VS (2004) The amniotic membrane in ophthalmology. Surv Ophthalmol 49(1):51–77. https://doi.org/10.1016/j.survophthal.2003.10.004

Fénelon M, Catros S, Meyer C, Fricain J-C, Obert L, Auber F et al (2021) Applications of human amniotic membrane for tissue engineering. Membranes 11(6):387. https://www.mdpi.com/2077-0375/11/6/387

Fernandez-Moure JS (2016) Lost in translation: the gap in scientific advancements and clinical application. Front Bioeng Biotechnol 4:43. https://doi.org/10.3389/fbioe.2016.00043

Hadi S, Sukmawan J (2014) Effect of lyophilized amniotic membrane, hyaluronic acid, and their combination in preventing adhesion after tendon repair in New Zealand white rabbits. Med J Indones 23:65. https://doi.org/10.13181/mji.v23i2.932

Hassan M, Prakasam S, Bain C, Ghoneima A, Liu SS (2017) A randomized split-mouth clinical trial on effectiveness of amnion-chorion membranes in alveolar ridge preservation: a clinical, radiologic, and morphometric study. Int J Oral Maxillofac Implants 32(6):1389–1398. https://doi.org/10.11607/jomi.5875

Higa K, Shimmura S, Shimazaki J, Tsubota K (2005) Hyaluronic acid-CD44 interaction mediates the adhesion of lymphocytes by amniotic membrane stroma. Cornea 24(2):206–212

Holtzclaw DJ, Toscano NJ (2013) Amnion–chorion allograft barrier used for guided tissue regeneration treatment of periodontal intrabony defects: a retrospective observational report. Clin Adv Periodontics 3(3):131–137. https://doi.org/10.1902/cap.2012.110110

Imamura K, Hamada Y, Yoshida W, Murakami T, Nakane-Koyachi S, Yoshikawa K, Saito A (2022) Investigating the effects of dehydrated human amnion-chorion membrane on periodontal healing. Biomol Ther 12(6):857. https://doi.org/10.3390/biom12060857

In 't Anker PS, Scherjon SA, Kleijburg-van der Keur C, de Groot-Swings GM, Claas FH, Fibbe WE, Kanhai HH (2004) Isolation of mesenchymal stem cells of fetal or maternal origin from human placenta. Stem Cells 22(7):1338–1345. https://doi.org/10.1634/stemcells.2004-0058

Insausti CL, Alcaraz A, García-Vizcaíno EM, Mrowiec A, López-Martínez MC, Blanquer M et al (2010) Amniotic membrane induces epithelialization in massive posttraumatic wounds. Wound Repair Regen 18(4):368–377. https://doi.org/10.1111/j.1524-475X.2010.00604.x

ISO 13485:2012 Medical devices—quality management systems—requirements for regulatory purposes (2016)

Jain A, Ganesh N, Venkatesh P (2018) Quality standards for medical devices. Int J Drug Regul Aff 2:19–24. https://doi.org/10.22270/ijdra.v2i4.149

Ji SZ, Xiao SC, Luo PF, Huang GF, Wang GY, Zhu SH et al (2011) An epidermal stem cells niche microenvironment created by engineered human amniotic membrane. Biomaterials 32(31): 7801–7811. https://doi.org/10.1016/j.biomaterials.2011.06.076

Jin CZ, Park SR, Choi BH, Lee KY, Kang CK, Min BH (2007) Human amniotic membrane as a delivery matrix for articular cartilage repair. Tissue Eng 13(4):693–702. https://doi.org/10.1089/ten.2006.0184

Kim JC, Tseng SC (1995) Transplantation of preserved human amniotic membrane for surface reconstruction in severely damaged rabbit corneas. Cornea 14(5):473–484

Kitagawa K, Yanagisawa S, Watanabe K, Yunoki T, Hayashi A, Okabe M, Nikaido T (2009) A hyperdry amniotic membrane patch using a tissue adhesive for corneal perforations and bleb leaks. Am J Ophthalmol 148(3):383–389. https://doi.org/10.1016/j.ajo.2009.03.030

Koob TJ, Lim JJ, Massee M, Zabek N, Denozière G (2014) Properties of dehydrated human amnion/chorion composite grafts: implications for wound repair and soft tissue regeneration. J Biomed Mater Res B Appl Biomater 102(6):1353–1362. https://doi.org/10.1002/jbm.b.33141

Krishnamurithy G, Shilpa PN, Ahmad RE, Sulaiman S, Ng CLL, Kamarul T (2011) Human amniotic membrane as a chondrocyte carrier vehicle/substrate: in vitro study. J Biomed Mater Res A 99A(3):500–506. https://doi.org/10.1002/jbm.a.33184

Kubo M, Sonoda Y, Muramatsu R, Usui M (2001) Immunogenicity of human amniotic membrane in experimental xenotransplantation. Invest Ophthalmol Vis Sci 42(7):1539–1546

Lauritsen KJ, Nguyen T (2009) Combination products regulation at the FDA. Clin Pharmacol Ther 85(5):468–470. https://doi.org/10.1038/clpt.2009.28

Leppanen OV, Karjalainen T, Goransson H, Hakamaki A, Havulinna J, Parkkinen J, Jokihaara J (2017) Outcomes after flexor tendon repair combined with the application of human amniotic membrane allograft. J Hand Surg Am 42(6):474.e471–474.e478. https://doi.org/10.1016/j.jhsa.2017.03.006

Lindenmair A, Nürnberger S, Stadler G, Meinl A, Hackl C, Eibl J et al (2014) Intact human amniotic membrane differentiated towards the chondrogenic lineage. Cell Tissue Bank 15(2): 213–225. https://doi.org/10.1007/s10561-014-9454-9

Liu Y, Peng Y, Fang Y, Yao M, Redmond RW, Ni T (2016) No midterm advantages in the middle term using small intestinal submucosa and human amniotic membrane in Achilles tendon transverse tenotomy. J Orthop Surg Res 11(1):125. https://doi.org/10.1186/s13018-016-0463-1

Liu C, Yu K, Bai J, Tian D, Liu G (2018) Experimental study of tendon sheath repair via decellularized amnion to prevent tendon adhesion. PLoS One 13(10):e0205811

Liu C, Tian S, Bai J, Yu K, Liu L, Liu G et al (2020) Regulation of ERK1/2 and SMAD2/3 pathways by using multi-layered electrospun PCL-amnion nanofibrous membranes for the prevention of post-surgical tendon adhesion. Int J Nanomedicine 15:927–942. https://doi.org/10.2147/IJN.S231538

Malhotra C, Jain AK (2014) Human amniotic membrane transplantation: different modalities of its use in ophthalmology. World J Transplant 4(2):111–121. https://doi.org/10.5500/wjt.v4.i2.111

Mandla S, Davenport Huyer L, Radisic M (2018) Review: multimodal bioactive material approaches for wound healing. APL Bioeng 2(2):021503. https://doi.org/10.1063/1.5026773

Mohr S, Portmann-Lanz CB, Schoeberlein A, Sager R, Surbek DV (2010) Generation of an osteogenic graft from human placenta and placenta-derived mesenchymal stem cells. Reprod Sci 17(11):1006–1015. https://doi.org/10.1177/1933719110377471

Morris ZS, Wooding S, Grant J (2011) The answer is 17 years, what is the question: understanding time lags in translational research. J R Soc Med 104(12):510–520. https://doi.org/10.1258/jrsm.2011.110180

Murphy SV, Skardal A, Song L, Sutton K, Haug R, Mack DL et al (2017) Solubilized amnion membrane hyaluronic acid hydrogel accelerates full-thickness wound healing. Stem Cells Transl Med 6(11):2020–2032. https://doi.org/10.1002/sctm.17-0053

Nicodemo MdC, Neves L Rd, Aguiar JC, Brito FS, Ferreira I, Sant'anna LB et al (2017) Amniotic membrane as an option for treatment of acute Achilles tendon injury in rats. Acta Cir Bras 32(2): 125–139

Niknejad H, Deihim T, Solati-Hashjin M, Peirovi H (2011) The effects of preservation procedures on amniotic membrane's ability to serve as a substrate for cultivation of endothelial cells. Cryobiology 63(3):145–151. https://doi.org/10.1016/j.cryobiol.2011.08.003

Okabe M, Kitagawa K, Yoshida T, Suzuki T, Waki H, Koike C et al (2014) Hyperdry human amniotic membrane is useful material for tissue engineering: physical, morphological properties, and safety as the new biological material. J Biomed Mater Res A 102(3):862–870. https://doi.org/10.1002/jbm.a.34753

Ozboluk S, Ozkan Y, Ozturk A, Gul N, Ozdemir RM, Yanik K (2010) The effects of human amniotic membrane and periosteal autograft on tendon healing: experimental study in rabbits. J Hand Surg Eur Vol 35(4):262–268. https://doi.org/10.1177/1753193409337961

Ozgenel GY (2004) The effects of a combination of hyaluronic and amniotic membrane on the formation of peritendinous adhesions after flexor tendon surgery in chickens. J Bone Joint Surg Br 86(2):301–307. https://doi.org/10.1302/0301-620x.86b2.14435

Peirovi H, Rezvani N, Hajinasrollah M, Mohammadi SS, Niknejad H (2012) Implantation of amniotic membrane as a vascular substitute in the external jugular vein of juvenile sheep. J Vasc Surg 56(4):1098–1104. https://doi.org/10.1016/j.jvs.2012.02.036

Portmann-Lanz CB, Ochsenbein-Kölble N, Marquardt K, Lüthi U, Zisch A, Zimmermann R (2007) Manufacture of a cell-free amnion matrix scaffold that supports amnion cell outgrowth in vitro. Placenta 28(1):6–13. https://doi.org/10.1016/j.placenta.2006.01.004

Prakash S, Kalra P, Dhal A (2020) Flexor tendon repair with amniotic membrane. Int Orthop 44(10):2037–2045. https://doi.org/10.1007/s00264-020-04752-1

Rao TV, Chandrasekharam V (1981) Use of dry human and bovine amnion as a biological dressing. Arch Surg 116(7):891–896. https://doi.org/10.1001/archsurg.1981.01380190029007

Ricci E, Vanosi G, Lindenmair A, Hennerbichler S, Peterbauer-Scherb A, Wolbank S et al (2013) Anti-fibrotic effects of fresh and cryopreserved human amniotic membrane in a rat liver fibrosis model. Cell Tissue Bank 14(3):475–488. https://doi.org/10.1007/s10561-012-9337-x

Rocha SCM, Baptista CJM (2015) Biochemical properties of amniotic membrane. In: Mamede AC, Botelho MF (eds) Amniotic membrane: origin, characterization and medical applications. Springer, Dordrecht, pp 19–40

Röck T, Bartz-Schmidt KU, Landenberger J, Bramkamp M, Röck D (2018) Amniotic membrane transplantation in reconstructive and regenerative ophthalmology. Ann Transplant 23:160–165. https://doi.org/10.12659/aot.906856

Saini NS, Mirakhur KK (2014) Ultrasonography of mid-metacarpal region following tendon allografting in equine. Indian J Anim Sci 71(2) https://epubs.icar.org.in/index.php/IJAnS/article/view/36415

Saini NS, Mirakhur KK, Singh M, Roy KS (1998) Homologous plasma-stored tendon grafting in equids—gross and histomorphological observations. Res Vet Sci 64(2):85–90. https://doi.org/10.1016/s0034-5288(98)90000-5

Sang R, Liu Y, Kong L, Qian L, Liu C (2020) Effect of acellular amnion with increased TGF-beta and bFGF levels on the biological behavior of tenocytes. Front Bioeng Biotechnol 8:446. https://doi.org/10.3389/fbioe.2020.00446

Shao C, Sima J, Zhang SX, Jin J, Reinach P, Wang Z, Ma J-x (2004) Suppression of corneal neovascularization by PEDF release from human amniotic membranes. Invest Ophthalmol Vis Sci 45(6):1758–1762. https://doi.org/10.1167/iovs.03-0882

Singh R, Chouhan U, Purohit S, Gupta P, Kumar P, Kumar A et al (2004) Radiation processed amniotic membranes in the treatment of non-healing ulcers of different etiologies. Cell Tissue Bank 5(2):129–134

Stern M (1913) The grafting of preserved amniotic membrane to burned and ulcerated surfaces, substituting skin grafts: a preliminary report. J Am Med Assoc 60(13):973–974. https://doi.org/10.1001/jama.1913.04340130021008

Stock SJ, Kelly RW, Riley SC, Calder AA (2007) Natural antimicrobial production by the amnion. Am J Obstet Gynecol 196(3):255.e251–255.e256. https://doi.org/10.1016/j.ajog.2006.10.908

Szekeres-Bartho J (2002) Immunological relationship between the mother and the fetus. Int Rev Immunol 21(6):471–495. https://doi.org/10.1080/08830180215017

Szukiewicz D, Alkhalayla H, Pyzlak M, Watroba M, Szewczyk G, Wejman J (2016) Human beta-defensin 1, 2 and 3 production by amniotic epithelial cells with respect to human papillomavirus (HPV) infection, HPV oncogenic potential and the mode of delivery. Microb Pathog 97:154–165. https://doi.org/10.1016/j.micpath.2016.06.010

Tan SL, Sulaiman S, Pingguan-Murphy B, Selvaratnam L, Tai CC, Kamarul T (2011) Human amnion as a novel cell delivery vehicle for chondrogenic mesenchymal stem cells. Cell Tissue Bank 12(1):59–70. https://doi.org/10.1007/s10561-009-9164-x

Tang K, Wu J, Xiong Z, Ji Y, Sun T, Guo X (2017) Human acellular amniotic membrane: a potential osteoinductive biomaterial for bone regeneration. J Biomater Appl 32(6):754–764. https://doi.org/10.1177/0885328217739753

Tarabah F (2015) 8—Good manufacturing practice (GMP) for biomaterials and medical devices in the EU and the USA. In: Amato SF, Ezzell RM (eds) Regulatory affairs for biomaterials and medical devices. Woodhead Publishing, pp 115–143

Title 21 Food and Drugs—Chapter I—Food and Drug Administration Department of Health and Human Services—Part 4 Current good manufacturing practice requirements for combination products: proposed rule (2016)

Title 21 Food and Drugs—Chapter I—Food and Drug Administration Department of Health and Human Services—Subchapter H—Medical Devices—Part 820 Quality System Regulation (2016)

Uchide N, Ohyama K, Bessho T, Takeichi M, Toyoda H (2012) Possible roles of proinflammatory and chemoattractive cytokines produced by human fetal membrane cells in the pathology of adverse pregnancy outcomes associated with influenza virus infection. Mediat Inflamm 2012: 270670. https://doi.org/10.1155/2012/270670

Visuthikosol V, Somna R, Nitiyanant P, Navikarn T (1992) The preparation of lyophylised of fetal membrane for biological dressing. J Med Assoc Thail 75(Suppl 1):52–59

Yang L, Shirakata Y, Shudou M, Dai X, Tokumaru S, Hirakawa S et al (2006) New skin-equivalent model from de-epithelialized amnion membrane. Cell Tissue Res 326(1):69–77. https://doi.org/10.1007/s00441-006-0208-2

Zare-Bidaki M, Sadrinia S, Erfani S, Afkar E, Ghanbarzade N (2017) Antimicrobial properties of amniotic and chorionic membranes: a comparative study of two human fetal sacs. J Reprod Infertil 18(2):218–224. https://pubmed.ncbi.nlm.nih.gov/28868246

Zarei H, Karimpour A, Reza Khalatbary A, Talebpour Amiri F (2019) Homing of adipose stem cells on the human amniotic membrane as a scaffold: a histological study. Int J Reprod Biomed 18(1): 21–32. https://doi.org/10.18502/ijrm.v18i1.6193

Zelen CM, Poka A, Andrews J (2013) Prospective, randomized, blinded, comparative study of injectable micronized dehydrated amniotic/chorionic membrane allograft for plantar fasciitis—a feasibility study. Foot Ankle Int 34(10):1332–1339. https://doi.org/10.1177/1071100713502179

Zukawa M, Okabe M, Osada R, Makino H, Nogami M, Seki S et al (2022) Effect of hyperdry amniotic membrane in preventing tendon adhesion in a rabbit model. J Orthop Sci 27(3): 707–712. https://doi.org/10.1016/j.jos.2021.03.005

Bioscaffolds and Cell Source in Cartilage Tissue Engineering

8

Kien Hui Chua and Siti Nurhadis Che Omar

Abstracts

Cartilage is a unique connective tissue that is found in the joints between bones, ribs, ears, nasal bridge, trachea and intervertebral discs (IVDs). It acts as a structural tissue and cushion between joints. Cartilage has limited regenerative capacity due to its avascular composition; any damage in the cartilage is mostly left unrepaired without surgical and therapeutic interventions. Cartilage damage often results from surgery due to tumour removal while joint cartilage damage is caused by sports-related injury. Currently, cartilage tissue engineering (CTE) appears to have potential in the treatment of cartilage damage. The use of specific cell sources in combination with various types of scaffolds could provide many potential solutions for cartilage repair. Differentiated cells (articular, nasal and auricular chondrocytes) and stem cells (mesenchymal stem cells) have been studied to evaluate their ability to repair cartilage damage. Successful CTE must consider factors such as efficient use of chondrogenic factors, sufficient supply of chondrogenic progenitors and heterogeneous environment for cell-friendly and biocompatible scaffolds. This chapter highlights the advancement of CTE in terms of cell sources, types of scaffold materials (natural, synthetic or biocomposites) used in CTE till date and how they accelerate the CTE process for clinical treatment.

Keywords

Cartilage repair · Chondrogenesis · Chondrocyte · MSCs · Biocomposite · Cartilage tissue engineering

K. H. Chua (✉) · S. N. Che Omar
Department of Physiology, Faculty of Medicine, Universiti Kebangsaan Malaysia, Kuala Lumpur, Malaysia
e-mail: ckienhui@ppukm.ukm.edu.my

W. S. Wan Kamarul Zaman, A. A. Nurul (eds.), *Sustainable Material for Biomedical Engineering Application*, https://doi.org/10.1007/978-981-99-2267-3_8

Abbreviations

3D	Three dimensional
ACI	Autologous chondrocyte implantation
ACs	Articular chondrocytes
ADSCs	Adipose-derived stem cells
AuCs	Auricular chondrocytes
BMSCs	Bone-Marrow stem cells
CPs	Chondroprogenitor cells
CTE	Cartilage Tissue Engineering
DMEM	Dulbecco's Modified Eagle Medium
ECM	Extracellular matrix
FCS	Fetal calf serum (FCS)
FDA	Food and Drug Administration
MSC	Mesenchymal stem cell
NCs	Nasal septum chondrocytes
PCL	Polycaprolactones
PEG	Polyethylene glycol
PGA	Polyglycolic acid
PLA	Polylactic acid
SDSCs	Synovium-derived stem cells
SVF	Stromal vascular fraction
TGF	Transforming growth factor
UCBSCs	Umbilical cord blood-derived stem cells

8.1 Introduction

Cartilage is a unique and smooth connective tissue. It consists of chondrocyte cells and extracellular matrix (ECM) that can be abundantly found in the ears, nasal bridge, trachea, joint cartilage of IVD and knees. Cartilage consists of elastic cartilage, fibrocartilage and hyaline, which can be differentiated by the composition of ECM. Hyaline cartilage at the joints functions as a joint cushion to absolve any force created during movements. Furthermore, it can form shape, support, allowing the well-ordered articulation of bones around each other and prolongation of bone growth (Alkaya et al. 2020). Cartilage damage can occur due to congenital problems like undeveloped ear (microtia), tissue that is removed by surgery due to tumour (nose and nasal septum) while joint cartilage damage can result from sport-related injuries. Some cartilage damage can be treated by surgical intervention, which is to replace it with prosthetic inert scaffold. However, synthetic material used has common problems such as poor mechanical property, inflexible and non-natural with limited bending abilities. Whereas for a total knee replacement, the transplanted foreign material requires replacement every 5–10 years. Besides that, there are possibilities of complications due to grafted tissue contraction such as degradation of tissue and donor source limitation and availability.

Tissue engineering is a multi-disciplinary approach to rebuild new tissues and replace damaged or diseased tissues through a combination of cells, growth factors and scaffolds. Similar to constructing a building, it requires meticulous selection of materials and is geometrically designed. Suitable planning on the construction process and sequence are important in which the cell culture must be carried out in the cGMP facility; by grafting it surgically as well as addressing the graft taking issues, etc. Previous studies have proven that the success rate of cartilage tissue engineering (CTE) is related to the efficacious use of chondrogenic factors that promote chondrogenesis, sufficient number of progenitors such as chondroblast, mesenchymal stem cells and the use of 3D biocompatible scaffold material for cell attachment and tissue remodelling (Green et al. 2015; Mostafa et al. 2019; Castro-Vinuelas et al. 2018; Rim et al. 2018).

Various cell sources which consist of stem cells, progenitor cells and adult tissue-derived cells have been studied to obtain a sufficient number of cells for tissue engineering. Of late, the mesenchymal stem cells (MSCs) have become one of the most popular cells due to their multipotent properties, including the potential to self-renew and differentiate into multiple-lineages such as cartilage, bone, muscle and fat. MSCs are abundantly found in bone marrow, adipose tissue, blood, amniotic fluid and connective soft tissues (Li et al. 2014). Studies have proved that MSCs are suitable for cartilage repair, where the formation of subchondral bone has better cell arrangement and integration than mature chondrocytes (Green et al. 2015; Pittenger et al. 2019; Gomez-Salazar et al. 2020). Therefore, cartilage remodelling through MSCs by seeding them in scaffolds to form cell aggregates in 3D has a better potential in the formation of new cartilage.

Scaffold materials can be found in natural, biological or synthetic origin and are classified as degradable or non-degradable. Scaffold is an artificial temporary platform that acts as the geometric design for cell attachment and growth. Most scaffolds can be degraded and replaced by the cells' extracellular matrices. Then, the newly formed tissue will go through a long remodelling process to optimise the tissue function towards maturation. The aim of scaffold usage in CTE is to facilitate the surgical process and make the surgery less invasive, simple, fast and cause less injury. Therefore, such hydrogel models are best to be practised with micro surgery and injection methods. In addition, treatment in an injectable form with the utilisation of a decellularised scaffold is an option for the aging population, particularly for the treatment of hip joints, ankle joints and intervertebral discs (IVDs). The combination of two or more matrix materials consisting of natural polymers and synthetic polymers is known as biocomposite. Biocomposites come in gel form, multi-layer 3D structure, beads, etc. Studies have shown that the biocomposites used in CTE would increase the rate of cell attachment and engagement, cell proliferation, extracellular matrix synthesis and remodelling of the tissue (Carletti et al. 2011). Biocomposites consist of various components such as collagen, proteins, polypeptides, glycosaminoglycan, amino acid and glycopeptides that are mixed in different ratios to promote cell growth and the synthesis of macromolecules in the new cartilage.

8.2 Cell Sources for Cartilage Tissue Engineering

8.2.1 Chondrocytes from Cartilage

The utilisation of cultured chondrocytes through the expansion in a chondrogenic medium has shown great potential for cartilage repair and regeneration. Cartilage is a smooth and elastic connective tissue, which consist of chondrocytes with low metabolic rate and extracellular matrix. Cartilage functions as a joint cushion that provides shape and support, well-ordered articulation of bones around each other and prolongation of bone growth. There are two main types of cartilage, namely hyaline cartilage (articular cartilage and nasal septum cartilage) and elastic cartilage (auricular cartilage). Hyaline cartilage is abundant in the joints between bones (Alkaya et al. 2020) and elastic cartilage makes up the outer ear (Goh et al. 2016). Cartilage damages that are caused by sports injury and osteoarthritis are the most common problems in orthopaedic clinics. Although cartilage donor tissue is extremely limited, there are areas of cartilage that can be harvested as donor tissue such as non-weight bearing articular cartilage at the knee area and surgically removed nasal septal cartilage. These cartilages do not play important roles in the overall body structure support; therefore, they are suitable to be harvested. These harvested cartilages can be digested using collagenase type II enzyme to release the chondrocytes as the seed cells for chondrocyte culture and subsequent CTE.

8.2.1.1 Articular Chondrocytes (ACs)

Articular chondrocytes (ACs) are the resident cell type in the articular cartilage of the bone as they are able to secrete ECM and maintain the cartilage. ACs can be isolated from the cartilage tissue harvested during the intraoperative procedure as well as from autopsies' tissue (Tew et al. 2008). ACs have the capability to proliferate in the monolayer culture. However, their expansion in the monolayer culture would lead to a gradual loss of their differentiated phenotype such as collagen type II in culture (Diaz-Roero et al. 2005; Jakob et al. 2001). Additionally, the limitation of redifferentiation potential of ACs after implantation (Passaretti et al. 2001; Pelttari et al. 2006; Dell'Accio et al. 2001) has been confirmed to affect new cartilage formation. However, the redifferentiation of ACs can be fixed when they are transferred into a 3-dimensional culture system and medium supplemented (Barbero et al. 2003) with specific regulatory molecules such as transforming growth factor beta 1 or 2 (TGF-β 1 or 2) and insulin growth factor 1 (IGF-1) (Yaeger et al. 1997). Therefore, medium formulation containing TGF-β family members, fibroblast growth factor 2 (FGF-2), IGF-1 and platelet-derived growth factor BB (PDGF-BB) are referred to as the standard chondrogenic medium in cartilage differentiation (Tew et al. 2008; Barbero et al. 2003).

8.2.1.2 Nasal Septum Chondrocytes (NCs)

Nasal septum cartilage is a hyaline cartilage which consists of matric extracellular that is produced by nasal septum chondrocytes (NCs) and exhibits a greater capability to restore the articular cartilage than the auricular chondrocytes themselves (Pelttari et al. 2017; Mumme et al. 2016). These chondrocytes are harvested from the nasoseptal cartilage biopsies through anaesthetic surgery and enzymatic digestion (Pelttari et al. 2017). NCs are shown to have higher proliferative capacity in the monolayer culture, (Rotter et al. 2002) similar to auricular chondrocytes. NCs exhibit chondrogenic genes' expression like collagen type II, but the expression would decrease in serial expansion (Homicz et al. 2002). They also have great potential to redifferentiate into chondrocytes and form hyaline cartilage in the 3-dimensional culture system with the addition of chondrogenic medium containing TGF-β and IGF-1 (Homicz et al. 2002). The advantages of NCs for clinical application are easy to harvest, high proliferation capacity, high potential in cartilage formation and less dependent on the age of the donor than ACs. Moreover, they also have the ability for self-renewal and high plasticity after implantation, (Pelttari et al. 2014) making them potential candidates for cartilage regeneration.

8.2.1.3 Auricular Chondrocytes (AuCs)

In general, ear or auricular cartilage is surrounded by the perichondrium layer cells with high proliferation rate, multi-lineages differentiation ability and high expression of CD44 and CD90 stem cell markers (Ishak et al. 2011). However, past studies conducted on the chondrocytes isolated from the auricular cartilage discovered that the perichondrium layer did not go through a specific separation from the chondrium layer and had cells characterised by cartilage reporter cells. Previous studies also had successfully proven that ear tissue from microtia patients were able to grow and reproduce through multi-layer development methods (Goh et al. 2016; Oseni et al. 2013). However, cartilage-like tissue obtained in the form of fibre was deemed suitable for autologous cartilage-based applications (ACI) but regarded as unsuitable for ear-shaped cartilage regeneration (Zhang et al. 2014). In addition, cartilage regeneration using only auricular chondrocytes were found to be insufficient due to the decreasing chondrogenic capacity of auricular chondrocytes once they reached the third generation of cultivation (Goh et al. 2016; Hong and Reddi 2013). Therefore, the stem cells used in cartilage tissue regeneration to replace or use with the auricular chondrocytes have attracted the attention of researchers due to their ability to renew themselves, good differentiation capacity and capability to overcome the problem of losing chondrogenic characteristics (Che Omar et al. 2018; Shi et al. 2017; Chiu et al. 2016).

8.2.2 Progenitors

8.2.2.1 Chondroprogenitor Cells (CPs)

Chondroprogenitors cells (CPs) are progenitor cells that can differentiate into chondrocytes and can be isolated from various types of tissues including cartilage, adipose, meniscus and synovium (Jayasuriya and Chen 2015; Dowthwaite et al. 2004). Chondroprogenitors can be isolated using an enzyme digestion method such as collagen type I and collagen type II, then purified using cell sorter to isolate those with stem cell surface markers (CD105, CD9, CD90, CD166 and CD146) (Su et al. 2015). CPs have the advantages in terms of clinical applications whereby they can differentiate into chondrogenic cells and its expansion does not alter its differentiation capacity when cultured in medium with 40% of fetal calf serum (FCS) and TGF-β 1 (Melero-Martin et al. 2006). However, the lack of autologous sources have limited their use in the clinical application (Jayasuriya and Chen 2015).

8.2.3 Mesenchymal Stem Cells (MSCs)

Mesenchymal stem cells (MSCs) are adult stem cells with multipotency and they are regarded as a suitable cell source for cartilage tissue engineering. MSCs can be derived from bone marrow, adipose tissue, umbilical cord blood as well as connective soft tissues (Gomez-Salazar et al. 2020; Carletti et al. 2011). MSCs have demonstrated self-renewal characteristics and are able to differentiate into specialised cells such as chondrocytes, osteoblasts and adipocytes under appropriate conditions (Dominici et al. 2006). MSCs can adhere to plastic dishes and express a high level of stem cells surface markers; CD73, CD90 and CD105 (Hamid et al. 2012). MSCs also have high regenerative potential, unlimited division capacity, self-renewal ability, easy access and hypo-immunogenicity. All these criteria have favoured their utilisation as progenitor cells in cartilage replacement. In lieu of this, the chondrogenic potential of MSCs has been applied in clinical applications for treatment of cartilage defects.

8.2.3.1 Bone-Marrow Stem Cells (BMSCs)

BMSCs are adult MSCs that are derived from the bone marrow and they are widely used in cartilage engineering. It can be harvested and used autologously as well as allogeneically. Studies have shown that approximately 100–150 million BMSC cells can be produced within 3 weeks from 25 ml of bone marrow aspirate (Zhao et al. 2017). In addition, the FGF2 supplementation in BMSCs culture has successfully produced a high number of cells with a lifespan of more than 70 doublings and its differentiation ability can increase up to 50 doublings (Tew et al. 2008). BMSCs also have shown greater differentiation in chondrogenic medium through pellet cultures which resemble condensation of embryonic mesenchymal pre-cartilage stage. The addition of bioactive signals such as TGF-β, dexamethasone, ascorbic acid-2-phosphate, proline and ITS premix (insulin, transferrin, selenium, linoleic acid) in the chondrogenic medium also promote the formation of hyaline tissue after 14 days (Mackay et al. 1998). However, the differentiation of BMSCs into chondrocytes is believed to exhibit hypertrophic characteristics, leading to the establishment of calcification (Bian et al. 2011a). Therefore, the use of co-culture methods between BMSCs and chondrocytes have been proposed to produce a better cartilage matrix synthesis (Bian et al. 2011a; Alves da Silva et al. 2015; Giovannini et al. 2010).

However, the utilisation of BMSCs for cartilage repair can overcome the issue of chondrocyte shortage, donor site morbidity and chondrocyte dedifferentiation.

8.2.3.2 Adipose-Derived Stem Cells (ADSCs)

Adipose tissue or stromal vascular fraction adipose tissue (SVF) is the second source of MSCs that can be harvested with a high abundance of cells and clonogenic potential (Halvorsen et al. 2000; Vishnubalaji et al. 2012). Adipose tissue is obtained through a liposuction procedure and it is known as lipoaspirate. Once digested with collagenase, the stromal vascular fraction (SVF) adipose tissue is believed to contain ADSCs, fibroblasts, white and red blood cells, endothelial cells, etc. (Pak et al. 2016). ADSCs have shown chondrogenic differentiation when they are cultured in Dulbecco's Modified Eagle Medium (DMEM) containing TGF-β3, albumin, dexamethasone, ascorbic acid, transferrin and insulin (Mumme et al. 2016; Stromps et al. 2014; Baptista et al. 2013) with 5% oxygen deprivation (Mumme et al. 2016; Merceron et al. 2010). ADSCs also have exhibited greater growth rate in monolayer culture. However, it is necessary to culture ADSCs in the pellet form or 3-dimensional scaffolds to overcome the problems of growth inhibition and promote chondrogenic differentiation that occurs in ADSCs (Xu et al. 2008). The co-culture between ADSCs and chondrocytes has demonstrated better development and differentiation (Goh et al. 2016). In addition, the application of mechanical forces onto the cell-loaded scaffold also has shown an increase of chondrogenesis during in vitro culture (González-Cruz and Darling 2013).

8.2.3.3 Umbilical Cord Blood-Derived Stem Cells (UCBSCs)

UCBSCs are allogeneic MSCs that are derived from the umbilical cord blood and they are widely commercialised due to their low immunogenicity after transplantation (Erices et al. 2000; Song et al. 2020). They have higher proliferation capacity, but lower isolation rate of the stem cell surface markers CD90 and CD105 as compared to BMSCs and ADSCs (Kern et al. 2006). In addition, they also have intrinsic chondrogenic potential and can be used to treat cartilage damage when they are seeded on the collagen sponge scaffolds under normoxic conditions in the presence of BMP-2 and TGF-β1 (Gómez-Leduc et al. 2016).

8.2.3.4 Synovium-Derived Stem Cells (SDSCs)

Synovium-derived stem cells (SDSCs) can be harvested from the synovium that are found in the knee joint (De Bari et al. 2001). Synovium is a thin membrane lining of the synovial joint cavity that produces fluid for joint lubrication and cartilage nutrition. SDSCs are obtained through minimally invasive and routine arthroscopic procedures (Garcia et al. 2016; Fernandes et al. 2018) with debris removal methods in arthroscopic surgery or infrapatellar fat pads operation. The main characterisation of SDSCs is by using the surface markers hyaluronic acid receptor (CD44) and stem cell surface markers (CD105, CD73, CD166, CD90, CD106, STRO-1) (Mochizuki et al. 2006; Li et al. 2020). Moreover, SDSCs have shown greater capacity of proliferation and chondrogenic potential than BMSCs and ADSCs (Kubosch et al. 2018) as the former in culture pellet forms have increased in size after 21 days

(Shirasawa et al. 2006). In addition, the chondrogenic differentiation ability of SDSCs is not tied to the age of the donor or site of collection (González-Cruz and Darling 2013; Kern et al. 2006). Studies have shown that the chondrogenesis of SDSCs can be increased by the addition of promoting factors such as bone morphogenetic protein (BMP-2), TGF-β and dexamethasone (Shirasawa et al. 2006). SDSCs also have exhibited an imprinted code for joint morphogenesis, whereby an expression of bone morphogenetic protein 7 (BMP-7) was detected in the skeletal muscle and ectopic formation of joint-like structures (Roelofs et al. 2017).

8.3 Type of Scaffolds for Cartilage Tissue Engineering

Scaffold is an artificial temporary platform that acts as the geometrical design structure for the cells to locate and attach. It should be degraded with time so that the cells will be able to replace it with their own extracellular matrices. Furthermore, it must allow tissue remodelling over time for tissue maturation and increase in function. Scaffolds can be divided into natural polymers, synthetic polymers or a combination of both as biocomposite scaffolds.

8.3.1 Natural Polymers

Natural polymers have been used extensively in cartilage tissue engineering, including collagen, fibrin, alginate, chitosan, polypeptide, glycopeptide, etc. These polymers are used widely due to their biocompatibility and biodegradability properties, good cellular interactions as they encourage cell adhesion as well as activate the chondrogenic pathways and minimise immune responses (Alkaya et al. 2020). Studies have shown that natural polymers are able to reduce the immune response when they are implanted in vivo (Mohammadinejad et al. 2020) as they have less immunogenic properties than synthetic polymers.

8.3.1.1 Collagen

Collagen is a protein with triple-helix structure. It can form a reversible gel and is able to assist in cell binding, proliferation and differentiation. Collagen has a rapid biodegradation rate, easy to absorb without causing any inflammation and has good tissue compatibility. Some of the advantages of collagen as a biocomposite in cartilage tissue engineering include its mechanical strength and swelling tendency (Turk et al. 2018). Collagen can be obtained from aquatic sources such as fish waste (bones, scales, skin), squid, jellyfish and bovine sources (Ge et al. 2012; Mahboob 2015; Silva et al. 2014). Studies have reported that the combination of collagen scaffold with stem cells or chondrocytes has exhibited good cartilage growth by promoting cell proliferation and mechanical strength (Pulkkinen et al. 2010), increasing the expression of cartilaginous matrix (Stoddart et al. 2009) and the formation of cartilage-like cells with lacuna structure (Chen et al. 2011). These findings have demonstrated that pure collagen scaffold or a combination of collagen with other types of material are able to create a biocompatible scaffold that is effective for cell seeding, promoting cartilage repair and regeneration (Wang et al. 2017a).

8.3.1.2 Fibrin

Fibrin is a blood clotting protein with biocompatibility and biodegradability properties. Fibrin products are recognised by the Food and Drug Administration (FDA) as medical products (Duarte Campos et al. 2012; Liu and Yu 2016). Recent studies have shown that fibrin-based scaffolds can produce high cell seeding, encourage proliferation, migration, chondrogenic differentiation, cartilage ECM secretion as well as cartilage-like tissue formation in vivo (Almeida et al. 2016; Park and Woo 2018). Moreover, fibrin scaffold also acts as a healing agent where it provides a surface for proliferation, cellular attachment, 3D fibrous support and a nano-textured surface for cell signalling and interaction (Rajangam and An 2013). Fibrin scaffolds are the pillars of ECM development in tissue engineering which support connective tissue development such as cartilage, tendons, bones and nerves formation.

8.3.1.3 Alginate

Alginate is a polysaccharide which can be found abundantly in the cell walls of brown algae. This scaffold supports the growth, migration and proliferation of cells through certain surface receptors (Duarte Campos et al. 2012). Alginate has the characteristics of cell friendly gelatin which is biocompatible, has low immunogenicity, is chondro-inductive and can be efficiently hydrated to expand. The Food and Drug Administration (FDA) has approved alginate as an artificial flavour and for utilisation in wound treatment as they are affordable, non-toxic and easy to form 3-dimensional scaffolds. Previous studies have shown that a variety of 3D alginate scaffolds that are seeded with stem cells/chondrocytes and cultured in vitro/in vivo have exhibited a development of cartilage-like tissue with characteristics of chondrocyte phenotype (Wang et al. 2012; Zeng et al. 2014). This is demonstrated by the occurrence of GAG/protein-protein electrostatic interactions (Arlov et al. 2014), chondrogenic gene expression (Bian et al. 2011b; Kerschenmeyer et al. 2017) and stable cartilage-like tissue formation (Ma et al. 2012).

8.3.1.4 Chitosan

Chitosan is a linear polysaccharide that consists of glucosamine which is non-toxic, biocompatible and biodegradable (Singh et al. 2017). Chitosan can be found in semi-crystalline polymers and can be extracted from arthropod exoskeletons. Chitosan is evidently used as an injectable and implantable periodontal and orthopaedic device, drug delivery carrier, healing agent and bioscaffold. Chitosan also can be incorporated with growth factors and ECM for cartilage regeneration. Studies have demonstrated that chitosan scaffolds or biocomposite of chitosan are able to repair damaged cartilage back into the hyaline cartilage (Gohil 2017). Chitosan biocomposite also has shown increased cell proliferation and collagen synthesis as well as vascularisation in the repaired tissue. Chitosan biocomposite has seen potential in the treatment of cartilage damage and cartilage regeneration due to its biodegradability and biocompatibility (Mallick et al. 2018). Therefore, chitosan has been utilised as a growth factor delivery system to promote cell proliferation and tissue regeneration (Tao et al. 2020).

8.3.2 Synthetic Polymers

The physical and chemical properties of synthetic polymers can be modified to exert suitable mechanical and degradation characteristics. The advantages of synthetic polymers include the flexibility in size and shape which can be predicted, controlled and processed to achieve desired chemical properties in tissue engineering applications. They also exhibit a slower degradation rate due to high mechanical strength. There are various types of synthetic polymers that have been utilised in cartilage tissue engineering, like polylactic acid (PLA), polycaprolactones (PCL), polyglycolic acid (PGA) and polyethylene glycol (PEG) due to their ability to produce cartilage-like cells in vitro and in vivo.

8.3.2.1 Polylactic Acid (PLA)

Polylactic acid (PLA) is a linear polyester with crystallinity and hydrophilicity (Ahmed and Hincke 2014). It also has a robust structure, good biocompatibility, slow degradation rate and low toxicity (Lopes et al. 2012; Li et al. 2015). PLA has been approved by the FDA and it is widely used in medicine, especially in orthopaedic treatment (Tyler et al. 2016). Studies have shown that PLA scaffolds or PLA-based composites that are seeded with cells/stem cells have exhibited good cell proliferation, viability, differentiation, particularly in hyaline cartilage tissue formation (Smieszek et al. 2019). According to Ahmed and Hincke, PLA scaffold that was seeded with MSCs exhibited successful proliferation with robust structure in the rabbit model as it formed hyaline-like cartilage tissue (Ahmed and Hincke 2014). Meanwhile, PLA-based composites with nanohydroxyapatite (nHAp) showed proliferation and differentiation of the progenitor cells (Smieszek et al. 2019). In addition, PLA can be moulded into a 3D scaffold material due to their low viscosity and good thermoplastic properties (Coppola et al. 2018; Dizon et al. 2018). The combination of *Pennisetum purpureum* with PLA matrices as the scaffold also showed high biodegradability and increased mechanical strength. PLA polymers are also used in medicine such as treatment for cartilage, bone and peripheral nerves (Suzuki et al. 2016).

8.3.2.2 Polyglycolic Acid (PGA)

Polyglycolic Acid (PGA) is a linear crystalline hydrophobic polyester that has been used in tissue engineering such as cartilage, tendons, skin, peripheral nerves and blood vessels (Nakao et al. 2017). Studies have shown that PGA scaffolds that are seeded with MSCs have successfully maintained the chondrogenic phenotype with hyaline cartilage characteristics and formed subchondral bone. In addition, PGA composites in sandwich structure have promoted auricular cartilage repair as they are able to form 3-dimension auricular cartilage regeneration and promote angiogenesis around the implant (Birru et al. 2018). However, PGA has certain disadvantages in which it produces acidic degradation products that would trigger counter-reactions and reduce its biocompatibility. Therefore, the establishment of PLA/PGA composites is deemed as a more effective technique for new cartilage regeneration (Venkatesan et al. 2020).

8.3.2.3 Polycaprolactones (PCL)

Polycaprolactones (PCL) is a combination of semi crystalline and aliphatic polymers. It is biocompatible, biodegradable, able to exert elasticity, hydrophobic and tough in terms of mechanical strength. PCL has received FDA approval due to its low immunogenic property and biodegradability. It can also mimic the viscoelastic and anisotropic biomechanical properties of articular cartilage (Venkatesan et al. 2020). Additionally, PCL by-products are harmless and can be metabolised through the tricarboxylic acid cycle (Moura et al. 2020). However, the hydrophobic nature of PCL has been found to inhibit cell adhesion and affect its use in cartilage tissue engineering. Therefore, PCL-based composites in the nanofibrous form and 3D structures that are seeded with MSCs have been tested for their usage in tissue regeneration. The findings have demonstrated an increase of chondrogenic gene expression, enhancement of MSC chondrogenic differentiation and favour the mechanical properties without causing inflammatory cytokines release (Ousema et al. 2012). Moreover, PCL also has exhibited successful repair of full-thickness cartilage defects in animal models (Shafiee et al. 2011).

8.3.2.4 Polyethylene Glycol (PEG)

Polyethylene glycol (PEG) is a polymer that can dissolve in water and cannot be identified by the immune system. It is biocompatible and possesses optimal physicochemical and biological characteristics. PEG exists in various structures including linear, 3-arm, 4-arm, dendritic-like and comb-like (Karim et al. 2016; Ding and Li 2017). PEG is hydrolytically degraded and minimally metabolised in the body by the liver or kidneys. Studies have shown that PEG-based bioscaffolds such as PEG hydrogels and PEG films that are seeded with chondrocytes or stem cells in animal models have demonstrated an increase in the mechanical strength and exhibited Young's modulus compressive feature that is similar to the cartilage-like tissue (Duarte Campos et al. 2012). Furthermore, these cells have chondrocyte-like characteristics with chondrocyte-associated marker expressions such as type I and II collagen, aggrecan and glycosaminoglycan (Scholz et al. 2010; Neumann et al. 2016; Wang et al. 2017b). Therefore, there is a possibility that a more effective cartilage tissue formation can be achieved using PEG bioscaffold (Fu et al. 2016).

8.3.3 Biocomposite Polymers

Biocomposites polymers are a combination of two or more matrix materials that consist of natural polymers and synthetic polymers or any other suitable combination to achieve and promote better tissue engineering. These biocomposite polymers can be fabricated in a gel form or multilayer form (sandwich structures).

8.3.3.1 Gel Form Biocomposite

Biocomposite in the gel form is a mixture of polymers in semi-solid form. It has sufficient fluidic nature that can encourage homogeneity and flexibility from liquid to semi-solid form with the change of temperature (Curvello et al. 2019). Hydrogel for example, can be developed from natural polymers, synthetic polymers or a combination of both polymers. The characteristics of hydrogel are similar to natural ECMs and they can control the morphology of the cell, cell growth and differentiation at the affected area. Previous studies have summarised that a single natural hydrogel like alginate is unsuitable for cartilage treatment due to its short-term stability and low cellular adhesiveness (Müller et al. 2017). Not only that, it has also exhibited low mechanical properties and must be improved with porous scaffolds that have good 3D structure which can be slowly degraded or bio-inert, totally. Therefore, synthetic polymers hydrogel like polyethylene glycol (PEG), etc. with highly tunable biocompatibility, biodegradability, biochemical characteristics and mechanical properties are deemed more effective than natural polymer hydrogel (Zhang et al. 2015). Besides that, the combination of hydrogel has provided additional possibilities for cartilage regeneration and clinical translation. Studies have shown that biocomposite hydrogel with synthetic polymer that consists of bacterial cellulose (BC) nanofiber network with PVA–poly (2-acrylamido-2-methyl-1-propanesulfonic acid sodium salt) (PAMPS) double-network hydrogel has exhibited good mechanical strengths (Yang et al. 2020). Another example of biocomposite hydrogel is gelatin/polycaprolactone–polyethylene glycol (Gel/PCEC) also has shown differentiation of ADSCs and high chondrogenesis genes expression such as collagen type II and aggrecan (Asadi et al. 2019).

8.3.3.2 Multilayer Form Biocomposite

Biocomposite in a layer form is known as sandwich structures that are formed from the outside to inside orientation (Fig. 8.1). Previous studies had exhibited improvement in ear and trachea reconstruction by sandwich structural biocomposite using cultured chondrocytes and non-degradable polymers (Lee et al. 2020). However, the grafted natural cartilage was found to shrink in size with the occurrence of stenosis in the transplanted trachea (Berg et al. 2014). In addition, a non-degradable polymer is also a non-union at the edge of the surgical site. This is due to the differences of physical and biochemical properties between the natural tissue and synthetic polymer (Lee et al. 2014). Therefore, the use of an inert core and a stable non-degradable material are crucial for structural support.

8.3.4 Medpor

Medpor is an inert non-degradable polymer with zero elastic properties. It can be composited with human chondrocytes and fibrin for ear cartilage tissue engineering

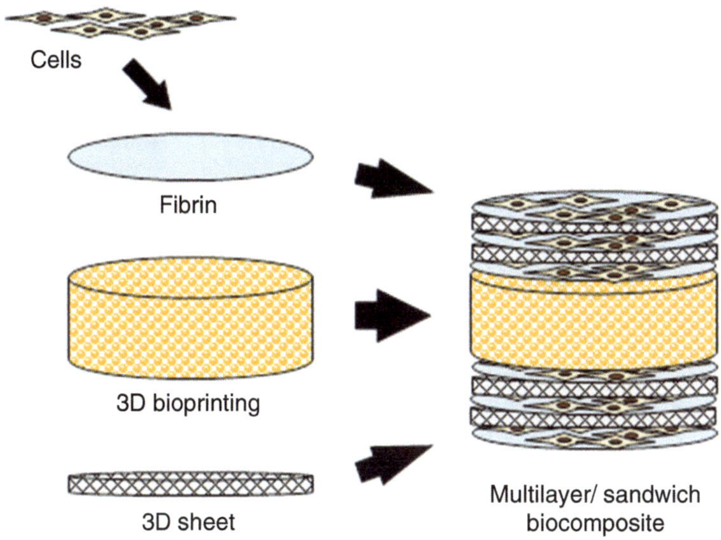

Fig. 8.1 Schematic illustration of multilayer/sandwich biocomposite

(Fig. 8.2). Medpor is coated with cultured chondrocytes that are mixed with stem cells to increase the cell number for faster tissue engraftment. Due to this, chondrocyte seeding is important to form less fibrotic tissue. With seeded chondrocytes in Medpor that are coated with fibrin, it can help the formation of elastic cartilage and fibrous connective tissue. This biocomposite also can sustain better and avoid Medpor from exposing out of the skin. Recently, the use of PLGA-Medpor biocomposite in ear reconstruction has shown an increase in the production of fibrous tissue than cartilage-like tissue due to low and inhomogeneous cell seeding throughout the scaffold (Griffon et al. 2005). To overcome these issues, the bioreactor method is used post static cells seeding in Medpor biocomposite to promote cells proliferation and attachment.

8.4 Conclusion

Till date, treating cartilage defects remains a big challenge due to the poor healing capability of the cartilage. Cartilage tissue engineering (CTE) appears as a promising technique in cartilage treatment. The combination of various cells, particularly the stem cells, together with polymer scaffolds promises a better outcome for cartilage repair. Studies on the suitability of polymers that consist of natural polymers, synthetic polymers and biocomposite that are seeded with stem cells have promoted better cell growth, differentiation and cell attachment as well as providing mechanical strength for the new cartilage formation in vivo. However, due to the lack of clinical evidence, the method is yet to be utilised as a standard treatment in surgery. Therefore, modification of scaffolds with biocomposite concept could potentially

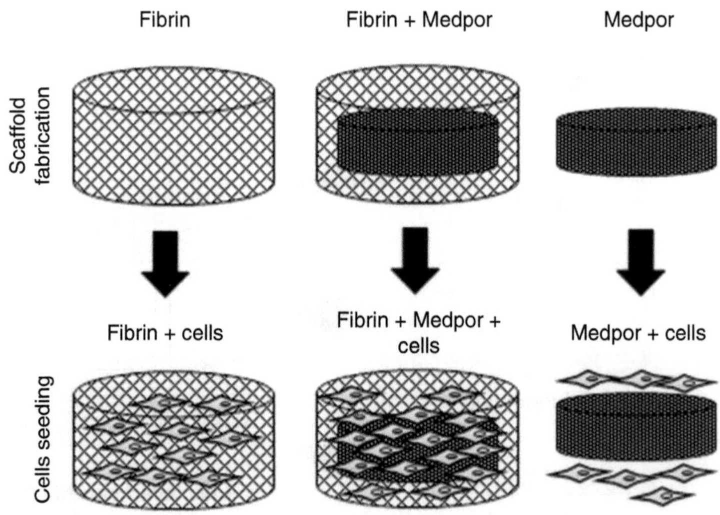

Fig. 8.2 Schematic illustration of Medpor biocomposite

provide a better environment for the new cartilage regeneration. Besides that, more studies in exploring the use of acellular technology such as secretome and exosome should be considered. The combination of biocomposite scaffold and high beneficial protein from cells by-product could be a new challenge and promising in the future option for cartilage repairs.

Acknowledgments This project was supported by Universiti Kebangsaan Malaysia (UKM), Malaysia.

Funding and Conflicts of Interest The authors declare no conflicts of interest.

References

Ahmed TA, Hincke MT (2014) Mesenchymal stem cell-based tissue engineering strategies for repair of articular cartilage. Histol Histopathol 29:669–689. https://doi.org/10.14670/HH-29.669

Alkaya D, Gurcan C, Kilic P, Yilmazer A, Gurman G et al (2020) Where is human-based cellular pharmaceutical R&D taking us in cartilage regeneration. Biotechnology 10:161. https://doi.org/10.1007/s13205-020-2134-2135

Almeida HV, Eswaramoorthy R, Cunniffe G, Kelly DJ et al (2016) Fibrin hydrogels functionalized with cartilage extracellular matrix and incorporating freshly isolated stromal cells as an injectable for cartilage regeneration. Acta Biomater 36:55–62

Alves da Silva ML, Costa-Pinto AR, Martins A, Correlo VM, Sol P, Bhattacharya M, Faria S, Reis RL, Neves NM et al (2015) Conditioned medium as a strategy for human stem cells chondrogenic differentiation. J Tissue Eng Regen Med 9:714

Arlov O, Aachmann FL, Sundan A, Espevik T, Skjak-Braek G et al (2014) Heparin-like properties of sulfated alginates with defined sequences and sulfation degrees. Biomacromolecules 15: 2744–2750. https://doi.org/10.1021/bm500602w

Asadi N, Alizadeh E, Rahmani A, Bakhshayesh D, Mostafavi E, Akbarzadeh A, Davaran S et al (2019) Fabrication and in vitro evaluation of nanocomposite hydrogel scaffolds based on gelatin/PCL-PEG-PCL for cartilage tissue engineering. ACS Omega 4:449–457

Baptista LS, Silva KR, Pedrosa CSG, Amaral RJFC, Belizário JV, Borojevic R, Granjeiro JM et al (2013) Bioengineered cartilage in a scaffold-free method by human cartilage-derived progenitor cells: a comparison with human adipose-derived mesenchymal stromal cells, vol 37. Artif Organs, p 1068

Barbero A, Ploegert S, Heberer M, Martin I et al (2003) Plasticity of clonal populations of dedifferentiated adult human articular chondrocytes. Arthritis Rheum 48:1315

Berg M, Ejnell H, Kovacs A et al (2014) Replacement of a tracheal stenosis with a tissue-engineered human trachea using autologous stem cells: a case report. Tissue Eng A 20:389–397

Bian L, Zhai DY, Mauck R, Burdick JA (2011a) Coculture of human mesenchymal stem cells and articular chondrocytes reduces hypertrophy and enhances functional properties of engineered cartilage. Tissue Eng A 17:1137–1145

Bian L, Zhai DY, Tous E, Rai R, Mauck RL, Burdick JA et al (2011b) Enhanced MSC chondrogenesis following delivery of TGF-beta3 from alginate microspheres within hyaluronic acid hydrogels in vitro and in vivo. Biomaterials 32:6425–6434. https://doi.org/10.1016/j.biomaterials.2011.05.033

Birru B, Mekala NK, Parcha SR (2018) Improved osteogenic differentiation of umbilical cord blood MSCs using custom made perfusion bioreactor. Biomed J 41:290–297. https://doi.org/10.1016/j.bj.2018.07.002

Carletti E, Motta A, Migliaresi C (2011) Scaffolds for tissue engineering and 3D cell culture. Methods Mol Biol 695:17–39

Castro-Vinuelas R, Sanjurjo-Rodriguez C, Pineiro-Ramil M, Hermida-Gomez T, Fuentes-Boquete IM, de Toro-Santos FJ et al (2018) Induced pluripotent stem cells for cartilage repair: current status and future perspectives. Eur Cell Mater 36:96–109. https://doi.org/10.22203/eCM.v036a08

Che Omar SN, Goh BS, Ubaidah MA, Khairoji KA, Sulaiman S, Saim L, Chua KH et al (2018) Transforming growth factor beta 3 induced human adipose-derived stem cells for auricular chondrogensis. Sains Malaysiana 47(10):2349–2358. https://doi.org/10.17576/jsm-2018-4710-11

Chen WC, Yao CL, Wei YH, Chu IM et al (2011) Evaluating osteochondral defect repair potential of autologous rabbit bone marrow cells on type II collagen scaffold. Cytotechnology 63:13–23

Chiu LLY, Bianco J, Giardini-rosa R, Collavino K, Waldman SD et al (2016) Direct and indirect co-culture of bone marrow stem cells and adipose-derived stem cells with chondrocytes in 3D scaffold-free culture. J Regen Med Tissue Eng 5:1–9

Coppola B, Cappetti N, Di Maio L, Scarfato P, Incarnato L et al (2018) 3D printing of PLA/clay nanocomposites: influence of printing temperature on printed samples properties. Materials (Basel) 11:1947. https://doi.org/10.3390/ma11101947

Curvello R, Raghuwanshi VS, Garnier G (2019) Engineering nanocellulose hydrogels for biomedical applications. Adv Colloid Interf Sci 267:47–61. https://doi.org/10.1016/j.cis.2019.03.002

De Bari C, Dell'Accio F, Tylzanowski P, Luyten FP et al (2001) Multipotent mesenchymal stem cells from adult human synovial membrane. Arthritis Rheum 3:P39

Dell'Accio F, De Bari C, Luyten FP (2001) Molecular markers predictive of the capacity of expanded human articular chondrocytes to form stable cartilage in vivo. Arthritis Rheum 44: 1608

Diaz-Roero J, Gaillard JP, Grogan SP, Nesic D, Trub T, Mainil-Varlet P et al (2005) Immuno phenotypic analysis of human articular chondrocytes: changes in surface markers associated with cell expansion in monolayer culture. J Cell Physiol 202:731

Ding C, Li Z (2017) A review of drug release mechanisms from nanocarrier systems. Mater Sci Eng C Mater Biol Appl 76:1440–1453. https://doi.org/10.1016/j.msec.2017.03.130

Dizon JRC, Espera AH, Chen QY, Advincula RC et al (2018) Mechanical characterization of 3D-printed polymers. Addit Manuf 20:4467. https://doi.org/10.1016/j.addma.2017.12.002

Dominici M, Le Blanc K, Mueller I, Slaper-Cortenbach I, Marini FC, Krause DS, Deans RJ, Keating A, Prockop DJ, Horwitz EM et al (2006) Minimal criteria for defining multipotent mesenchymal stromal cells. The International Society for Cellular Therapy position statement. Cytotherapy 8(4):315–317

Dowthwaite GP, Bishop JC, Redman SN, Khan IM, Rooney P, Evans DJR, Haughton L, Bayram Z, Boyer S, Thomson B et al (2004) The surface of articular cartilage contains a progenitor cell population. J Cell Sci 117:889

Duarte Campos DF, Drescher W, Rath B, Tingart M, Fischer H et al (2012) Supporting biomaterials for articular cartilage repair. Cartilage 3:205–221

Erices A, Conget P, Minguell JJ (2000) Mesenchymal progenitor cells in human umbilical cord blood. Br J Haematol 109:235

Fernandes TL, Kimura HA, Pinheiro CCG, Shimomura K, Nakamura N, Ferreira JR, Gomoll AH, Hernandez AJ, Bueno DF et al (2018) Human synovial mesenchymal stem cells good manufacturing practices for articular cartilage regeneration. Tissue Eng Part C Methods 24:709

Fu N, Liao J, Lin S, Sun K, Tian T, Zhu B et al (2016) PCL-PEG-PCL film promotes cartilage regeneration in vivo. Cell Prolif 49:729–739. https://doi.org/10.1111/cpr.12295

Garcia J, Wright K, Roberts S, Kuiper JH, Mangham C, Richardson J, Mennan C et al (2016) Characterisation of synovial fluid and infrapatellar fat pad derived mesenchymal stromal cells: the influence of tissue source and inflammatory stimulus. Sci Rep 6:24295

Ge Z, Li C, Heng BC, Cao G, Yang Z et al (2012) Functional biomaterials for cartilage regeneration. J Biomed Mater Res A 100:2526–2536. https://doi.org/10.1002/jbm.a.34147

Giovannini S, Diaz-Romero J, Aigner T, Heini P, Mainil-Varlet P, Nesic D et al (2010) Micromass co-culture of human articular chondrocytes and human bone marrow mesenchymal stem cells to investigate stable neocartilage tissue formation in vitro. Eur Cell Mater 20:245

Goh BS, Che Omar SN, Ubaidah MA, Saim L, Sulaiman S, Chua KH et al (2016) Chondrogenesis of human adipose derived stem cells for future microtia repair using co-culture technique. Acta Otolaryngol 137:432. https://doi.org/10.1080/00016489.2016.1257151

Gohil SV (2017) 7-chitosan-based scaffolds for growth factor delivery A2. In: Bumgardner JD (ed) Chitosin based biomaterials, vol 2. Woodhead Publishing, pp 175–207

Gómez-Leduc T, Hervieu M, Legendre F, Bouyoucef M, GruchybN PL, De Vienne C, Herlicoviez M, Demoor M, Galéra P et al (2016) Chondrogenic commitment of human umbilical cord blood-derived mesenchymal stem cells in collagen matrices for cartilage engineering, vol 6. Sci Rep, p 32786

Gomez-Salazar M, Gonzalez-Galofre ZN, Casamitjana J, Crisan M, James AW, Peault B et al (2020) Five decades later, are mesenchymal stem cells still relevant? Front Bioeng Biotechnol 8: 148. https://doi.org/10.3389/fbioe.2020.00148

González-Cruz RD, Darling EM (2013) Adipose-derived stem cell fate is predicted by cellular mechanical properties, vol 2. Adipocyte, p 87

Green JD, Tollemar V, Dougherty M, Yan Z, Yin L, Ye J et al (2015) Multifaceted signaling regulators of chondrogenesis: implications in cartilage regeneration and tissue engineering. Genes Dis 2:307–327. https://doi.org/10.1016/j.gendis.2015.09.003

Griffon DJ, Sedighi MR, Sendemir-Urkmez A, Stewart AA, Jamison R et al (2005) Evaluation of vacuum and dynamic cell seeding of polyglycolic acid and chitosan scaffolds for cartilage engineering. Am J Vet Res 66:599–605

Halvorsen YC, Wilkison WO, Gimble JM (2000) Adipose-derived stromal cells—their utility and potential in bone formation. Int J Obes 24:S41

Hamid AA, Ruszymah BHI, Aminuddin S, Somasumdaram S, Chua KH et al (2012) Characterization of human adipose-derived stem cells and expression of chondrogenic genes during induction of cartilage differentiation. Clinics 67:99–106

Homicz MR, Schumacher BL, Sah RL, Watson D et al (2002) Effects of serial expansion of septal chondrocytes on tissue-engineered neocartilage composition, vol 127. Otolaryngology Head Neck Surg, p 398

Hong E, Reddi AH (2013) Dedifferentiation and re- differentiation of articular chondrocytes from surface and middle zones: changes in microRNAs-221/-222, -140, and –143/145 expression. Tissue Eng A 19:1015

Ishak MF, Chua KH, Asma A, Saim L, Aminuddin BS, Ruszymah BHI et al (2011) Stem cell genes are poorly expressed in chondrocytes from microtic cartilage. Int J Pediatr Otorhinolaryngol 75: 835–840

Jakob M, Démarteau O, Schäfer D, Hintermann B, Dick W, Heberer M, Martin I et al (2001) Specific growth factors during the expansion and redifferentiation of adult human articular chondrocytes enhance chondrogenesis and cartilaginous tissue formation in vitro. J Cell Biochem 81:368

Jayasuriya CT, Chen Q (2015) Potential benefits and limitations of utilizing chondroprogenitors in cell-based cartilage therapy. Connect Tissue Res 56:265

Karim AA, Dou Q, Li Z, Loh XJ et al (2016) Emerging supramolecular therapeutic carriers based on host-guest interactions. Chem Asian J 11:13001321. https://doi.org/10.1002/asia.201501434

Kern S, Eichler H, Stoeve J, Klüter H, Bieback K et al (2006) Comparative analysis of mesenchymal stem cells from bone marrow, umbilical cord blood, or adiposetissue. Stem Cells 24:1294

Kerschenmeyer A, Arlov O, Malheiro V, Steinwachs M, Rottmar M, Maniura Weber K et al (2017) Anti-oxidant and immune-modulatory properties of sulfated alginate derivatives on human chondrocytes and macrophages. Biomater Sci 5:1756–1765. https://doi.org/10.1039/c7bm00341b

Kubosch EJ, Lang G, Furst D, Kubosch D, Izadpanah K, Rolauffs B, Sudkamp NP, Schmal H et al (2018) The potential for synovium-derived stem cells in cartilage repair. Curr Stem Cell Res Ther 13:174

Lee JS, Hong JM, Jung JW, Shim JH, Oh JH, Cho DW et al (2014) 3D printing of composite tissue with complex shape applied to ear regeneration. Biofabrication 6:2

Lee JY, Park JH, Ahn MJ, Kim SSW, Cho DW et al (2020) Long-term study on off-the-shelf tracheal graft: a conceptual approach for urgent implantation. Mater Des 185:108218. https://doi.org/10.1016/j.matdes.2019.108218

Li YY, Cheng HW, Cheung KM, Chan D, Chan BP et al (2014) Mesenchymal stem cell-collagen microspheres for articular cartilage repair: cell density and differentiation status. Acta Biomater 10:1919–1929. https://doi.org/10.1016/j.actbio.2014.01.002

Li RY, Liu ZG, Liu HQ, Chen L, Liu JF, Pan YH et al (2015) Evaluation of biocompatibility and toxicity of biodegradable poly (DL-lactic acid) films. Am J Transl Res 7:1357–1370

Li N, Gao J, Mi L, Zhang G, Zhang L, Zhang N, Huo R, Hu J, Xu K et al (2020) Synovial membrane mesenchymal stem cells: past life, current situation, and application in bone and joint diseases. Stem Cell Res Ther 11:1–12

Liu W, Yu X (2016) Injectable natural biomaterials for cartilage tissue engineering. Clin Surg 1: 1103

Lopes MS, Jardini AL, Filho RM (2012) Poly (lactic acid) production for tissue engineering applications. Procedia Eng 42:1402–1413. https://doi.org/10.1016/j.proeng.2012.07.534

Ma K, Titan AL, Stafford M, Zheng C, Levenston ME et al (2012) Variations in chondrogenesis of human bone marrow-derived mesenchymal stem cells in fibrin/alginate blended hydrogels. Acta Biomater 8:3754–3764. https://doi.org/10.1016/j.actbio.2012.06.028

Mackay AM, Beck SC, Murphy JM, Barry FP, Chichester CO, Pittenger MF et al (1998) Chondrogenic differentiation of cultured human mesenchymal stem cells from marrow. Tissue Eng 4:415

Mahboob S (2015) Isolation and characterization of collagen from fish waste material skin, scales and fins of Catla catla and Cirrhinus mrigala. J Food Sci Technol 52:4296–4305. https://doi.org/10.1007/s13197-014-1520-1526

Mallick SP, Singh BN, Rastogi A, Srivastava P et al (2018) Design and evaluation of chitosan/poly (l-lactide)/pectin based composite scaffolds for cartilage tissue regeneration. Int J Biol Macromol 112:909–920. https://doi.org/10.1016/j.ijbiomac.2018.02.049

Melero-Martin JM, Dowling MA, Smith M, Al-Rubeai M (2006) Optimal in-vitro expansion of chondroprogenitor cells in monolayer culture. Biotechnol Bioeng 93:519

Merceron C, Vinatier C, Portron S, Masson M, Amiaud J, Guigand L, Chérel Y, Weiss P, Guicheux J et al (2010) Differential effects of hypoxia on osteochondrogenic potential of human adipose-derived stem cells. Am J Physiol Cell Physiol 298:C355

Mochizuki T, Muneta T, Sakaguchi Y, Nimura A, Yokoyama A, Koga H, Sekiya I et al (2006) Higher chondrogenic potential of fibrous synovium- and adipose synovium-derived cells compared with subcutaneous fat-derived cells: distinguishing properties of mesenchymal stem cells in humans. Arthritis Rheum 54(3):843–853

Mohammadinejad R, Kumar A, Ranjbar-Mohammadi M, Ashrafizadeh M, Han SS, Khang G et al (2020) Recent advances in natural gum-based biomaterials for tissue engineering and regenerative medicine: a review. Polymers (Basel) 12:176. https://doi.org/10.3390/polym12010176

Mostafa S, Pakvasa M, Coalson E, Zhu A, Alverdy A, Castillo H et al (2019) The wonders of BMP9: from mesenchymal stem cell differentiation, angiogenesis, neurogenesis, tumorigenesis, and metabolism to regenerative medicine. Genes Dis 6:201–223. https://doi.org/10.1016/j.gendis.2019.07.003

Moura CS, Silva JC, Faria S, Fernandes PR, da Silva CL, Cabral JMS et al (2020) Chondrogenic differentiation of mesenchymal stem/stromal cells on 3D porous poly (epsilon-caprolactone) scaffolds: effects of material alkaline treatment and chondroitin sulfate supplementation. J Biosci Bioeng 129:756–764. https://doi.org/10.1016/j.jbiosc.2020.01.004

Müller M, Ozturk E, Arlov O et al (2017) Alginate sulfate nanocellulose bioinks for cartilage bioprinting applications. Ann Biomed Eng 45:210–223

Mumme M, Barbero A, Miot S, Wixmerten A, Feliciano S, Wolf F, Asnaghi AM, Baumhoer D, Bieri O, Kretzschmar M et al (2016) Nasal chondrocyte-based engineered autologous cartilage tissue for repair of articular cartilage defects: An observational first-in-human trial. Lancet 388:1985

Nakao H, Jacquet RD, Shasti M, Isogai N, Murthy AS, Landis WJ et al (2017) Long-term comparison between human normal conchal and microtia chondrocytes regenerated by tissue engineering on nanofiber polyglycolic acid scaffolds. Plast Reconstr Surg 139:911e–921e

Neumann AJ, Quinn T, Bryant SJ (2016) Nondestructive evaluation of a new hydrolytically degradable and photo-clickable PEG hydrogel for cartilage tissue engineering. Acta Biomater 39:1–11

Oseni AO, Butler PE, Seifalian AM (2013) Optimization of chondrocyte isolation and characterization for large-scale cartilage tissue engineering. J Surg Res 181(1):41–48

Ousema PH, Moutos FT, Estes BT, Caplan AI, Lennon DP, Guilak F et al (2012) The inhibition by interleukin 1 of MSC chondrogenesis and the development of biomechanical properties in biomimetic 3D woven PCL scaffolds. Biomaterials 33:8967–8974. https://doi.org/10.1016/j.biomaterials.2012.08.045

Pak J, Lee JH, Kartolo WA, Lee SH (2016) Cartilage regeneration in human with adipose tissue-derived stem cells: current status in clinical implications. Biomed Res Int 2016:4702674

Park CH, Woo KM (2018) Fibrin-based biomaterial applications in tissue engineering and regenerative medicine. Adv Exp Med Biol 1064:253–261

Passaretti D, Silverman RP, Huang W, Kirchhoff CH, Ashiku S, Randolph MA, Yaremchuk MJ et al (2001) Cultured chondrocytes produce injectable tissue-engineered cartilage in hydrogel polymer. Tissue Eng 7:805

Pelttari K, Winter A, Steck E, Goetzke K, Hennig T, Ochs BG, Aigner T, Richter W et al (2006) Premature induction of hypertrophy during in vitro chondrogenesis of human mesenchymal stem cells correlates with calcification and vascular invasion after ectopic transplantation in SCID mice. Arthritis Rheum 54:3254

Pelttari K, Pippenger B, Mumme M, Feliciano S, Scotti C, Mainil-Varlet P, Procino A, Von Rechenberg B, Schwamborn T, Jakob M et al (2014) Adult human neural crest-derived cells for articular cartilage repair. Sci Transl Med 6:251ra119

Pelttari K, Mumme M, Barbero A, Martin I et al (2017) Nasal chondrocytes as a neural crest-derived cell source for regenerative medicine. Curr Opin Biotechnol 47:1–6

Pittenger MF, Discher DE, Peault BM, Phinney DG, Hare JM, Caplan AI et al (2019) Mesenchymal stem cell perspective: cell biology to clinical progress. NPJ Regen Med 4:22. https://doi.org/10.1038/s41536-019-0083-86

Pulkkinen HJ, Tiitu V, Valonen P, Jurvelin JS, Lammi MJ et al (2010) Engineering of cartilage in recombinant human type II collagen gel in nude mouse model in vivo. Osteoarthr Cartil 18:1077–1087

Rajangam T, An SSA (2013) Fibrinogen and fibrin based micro and nano scaffolds incorporated with drugs, proteins, cells and genes for therapeutic biomedical applications. Int J Nanomedicine 8:3641–3662

Rim YA, Nam Y, Park N, Lee J, Park SH, Ju H et al (2018) Repair potential of non surgically delivered induced pluripotent stem cell-derived chondrocytes in a rat osteochondral defect model. J Tissue Eng Regen Med 12:1843–1855. https://doi.org/10.1002/term.2705

Roelofs AJ, Zupan J, Riemen AHK, Kania K, Ansboro S, White N, Clark SM, De Bari C et al (2017) Joint morphogenetic cells in the adult mammalian synovium. Nat Commun 8:15040

Rotter N, Bonassar LJ, Tobias G, Lebl M, Roy AK, Vacanti CA et al (2002) Age dependence of biochemical and biomechanical properties of tissue-engineered human septal cartilage. Biomaterials 23:3087

Scholz B, Kinzelmann C, Benz K, Mollenhauer J, Wurst H et al (2010) Suppression of adverse angiogenesis in an albumin-based hydrogel for articular cartilage and intervertebral disc regeneration. Eur Cell Mater 20:24–36

Shafiee A, Soleimani M, Chamheidari GA, Seyedjafari E, Dodel M, Atashi A et al (2011) Electro spun nanofiber-based regeneration of cartilage enhanced by mesenchymal stem cells. J Biomed Mater Res A 99:467–478. https://doi.org/10.1002/jbm.a.33206

Shi J, Liang J, Guo B, Zhang Y, Hui Q, Chang P, Tao K et al (2017) Adipose-derived stem cells cocultured with chondrocytes promote the proliferation of chondrocytes. Stem Cells Int 2017:1709582

Shirasawa S, Sekiya I, Sakaguchi Y, Yagishita K, Ichinose S, Muneta T et al (2006) In vitro chondrogenesis of human synovium-derived mesenchymal stem cells: optimal condition and comparison with bone marrow-derived cells. J Cell Biochem 97:84

Silva TH, Moreira-Silva J, Marques AL, Domingues A, Bayon Y, Reis RL et al (2014) Marine origin collagens and its potential applications. Mar Drugs 12:5881–5901. https://doi.org/10.3390/md12125881

Singh R, Shitiz K, Singh A (2017) Chitin and chitosan: biopolymers for wound management. Int Wound J 14:1276–1289

Smieszek A, Marycz K, Szustakiewicz K, Kryszak B, Targonska S, Zawisza K et al (2019) New approach to modification of poly (l-lactic acid) with nanohydroxyapatite improving functionality of human adipose-derived stromal cells (hASCs) through increased viability and enhanced mitochondrial activity. Mater Sci Eng C Mater Biol Appl 98:213–226. https://doi.org/10.1016/j.msec.2018.12.099

Song JS, Hong KT, Kim NM, Jung JY, Park HS, Lee SH, Cho YJ, Kim SJ et al (2020) Implantation of allogenic umbilical cord blood-derived mesenchymal stem cells improves knee osteoarthritis outcomes: two-year follow-up, vol 14. Regen Ther, p 32

Stoddart MJ, Grad S, Eglin D, Alini M et al (2009) Cells and biomaterials in cartilage tissue engineering. Regen Med 4:81–98

Stromps JP, Paul NE, Rath B, Nourbakhsh M, Bernhagen J, Pallua N et al (2014) Chondrogenic differentiation of human adiposederived stem cells: a new path in articular cartilage defect management? Biomed Res Int 2014:740926

Su X, Wu Z, Chen J, Wu N, Ma P, Xia Z, Jiang C, Ye Z, Liu S, Liu J et al (2015) CD146 as a new marker for an increased chondroprogenitor cell sub-population in the later stages of osteoarthritis. J Orthop Res 33:84

Suzuki H, Araki K, Matsui T, Tomifuji M, Yamashita T, Kobayashi Y et al (2016) Value of a novel PGA-collagen tube on recurrent laryngeal nerve regeneration in a rat model. Laryngoscope 126: E233–E239. https://doi.org/10.1002/lary.25750

Tao F, Cheng Y, Shi X, Zheng H, Du Y, Xiang W, Deng H et al (2020) Applications of chitin and chitosan nanofibers in bone regenerative engineering. Carbohydr Polym 230:115658

Tew SR, Murdoch AD, Rauchenberg RP, Hardingham TE et al (2008) Cellular methods in cartilage research: primary human chondrocytes in culture and chondrogenesis in human bone marrow stem cells. Methods 45:2–9

Turk S, Altinsoy I, Celebi Efe G, Ipek M, Ozacar M, Bindal C et al (2018) 3D porous collagen/ functionalized multi-walled carbon nanotube/chitosan/hydroxyapatite composite scaffolds for bone tissue engineering. Mater Sci Eng C Mater Biol Appl 92:757–768. https://doi.org/10.1016/ j.msec.2018.07.020

Tyler B, Gullotti D, Mangraviti A, Utsuki T, Brem H et al (2016) Polylactic acid (PLA) controlled delivery carriers for biomedical applications. Adv Drug Deliv Rev 107:163–175. https://doi.org/ 10.1016/j.addr.2016.06.018

Venkatesan JK, Falentin-Daudre C, Leroux A, Migonney V, Cucchiarini M et al (2020) Biomaterial-guided recombinant adeno-associated virus delivery from poly(sodium styrene sulfonate)-grafted poly(varepsilon-caprolactone) films to target human bone marrow aspirates. Tissue Eng Part A 26:450–459. https://doi.org/10.1089/ten.TEA.2019.0165

Vishnubalaji R, Al-Nbaheen M, Kadalmani B, Aldahmash A, Ramesh T et al (2012) Comparative investigation of the differentiation capability of bone-marrow- and adipose-derived mesenchymal stem cells by qualitative and quantitative analysis. Cell Tissue Res 347:419

Wang CC, Yang KC, Lin KH, Liu YL, Liu HC et al (2012) Cartilage regeneration in SCID mice using a highly organized three-dimensional alginate scaffold. Biomaterials 33:120–127

Wang J, Sun B, Tian L, He X, Gao Q et al (2017a) Evaluation of the potential of rhTGF- ß3 encapsulated P(LLA-CL)/collagen nanofibers for tracheal cartilage regeneration using mesenchymal stems cells derived from Wharton's jelly of human umbilical cord. Mater Sci Eng C Mater Biol Appl 70:637–645

Wang J, Zhang F, Tsang WP, Wan C, Wu C et al (2017b) Fabrication of injectable high strength hydrogel based on 4-arm star PEG for cartilage tissue engineering. Biomaterials 120:11–21

Xu J, Wang W, Ludeman M, Cheng K, Hayami T, Lotz JC, Kapila S et al (2008) Chondrogenic differentiation of human mesenchymal stem cells in three-dimensional alginate gels. Tissue Eng Part A 14:667

Yaeger PC, Masi TL, De Ortiz JLB, Binette F, Tubo R, McPherson JM et al (1997) Synergistic action of transforming growth factor-β and insulin-like growth factor -I induce expression of type II collagen and aggrecan genes in adult human articular chondrocytes. Exp Cell Res 237: 318

Yang F, Zhao J, Koshut WJ, Watt J, Riboh JC, Gall K, Wiley BJ et al (2020) A synthetic hydrogel composite with the mechanical behavior and durability of cartilage. Adv Funct Mater 30:1–8

Zeng L, Yao Y, Wang DA, Chen X et al (2014) Effect of microcavitary alginate hydrogel with different pore sizes on chondrocyte culture for cartilage tissue engineering. Mater Sci Eng C Mater Biol Appl 34:168–175

Zhang L, He A, Yin Z, Yu Z, Luo X, Liu W, Zhang W et al (2014) Regeneration of human-ear-shaped cartilage by co-culturing human microtia chondrocytes with BMSCs. Biomaterials 35(18):4878–4887

Zhang X, Xu B, Puperi DS, Yonezawa AL, Wu Y, Tseng H et al (2015) Integrating valve-inspired design features into poly(ethylene glycol) hydrogel scaffolds for heart valve tissue engineering. Acta Biomater 14:11–21

Zhao X, Hwang NS, Bichara DA, Saris DB, Malda J, Vacanti JP, Pomerantseva I, Sundback CA, Langer R, Anderson DG et al (2017) Chondrogenesis by bone marrow-derived mesenchymal stem cells grown in chondrocyte-conditioned medium for auricular reconstruction. J Tissue Eng Regen Med 11:2763

Lipase Synthesis Using Palm Oil Mill Effluent for Polycaprolactone Production

9

Jegalakshimi Jewaratnam and Paveethra Thegarathah

Abstract

Polycaprolactone (PCL) is a popular biopolymer utilized in the medical field for tissue engineering, drug delivery, and wound healing. Main factors of concern when it comes to material applications in the human body are toxicity and biodegradability. It must be biocompatible. One of the methods employed in the polymerization of ε-caprolactone (ε-CL) is ring opening polymerization using lipase as the catalyst. This is clearly due to lipase's characteristics like high stability, specificity, and thermostability. Biocatalysts have remained a major topic of discussion in the industrial sector due to their value in a wide range of applications. The enzyme industry has seen billion-dollar growth over the years, and it is anticipated that it will continue to develop significantly in the near future. However, commercial lipase is rather expensive. This issue leads the industry to utilize chemical substances in polymerization as an alternative for cheaper PCL production with toxicity as the major consequence. Therefore, this chapter suggests a sustainable way to produce cost-effective lipase by employing palm oil mill effluent (POME) as a fungal growth medium for lipase production. Materializing this research will enable a sustainable, biocompatible, and cheaper local PCL synthesis.

Keywords

Polycaprolactone · Cost-effective lipase · Palm oil mill effluent · Fermentation · Ring opening polymerization

J. Jewaratnam (✉) · P. Thegarathah
Department of Chemical Engineering, Faculty of Engineering, Center for Separation Science & Technology (CSST), Universiti Malaya, Kuala Lumpur, Malaysia
e-mail: jegalaxmi24@um.edu.my

W. S. Wan Kamarul Zaman, A. A. Nurul (eds.), *Sustainable Material for Biomedical Engineering Application*, https://doi.org/10.1007/978-981-99-2267-3_9

Abbreviations

BOD	Biological oxygen demand
COD	Chemical oxygen demand
ε-CL	ε-caprolactone
FDA	Food and Drug Administration
K	Potassium
MW	Molecular weight
N	Nitrogen
P	Phosphorus
PCL	Polycaprolactone
PDI	Polydispersity index
POME	Palm oil mill effluent
ROP	Ring-opening polymerization
SmF	Submerged fermentation
TOC	Total organic carbon
TSS	Total suspended solid

9.1 Polycaprolactone (PCL): An Excellent Biomaterial for Biomedical Applications

PCL, a hydrophobic semi-crystalline aliphatic polyester, is a well-known biocompatible and biodegradable polymer. PCL was widely used in the medical field since 1970s–1980s when the resorbable-polymer-bloom took place. It was commonly used in the applications of tissue engineering and drug release control (Nhi et al. 2016; Kotcharat et al. 2021). According to (Woodruff and Hutmacher 2010), PCL has a magnificent shaping characteristic that allows it to be shaped into appropriate pore sizes that are suited for the in-growth of tissue and regulated drug distribution within the matrix. However, PCL was almost forgotten for 20 years. Synthetically produced polymers are often used in this area due to their processability, high mechanical performance, and degradability (Jouibary et al. 2022) The recent urge of producing biomaterials has brought PCL into production again (Woodruff and Hutmacher 2010). Since conventional petroleum-based materials pose a hazard to the environment, several investigations have been done to find an alternative, ecologically friendly biomaterial.

Polycaprolactone (PCL) is one such biomaterial that has become increasingly popular due to its eco-friendliness, biocompatibility, FDA approval for internal medicinal usage, and low immunogenicity (Dhanasekaran et al. 2022). It possesses a large specific area, high porosity, and small pore size due to its nanofibrous nature (Van der Schueren et al. 2011). PCL is used in the medical field for several compelling reasons as shown in Fig. 9.1.

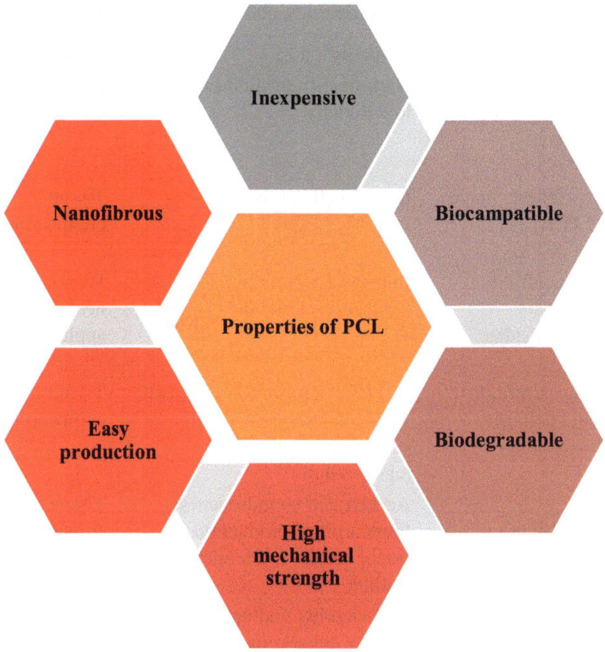

Fig. 9.1 Properties of PCL

Table 9.1 shows a review of recent applications and modifications done to PCL to improvise its performances in tissue engineering, drug delivery, and wound healing. Biomaterials are crucial in sustaining health care level (El-Naggar et al. 2021). It is clear that nanofibrous scaffolds are receiving a lot of attention in tissue engineering (Nooeaid et al. 2020) for their qualities like processability, great mechanical performances, and biodegradability (Jouibary et al. 2022; Altun et al. 2019). The properties of PCL like viscoelasticity, solubility in several solvents, and easy fabrication make PCL a suitable material for wound healing application (El-Naggar et al. 2021).

9.2 Biocatalyst for Ring Opening Polymerization of Polycaprolactone

The literature describes two basic approaches to produce polycaprolactone which are: polycondensation of the 6-hydroxyhexanoic acid and ring opening polymerization (ROP) of the lactone monomers (ε-CL). ROP is preferable compared to polycondensation because ROP yields a polymer with lower polydispersity index (PDI) and higher molecular weight (MW) (molecular weight). There are four general mechanisms for ROP of lactones which are: anionic, cationic, radical, and coordination polymerization mechanisms. These mechanisms are influenced by the

Table 9.1 Review of recent advances, properties, and applications of PCL

Recent advances	Properties	Application	References
Macrophage membrane-biomimetic platform for nano-delivery of PCL nano-camptothecin	High-quality medicines with low toxicity	Drug delivery	Ying et al. (2023)
Fabricated skin scaffolds made of collagen and PCL	Higher cell survival	Tissue engineering	Jouibary et al. (2022)
Nanofibrous PCL scaffolds electrospun with binary hydroxyapatite/Pd/vanadate ions	Improved wound healing and reduced bacterial development	Wound healing application	El-Morsy et al. (2022)
A bioactive porous scaffold made of PCL, and collagen treated with phosphorous	Scaffold that can facilitate cell attachment and better cell proliferation for bone regeneration	Tissue engineering	Safari et al. (2022)
Electrospun PCL/hydroxyapatite/ZnO films	Accelerated chondrogenesis, fibrocartilage production, osteogenesis, and tendon healing	Tissue engineering	He et al. (2021)
Scaffolds made of PCL nanofibers and laser-ablated Au@Se core shells	Cell viability and antibacterial activity enhancement	Wound healing application	Al-Mogbel et al. (2021)
Simple addition of hydroxyapatite doped with thallium ions to PCL nanofibers	Efficient cell viability	Wound healing application	El-Naggar et al. (2021)
In-situ sol-gel produced PCL/fluoride substituted-hydroxyapatite (PCL/FHA) nanocomposite coatings	Compact and crack free coating	Dental implant	Ansari et al. (2020)
Blend biopolymeric nanofibrous scaffolds made of cellulose acetate and PCL that incorporate laser-cut metallic nanoparticles	Biocompatible and improvised antibacterial activity	Wound healing application	Ahmed et al. (2020)
Pulsed laser deposition for the manipulation of AgNPs coated on a selenium, carbonated hydroxyapatite, and PCL nano-fibrous	Increases bacterial inhibition and cell growth	Wound healing application	Menazea et al. (2020)
Electrospun PCL/silk fibroin nanofibrous bioactive scaffold	Non-toxic and enhanced cell proliferation	Tissue engineering	Nazeer et al. (2019)
Three-dimensionally printed biodegradable PCL mesh	Dependable, low-complication stabilization of orbital wall deficiencies	Tissue engineering	Kim (2019)
Incorporating magnesium metallic particles into a PCL nanofiber mesh	Improved tissue healing	Tissue engineering	Adhikari et al. (2019)

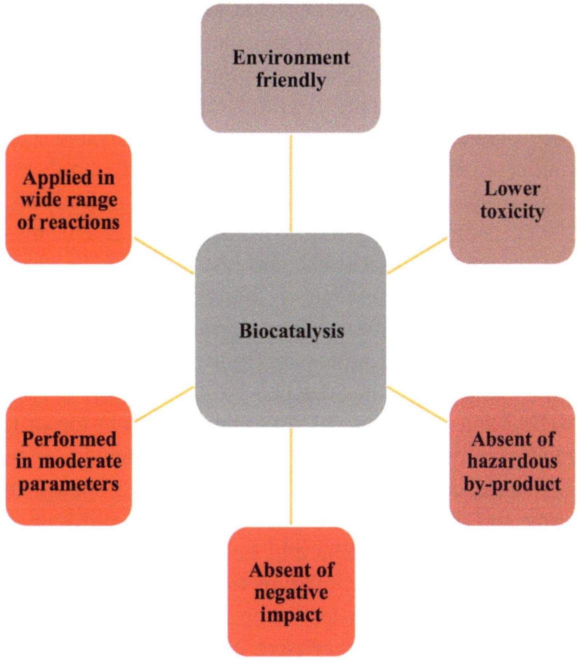

Fig. 9.2 Advantages of utilizing biocatalyst

copolymer's structure, molecular weight distribution, and chemical makeup of the end groups. Different catalytic systems, including organic, enzymatic, and metal-based ones, can speed up the ROP process (Labet and Thielemans 2009; Mahmoud Salehi et al. 2021).

The traditional method uses metals, acids, and alkalis as catalysts at high or low temperatures, together with organic solvents and chemicals, making these processes hazardous and posing a severe threat to the environment. Metallic catalysts are useful but may also be harmful and produce more by-products when utilized (Lu et al. 2019). Enzymes, in contrast, are non-toxic, biocompatible, and biodegradable, which dissipates worries about downstream pollution (Jegannathan and Nielsen 2013). Enzymes serve as an appropriate, environmentally acceptable replacement for these conventional techniques (Vivek et al. 2022). Hence, enzymes are biocatalysts that are often favorable to the environment and can be used in a range of bioprocesses (Vivek et al. 2022; Shi et al. 2020). On top of that, due to their lower toxicity and ability to function in mild settings, enzyme-based polymerization is seen as an alternative to traditional polymerization (Pakalapati et al. 2019). Enzymes provide more efficient production than the chemical process used, with few to no adverse effects and no generation of harmful byproducts (Vivek et al. 2022). Figure 9.2 summarizes the strength of biocatalyst.

Lipase is one of the biocatalysts with the highest demand as it is being widely applied in various industries like pharmaceuticals, food, detergent, cosmetics, and textiles (Phukon et al. 2020; Bharathi and Rajalakshmi 2019; Martinez-Ruiz et al.

2018; Navvabi et al. 2018). Due to their wide range of catalytic reactions, lipases catalyze numerous biotransformation synthesis reactions in both aqueous and non-aqueous media (Nimkande and Bafana 2022), including esterification, inter-esterification, and trans-esterification reactions (Knappmann et al. 2017; Xin et al. 2017). On top of that, their excellent specificity and stability, lipases are favored over chemical catalysts in many applications. The production of biodiesel (Selvakumar and Sivashanmugam 2017), meat, dairy, fats and oils, surfactants (Hemlata et al. 2016), tanneries, cosmetics (Mouad et al. 2016), biosensors (Narwal and Pundir 2017), and pharmaceuticals are just a few of the processes and industries that use microbial lipases, particularly bacterial and fungal lipases (Nimkande and Bafana 2022). It is the third largest group of enzymes in the enzyme market after carbohydrases and proteases (Martinez-Ruiz et al. 2018). The global market of industrial enzymes has reached 7 billion dollars in 2017 and it is expected to reach 10.5 billion dollars by 2024 (Anon 2018). On top of this, application of biological catalyst in industrial processes is highly beneficial to the environment (Atalah et al. 2019). Hence, microbial lipases are receiving special attention. It is also due to their properties like enantio-selectivity, broad substrate specificity, and stability in extreme conditions (Bharathi and Rajalakshmi 2019; Atalah et al. 2019; Thapa et al. 2019). However, many researchers are seeking cheaper lipase as commercial lipases are sold at high cost comparatively (Costa et al. 2017). This is mainly because lipase production is expensive and being the main hindrance for using lipase in various industrial applications (Luna et al. 2018).

9.3 Palm Oil Mill Effluent (POME) as a Fermentation Medium

The need for food, fuel, and other necessities is rising as the population grows. In this setting, the agricultural, pastoral, and agro-industrial sectors began to develop quickly, which in turn caused the amount of waste produced by each of their individual operations to rise (Cremonez et al. 2021). A study by (Dahiya et al. 2018) says that agro-industry produces the most residues among the other industries. During its production, handling, storage, processing, distribution, or consumption, almost 1.3 billion tonnes of food are wasted annually. Composting, incineration, and landfilling are the most popular ways to dispose of these wastes. However, the main issues with these conventional technologies are their high administration costs and damaging environmental effects from the release of greenhouse gases (Gil et al. 2015; Greses et al. 2020). This rise in wastes results in a variety of severe environmental problems (Mo et al. 2018) without proper treatment (Anastopoulos et al. 2017).

Therefore, the search for decentralized, alternative, and clean energy sources is expanding at the same time as there is significant potential for using leftover biomass from agro-industrial activities as a raw material for obtaining energy, lowering environmental risks, and enhancing those already established value chains (Cremonez et al. 2021). Wastes and by-products from the agro-industry are highly potential for bioconversion. In light of the possibility of using a variety of organic compounds as feedstock to support a more sustainable and economical method of handling agricultural wastes, biological processes are gaining a lot of attention (Strazzera et al. 2018).

With palm oil and palm kernel oil accounting for around 33% of the world's production of oils and fats, oil palm plays a significant role in the vegetable oil sectors (Tan and Lim 2019). Figures say that 64 million tonnes of palm oil were produced in Indonesia, Malaysia, and Thailand in 2017 (Cheng et al. 2019a). Then, the demand for palm oil reached 71.48 million tonnes globally in 2019, and from 2020 to 2027, it is projected to expand at a 2.3% annual pace. Palm oil can lower cholesterol levels, improve brain health, lower oxidative stress, halt the progression of disease, promote vitamin A status, and enhance the health of the skin and hair (Low et al. 2021).

The increase in demand and supply for palm oil results in an increase of palm oil mill effluent (POME) released due to the growth of the oil palm agribusiness (Ratnasari et al. 2022). According to estimates, every tonne of crude palm oil (CPO) that is processed industrially can result in 2.5–3.8 tonnes of POME waste (Cheng et al. 2019b). With a production volume of 72 million tonnes of crude palm oil in 2018, 180 to 270 million tonnes of POME were produced, the majority of which was released into the environment without sufficient treatment. The presence of POME in the environment has multiple detrimental effects on human health, aquatic life, water quality, ground water, and soil (Jasni et al. 2020; Zulfahmi et al. 2021). Due to its 100 times greater pollution than municipal trash, POME is said to be the world's most polluted wastewater (Norfadilah et al. 2016; Wee et al. 2022).

POME is a complex effluent that is not hazardous but is yet highly damaging because of its high organic content. Therefore, before releasing it into rivers or the ocean, appropriate treatment or utilization is required (Garritano et al. 2018). The sustainability of oil palms must be carefully considered. A range of POME utilization strategies has been investigated and put into practice to provide alternative renewable energy to the palm oil milling operations that can reduce biomass fuel consumption and mill expenses (Tan and Lim 2019; Harsono et al. 2014).

The majority of POME's components are lipids, carbohydrates (Maaroff et al. 2019), minerals, proteins, and nitrogenous substances, making it suitable for biological handling techniques (Tan and Lim 2019). Besides, it also contains significant amounts of suspended particles, oil and grease, and organic material (Garritano et al. 2018). Hence, it acts as a suitable cheap substrate for bioconversion. Table 9.2. shows the revalorization of POME for value-added products in recent times.

Information gathered shows that POME has been serving as a great fermentation substrate for biogas production. By virtue of its extensive applicability, adaptable bioproduct creation, low energy requirement, and effective degradation, POME bioconversion is the most practical valorization strategy. By using foreign microorganisms to speed up biodegradation or promote the bioproduction of a variety of metabolites, POME bioconversions are marketed as the more sophisticated form of biological-based open ponding treatment (which depends on native microorganisms) (Cheng et al. 2019a).

Table 9.2 Revalorisation of raw POME through fermentation

Raw material	Microbial strain	Fermentation approach	Value added product	Reference
POME	*Thermoanaerobacterium thermosaccharolyticum*	Dark fermentation	Biohydrogen	Arisht et al. (2022)
POME and crude glycerol	*Sandarakinotalea* sp., *Lachnobacterium* sp., *Blautia* sp., *Marinobacter* sp., *Anaerotruncus* sp., *Clostridium* sp., *Enterobacter* sp., and *Acinetobacter* sp.	Mesophilic anaerobic digestion	Biomethane	Prasertsan et al. (2021)
POME	*Clostridium* sensu stricto *1* and *Lactobacillus*	Dark fermentation	Biohydrogen	Akhbari et al. (2021)
POME	*Yarrowia lipolytica* TISTR 5151	Aerobic fermentation and anaerobic digestion	Biomethane	Louhasakul et al. (2021)
POME	*Anaerobic consortium of bacteria: Bacillus toyonensis* strain BCT-71120 and *Stenotrophomonas rhizophila* strain e-p10	Aerobic fermenttaion	Biomethane	Said et al. (2021)
POME	Two microbial consortia and a pure culture of *Clostridium beijerinckii* (ATCC 8260)	Dark fermentation	Biohydrogen	Rosa et al. (2020)
POME	*Clostridium* sp. and *Methanosphaera* sp.	Thermophilic fermentation	Biohythane	Mamimin et al. (2019)
POME	*Chlorella sorokiniana* CY-1	Microalgae cultivation	Lipid	Cheah et al. (2018)

Simultaneous biodegradation of complex organics and absorption of simple metabolites might support the growth of microorganisms in POME wastewater, which contains large amounts of organics substances (COD, BOD5, TSS, & TOC) and nutrients (N, P, & K). POME is also usually acknowledged as an affordable alternative to expensive carbon sources (glucose, sucrose, and wheat flour) and nitrogen supplies as a microbial substrate (yeast extract, urea, and ammonium nitrate) (Aljuboori et al. 2014). Revalorization of POME promotes the growth of microorganisms to produce bio-sludge for wastewater treatment and biocarriers of extracellular metabolites like gas and suspended or extractable solutes. Employing biological agents to deplete organics is environmentally beneficial regardless of the presence of external substrates (Halim et al. 2016). In some studies, nitrogen supplements like fertilizer and urea, and enzyme inducers namely cellulose, peptone, and olive oil are added in trace amounts. Addition of these substances assists microorganisms to adapt in POME (Mishra et al. 2016), enhances their growth (Prasertsan and Binmaeil 2018), and improves enzyme production (cellulase, lipase and protease) (Wu et al. 2009).

In addition, POME being a thick liquid provides a great medium for submerged fermentation (SmF). SmF provides benefits like simplicity of controlling the factors and enzyme isolation (Musatti et al. 2017). Majority of extracellular microbial enzymes commercial productions employ submerged fermentation techniques (et al. 2022).

9.4 Extracellular Fungal Lipase Production in POME

Numerous products have shown POME to be an effective raw material, and many processes have been developed to turn POME into value-added products. The manufacture of lipase from this cheap raw material could increase the competitiveness of industrial operations and lower the cost of the enzyme. In order to produce lipase using POME, it would be highly appropriated to use fungal strains. This is because these strains are easy to produce and have a shorter generation time in fermentation and molecular manipulation. Fungal lipases have garnered a lot of attention due to their higher stability and substrate selectivity (Geoffry and Achur 2018a). There are more than 1.5 million different kinds of fungi in the world. By adapting to different environments and developing a wide range of survival mechanisms, fungi can produce a huge variety of lipases (et al. 2022). Lipases have been found to be released into the growing medium by a variety of fungal strains, including *Penicillium cyclopium* (Chahinian et al. 2000; Ibrik et al. 1998), *Penicillium simplicissimum* (Sztajer et al. 1992), *Aspergillus niger* (Namboodiri and Chattopadhyaya 2000), *Aspergillus oryzae* (Toida et al. 1998), *Botrytis cinerea* (Commenil et al. 1995), *Trichosporon asteroids* (Dharmsthiti and Ammaranond 1997), *Trichosporon laibachii* (Liu et al. 2004), *Rhizopus* sp., *R. miehei, Geotrichum candidum* (Sidebottom et al. 1991; Jacobsen and Poulsen 1995), *Pichia burtonii* (Sugihara et al. 1995), *Candida cylidracae* (Muralidhar et al. 2001) and *Fusarium solani* (Knight et al. 2000). Filamentous fungi such as *Aspergillus niger, Colletotrichum gloeosporioides, Rhizopus oryzae, Trichoderma viride,* and *Yarrowia lipolytica* (Ramos Sanchez 2015; Kumar et al. 2011) are among the best extracellular lipase producers, and the extraction and purification procedure are comparatively easier. They are often popular for industrial use due to their excellent stability and wide substrate specificity (Geoffry and Achur 2018b). Fungal lipases play a role in the synthesis of various monomers, oligomers, and polymers.

Plants, animals, and microbes all manufacture lipases (Li et al. 2017; Das et al. 2017). Most of the commercial enzymes produced in recent years using microorganisms produce different types of extracellular lipases (Gupta et al. 2015; Taskin et al. 2016). Different sources of lipases have different characteristics in terms of positional specificity, fatty acid specificity, thermostability, pH optimum, and other factors. Despite notable progress in recent years, the need to identify and isolate novel fungus strains that produce lipases with catalytic properties suitable for industrial processes remains a pressing concern. A barrier in large-scale commercial production continues to be the understanding of medium composition, which has a significant impact on the development and synthesis of metabolites in

Table 9.3 Review of fungal lipase production using agro-industrial wastes as substrate

Fungal strain	Growth medium	Lipase activity	Reference
Sarocladium strictum	Crude glycerol	73.2 U/mg	Maidana Serpa et al. (2022)
Metarhizium anisopliae	Babbasu coconut bagasse	–	Diniz da Silva et al. (2022)
Fusarium sp. (GFC)	Crambe oil supplemented with Triton X-100, ammonium sulphate and yeast extract	2.88 U/mL	de Oliveira et al. (2020)
Fusarium sp. (GFC)	Crambe cake in phosphate buffer	74.00 U/g	de Oliveira et al. (2020)
Aspergillus niger MTCC 872	Tri-substrate mixture of *Prosopis juliflora* (PJ), red gram husk (RGH) and cotton seed cake (CSC)	269.87 ± 8.09 U/g	Mandari et al. (2020)
Meyerozyma guilliermondii	Cheese whey in Vogel's medium (sodium citrate, mineral salts, and biotin)	285.82 U/mL	Knob et al. (2020)
Beauveria bassiana lipase 1 expressed in Aspergillus nidulans A773	Different agroindustry inputs (corn meal, corn syrup, sorghum seed and wheat bran)	33.01; 42.93; 45.12; and 47.53 U/mL, respectively	Cerioni Spiropulos Gonçalves et al. (2020)
Aspergillus niger	Rice bran and Jatropha seed cake	282 U/mL	Putri et al. (2020)

microorganisms as well as the high cost of producing lipases (Geoffry and Achur 2018b).

There are very few studies that have utilised fungal lipase in ring opening polymerization of ε-CL in recent years. Knowing the qualities of fungal lipase and their ability to grow on a wide range of substrate to produce stable high yield lipase, more studies should be conducted to explore in order to obtain sources of cost-effective lipase biocatalyst for the synthesis of PCL.

A review (Table 9.3) of studies of fungal lipase production in agro-industrial waste in recent years has been done to have an insight of the current trend in types of fungal species employed. *A. niger* is showing the highest lipase activity which is above 250 U/mL. *A. niger* is a filamentous fungus, growing conidiophore protrusions, and producing conidial spores. It is found abundantly in soil, thrives in decaying organic matter, and can withstand extremely low and high temperatures (Cortesão et al. 2020). They also discovered that *A. niger* can also withstand radiation extremities in space such as solar flares, UV radiation, and X-rays (Li et al. 2019). This finding explains that *A. niger* could survive in a wide range of environments and could be one example of a good lipase producing agent.

Finally, indigenous fungal strains found in POME are *Aspergillus fumigatu (*Bala et al. 2018*;* Rupani et al. 2022*), Aspergillus nomius* 108PF, *Aspergillus niger* 109PF, *Meyerozyma guilliermondii* 110PF (Bala et al. 2018). This information reveals that most of them are from the Aspergillus genus. This discovery strengthens the idea of utilizing *Aspergillus* species for lipase production in POME.

9.5 Bioethics and Clinical Trials

Utilisation of fungal lipase synthesized using POME as a medium raises bioethics concern. This is because the use of PCL is mainly in biomedical applications. Toxicity tests of POME, lipase, and PCL will have to be carried out prior to the modification of the material. Given that biomedical research methodologies are complex and varied, including surgery and cells, it is crucial to look for iterative value. Similarly, ethical issues that occur in biomedical research could also be complicated. Thus, it requires serious consideration (Baker et al. 2016).

9.6 Conclusion: Cost Effective Fungal Lipase Application for Sustainable Local PCL Production

POME could be a great substrate as there are traces of palm oil in it. Palm oil acts as an inducer for lipase production. On top of that, POME is rich in nutrients which act as a carbon source, nitrogen source, and surfactant for lipase production (Geoffry and Achur 2018a). Hence, Fig. 9.3 suggests a framework for a sustainable and biocompatible PCL synthesis. Currently, this research is at the stage of fungal lipase purification to produce lipase at higher activity and stability. Purified lipase is then expected to be immobilised to be applied in ring-opening polymerization of ε-caprolactone. PCL produced through this procedure will then be tested for its toxicity to verify the biosafety level.

It is expected that lipase production using low-cost substrates like POME within countries which are being global palm oil producers would allow competitive pricing

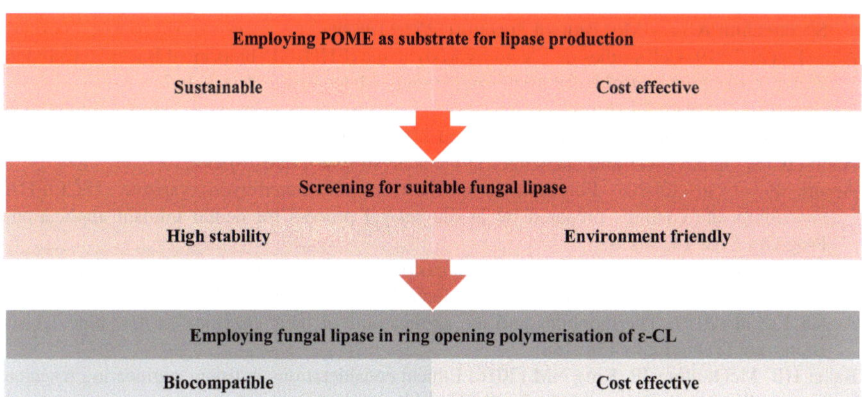

Fig. 9.3 Proposed idea for sustainable and cost-effective local PCL production

of the enzyme for industrial applications. Utilization of agro-industrial waste, POME, will ensure the realization of enzyme production for development of bio-industrial processes for economic mobility through biotechnology.

Acknowledgement The authors gratefully acknowledge the financial support from the Ministry of Science Technology and Innovation (MOSTI), Malaysia; FRGS FP030-2020.

Statements and Declaration *Funding*

This work was funded by FRGS FP030-2020. Author Jegalakshimi Jewaratnam has received research support from the Ministry of Science Technology and Innovation (MOSTI), Malaysia.

Competing interests

The authors declare no conflicts of financial interests or personal relationship. The authors alone are accountable for the content and writing of this article.

Authors contribution

Jegalakshimi Jewaratnam: Conceptualization, Resources, Writing—Review and Editing, Supervision, Project administration, Funding acquisition; Paveethra Thegarathah: Writing—Original Draft.

References

Adhikari U et al (2019) Embedding magnesium metallic particles in polycaprolactone nanofiber mesh improves applicability for biomedical applications. Acta Biomater 98:215–234

Afifa et al (2022) Bioprospecting fungal-derived value-added bioproducts for sustainable pharmaceutical applications. Sustain Chem Pharm 29:100755

Ahmed MK, Menazea AA, Abdelghany AM (2020) Blend biopolymeric nanofibrous scaffolds of cellulose acetate/epsilon-polycaprolactone containing metallic nanoparticles prepared by laser ablation for wound disinfection applications. Int J Biol Macromol 155:636–644

Akhbari A, Chuen OC, Ibrahim S (2021) Start-up study of biohydrogen production from palm oil mill effluent in a lab-scale up-flow anaerobic sludge blanket fixed-film reactor. Int J Hydrog Energy 46(17):10191–10204

Aljuboori AH et al (2014) Production of a bioflocculant from Aspergillus niger using palm oil mill effluent as carbon source. Bioresour Technol 171:66–70

Al-Mogbel MS et al (2021) Conditions adjustment of polycaprolactone nanofibers scaffolds encapsulated with core shells of Au@Se via laser ablation for wound healing applications. Spectrochim Acta A Mol Biomol Spectrosc 259:119899

Altun E et al (2019) Cell studies on Electrohydrodynamic (EHD)-3D-bioprinted Bacterial Cellulose \Polycaprolactone scaffolds for tissue engineering. Mater Lett 234:163–167

Anastopoulos I et al (2017) A review on waste-derived adsorbents from sugar industry for pollutant removal in water and wastewater. J Mol Liq 240:179–188

Anon (2018) On advances and challenges in biocatalysis. Nat Catal 1:635–636

Ansari Z et al (2020) Polycaprolactone/fluoride substituted-hydroxyapatite (PCL/FHA) nanocomposite coatings prepared by in-situ sol-gel process for dental implant applications. Prog Org Coat 147:105873

Arisht SN et al (2022) Enhancing biohydrogen gas production in anaerobic system via comparative chemical pre-treatment on palm oil mill effluent (POME). J Environ Manag 321:115892

Atalah J et al (2019) Thermophiles and the applications of their enzymes as new biocatalysts. Bioresour Technol 280:478–488

Baker HB, McQuilling JP, King NM (2016) Ethical considerations in tissue engineering research: case studies in translation. Methods 99:135–144

Bala JD et al (2018) Microbiota of palm oil mill wastewater in Malaysia. Trop Life Sci Res 29(2): 131–163

Bharathi D, Rajalakshmi G (2019) Microbial lipases: an overview of screening, production and purification. Biocatal Agric Biotechnol 22:101368

Cerioni Spiropulos Gonçalves E et al (2020) Potential biodiesel production from Brazilian plant oils and spent coffee grounds by Beauveria bassiana lipase 1 expressed in Aspergillus nidulans A773 using different agroindustry inputs. J Clean Prod 256:120513

Chahinian H et al (2000) Production of extracellular lipases by Penicillium cyclopium purification and characterization of a partial acylglycerol lipase. Biosci Biotechnol Biochem 64(2):215–222

Cheah WY et al (2018) Microalgae cultivation in palm oil mill effluent (POME) for lipid production and pollutants removal. Energy Convers Manag 174:430–438

Cheng YW et al (2019a) Harnessing renewable hydrogen-rich syngas from valorization of palm oil mill effluent (POME) using steam reforming technique. Renew Energy 138:1114–1126

Cheng YW et al (2019b) Hydrogen-rich syngas production via steam reforming of palm oil mill effluent (POME) – a thermodynamics analysis. Int J Hydrog Energy 44(37):20711–20724

Commenil P et al (1995) Purification and properties of an extracellular lipase from the fungus Botrytis cinerea. Lipids 30(4):351–356

Cortesão M et al (2020) Aspergillus niger spores are highly resistant to space radiation. Front Microbiol 11:560

Costa TM et al (2017) Lipase production by Aspergillus Niger grown in different agro-industrial wastes by solid-state fermentation. Braz J Chem Eng 34(2):419–427

Cremonez PA et al (2021) Two-stage anaerobic digestion in agroindustrial waste treatment: a review. J Environ Manag 281:111854

Dahiya S et al (2018) Food waste biorefinery: sustainable strategy for circular bioeconomy. Bioresour Technol 248(Pt A):2–12

Das A et al (2017) Coconut oil induced production of a surfactant-compatible lipase from Aspergillus tamarii under submerged fermentation. J Basic Microbiol 57(2):114–120

de Oliveira BH et al (2020) Biocatalytic benefits of immobilized Fusarium sp. (GFC) lipase from solid state fermentation on free lipase from submerged fermentation. Ind Crop Prod 147:112235

Dhanasekaran NPD, Muthuvelu KS, Arumugasamy SK (2022) Recent advancement in biomedical applications of polycaprolactone and polycaprolactone-based materials. In: Encyclopedia of materials: plastics and polymers, pp 795–809

Dharmsthiti S, Ammaranond P (1997) Purification and characterization of lipase from a raw-milk yeast (Trichosporon asteroides). Biotechnol Appl Biochem 26(2):111–116

Diniz da Silva MP et al (2022) A two-phase model for simulation of water transfer during lipase production by solid-state cultivation in a tray bioreactor using babassu residues as substrate. Chem Eng Process 177:108981

El-Morsy MA et al (2022) Electrospun nanofibrous scaffolds of polycaprolactone containing binary ions of Pd/vanadate doped hydroxyapatite for biomedical applications. J Drug Delivery Sci Technol 70:103153

El-Naggar ME et al (2021) Facile modification of polycaprolactone nanofibers with hydroxyapatite doped with thallium ions for wound and mucosal healing applications. J Mater Res Technol 15: 2909–2917

Garritano AN et al (2018) Palm oil mill effluent (POME) as raw material for biohydrogen and methane production via dark fermentation. Renew Sust Energ Rev 92:676–684

Geoffry K, Achur RN (2018a) Optimization of novel halophilic lipase production by Fusarium solani strain NFCCL 4084 using palm oil mill effluent. J Genet Eng Biotechnol 16(2):327–334

Geoffry K, Achur RN (2018b) Screening and production of lipase from fungal organisms. Biocatal Agric Biotechnol 14:241–253

Gil A et al (2015) Mixture optimization of anaerobic co-digestion of tomato and cucumber waste. Environ Technol 36(20):2628–2636

Greses S, Tomas-Pejo E, Gonzalez-Fernandez C (2020) Agroindustrial waste as a resource for volatile fatty acids production via anaerobic fermentation. Bioresour Technol 297:122486

Gupta R et al (2015) Molecular and functional diversity of yeast and fungal lipases: their role in biotechnology and cellular physiology. Prog Lipid Res 57:40–54

Halim FTA et al (2016) Sustainable microalgae-based palm oil mill effluent treatment process with simultaneous biomass production. Can J Chem Eng 94(10):1848–1854

Harsono SS, Grundmann P, Soebronto S (2014) Anaerobic treatment of palm oil mill effluents: potential contribution to net energy yield and reduction of greenhouse gas emissions from biodiesel production. J Clean Prod 64:619–627

He X et al (2021) Electrospun polycaprolactone/hydroxyapatite/ZnO films as potential biomaterials for application in bone-tendon interface repair. Colloids Surf B Biointerfaces 204:111825

Hemlata B, Uzma Z, Tukaram K (2016) Substrate kinetics of thiol activated hyperthermostable alkaline lipase of Bacillus sonorensis 4R and its application in bio-detergent formulation. Biocatal Agric Biotechnol 8:104–111

Ibrik A et al (1998) Biochemical and structural characterization of triacylglycerol lipase from Penicillium cyclopium. Lipids 33(4):377–384

Jacobsen T, Poulsen OM (1995) Comparison of lipases from different strains of the fungus Geotrichum candidum. Biochim Biophys Acta 1257(2):96–102

Jasni J et al (2020) Comparative toxicity effect of organic and inorganic substances in palm oil mill effluent (POME) using native microalgae species. J Water Process Eng 34:101165

Jegannathan KR, Nielsen PH (2013) Environmental assessment of enzyme use in industrial production – a literature review. J Clean Prod 42:228–240

Jouibary YM, Rezvanpour A, Akbari B (2022) Fabrication of polycaprolactone/collagen scaffolds reinforced by graphene oxide for tissue engineering applications. Mater Lett 322:132477

Kim SY (2019) Application of the three-dimensionally printed biodegradable polycaprolactone (PCL) mesh in repair of orbital wall fractures. J Craniomaxillofac Surg 47(7):1065–1071

Knappmann I et al (2017) Lipase-catalyzed kinetic resolution as key step in the synthesis of enantiomerically pure sigma ligands with 2-benzopyran structure. Bioorg Med Chem 25(13): 3384–3395

Knight K et al (2000) Immobilization of lipase from Fusarium solani FS1. Braz J Microbiol 31(3): 219–221

Knob A et al (2020) A novel lipolytic yeast Meyerozyma guilliermondii: efficient and low-cost production of acid and promising feed lipase using cheese whey. Biocatal Agric Biotechnol 24: 101565

Kotcharat P et al (2021) Development of bacterial cellulose and polycaprolactone (PCL) based composite for medical material. Sustain Chem Pharm 20:100404

Kumar S et al (2011) Use of evolutionary operation (EVOP) factorial design technique to develop a bioprocess using grease waste as a substrate for lipase production. Bioresour Technol 102(7): 4909–4912

Labet M, Thielemans W (2009) Synthesis of polycaprolactone: a review. Chem Soc Rev 38(12): 3484–3504

Li X et al (2017) Effective detoxification of patulin from aqueous solutions by immobilized porcine pancreatic lipase. Food Control 78:48–56

Li Q et al (2019) Improving the thermostability and pH stability of Aspergillus niger xylanase by site-directed mutagenesis. Appl Biochem Microbiol 55(2):136–144

Liu J et al (2004) Kinetic resolution of ketoprofen ester catalyzed by lipase from a mutant of CBS 5791. J Ind Microbiol Biotechnol 31(11):495–499

Louhasakul Y et al (2021) Valorization of palm oil mill wastewater for integrated production of microbial oil and biogas in a biorefinery approach. J Clean Prod 296:126606

Low SS et al (2021) Microalgae cultivation in palm oil mill effluent (POME) treatment and biofuel production. Sustainability 13(6):1–17

Lu Y et al (2019) Immobilized Candida Antarctica lipase B catalyzed synthesis of biodegradable polymers for biomedical applications. Biomater Sci 7(12):4963–4983

Luna C et al (2018) Evaluation of lipases from wild microbial strains as biocatalysts in biodiesel production. Separations 5(4):53

Maaroff RM et al (2019) Biohydrogen production from palm oil mill effluent (POME) by two stage anaerobic sequencing batch reactor (ASBR) system for better utilization of carbon sources in POME. Int J Hydrog Energy 44(6):3395–3406
Mahmoud Salehi AO et al (2021) Use of polycaprolactone in corneal tissue engineering: a review. Mater Today Commun 27:102402
Maidana Serpa JD et al (2022) Sarocladium strictum lipase (LipSs) produced using crude glycerol as sole carbon source: a promising enzyme for biodiesel production. Biocatal Agric Biotechnol 40:102299
Mamimin C et al (2019) Enhancement of biohythane production from solid waste by co-digestion with palm oil mill effluent in two-stage thermophilic fermentation. Int J Hydrog Energy 44(32): 17224–17237
Mandari V, Nema A, Devarai SK (2020) Sequential optimization and large scale production of lipase using tri-substrate mixture from Aspergillus niger MTCC 872 by solid state fermentation. Process Biochem 89:46–54
Martinez-Ruiz A et al (2018) Continuous ethyl oleate synthesis by lipases produced by solid-state fermentation by Rhizopus microsporus. Bioresour Technol 265:52–58
Menazea AA, Abdelbadie SA, Ahmed MK (2020) Manipulation of AgNPs coated on selenium/carbonated hydroxyapatite/ε-polycaprolactone nano-fibrous via pulsed laser deposition for wound healing applications. Appl Surf Sci 508:145299
Mishra P et al (2016) Enhanced hydrogen production from palm oil mill effluent using two stage sequential dark and photo fermentation. Int J Hydrog Energy 41(41):18431–18440
Mo J et al (2018) A review on agro-industrial waste (AIW) derived adsorbents for water and wastewater treatment. J Environ Manag 227:395–405
Mouad AM et al (2016) Aminolysis of linoleic and salicylic acid derivatives with Candida Antarctica lipase B: a solvent-free process to obtain amphiphilic amides for cosmetic application. J Mol Catal B Enzym 126:64–68
Muralidhar RV, Chirumamilla RR, Ramachandran VN, Marchant R, Nigam P (2001) Racemic resolution of RS-baclofen using lipase from Candida cylindracea. Mededelingen 66:227–232
Musatti A et al (2017) Use of solid digestate for lignocellulolytic enzymes production through submerged fungal fermentation. J Environ Manag 199:1–6
Namboodiri VM, Chattopadhyaya R (2000) Purification and biochemical characterization of a novel thermostable lipase from Aspergillus niger. Lipids 35(5):495–502
Narwal V, Pundir CS (2017) An improved amperometric triglyceride biosensor based on co-immobilization of nanoparticles of lipase, glycerol kinase and glycerol 3-phosphate oxidase onto pencil graphite electrode. Enzym Microb Technol 100:11–16
Navvabi A et al (2018) Novel lipases discovery specifically from marine organisms for industrial production and practical applications. Process Biochem 70:61–70
Nazeer MA, Yilgor E, Yilgor I (2019) Electrospun polycaprolactone/silk fibroin nanofibrous bioactive scaffolds for tissue engineering applications. Polymer 168:86–94
Nhi TT et al (2016) Fabrication of electrospun polycaprolactone coated withchitosan-silver nanoparticles membranes for wound dressing applications. J Mater Sci Mater Med 27(10):156
Nimkande VD, Bafana A (2022) A review on the utility of microbial lipases in wastewater treatment. J Water Process Eng 46:102591
Nooeaid P et al (2020) Polylactic acid microparticles embedded porous gelatin scaffolds with multifunctional properties for soft tissue engineering. J Sci Adv Mater Devices 5(3):337–345
Norfadilah N et al (2016) Bio-hydrogen production from palm oil mill effluent (POME): a preliminary study. Int J Hydrog Energy 41(28):11960–11964
Pakalapati H, Tariq MA, Arumugasamy SK (2019) Optimization and modelling of enzymatic polymerization of epsilon-caprolactone to polycaprolactone using Candida Antartica lipase B with response surface methodology and artificial neural network. Enzym Microb Technol 122: 7–18

Phukon LC, Chourasia R, Kumari M, Godan TK, Sahoo D, Parameswaran B, Rai AK (2020) Production and characterisation of lipase for application in detergent industry from a novel pseudomonas helmanticensis HS6. Bioresour Technol 309:8
Prasertsan P, Binmaeil H (2018) Treatment of palm oil mill effluent by thermotolerant polymer-producing fungi. J Water Environ Technol 16(3):127–137
Prasertsan P et al (2021) Enhanced biogas production by co-digestion of crude glycerol and ethanol with palm oil mill effluent and microbial community analysis. Biomass Bioenergy 148:106037
Putri DN et al (2020) Optimization of Aspergillus niger lipase production by solid state fermentation of agro-industrial waste. Energy Rep 6:331–335
Ramos Sanchez LB (2015) Fungal lipase production by solid-state fermentation. J Bioprocess Biotech 05(02):9
Ratnasari A et al (2022) Advances in pretreatment technology for handling the palm oil mill effluent: challenges and prospects. Bioresour Technol 344(Pt B):126239
Rosa D et al (2020) Biological hydrogen production from palm oil mill effluent (POME) by anaerobic consortia and Clostridium beijerinckii. J Biotechnol 323:17–23
Rupani PF et al (2022) Changes in the microbiota during biological treatment of palm oil mill waste: a critical review. J Environ Manag 320:115772
Safari B et al (2022) A bioactive porous scaffold containing collagen/phosphorous-modified polycaprolactone for osteogenesis of adipose-derived mesenchymal stem cells. Eur Polym J 171:111220
Said M et al (2021) Production of methane as bio-fuel from palm oil mill effluent using anaerobic consortium bacteria. J Clean Prod 282:124424
Selvakumar P, Sivashanmugam P (2017) Optimization of lipase production from organic solid waste by anaerobic digestion and its application in biodiesel production. Fuel Process Technol 165:1–8
Shi K et al (2020) Enzymatic hydrolysis of polyester: degradation of poly(epsilon-caprolactone) by Candida Antarctica lipase and fusarium solani cutinase. Int J Biol Macromol 144:183–189
Sidebottom CM et al (1991) Geotrichum candidum produces several lipases with markedly different substrate specificities. Eur J Biochem 202(2):485–491
Strazzera G et al (2018) Volatile fatty acids production from food wastes for biorefinery platforms: a review. J Environ Manag 226:278–288
Sugihara A et al (1995) Purification and characterization of a lipase from Pichia burtonii. Appl Microbiol Biotechnol 43(2):277–281
Sztajer H et al (1992) Purification and properties of lipase from Penicillium simplicissimum. Biochim Biophys Acta 1124(3):253–261
Tan YD, Lim JS (2019) Feasibility of palm oil mill effluent elimination towards sustainable Malaysian palm oil industry. Renew Sust Energ Rev 111:507–522
Taskin M et al (2016) Lipase production with free and immobilized cells of cold-adapted yeast Rhodotorula glutinis HL25. Biocatal Agric Biotechnol 8:97–103
Thapa S et al (2019) Biochemical characteristics of microbial enzymes and their significance from industrial perspectives. Mol Biotechnol 61(8):579–601
Toida J et al (1998) Purification and characterization of triacylglycerol lipase from Aspergillus oryzae. Biosci Biotechnol Biochem 62(4):759–763
Van der Schueren L et al (2011) An alternative solvent system for the steady state electrospinning of polycaprolactone. Eur Polym J 47(6):1256–1263
Vivek K, Sandhia GS, Subramaniyan S (2022) Extremophilic lipases for industrial applications: a general review. Biotechnol Adv 60:108002
Wee ANCH et al (2022) Techno-economic assessment of hydrogen production via steam reforming of palm oil mill effluent. Sustain Energy Technol Assess 53:102575
Woodruff MA, Hutmacher DW (2010) The return of a forgotten polymer—Polycaprolactone in the 21st century. Prog Polym Sci 35(10):1217–1256

Wu TY et al (2009) A holistic approach to managing palm oil mill effluent (POME): biotechnological advances in the sustainable reuse of POME. Biotechnol Adv 27(1):40–52
Xin J et al (2017) Effect of post weld heat treatment on the microstructure and mechanical properties of ITER-grade 316LN austenitic stainless steel weldments. Cryogenics 83:1–7
Ying K et al (2023) Macrophage membrane-biomimetic adhesive polycaprolactone nanocamptothecin for improving cancer-targeting efficiency and impairing metastasis. Bioact Mater 20:449–462
Zulfahmi I et al (2021) Phytoremediation of palm oil mill effluent (POME) using water spinach (Ipomoea aquatica Forsk). Environ Technol Innov 21:101260

Engineered Microbial Sensing Element-Based Biosensor for Sustainable Biomedical Engineering Application

10

Salmah Binti Karman, Suriani Binti Ibrahim, and Chan Chow Khuen

Abstract

This chapter discusses engineered microbial-based biosensors. The chapter focuses on explaining the key components, basic principles, and applications in the medical field. The development of low manufacturing cost, and highly sensitive, selective, and stable microbial biosensors is the main goal of this research. Their performance would be mostly affected by the microbial-incorporated sensing element, which is the main component of these biosensors. There are hundreds of microbes incorporated as bio-sensing elements for detecting chemical and biological substances. There are electrochemical and optical-based microbial biosensors that are mostly used for medical application, which represent their transducer's working principles. A deep knowledge on the sensing elements as well as transducer principles is very important in order to optimize the performance of these devices.

Keywords

Biosensor · SPR · Engineered microbial · Electrochemical · Optical · Genetically encoded protein · Whole-cell based

S. B. Karman (✉) · C. C. Khuen
Department of Biomedical Engineering, Faculty of Engineering, Universiti Malaya, Kuala Lumpur, Malaysia
e-mail: salmah_karman@um.edu.my

S. B. Ibrahim
Department of Mechanical Engineering, Faculty of Engineering, Universiti Malaya, Kuala Lumpur, Malaysia

W. S. Wan Kamarul Zaman, A. A. Nurul (eds.), *Sustainable Material for Biomedical Engineering Application*, https://doi.org/10.1007/978-981-99-2267-3_10

Abbreviations

AFM	Atomic force microscopy
BW	Bioelectrochemical wire
CP	Conducting polymer
DET	Direct electron transfer
DNA	Deoxyribonucleic acid
ECM	Electrochemical machining
EET	Extracellular electron transfer
ELISA	Enzyme-linked immunosorbent assay
GBP	Gold-binding peptide
GEM	Genetically engineered microorganisms
Gfp	Green fluorescent protein
GoIS	Genes of interest
IMBED	Ingestible micro-bio-electronic device
Luc	Luciferase
lucZ	Beta-galactocidase
Lux	Licberase
MES	Electrochemical system
MET	Mediated electron transfer
METC	Microbial electron transport chain
MFC	Microbial fuel cell
MT	Metallothionein
PCR	Polymerase chain reaction
QSM	Quorum-sensing molecule
RNA	Ribonucleic acid
scFv	Single chain variable fragment
SPE	Screen-printed carbon electrode
STM	Scanning tunnelling microscopy

10.1 Introduction

The first biosensor was developed by Clark and Lyon in 1962 for the detection of glucose. From then onward, other researchers have extended the study and explored the potential of biosensors to be used and applied in areas, such as healthcare, environmental, pharmaceutical, and food industry (Amine et al. 2006). A biosensor is a device that detects, captures, analyses, and interprets the biological or chemical responses from the surroundings, and then transforms it to readable signals, such as current, voltage, or colour signals (Haleem et al. 2021). The biological or chemical reactions could be produced from the oxidation-reduction process done by the biological components, such as enzymes, cells, or RNA. The biological recognition element, which is a part of the biosensors, is an important component to identify and detect a signal within the target compound or analyte sample (Daunert et al. 2000; Neethirajan et al. 2018). The biological response obtained by this recognition element is then measured and converted to readable electrical signals by the transducer through conversion principles, such as electrochemical, optical, acoustical, mechanical, calorimetric, or electronical approaches (Byfield and Abuknesha 1994; Cunningham 1998; Eggins 2002; Wilson and Gifford 2005).

The typical output of the transducer used in a biosensor is presented in electric forms, and are submitted to the supplementary electronic device for result reading by the user (Haleem et al. 2021; Bilal and Iqbal 2019). This electronic component is usually integrated with the signal processors, and is capable to process and exploit the data for an extended data analysis. The measurable signal from the transducer that is equipped with an electronic device represents the status of analytes' output, such as concentration, types of products, and others (Cunningham 1998; Eggins 2002; Wilson and Gifford 2005; Wilson 2004; Sadana 2001; Lei et al. 2006; Belkin 2003; Su et al. 2011). The microbial biosensor is critically used to track and monitor the biological changes in the environmental, medical, agricultural, security, and water management. In medical applications, the biosensor is employed to monitor the general healthcare, disease diagnosis, infection screening, clinical analysis, as well as trackers, such as in wearable devices (Mowbray and Amiri 2019; Kucherenko et al. 2019). The glucose biosensors are highly important in measuring the blood glucose levels, thus assisting people to control their blood sugar to normal levels (Bahl et al. 2020; Kamel and Khattab 2020).

10.2 Microbial Biosensor

The integration of microbes as a biological sensing element has allowed biosensors to detect the target substance through the biological reaction process (Su et al. 2011). The typical biosensing platforms always incorporate biological materials, including cells, antibodies, and enzymes. In microbial biosensors, the sensing elements consist of parts of microbes or entire cells. The natural capability of living cells to detect various signals from the surroundings has attracted and inspired engineers to develop the bioinspired sensing device. Microbes used as sensing elements should fulfill all the criteria of low cost, and good portability, user friendly, and stability in various conditions. The microbes possess a simpler preparation method, skipping the purification process required, such as in enzymes preparation process.

The advanced technology of synthetic biology nowadays has provided an engineered microbial production that ultimately supports the microbial biosensors with specific capability and high efficiency. Due to the mass production through synthetic biological technology and cell culture, the engineered microbes of algae, bacteria, and yeast will be the most selected for the fabrication of biosensors due to their good endurance in vitro (Byfield and Abuknesha 1994; Su et al. 2011; Brennan et al. 2016; Liu et al. 2014). Incorporating this type of microbes will help in simplifying the fabrication steps for the sensing platform, and improving the performance of biosensors. By taking advantage of genetically engineered microorganisms

(GEMs) technology to manipulate the metabolic process of living microbes, the highly selective and sensitive microbial biosensors to a number of chemicals at the DNA level can be potentially achieved (Belkin 2003; Paul et al. 2005; Urgun-Demirtas et al. 2006). This can be done by blocking or inducing any metabolic, microbial approach when responding to the target substrate. Furthermore, the engineered cells are able to produce the precursors for the subsequent synthesis of materials, thus increasing the specificity (Burgos-Morales et al. 2021). By exquisitely attuning their metabolism to their environment, the engineered cells could appropriately capture, process, and modulate the target signal.

10.3 Development of Microbial Based Biosensor for Medical Application

The current strategies of polymerase chain reaction (PCR) and enzyme-linked immunosorbent assay (ELISA) are the gold standard for pathogens' nucleic acids detection, but they are highly costly, time consuming and complicated, and requiring expert operators (Yager et al. 2008; Fu et al. 2011). Due to the resource-limited settings, most of the testing for infectious diseases is performed by microscopy or agglutination tests, thus reducing the efficiency (Chang et al. 2017).

The demand for portable, on-site, and real-time screening devices nowadays has increased the biosensors' popularity among scientists. The biosensors research has obtained a major impact recently, contributing to many disciplines of applications, such as environmental monitoring (Cai et al. 2018; Gao et al. 2017; Vopálenská et al. 2015; Yang et al. 2016), clinical diagnosis (Jiang et al. 2015; Pandey et al. 2018), food safety (Jiang et al. 2015; Zou et al. 2016), geo-exploration (Zammit et al. 2013), and others. The types of microbes used for sensing elements would influence the functional strategy of the device to detect the target analyte due to the response time, sensitivity, and specificity (Kumar et al. 2017; Singh et al. 2012). The existing enzymatic sensing platform has been proven to yield high specificity in responding to the target analytes (Kumar et al. 2017; Singh et al. 2012; Pandey et al. 2017). However, due to the high cost and limitation of enzymes in certain conditions, living cells become an alternative to these molecular-based sensing elements. Due to the wide variety of molecules in different proportions as well as advances in synthetic biology, the living cells or microbes offer quantitative response to specific stimuli of single or multi-types of analytes with less effort and expenses (Corcoran and Rechnitz 1985; Courbet et al. 2015). The ability to sense the multi-types of analytes is also encouraged by the incorporation of microfluidics technology in microbial-based biosensor development (Gupta et al. 2019; Pasternak et al. 2017; Shukla et al. 2018; Syal et al. 2016; Tucci et al. 2019). The incorporation of microbial cells has enabled the sensor to continuously monitor the chemical interaction, active agent's effect, and environmental conditions, all of which are highly beneficial for pharmaceutical application (Liu et al. 2014; Cevenini et al. 2016; Pan et al. 2015).

The biological reaction outputs from the diagnosis programmes, such as cancer screening, are commonly measured by using the electrochemical or optical-based biosensors (Jin et al. 2020; Chen and Shamsi 2017). In cancer diagnosis, the microbial biosensor is used to detect cancer biomarkers, which are the significant indicators of tumor growth, hence providing a prognostic approach to treatment. Nowadays, the microbial biosensor technology has led to the discovery of new precise and potent analytical sensors that can efficiently work in detecting the target substance in sensitive bioelements, such as tissues, organelles, cells, enzymes, antibodies, and nucleic acids (Haleem et al. 2021).

10.4 Transducer

The living cell-incorporated biosensor is sometimes known as metabolite biosensors (Liu et al. 2015), which involve genetically-encoded protein or RNA. When this genetically-encoded protein interacts with a metabolite, a reaction process occurs in the host cell that has the target analyte, generating a readable output signal. The output generated by the protein expression is controlled by three main parameters: modulating transcription rates, translation rates, and post-translational rate.

The typical electrochemical transducer reads out the signal from the redox reaction process in the microbes' metabolism when the target analyte is presented (Su et al. 2011). The performance of electrochemical transducer-based biosensors mostly relies on the electron transfer activity instead of the type of microbes. In fact, the electron transfer across the electrode of an electrochemical transducer can be improved by integrating the redox active mediators externally. However, electrochemical systems can be easily interfered with by the interfering species, hence reducing its performance. The performance of the optical transducer-based biosensor strongly relies on the microbes' genetics. The genetic modification in the microbes will influence the quality of visual output in responding to the target analyte. The analyte-specific promoter and reporter genes are always involved in the genetic modifications of the microbial cells that act as a biosensor's sensing element

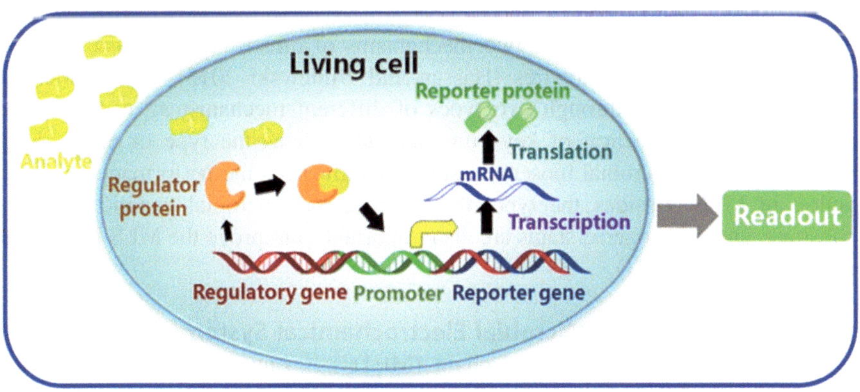

Fig. 10.1 A schematic diagram representing a typical microbial biosensor. (Reproduced under the Creative Common Licence from (Gui et al. 2017))

(Fig. 10.1). When the sensing element's cell probe is exposed to the target compound in the sample, the promoter gene will be activated, hence inducing the reporter genes to express the fluorescent proteins. The optical biosensors could offer a higher specificity than the electrochemical-based biosensor could. However, the manufacturing cost of optical-based biosensors could be more expensive due to the gene modification requirement. Even though the fluorescent protein-based optical biosensor could offer high specificity, it may provide low reliability and stability in terms of sensitivity among the different batches of the cell populations (Gupta et al. 2019). This has been indicated in a recent study on the detection and measurement of reactions between analyte and immobilized microbes on sensing platforms (Su et al. 2011).

10.4.1 Electrochemical Techniques

Electrochemical principle is a common technology utilized in the development of microbial biosensors. The technology that integrates the microbial systems, electrochemical principle, and sensing device application could be defined as an electrochemical system (MES) (Selim et al. 2017). MES exploits the biocatalytic activity, which involves electron transfer from a microorganism cell to the working electrode, acting as an electron donor and electron acceptor, respectively (Khater et al. 2015). The ability of electrons to transfer effectively from one part to the other part is crucial to obtain a high-performing electrochemical-based biosensor. There are two types of interaction involved in the MES system. Based on the nature of these interactions, MESs can use processes that involve:

1. Direct interaction between electrode and cell, involving the extracellular electron transfer.
2. Indirect interaction represented by the electrochemical interaction among the microbial environment such as pH level.

More specifically, there are two mechanisms of direct electron transfer: cell's outer membrane and nanowires (Hassan and Bilitewski 2013). Meanwhile, the indirect transfer runs through two types of different mechanisms: mediators and catabolites. The selection of key components, such as the type of electrode in constructing the microbial biosensor, strongly influences the performance of interaction in MES. Besides the type of electrode, the microbial genetic, types of substrate, and electrogenic status are also important to improve the MES interaction (Mahmoud et al. 2021).

10.4.1.1 Principles of Microbial Electrochemical Systems (MES)

The microbial electron transport chain (METC) is one of the basic principles incorporated in MES. The principles involve the reduction-oxidation reaction process of organic substrates (Wesolowski et al. 2008). The cellular activity can be used as a main indicator of electron transport efficiency in microbial respiration (Santoro et al. 2017). One of the examples of direct electron transfer can be observed in the measurement of dissolved oxygen consumption by microbes, with the electrical current generated indicating the concentration of viable cells (Holder et al. 2017). On the other hand, the number of non-viable microbial cells could be known due to the zero output of electrochemical signals. Here, strong correlation exists between electrochemical response and redox reaction, anodic respiration, and cell-electrode interaction (Chouler et al. 2018; Sedki et al. 2019). This capability has allowed further development of high-performance microbial biosensors to be effectively used in the environmental and biomedical field.

10.4.1.2 Extracellular Electron Transfer Based Biosensor

The direct transfer of electrons from the cell to the electrode via the cytoplasmic membranes as an external electron transport chain comprises an extracellular electron transfer (EET) (Hassan and Bilitewski 2011; Zhao et al. 2021). The performance of this kind of biosensors is strongly influenced by the electron transfer, which relies on the microbial cell wall and electron transporter membrane. The modification on the proposed cell to be used as the biosensor's probe would be important in device optimization.

10.4.1.3 Direct Electron Transfer (DET) Via Electroactive Microbes

The exoelectrogen microbes, which are the electroactive species, play an important role in transferring the electron directly to conductive materials without the assistance from mediators (Logan et al. 2019; Yang et al. 2012; Kumar et al. 2016). As an electroactive organism, the exoelectrogens perform the redox process at anode (Kracke et al. 2015), transferring the electron to the surface electrode through physical contact (Shi et al. 2016).

10.4.1.4 Mediated Electron Transfer

The mediated electron transfer (MET) process is required when the artificial redox mediators are used to enable the electron transfer from an inactive electrochemical microbe to the electrode (Pankratova and Gorton 2017; Hernandez and Osma 2020; Simoska et al. 2021; Wu et al. 2014a; Schmitz and Rosenbaum 2018). The application of known artificial redox mediators in electron transfer process and the resulting microbe-electrode interaction behaviour are well documented, thus having potential to be used as a rapid detection application (Hassan and Wollenberger 2016; Fang et al. 2020; Kwon et al. 2018; Li et al. 2013). The examples of single and double electron mediators are 2,6-dichlorophenolindophenol (Hassan and Bilitewski 2011), and 2,3,5,6-Tetramethyl-p-phenylenediamine, respectively. Another example of a double electron mediator is the combination of menadione and hexacyaanoferrate (III) (Rawson et al. 2014; Zhao et al. 2007). In one of the examples, the double-mediator system was used to amplify the electron activity between NAD(P)H: quinone interaction (Leung et al. 2013). In the mediated electron transfer principle, both anode and cathode are involved in the microorganism metabolism process for oxidation reaction and reduction reaction, respectively. From this metabolism-based redox process, the electron will be released by the anode, and then transferred to the cathode, which acts as an acceptor (Logan et al. 2019).

10.4.1.5 Bioelectrochemistry of Biofilms Based Biosensor

The formation of electroactive biofilm plays an important role as a medium for direct communication between metabolic-active microorganisms with the electrode in non-mediated electrochemical systems (Mahmoud et al. 2021, 2020, 2018; Cong et al. 2013). The biofilms are developed by mixing the heterogenous microbial cells with exopolysaccharides matrix, as well as containing the proteins and extracellular DNA (Saratale et al. 2017; Flemming and Wingender 2010; Azeredo et al. 2017). It has been proven that it can be used as the best substrate for living cells. The performance of this biofilm strongly relies on bacterial species and environmental conditions. The performance of biofilm-based microbial electrochemical systems can be improved by selecting the types of biofilm formation (Halan et al. 2012), and different electrode modifiers (Mahmoud et al. 2020; Cornejo et al. 2015).

10.4.1.6 Microbial Fuel Cells Based Biosensor

The device that converts chemical energy to electrical energy is commonly known as fuel cells. The device that exploits the redox process reactions from microbial respiration and generates the electron transfer through the electrochemical system could be defined as microbial fuel cells (Dolch et al. 2014). Incorporating the principles of electrochemical systems, the microbial fuel cells have key components that include anode and cathode electrode, external circuit, electrogenic cell and its culture medium, and biofilm (Santoro et al. 2017). The cells' regulator factor, biofilm structure adaptation, and type of microbials strongly affect the metabolic reaction between cell and electrode. Thus, the extended research on these key components is important for device optimization (Katuri et al. 2020).

10.4.1.7 Impact of the Electrode Materials on the Performance of MESs

As discussed in the previous paragraph, the electrode is a place where the microbial is attached and performs a redox process through respiration activity, generating and transferring the electron from anode to cathode whether using a mediator or not. The conventional carbon-based electrochemical electrodes built from materials, such as graphite granules, graphite felt and others, have the limitation due to the small surface area (Sharma et al. 2019; Zhou et al. 2011). The modification on the electrode material and design, as well as the improvement on fabrication step will help enhance the electrochemical system's performance. The use of the nanostructured materials for electrode fabrication will increase the performance of the electrode due to the high surface area, conductivity, and catalytic activity (Dige et al. 2007; Liu et al. 2020). Furthermore, using innovative carbon materials, such as carbon fibres and carbon nanotube, will expand the surface area of the electrochemical active, hence improving the device performance (Qiao et al. 2007; Sharma et al. 2008).

10.4.2 Optical Techniques

Optical transducer could be defined as a device that converts any reaction responses to visible output, such as luminescent, fluorescence, and colour signals. The optical-based microbial biosensor is a device which is developed by incorporating an optical transducer, measuring the microorganism-analyte reaction captured by the bio-sensing platform and expressing this response through the visible signal. The concentration of target analytes corresponds to the observed optical signal. The performance of the device in expressing the output strongly relies on the microorganism's gene, especially involved in the optical output signal. The current technology of synthetic biology has contributed to the major successful factors, with capability to genetically modify microorganisms as requested. Normally, optically active microorganisms always consist of the optically active gene inside their RNA, also known as the inducible reporter gene. When the target compound is detected, the reporter gene will be regulated by the regulator protein to induce the expression of a measurable optical signal (Sharma et al. 2019).

In bioluminescence-based biosensors, the reaction between the target analyte and the microorganism used in the sensing platform will induce the microorganism's reporter gene to express the luminescent light. In colorimetric sensing technique, the chromogen substrate is converted to the colourful material to identify the presence of target analyte. The colorimetric technique is commonly used as a cost-effective biosensor due to its simple and inexpensive fabrication process.

10.4.2.1 Reporter Gene

A reporter gene is a gene attached to a regulatory sequence (promoter) of another gene of interest (from bacteria, virus, animal genome, etc.). The reporter gene coding for an oxidative enzyme (like Luciferase) could convert a substrate into a product that can be quantified by measuring the released light. The reporter gene is capable of tracing the physical location of the DNA segment to monitor the gene expression (Clark et al. 2019).

The types of reporter gene strongly determine the performance of the microbial biosensor in measuring the target analyte. The sensitivity and selectivity of the device is highly affected by the performance of a reporter gene in converting a biological response into the readable signal. Among the reporter genes that can successfully yield high-performing biosensors are bacterial licberase (lux), firefly luciferase (luc), green fluorescent protein (gfp), and Beta-galactocidase (lucZ). Gfp does not require an ATP to emit fluorescent light (Singh et al. 2000). However, using the gfp for certain host cells will make the biosensor less sensitive due to interference. The lux and lacZ-based biosensors usually offer higher sensitivity compared to gfp (Stocker et al. 2003). However, lux is weak to thermal and protein interference, and unsuitable to be used in mammal cells (Naylor 1999). The luc reporter gene has an advantage compared to lux and gfp due to its sensitivity to a wide range of responses (Joyeux et al. 1997; Welsh and Kay 1997). The enzymatic lucZ material is excellent and efficient in monitoring transfection, making it attractive to be used in molecular biology sensing devices. It has a great capability in colorimetric and fluorescent expression, with potential for rapid and simple biosensors (Jain and Magrath 1991). LucZ reacts to widely available chemiluminescent and electrochemical substrates, making it highly sensitive to low limit of detection and dynamic range of samples (Biran et al. 1999; Karttunen and Shastri 1991).

10.4.2.2 Transcriptor

A transcription factor is a protein responsible for controlling the transcription activity of genes from DNA to RNA. The transcription process involves the enzymatic catalyzation in the RNA synthesis process using the template origin from the gene's DNA. During the transcription process, the transcription factors bind to the regulatory gene also known as uncode DNA. After the binding, the code from the coding DNA is then expressed to the uncode DNA. The transcription process will not happen if there is no binding between the transcription factor and the regulatory gene.

Transcription factors are the key components in determining the specificity level of the device in responding to environmental conditions. These factors play an important role in expressing or restricting the reaction to the specific tissue. The importance of transcription factors on the structural and functional features has been reported previously (Molnár 2016; Ishihama 2013).

10.4.2.3 Regulatory Proteins

The regulatory protein, which is the key component in controlling the transcription process, has a high impact on the specificity and sensitivity of microbial biosensors in responding to target analytes. The GoIS protein is a regulatory protein that is highly selective to Au (Pontel et al. 2007). The combination of GoIS with the single amino acid has improved its selectivity to the greater range of target analytes (Cerminati et al. 2015), such as mercury, lead, and cadmium. Metallothionein (MT) protein is a regulatory protein that is highly selective to various heavy metals (Shang et al. 2002; Diaz et al. 2007). The uses of MT origin of MTT5 and MTT1 promoter genes in biosensors have increased the sensitivity and reduced the time response of the device to screen heavy metal contaminations (Xie et al. 2004). The presence of chpA promoter from utilization of *E. coli* as a sensing element has enabled the development of a highly sensitive device in detecting the CPF (Whangsuk et al. 2016; Zhang et al. 2012). The use of ars and crtA reporter genes has enabled the production of highly sensitive microbial biosensors in detecting very small samples of arsenite, as small as 5 μg/L (Fujimoto et al. 2006). The study has shown that these reporter genes have expressed colour change from yellow to red, representing the changes of arsenite concentration (Yeliseev et al. 1996; Wu and Rosen 1991).

10.4.2.4 Host Cells

The microbes used as a sensing element of the biosensor are commonly known as host cells. The host cell is the most important component in the development of biosensors since it detects and measures the target analytes through the involvement of the cell's components, such as cell wall, gene, protein, and enzymes. As described in previous sections, the types of the host cell greatly impact the performance of biosensors in terms of specificity, sensitivity, and time response. The examples of host cell use for microbial biosensors are *Pseudomonas putida* DOT-T1E (Espinosa-Urgel et al. 2015), *E. coli* (Zhang et al. 2012), and *Deinococcus deseri* (Brutesco et al. 2017). The excellent properties of ADP1 as a host cell include the high sensitivity of biosensors in detecting the alkanes and alkenes in water and soil samples (Gui et al. 2017; Zhang et al. 2012; Throne-Holst et al. 2007; Ratajczak et al. 1998).

10.5 Design and Principle of Microbial-Based Biosensor

The microbial-based biosensors are classified according to the sensing element principles, which incorporate the microbial elements, including the types of cells, molecules, and tissues (Lei et al. 2006; Pancrazio et al. 1999). The biosensor classified as the molecular type of sensing element uses an active substance, such as enzyme, DNA, and antigen for their sensing platform (Bousse 1996). While the enzymatic molecular-based biosensor offers high selectivity in sensing activity (Du et al. 2005), the steep cost to prepare enzymes, low detection, and short lifespan of enzyme molecules limit its application (Daunert et al. 2000; Gui et al. 2017). The whole-cell-based biosensor could be an alternative to the molecular-based biosensor to compensate for the limitation. The whole-cell-based biosensor is developed by incorporating living cells on the sensing element that is coupled with the transducer, generating the quantifiable signal that responds to the specific analytes (Gui et al. 2017).

Two main parts make up the microbial biosensor. The first is the primary transducer, made up from living cells. It is the sensing element to receive and produce signals. The second is the secondary transducer, which converts physiological signals to electrical signals. In brief, a typical whole-cell-based biosensor recognizes a type of analyte, amplifies, and converts the identified signal into an optical or electrical signal through a processor. Therefore, the bacteria or living cells are the elements that enables recognition of molecules as part of the readout mechanism (Wang and Liu 2010).

To improve the efficiency of a whole-cell-based biosensor, the type of regulatory proteins and reporter genes of the sensing element are important. Real-time portable devices can be developed through improved incorporation of cellular sensors with nano and micro-scaled chips, benefitting expansive applications from remote environmental toxin detection to bedside diagnostics. Through the genetic modification, the working cell could be engineered to function over an increased range of parameters, such as pH values and temperature (Ben-Yoav et al. 2009; Tian et al. 2017; Behzadian et al. 2011), thus increasing the capability, sensitivity, and selectivity for high-throughput in situ detection (Xie et al. 2004; Raut et al. 2012). Thus, its potential is significant for the future application in environmental monitoring and medical diagnostics.

10.6 Immobilization Strategies

For a biochemical response to be effectively transformed into a physical signal, the recognition element or microbial cells need to be intimately linked with the transducer. Consequently, the immobilization of microorganisms on transducers is critical to develop microbial biosensors. This immobilization can be done via cross-linking, covalent binding, entrapment, encapsulation, adsorption, and others (Cunningham 1998; Eggins 2002; Lei et al. 2006; Belkin 2003; Bjerketorp et al. 2006; D'Souza 2001a,b; Ding et al. 2008; Xu et al. 2006). Improvement of microbial immobilization can be obtained through a silicate network obtained from the sol-gel owing to the ability to adjust the pore size of the entrapment materials, increasing the mass transport of analytes (Song et al. 2005; Yu et al. 2005).

Due to the electrochemical characteristics of conducting polymer (CP), CP has gained traction for microbial immobilization (Su et al. 2011; Ahuja et al. 2007; Malhotra et al. 2006). The attachment of living cells onto detecting platforms or transducers has been tested via multiple immobilization strategies: covalent linkage, hydrogel entrapment or encapsulation, immobilization using electrochemical machining (ECM) components, and direct immobilization of on electrode or chip. Generally, determining the immobilization strategy involves several factors: the biosensor application, requirements for cell growth and viability, substrate toxicity, type of material or substrate for cell immobilization, and cell types utilized as bioreceptors (Gupta et al. 2019).

10.6.1 Passive Immobilization

Passive immobilization techniques depend on the cells' capability to attach to surfaces. The example techniques are those involving functional strategies that do not influence the cell viability, such as physical adsorption (Fang et al. 2016; Yamashita et al. 2016), hydrogel or sol gel entrapment or encapsulation, (Gao et al. 2017, 2016; Axelrod et al. 2016) and biofilm formation (Chouler et al. 2018; Catal et al. 2018; Webster et al. 2014). The microbial fuel cells (MFCs), cellulose nitrate, and polycarbonate films are among the biofilms applied recently to efficiently immobilize cells (Chee 2016; Qiao et al. 2015; Wei et al. 2017). The silicone beads and packed bed reactors have been utilized to form and adsorb biofilm for highly efficient immobilization of different green algae species (Haigh-Flórez et al. 2014; Tahirbegi et al. 2017).

Applying physical adsorption and biofilm formation have other benefits, such as direct contact between the cells and transducing surface, increasing sensitivity. A notable cell immobilization technique is entrapment in sol gel or hydrogel. Chitosan, agarose, and calcium alginate (or alginate-starch) are some of the most commonly utilized gel matrices. Besides maintaining viability, the matrices act as a 3D tissue-like environment so that cells can grow and proliferate. To enhance the signal transduction in entrapment immobilization technique, the incorporation with supportive materials is important. Those supportive materials are conductive materials, such as graphene oxide, carbon nanodots, and others (Fang et al. 2016; Jiang et al. 2017); nanomaterials, such as Magnetic Fe_3O_4 nanoparticles for labelling *Bacillus subtilis* cells (Wang et al. 2017); nanoporous gold, such as for selective detection of sulfide (Liu et al. 2017); and silicon-based nanostructures to, among others, immobilize breast cancer cells (Shashaani et al. 2016; Zanganeh et al. 2016).

10.6.2 Active Immobilization

Active immobilization requires chemical linkers or other materials to immobilize living cells onto an appropriate substrate, such as glutaraldehyde crosslinking. Active immobilization can be efficiently used instead of passive immobilization. Tight attachment of cells to the transducer surface leads to increased sensitivity. In active immobilization, bacterial cells are immobilized on various substrates using biomolecules that bind to receptors and other moieties occurring on the cell surface (Syal et al. 2016; Leonard et al. 2017; Safavieh et al. 2017).

10.6.3 Microfluidic/Lab on Chip Integration

Following the incorporation of microfluidics and biological sciences, specialized microfluidic chips have been developed to culture living cells within micro-sized wells, allowing cell proliferation (Popova et al. 2016). Consequently, specialized immobilization techniques are necessary for the microfluidic culture chips.

Microbial and animal cells, such as cancer, bacterial, and algal cells, cultured and grown in microfluidic chips have been numerously reported. Moreover, many kinds of transducers have been incorporated with these chips to capture responses to particular parameters and bioactive analytes, developing cell microarrays for high-throughput molecular detection (Weingeist et al. 2013; Wu et al. 2014b).

The lab-on-chip concept to analyse and monitor numerous cellular parameters allows platforms to achieve single cell resolution (Shukla et al. 2018; Wang et al. 2018). Recent efforts are spent on developing conductive hydrogels and 'soft' substrates that have both conductive electrodes for transduction and matrices for immobilization.

10.7 Microbially-Derived Whole-Cell Based Biosensors in Medical Application

Microbially-derived, whole-cell-based biosensors are widely used in healthcare. The engineered bacteria are used for in vitro diagnosis of infection by quorum molecules (Kumari et al. 2008). Synthetic genetic-engineered circuits are used to detect, amplify, and analyse pathological biomarkers in clinical samples, such as irregular glucosuria in the urine of diabetic patients (Courbet et al. 2015; Pham et al. 2022). The engineered *E. coli* derived whole-cell-based biosensor has been successfully implemented in in vivo diagnosis (Slomovic et al. 2015) to detect liver metastasis (Danino et al. 2015; Piñero-Lambea et al. 2015). Highly sensitive POC technology for biosensing devices is developed using yeast-based antibody display (Colby et al. 2004; Venkatesh et al. 2015; Aronoff-Spencer et al. 2016). The antibody display was also used to detect the *Salmonella* or Hepatitis C virus (Chang et al. 2017). Pathological bile salts as biomarkers in human serum samples are used as substrates in biosensing devices, allowing for detection with high specificity and sensitivity within 2 h (Chang et al. 2018, 2021).

The two-input zinc-sensing, engineered whole-cell-based biosensor incorporates the isopropyl ß-D-1-thiogalactopyranoside-inducible elements in *E. coli* and zinc-responsive transcription elements (McNerney et al. 2019), generating coloured pigments that correspond to different concentrations of zinc. Moreover, the measurement of zinc concentration at hour 12 is possible because the sensing cells sustain strong activity after rehydration and lyophilization in human serum samples. Besides that, microbial sensors and sensing devices can screen biomarkers in inaccessible body niches, such as the gastrointestinal (GI) tract.

The engineered cells also could be incorporated in designing microbial cells used for live diagnostics tools for human biomarkers and microbiome in the GI tract, including nitrate and tetrathionate (Daeffler et al. 2017; Riglar et al. 2017; Woo et al. 2020). The microbial biosensor that can simultaneously detect many indicators like oxygen, lactate, and pH has been developed (Chien et al. 2022). In the future, this device can be used so that cells can deliver drugs at localized and specific areas in the GI tract. Moreover, core bacteria sensors embedded within miniaturized ingestible micro-bio-electronic devices (IMBED) allow wireless transmission of the signal from the mammalian gut exposed elevated heme levels and to monitor internal organs (Mimee et al. 2018).

The whole-cell-based biosensors can constantly analyse and monitor various physiological parameters impacted by external factors, for example differences in the cell's action potential, impedance, and metabolism (Wolf and Mascher 2016). Consequently, at a cellular level, the living cell-based sensors are helpful to comprehend biological metabolic disorders and other diseases, hence improving health diagnosis, drugs evaluation, and cellular physiological analysis. Using immobilized bacteriophages as recognition receptors, a novel whole-cell-biosensor technique can specifically, rapidly, accurately, and directly detect bacteria (Shabani et al. 2008). The bacteriophages are covalently attached to a microarray of functionalized screen-printed carbon electrode (SPE) to directly detect *E. coli.*

The whole-cell-based biosensors are non-invasive, evaluating bacteria quorum-sensing molecules (QSMs) in samples obtained from patients suffering from bacterial gastrointestinal disorders (Winson et al. 1998; Kumari et al. 2006). Improving the biosensors to detect low levels of QSMs at the nanomolar of pathogens (Venkatesh et al. 2015; Pearson et al. 1995) has helped to detect non-typhoid Salmonella (iNTS) and reduce mortality rate (Morpeth et al. 2009). This whole-cell-based biosensor was developed by incorporating the electrochemical immunosensor principle and hosted in yeast (Venkatesh et al. 2015), which has been genetically modified to display both gold-binding peptide (GBP) and single chain variable fragment (scFv) antibodies on their surfaces. Besides sensitive and accurate detection in low level of samples, this type of biosensors also offers simple and fast detection for diagnosis of various diseases.

The engineered microbial-derived whole-cell-based biosensor is used to detect hydroxyphenylacetic acid, auto-inducer I, and cholera autoinducer-1 from *Candida albicans*, *Staphylococcus aureus,* and *Vibrio cholerae*, respectively, in clinical samples. The engineered *E. coli* cell-based biosensor is applied to perceive N-acylhomoserine lactone to discern and alert the intrusion of *Burkholderia pseudomallei* and *Pseudomonas aeruginosa* (Pham et al. 2022; Wu et al. 2021).

The use of microbial electrochemical sensors in pathogen diagnostics has been reported previously (Chen et al. 2020; Amiri et al. 2018; Karbelkar and Furst 2020; Simoska and Stevenson 2019). In this design, the pathogenic bacteria occurring in clinical samples are the catalysts due to the formation of redox-active mediators as secondary metabolites, creating readable signals only during the presence of the pathogen of interest. During infection stages, the human pathogen, *Pseudomonas aeruginosa*, redox-active generates phenazine molecules, which are detected in real-time by the microbial electrochemical sensors designed with unique array-based geometries (Simoska et al. 2021, 2019; Simoska and Stevenson 2019; Elliott et al. 2017; Jarosova et al. 2019).

The microbial-based bioelectrochemical biosensor for detecting riboflavin has been developed by incorporating the amperometric principle (Si et al. 2016). The biosensor has been made of bioelectrochemical wire (BW) involving the *Shewanella oneidensis* MR-1. Following the addition of riboflavin into the BW system, an electrochemical response was generated. The monitoring of riboflavin is important to ensure the correct consumption by humans to avoid the risks. The use of microbial electrochemical systems (MESs) in measuring cell viability has been reported (Selim et al. 2017; Hassan and Wollenberger 2019). Using this device, the living cells could be distinguished from the dead cells by characterizing the intracellular redox functions of living cells through electrochemically active biofilms.

The fast detection of paper-based whole cell biosensor was successfully developed by exposing it to the target pathogens and utilizing lycopene as an output that can be visualized (Wu et al. 2021). This biosensor to detect pathogens becomes an alternative to nucleic acid-based PCR assays due to its simplification of the sensing workflow.

10.8 Conclusions and Future Perspectives

Microbial biosensors are mostly classified based on the transducers, such as electrochemical and optical. Better understanding of the genetic information can improve the microbial-based biosensor for shorter response time, high sensitivity and high selectivity. Nevertheless, the limitation of genetically engineering bacteria and integration devices become the challenges in the development of biosensors.

In this regard, future work should be conducted to improve the performance of microbial biosensors. Those suggestions include (1) adapting the microbes for specific substrate detection using specific enzyme expression, (2) using the electroactive (intra or extracellular) species in the molecular structure and electron transfer mechanism to improve the microbes-electrodes interaction, (3) conducting the microbial biosensor optimization on expression or suppression of preferred or unwanted metabolic pathways to improve the specificity, (4) fabricating multifunctional and specific biosensors for real-time and rapid detection within extreme environments, (5) conducting the improvement on power generation, stability and component integration for low-cost device, (6) initiating biosensor integration with existing established system, such as scanning tunnelling microscopy (STM) and/or atomic force microscopy (AFM) to fulfil the need of application, and (7) enhancing the acceptability and feasibility of biosensor to be implemented. Finally, increasing endeavors are critical to realize the potential of cell-based biosensors, so that, in the future, it may have a positive impact on health care diagnosis and physiological monitoring systems. Overall, this could also ensure sustainability in biosensors development for various biomedical applications.

Acknowledgments This work was supported by the Malaysia Ministry of Higher Education (MOHE) under Fundamental Research Grant Scheme (FRGS), with Grant no of FP114-2020 (ref code: FRGS/1/2020/SKK06/UM/02/6).

Conflict of Interest Declaration All authors declared that there is no conflict of interest in this work.

References

Ahuja T, Mir IA, Kumar D (2007) Biomolecular immobilization on conducting polymers for biosensing applications. Biomaterials 28(5):791–805

Amine A et al (2006) Enzyme inhibition-based biosensors for food safety and environmental monitoring. Biosens Bioelectron 21(8):1405–1423

Amiri M et al (2018) Electrochemical methodologies for the detection of pathogens. ACS Sens 3(6):1069–1086

Aronoff-Spencer E et al (2016) Detection of hepatitis C core antibody by dual-affinity yeast chimera and smartphone-based electrochemical sensing. Biosens Bioelectron 86:690–696

Axelrod T, Eltzov E, Marks RS (2016) Bioluminescent bioreporter pad biosensor for monitoring water toxicity. Talanta 149:290–297

Azeredo J et al (2017) Critical review on biofilm methods. Crit Rev Microbiol 43(3):313–351

Bahl S et al (2020) Biosensors applications in fighting COVID-19 pandemic. Apollo Med 17(3):221

Behzadian F et al (2011) Construction and characterization of Escherichia coli whole-cell biosensors for toluene and related compounds. Curr Microbiol 62(2):690–696

Belkin S (2003) Microbial whole-cell sensing systems of environmental pollutants. Curr Opin Microbiol 6(3):206–212

Ben-Yoav H et al (2009) A whole cell electrochemical biosensor for water genotoxicity bio-detection. Electrochim Acta 54(25):6113–6118

Bilal M, Iqbal HMN (2019) Microbial-derived biosensors for monitoring environmental contaminants: recent advances and future outlook. Process Saf Environ Prot 124:8–17

Biran I et al (1999) On-line monitoring of gene expression. Microbiology 145(8):2129–2133

Bjerketorp J et al (2006) Advances in preservation methods: keeping biosensor microorganisms alive and active. Curr Opin Biotechnol 17(1):43–49
Bousse L (1996) Whole cell biosensors. Sensors Actuators B Chem 34(1):270–275
Brennan LM et al (2016) Preparation and testing of impedance-based fluidic biochips with RTgill-W1 cells for rapid evaluation of drinking water samples for toxicity. J Vis Exp 109:e53555
Brutesco C et al (2017) Bacterial host and reporter gene optimization for genetically encoded whole cell biosensors. Environ Sci Pollut Res 24(1):52–65
Burgos-Morales O et al (2021) Synthetic biology as driver for the biologization of materials sciences. Mater Today Bio 11:100115
Byfield MP, Abuknesha RA (1994) Biochemical aspects of biosensors. Biosens Bioelectron 9(4–5):373–399
Cai S et al (2018) Engineering highly sensitive whole-cell mercury biosensors based on positive feedback loops from quorum-sensing systems. Analyst 143(3):630–634
Catal T et al (2018) Monitoring of neomycin sulfate antibiotic in microbial fuel cells. Bioresour Technol 268:116–120
Cerminati S, Soncini FC, Checa SK (2015) A sensitive whole-cell biosensor for the simultaneous detection of a broad-spectrum of toxic heavy metal ions. Chem Commun 51(27):5917–5920
Cevenini L et al (2016) Smartphone-interfaced 3D printed toxicity biosensor integrating bioluminescent "sentinel cells". Sensors Actuators B Chem 225:249–257
Chang HJ et al (2017) Microbially derived biosensors for diagnosis, monitoring and epidemiology. Microb Biotechnol 10(5):1031–1035
Chang H-J et al (2018) A modular receptor platform to expand the sensing repertoire of bacteria. ACS Synth Biol 7(1):166–175
Chang H-J et al (2021) Programmable receptors enable bacterial biosensors to detect pathological biomarkers in clinical samples. Nat Commun 12(1):5216
Chee G-J (2016) A novel whole-cell biosensor for the determination of trichloroethylene. Sensors Actuators B Chem 237:836–840
Chen S, Shamsi MH (2017) Biosensors-on-chip: a topical review. J Micromech Microeng 27(8):083001
Chen H et al (2020) Fundamentals, applications, and future directions of bioelectrocatalysis. Chem Rev 120(23):12903–12993
Chien T et al (2022) Enhancing the tropism of bacteria via genetically programmed biosensors. Nat Biomed Eng 6(1):94–104
Chouler J et al (2018) A screen-printed paper microbial fuel cell biosensor for detection of toxic compounds in water. Biosens Bioelectron 102:49–56
Clark DP, Pazdernik NJ, McGehee MR (2019) Chapter 21 - Analysis of gene expression. In: Clark DP, Pazdernik NJ, McGehee MR (eds) Molecular biology, 3rd edn. Academic Cell, pp 651–690
Colby DW et al (2004) Engineering antibody affinity by yeast surface display. Methods Enzymol 388:348–358
Cong Y et al (2013) Efficient electrochemically active biofilm denitrification and bacteria consortium analysis. Bioresour Technol 132:24–27
Corcoran C, Rechnitz G (1985) Cell-based biosensors. Trends Biotechnol 3(4):92–96
Cornejo JA et al (2015) Surface modification for enhanced biofilm formation and electron transport in shewanella anodes. J Electrochem Soc 162(9):H597
Courbet A et al (2015) Detection of pathological biomarkers in human clinical samples via amplifying genetic switches and logic gates. Sci Transl Med 7(289):289ra83
Cunningham AJ (1998) Introduction to bioanalytical sensors. Wiley
Daeffler KN-M et al (2017) Engineering bacterial thiosulfate and tetrathionate sensors for detecting gut inflammation. Mol Syst Biol 13(4):923
Danino T et al (2015) Programmable probiotics for detection of cancer in urine. Sci Transl Med 7(289):289ra84
Daunert S et al (2000) Genetically engineered whole-cell sensing systems: coupling biological recognition with reporter genes. Chem Rev 100(7):2705–2738

Diaz S et al (2007) Tetrahymena metallothioneins fall into two discrete subfamilies. PLoS One 2(3): e291
Dige I et al (2007) In situ identification of streptococci and other bacteria in initial dental biofilm by confocal laser scanning microscopy and fluorescence in situ hybridization. Eur J Oral Sci 115(6):459–467
Ding L et al (2008) Trends in cell-based electrochemical biosensors. Curr Med Chem 15(30): 3160–3170
Dolch K et al (2014) Characterization of microbial current production as a function of microbe–electrode-interaction. Bioresour Technol 157:284–292
D'Souza SF (2001a) Microbial biosensors. Biosens Bioelectron 16(6):337–353
D'Souza SF (2001b) Immobilization and stabilization of biomaterials for biosensor applications. Appl Biochem Biotechnol 96(1-3):225–238
Du H et al (2005) Sensitivity and specificity of metal surface-immobilized "molecular beacon" biosensors. J Am Chem Soc 127(21):7932–7940
Eggins BR (2002) Chemical sensors and biosensors, vol 2. Wiley, London
Elliott J et al (2017) Transparent carbon ultramicroelectrode arrays for the electrochemical detection of a bacterial warfare toxin, pyocyanin. Anal Chem 89(12):6285–6289
Espinosa-Urgel M et al (2015) Engineering biological approaches for detection of toxic compounds: a new microbial biosensor based on the Pseudomonas putida TtgR repressor. Mol Biotechnol 57(6):558–564
Fang D et al (2016) A reagentless electrochemical biosensor based on thionine wrapped E. coli and chitosan-entrapped carbon nanodots film modified glassy carbon electrode for wastewater toxicity assessment. Electrochim Acta 222:303–311
Fang D et al (2020) Redox mediator-based microbial biosensors for acute water toxicity assessment: a critical review. ChemElectroChem 7(12):2513–2526
Flemming H-C, Wingender J (2010) The biofilm matrix. Nat Rev Microbiol 8(9):623–633
Fu E et al (2011) Perspective on diagnostics for global health. IEEE Pulse 2(6):40–50
Fujimoto H et al (2006) Whole-cell arsenite biosensor using photosynthetic bacterium Rhodovulum sulfidophilum. Appl Microbiol Biotechnol 73(2):332–338
Gao G et al (2016) Development of a mediated whole cell-based electrochemical biosensor for joint toxicity assessment of multi-pollutants using a mixed microbial consortium. Anal Chim Acta 924:21–28
Gao G et al (2017) A double-mediator based whole cell electrochemical biosensor for acute biotoxicity assessment of wastewater. Talanta 167:208–216
Gui Q et al (2017) The application of whole cell-based biosensors for use in environmental analysis and in medical diagnostics. Sensors (Basel) 17(7):1623
Gupta N et al (2019) Cell-based biosensors: recent trends, challenges and future perspectives. Biosens Bioelectron 141:111435
Haigh-Flórez D et al (2014) Microalgae dual-head biosensors for selective detection of herbicides with fiber-optic luminescent O2 transduction. Biosens Bioelectron 54:484–491
Halan B, Buehler K, Schmid A (2012) Biofilms as living catalysts in continuous chemical syntheses. Trends Biotechnol 30(9):453–465
Haleem A et al (2021) Biosensors applications in medical field: a brief review. Sensors Int 2:100100
Hassan RYA, Bilitewski U (2011) A viability assay for Candida albicans based on the electron transfer mediator 2,6-dichlorophenolindophenol. Anal Biochem 419(1):26–32
Hassan RYA, Bilitewski U (2013) Direct electrochemical determination of Candida albicans activity. Biosens Bioelectron 49:192–198
Hassan RYA, Wollenberger U (2016) Mediated bioelectrochemical system for biosensing the cell viability of Staphylococcus aureus. Anal Bioanal Chem 408(2):579–587
Hassan RY, Wollenberger U (2019) Direct determination of bacterial cell viability using carbon nanotubes modified screen-printed electrodes. Electroanalysis 31(6):1112–1117
Hernandez CA, Osma JF (2020) Microbial electrochemical systems: deriving future trends from historical perspectives and characterization strategies. Front Environ Sci 8:44

Holder SL, Lee C-H, Popuri SR (2017) Simultaneous wastewater treatment and bioelectricity production in microbial fuel cells using cross-linked chitosan-graphene oxide mixed-matrix membranes. Environ Sci Pollut Res 24(15):13782–13796

Ishihama A (2013) Transcription factors and transcriptional apparatus in bacteria. In: Dubitzky W et al (eds) Encyclopedia of systems biology. Springer New York, New York, NY, pp 2227–2235

Jain VK, Magrath IT (1991) A chemiluminescent assay for quantitation of β-galactosidase in the femtogram range: application to quantitation of β-galactosidase in lacZ-transfected cells. Anal Biochem 199(1):119–124

Jarosova R et al (2019) Inkjet-printed carbon nanotube electrodes for measuring pyocyanin and uric acid in a wound fluid simulant and culture media. Anal Chem 91(14):8835–8844

Jiang D et al (2015) Fluorescent magnetic bead-based mast cell biosensor for electrochemical detection of allergens in foodstuffs. Biosens Bioelectron 70:482–490

Jiang D et al (2017) Preliminary study on an innovative, simple mast cell-based electrochemical method for detecting foodborne pathogenic bacterial quorum signaling molecules (N-acyl-homoserine-lactones). Biosens Bioelectron 90:436–442

Jin X et al (2020) Artificial intelligence biosensors: challenges and prospects. Biosens Bioelectron 165:112412

Joyeux A et al (1997) Engineered cell lines as a tool for monitoring biological activity of hormone analogs. Anal Biochem 249(2):119–130

Kamel S, Khattab TA (2020) Recent advances in cellulose-based biosensors for medical diagnosis. Biosensors 10(6):67

Karbelkar AA, Furst AL (2020) Electrochemical diagnostics for bacterial infectious diseases. ACS Infect Dis 6(7):1567–1571

Karttunen J, Shastri N (1991) Measurement of ligand-induced activation in single viable T cells using the lacZ reporter gene. Proc Natl Acad Sci U S A 88(9):3972–3976

Katuri KP et al (2020) Electroactive biofilms on surface functionalized anodes: the anode respiring behavior of a novel electroactive bacterium, Desulfuromonas acetexigens. Water Res 185: 116284

Khater DZ et al (2015) Development of bioelectrochemical system for monitoring the biodegradation performance of activated sludge. Appl Biochem Biotechnol 175(7):3519–3530

Kracke F, Vassilev I, Krömer JO (2015) Microbial electron transport and energy conservation – the foundation for optimizing bioelectrochemical systems. Front Microbiol 6:575

Kucherenko I, Topolnikova YV, Soldatkin O (2019) Advances in the biosensors for lactate and pyruvate detection for medical applications: a review. TrAC Trends Anal Chem 110:160–172

Kumar R, Singh L, Zularisam AW (2016) Exoelectrogens: recent advances in molecular drivers involved in extracellular electron transfer and strategies used to improve it for microbial fuel cell applications. Renew Sust Energ Rev 56:1322–1336

Kumar S et al (2017) Protein functionalized nanostructured zirconia based electrochemical immunosensor for cardiac troponin I detection. J Mater Res 32(15):2966–2972

Kumari A et al (2006) Biosensing systems for the detection of bacterial quorum signaling molecules. Anal Chem 78(22):7603–7609

Kumari A, Pasini P, Daunert S (2008) Detection of bacterial quorum sensing N-acyl homoserine lactones in clinical samples. Anal Bioanal Chem 391(5):1619–1627

Kwon J et al (2018) Rapid and sensitive detection of Aspergillus niger using a single-mediator system combined with redox cycling. Anal Chem 90(22):13491–13497

Lei Y, Chen W, Mulchandani A (2006) Microbial biosensors. Anal Chim Acta 568(1):200–210

Leonard H et al (2017) Unraveling antimicrobial susceptibility of bacterial networks on micropillar architectures using intrinsic phase-shift spectroscopy. ACS Nano 11(6):6167–6177

Leung KM et al (2013) Shewanella oneidensis MR-1 bacterial nanowires exhibit p-type, tunable electronic behavior. Nano Lett 13(6):2407–2411

Li J et al (2013) The benzoquinone-mediated electrochemical microbial biosensor for water biotoxicity assay. Electrochim Acta 97:52–57

Liu F et al (2014) A lab-on-chip cell-based biosensor for label-free sensing of water toxicants. Lab Chip 14(7):1270–1280
Liu D, Evans T, Zhang F (2015) Applications and advances of metabolite biosensors for metabolic engineering. Metab Eng 31:35–43
Liu Z et al (2017) Nanoporous gold-based microbial biosensor for direct determination of sulfide. Biosens Bioelectron 98:29–35
Liu Y et al (2020) Microbial fuel cells: nanomaterials based on anode and their application. Energ Technol 8(9):2000206
Logan BE et al (2019) Electroactive microorganisms in bioelectrochemical systems. Nat Rev Microbiol 17(5):307–319
Mahmoud RH et al (2018) Assisting the biofilm formation of exoelectrogens using nanostructured microbial fuel cells. J Electroanal Chem 824:128–135
Mahmoud RH et al (2020) Biosensing of algal-photosynthetic productivity using nanostructured bioelectrochemical systems. J Chem Technol Biotechnol 95(4):1028–1037
Mahmoud RH et al (2021) Formation of electroactive biofilms derived by nanostructured anodes surfaces. Bioprocess Biosyst Eng 44(4):759–768
Malhotra BD, Chaubey A, Singh SP (2006) Prospects of conducting polymers in biosensors. Anal Chim Acta 578(1):59–74
McNerney MP et al (2019) Dynamic and tunable metabolite control for robust minimal-equipment assessment of serum zinc. Nat Commun 10(1):5514
Mimee M et al (2018) An ingestible bacterial-electronic system to monitor gastrointestinal health. Science 360(6391):915–918
Molnár F (2016) Transcription factors. pp 57–73
Morpeth SC, Ramadhani HO, Crump JA (2009) Invasive non-Typhi salmonella disease in Africa. Clin Infect Dis 49(4):606–611
Mowbray SE, Amiri AM (2019) A brief overview of medical fiber optic biosensors and techniques in the modification for enhanced sensing ability. Diagnostics 9(1):23
Naylor LH (1999) Reporter gene technology: the future looks bright. Biochem Pharmacol 58(5): 749–757
Neethirajan S et al (2018) Biosensors for sustainable food engineering: challenges and perspectives. Biosensors 8(1):23
Pan T et al (2015) High-throughput screening assay for the environmental water samples using cellular response profiles. Ecotoxicol Environ Saf 114:134–142
Pancrazio JJ et al (1999) Development and application of cell-based biosensors. Ann Biomed Eng 27(6):697–711
Pandey CM et al (2017) Highly sensitive electrochemical immunosensor based on graphene-wrapped copper oxide-cysteine hierarchical structure for detection of pathogenic bacteria. Sensors Actuators B Chem 238:1060–1069
Pandey CM et al (2018) Microfluidics based point-of-care diagnostics. Biotechnol J 13(1):1700047
Pankratova G, Gorton L (2017) Electrochemical communication between living cells and conductive surfaces. Curr Opin Electrochem 5(1):193–202
Pasternak G, Greenman J, Ieropoulos I (2017) Self-powered, autonomous biological oxygen demand biosensor for online water quality monitoring. Sensors Actuators B Chem 244:815–822
Paul D et al (2005) Accessing microbial diversity for bioremediation and environmental restoration. Trends Biotechnol 23(3):135–142
Pearson JP et al (1995) A second N-acylhomoserine lactone signal produced by Pseudomonas aeruginosa. Proc Natl Acad Sci 92(5):1490–1494
Pham HL, Ling H, Chang MW (2022) Design and fabrication of field-deployable microbial biosensing devices. Curr Opin Biotechnol 76:102731
Piñero-Lambea C et al (2015) Programming controlled adhesion of E. coli to target surfaces, cells, and tumors with synthetic Adhesins. ACS Synth Biol 4(4):463–473

Pontel LB et al (2007) GolS controls the response to gold by the hierarchical induction of Salmonella-specific genes that include a CBA efflux-coding operon. Mol Microbiol 66(3): 814–825

Popova AA et al (2016) Droplet-microarray on superhydrophobic-superhydrophilic patterns for high-throughput live cell screenings. RSC Adv 6(44):38263–38276

Qiao Y et al (2007) Carbon nanotube/polyaniline composite as anode material for microbial fuel cells. J Power Sources 170(1):79–84

Qiao L et al (2015) A novel pungency biosensor prepared with fixing taste-bud tissue of rats. Biosens Bioelectron 68:454–461

Ratajczak A, Geissdörfer W, Hillen W (1998) Expression of alkane hydroxylase from Acinetobacter sp. strain ADP1 is induced by a broad range of n-alkanes and requires the transcriptional activator AlkR. J Bacteriol 180(22):5822–5827

Raut N et al (2012) Engineered cells as biosensing systems in biomedical analysis. Anal Bioanal Chem 402(10):3147–3159

Rawson FJ, Downard AJ, Baronian KH (2014) Electrochemical detection of intracellular and cell membrane redox systems in Saccharomyces cerevisiae. Sci Rep 4(1):5216

Riglar DT et al (2017) Engineered bacteria can function in the mammalian gut long-term as live diagnostics of inflammation. Nat Biotechnol 35(7):653–658

Sadana A (2001) Engineering biosensors: kinetics and design applications. Academic Press

Safavieh M et al (2017) Rapid real-time antimicrobial susceptibility testing with electrical sensing on plastic microchips with printed electrodes. ACS Appl Mater Interfaces 9(14):12832–12840

Santoro C et al (2017) Microbial fuel cells: from fundamentals to applications. A review. J Power Sources 356:225–244

Saratale GD et al (2017) A comprehensive overview on electro-active biofilms, role of exo-electrogens and their microbial niches in microbial fuel cells (MFCs). Chemosphere 178: 534–547

Schmitz S, Rosenbaum MA (2018) Boosting mediated electron transfer in bioelectrochemical systems with tailored defined microbial cocultures. Biotechnol Bioeng 115(9):2183–2193

Sedki M et al (2019) Online-monitoring of biofilm formation using nanostructured electrode surfaces. Mater Sci Eng C 100:178–185

Selim HMM et al (2017) Bioelectrochemical systems for measuring microbial cellular functions. Electroanalysis 29(6):1498–1505

Shabani A et al (2008) Bacteriophage-modified microarrays for the direct impedimetric detection of bacteria. Anal Chem 80(24):9475–9482

Shang Y et al (2002) A robust inducible-repressible promoter greatly facilitates gene knockouts, conditional expression, and overexpression of homologous and heterologous genes in Tetrahymena thermophila. Proc Natl Acad Sci U S A 99(6):3734–3739

Sharma T et al (2008) Development of carbon nanotubes and nanofluids based microbial fuel cell. Int J Hydrog Energy 33(22):6749–6754

Sharma M et al (2019) Electrode material properties for designing effective microbial electrosynthesis systems. J Mater Chem A 7(42):24420–24436

Shashaani H et al (2016) Silicon nanowire based biosensing platform for electrochemical sensing of Mebendazole drug activity on breast cancer cells. Biosens Bioelectron 85:363–370

Shi L et al (2016) Extracellular electron transfer mechanisms between microorganisms and minerals. Nat Rev Microbiol 14(10):651–662

Shukla VC et al (2018) Lab-on-a-Chip platforms for biophysical studies of cancer with single-cell resolution. Trends Biotechnol 36(5):549–561

Si R-W et al (2016) Wiring bacterial electron flow for sensitive whole-cell amperometric detection of riboflavin. Anal Chem 88(22):11222–11228

Simoska O, Stevenson KJ (2019) Electrochemical sensors for rapid diagnosis of pathogens in real time. Analyst 144(22):6461–6478

Simoska O et al (2019) Electrochemical monitoring of the impact of polymicrobial infections on Pseudomonas aeruginosa and growth dependent medium. Biosens Bioelectron 142:111538

Simoska O et al (2021) Recent trends and advances in microbial electrochemical sensing technologies: an overview. Curr Opin Electrochem 30:100762
Singh SK et al (2000) Unfolding and internalization of proteins by the ATP-dependent proteases ClpXP and ClpAP. Proc Natl Acad Sci U S A 97(16):8898–8903
Singh R et al (2012) Nanostructured platform for the detection of Neisseria gonorrhoeae using electrochemical impedance spectroscopy and differential pulse voltammetry. Microchim Acta 177(1):201–210
Slomovic S, Pardee K, Collins JJ (2015) Synthetic biology devices for in vitro and in vivo diagnostics. Proc Natl Acad Sci U S A 112(47):14429–14435
Song SH et al (2005) Enzyme and microbial technology, vol 37. pp 567
Stocker J et al (2003) Development of a set of simple bacterial biosensors for quantitative and rapid measurements of arsenite and arsenate in potable water. Environ Sci Technol 37(20):4743–4750
Su L et al (2011) Microbial biosensors: a review. Biosens Bioelectron 26(5):1788–1799
Syal K et al (2016) Antimicrobial susceptibility test with plasmonic imaging and tracking of single bacterial motions on nanometer scale. ACS Nano 10(1):845–852
Tahirbegi IB et al (2017) Fast pesticide detection inside microfluidic device with integrated optical pH, oxygen sensors and algal fluorescence. Biosens Bioelectron 88:188–195
Throne-Holst M et al (2007) Identification of novel genes involved in long-chain n-alkane degradation by Acinetobacter sp. strain DSM 17874. Appl Environ Microbiol 73(10):3327–3332
Tian Y et al (2017) Construction and comparison of yeast whole-cell biosensors regulated by two RAD54 promoters capable of detecting genotoxic compounds. Toxicol Mech Methods 27(2): 115–120
Tucci M et al (2019) Microbial amperometric biosensor for online herbicide detection: photocurrent inhibition of Anabaena variabilis. Electrochim Acta 302:102–108
Urgun-Demirtas M, Stark B, Pagilla K (2006) Use of genetically engineered microorganisms (GEMs) for the bioremediation of contaminants. Crit Rev Biotechnol 26(3):145–164
Venkatesh AG et al (2015) Yeast dual-affinity biobricks: progress towards renewable whole-cell biosensors. Biosens Bioelectron 70:462–468
Vopálenská I, Váchová L, Palková Z (2015) New biosensor for detection of copper ions in water based on immobilized genetically modified yeast cells. Biosens Bioelectron 72:160–167
Wang P, Liu Q (2010) Cell-based biosensors principles and applications. Artech House, Boston
Wang J et al (2017) Ultramicroelectrode array modified with magnetically labeled Bacillus subtilis, palladium nanoparticles and reduced carboxy graphene for amperometric determination of biochemical oxygen demand. Microchim Acta 184(3):763–771
Wang X et al (2018) Three-dimensional graphene biointerface with extremely high sensitivity to single cancer cell monitoring. Biosens Bioelectron 105:22–28
Webster DP et al (2014) An arsenic-specific biosensor with genetically engineered Shewanella oneidensis in a bioelectrochemical system. Biosens Bioelectron 62:320–324
Wei L et al (2017) A kinetic study of bitter taste receptor sensing using immobilized porcine taste bud tissues. Biosens Bioelectron 92:74–80
Weingeist DM et al (2013) Single-cell microarray enables high-throughput evaluation of DNA double-strand breaks and DNA repair inhibitors. Cell Cycle 12(6):907–915
Welsh S, Kay SA (1997) Reporter gene expression for monitoring gene transfer. Curr Opin Biotechnol 8(5):617–622
Wesolowski J et al (2008) Sensing of oxygen in microtiter plates: a novel tool for screening drugs against pathogenic yeasts. Anal Bioanal Chem 391(5):1731–1737
Whangsuk W et al (2016) Specific detection of the pesticide chlorpyrifos by a sensitive genetic-based whole cell biosensor. Anal Biochem 493:11–13
Wilson JS (2004) Sensor technology handbook. Elsevier
Wilson GS, Gifford R (2005) Biosensors for real-time in vivo measurements. Biosens Bioelectron 20(12):2388–2403

Winson MK et al (1998) Construction and analysis of luxCDABE-based plasmid sensors for investigating N-acyl homoserine lactone-mediated quorum sensing. FEMS Microbiol Lett 163(2):185–192

Wolf D, Mascher T (2016) The applied side of antimicrobial peptide-inducible promoters from Firmicutes bacteria: expression systems and whole-cell biosensors. Appl Microbiol Biotechnol 100(11):4817–4829

Woo S-G et al (2020) A designed whole-cell biosensor for live diagnosis of gut inflammation through nitrate sensing. Biosens Bioelectron 168:112523

Wu J, Rosen B (1991) The ArsR protein is a trans-acting regulatory protein. Mol Microbiol 5(6): 1331–1336

Wu Y et al (2014a) Exogenous electron shuttle-mediated extracellular electron transfer of Shewanella putrefaciens 200: electrochemical parameters and thermodynamics. Environ Sci Technol 48(16):9306–9314

Wu X et al (2014b) A breast cancer cell microarray (CMA) as a rapid method to characterize candidate biomarkers. Cancer Biol Ther 15(12):1593–1599

Wu Y et al (2021) A whole-cell biosensor for point-of-care detection of waterborne bacterial pathogens. ACS Synth Biol 10(2):333–344

Xie X et al (2004) Development of an ultramicroelectrode arrays (UMEAs) sensor for trace heavy metal measurement in water. Sensors Actuators B Chem 97(2–3):168–173

Xu Z, Chen X, Dong S (2006) Electrochemical biosensors based on advanced bioimmobilization matrices. Trends Analyt Chem 25(9):899–908

Yager P, Domingo GJ, Gerdes J (2008) Point-of-care diagnostics for global health. Annu Rev Biomed Eng 10:107–144

Yamashita T et al (2016) A novel open-type biosensor for the in-situ monitoring of biochemical oxygen demand in an aerobic environment. Sci Rep 6(1):38552

Yang Y et al (2012) Bacterial extracellular electron transfer in bioelectrochemical systems. Process Biochem 47(12):1707–1714

Yang W et al (2016) Fast and sensitive water quality assessment: a μL-scale microbial fuel cell-based biosensor integrated with an air-bubble trap and electrochemical sensing functionality. Sensors Actuators B Chem 226:191–195

Yeliseev AA, Eraso JM, Kaplan S (1996) Differential carotenoid composition of the B875 and B800-850 photosynthetic antenna complexes in Rhodobacter sphaeroides 2.4. 1: involvement of spheroidene and spheroidenone in adaptation to changes in light intensity and oxygen availability. J Bacteriol 178(20):5877–5883

Yu D et al (2005) Aqueous sol-gel encapsulation of genetically engineered Moraxella spp. cells for the detection of organophosphates. Biosens Bioelectron 20(7):1433–1437

Zammit CM et al (2013) A whole cell biosensor for the detection of gold. PLoS One 8(8):e69292

Zanganeh S et al (2016) Electrochemical approach for monitoring the effect of anti tubulin drugs on breast cancer cells based on silicon nanograss electrodes. Anal Chim Acta 938:72–81

Zhang D et al (2012) Whole-cell bacterial bioreporter for actively searching and sensing of alkanes and oil spills. Microb Biotechnol 5(1):87–97

Zhao J et al (2007) The different behaviors of three oxidative mediators in probing the redox activities of the yeast Saccharomyces cerevisiae. Anal Chim Acta 597(1):67–74

Zhao J et al (2021) Microbial extracellular electron transfer and strategies for engineering electroactive microorganisms. Biotechnol Adv 53:107682

Zhou M et al (2011) An overview of electrode materials in microbial fuel cells. J Power Sources 196(10):4427–4435

Zou L et al (2016) Detection of diarrhetic shellfish poisoning toxins using high-sensitivity human cancer cell-based impedance biosensor. Sensors Actuators B Chem 222:205–212

Progress in Biomedical Applications Using Sustainable Nanoparticles

11

Lidawani Lambuk, Suhana Ahmad, Siti Khatijah Abdul Razak, Ros Akmal Mohd Idris, and Rohimah Mohamud

Abstract

Over the past decades, there has been a dramatic increase in the application of nanotechnology in diverse areas of biomedical research and the treatment of diseases. In nanotechnological applications, nanoparticles have high potential as robust and versatile tools and have been successfully employed, notably in drug formulations, targeted drug delivery, and biomedical imaging. Nonetheless, such advancements pose toxic elements for the environment and human health. Given the high production and general feasibility of nanomaterials, contamination by hazardous substances in environmental matrices is highly possible. Herein, the authors attempt to highlight recent updates on nanoparticles with an emphasis on the benefits and risks imposed on the environment and living beings. In particular, the chapter presents the assessment of green synthesized nanoparticles from a sustainability viewpoint.

Keywords

Nanoparticles · Sustainability · Green technology · Physicochemical properties · Biomedical application · Toxicology

L. Lambuk · S. K. Abdul Razak · R. A. Mohd Idris · R. Mohamud (✉)
Present Address: Department of Immunology, School of Medical Sciences, Universiti Sains Malaysia, Kota Bharu, Malaysia
e-mail: lidawani@usm.my; rohimahm@usm.my

S. Ahmad
Present Address: Department of Health, Medicine & Caring Sciences, Linköping University, Linköping, Sweden
e-mail: suhana.ahmad@liu.se

W. S. Wan Kamarul Zaman, A. A. Nurul (eds.), *Sustainable Material for Biomedical Engineering Application*, https://doi.org/10.1007/978-981-99-2267-3_11

Abbreviations

Ag	Silver
Ag_2S	Silver sulfide
Au	Gold
CO	Cobalt
-COOH	Carboxyl group
COVID-19	Coronavirus disease
CuO	Copper oxide
DDS	Drug delivery system
DNA	Deoxyribonucleic acid
DOX	Doxorubicin
FDA	The United States Food and Drug Administration
Fe	Iron
GO	Graphene oxide
IL-8	Interleukin 8
LC50	Lethal dose concentration 50
MDRS	Multidrug resistance strain
MRI	Magnetic resonance imaging
Ni	Nickel
NP	Nanoparticle
PEG	Polyethylene glycol
Pt	Platinum
PVC	Polyvinyl chloride
R&D	Research and development
ROS	Reactive oxygen species
-SH	Thiol group
SiO_2	Silicon oxide
TiO_2	Titanium oxide
WO_3	Tungsten oxide
ZnO	Zinc oxide

11.1 Introduction

Nanotechnology has been implemented across various application areas as a potential tool for positive sustainability by supporting waste and energy management solutions. In 1959, Richard Feynman, a well-known physicist, laid out an excellent concept of nanotechnology (Feynman 2018). Later, in 1974, the word "nanotechnology" was first adopted by Norio Taniguchi (Taniguchi 1974). He stated that "nanotechnology mainly consists of processing separation, consolidation, and deformation of materials by one atom or molecule". Significant advances in material science have confirmed Feynman's concept of influencing material characteristics at the atomic level. The popularity of nanotechnology, the development of various nanomaterials, and its incorporation into a broad array of consumer products, has expanded substantially over the past decades.

It has been documented that nanoscale materials in the form of nanoparticles (NPs) are beneficial for a plethora of specialties and have potent usage in medicine. The technology allows the development and control of the properties of materials into the desired nanoscale (Tolfree 2006). There is no doubt that this highly flexible technology has now advanced for a wide range of biomedical applications, which primarily improve or transform the remedial and treatment regimen, as well as prophylaxis at the cellular level, due to nanotechnology. According to BCC Research (https://www.bccresearch.com/market-research/healthcare/nanotechnology-medical-applications-market.html), the global market for nanotechnology used in biomedical applications is estimated to grow from $242.6 billion in 2021 to $493.5 billion by 2026 with a compound annual growth rate (CAGR) of 15.3% for the period of 2021–2026, citing market share based on therapeutic areas (e.g., anticancer, anti-inflammatories, cardiovascular, and others) and geography segments including Europe, North America, and Asia-Pacific. Based on these data, many perceive the introduction of nanotechnologies as a watershed moment in biomedical advancement, as proven by the attention on the inorganic particles used in the production of antimicrobial agents, biological markets for research and diagnosis, biomagnetic separation, drug carriers, MRI contrast medium, and in orthopedic devices. According to the European Technology Platform on Nanomedicine (ETPN), an association of the European Commission set up to address the application of nanotechnology for medical purposes, nanomedicine has significantly contributed to the early diagnosis, treatment, prevention, and follow-up of numerous diseases. Overall, the ETPN has identified over 70 nanotechnology clinical trial products that cover diseases such as cancer, cardiovascular, neurological, and inflammatory disorders, enabling advancements in all healthcare areas. As a matter of fact, the use of nanotechnology has already yielded innumerable benefits to society. There are many other potentials to implement technological advances, along with countless others waiting to be uncovered. Many biomedical applications are emerging, particularly for diagnostics, bioimaging, and drug delivery systems.

Significant advantages of this technology, for example, to develop a medicine to be placed within a nanoparticulate as the carrier, including improved bioavailability, rapid onset of action, dose consistency, and medications in smaller and more stable dose forms that could not previously be produced commonly (Singh and Amiji 2022). The unique features of NPs are impressive from a physicochemical standpoint. Nevertheless, they still raise concerns about potential detrimental impacts on human health and biological systems, particularly the nanoscale particles and materials that may be detrimental to human health because of their extreme reactivity, systemic availability, and environmental occurrence. Like any new invention, identifying prospective health risks is paramount for practically evaluating the usefulness and safety of any substances, materials, or products that may be manufactured.

Rather than being an exhaustive review, the purpose of this chapter is to give readers a brief introduction to the NPs utilized in medicine that have either undergone in vivo and/or in vitro research or have already been approved for clinical use. Importantly, in this chapter, the relationship between NPs and their health and environmental consequences are examined. It should be emphasized that this chapter exclusively discusses human and animal biological applications. Considering broad categories of medical diagnostic/imaging, therapeutics, and drug delivery systems, the purpose of this chapter was also to augment the recent body of work and advancements in nanotechnology for biomedical applications employing NPs with sustainable qualities. Wherever relevant, the readers will be directed to the several other in-depth reviews available in each application area. Finally, prospects and challenges for the sustainability of managing the health and environmental impact of NPs in indoor and large-scale manufacturing are also described.

11.2 Nanoparticles

Nanoparticles feature at least one dimension on the nanoscale and have distinctive characteristics compared to their non-nanoscale counterparts. The particles offer remarkable and novel properties that cannot simply be termed an intermediate state between the atomic or molecular/bulk state. NPs are known by many other names, including nanospheres, nanotubes, nanorods, nanochains, nanometers, nanostars, nanowires, and nanofibers (Kralj and Makovec 2015). Indeed, nanoscale particles have been used in medicine as tuneable macromolecular components whereby an active agent (drug or substance) is encapsulated, dissolved, or entrapped, or to which the agent is adsorbed or attached within the particles. The use of NPs for cancer and other disease diagnostics and therapeutics has been a rapidly advancing and evolving field.

Nanoparticles have a wide range of biomedical applications (Fig. 11.1), including diagnostics and imaging (Singh and Amiji 2022), drug delivery (Mitchell et al. 2021), vaccine development (Kheirollahpour et al. 2020), genetic screening (Francia et al. 2020), as well as treatments for diseases (Yetisgin et al. 2020) such as cancer, neurological disorders, heart diseases, and organ implants (Yao and Martins 2020). The unique physicochemical properties of NPs (Fig. 11.2) are being exploited in medical research, where they have been discovered to improve pharmacokinetic and targeting capabilities, along with size, shape, material, and structural surface properties (Mitchell et al. 2021). Based on structural surface properties, for example, the surface area of the materials is considerably enhanced at the nanoscale, contributing to greater reactivity and stability. The diminutive size of the nanomaterial also implies less energy required for functioning, thereby encouraging sustainability. Because of their small size, nanoscale materials can access previously inaccessible areas, allowing them to permeate biological structures and disrupt biomolecular interactions (e.g., intracellular trafficking, cellular component synthesis and degradation, neurotransmission, receptor-ligand complexes, and initiating

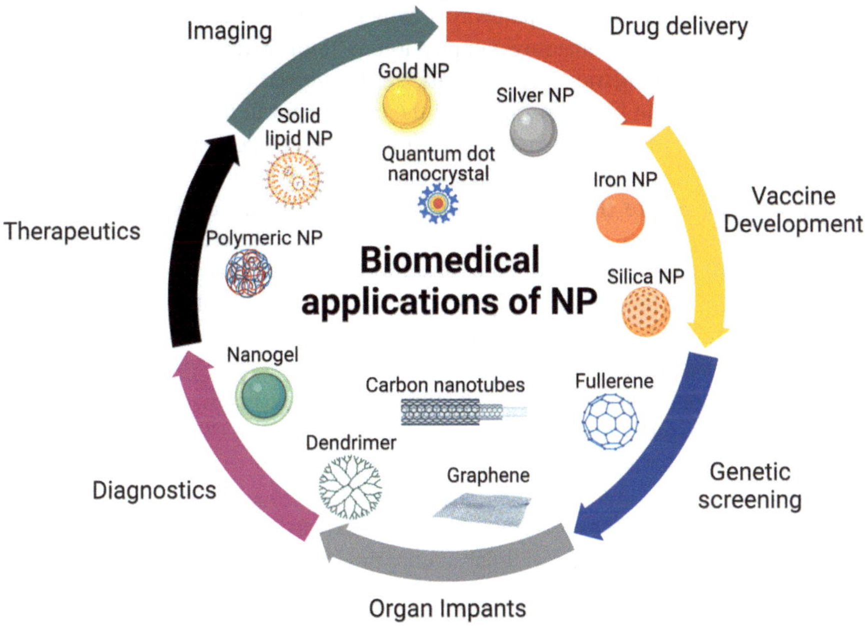

Fig. 11.1 Schematic showing some of the leading biomedical applications of various types of nanoparticles. *NP* nanoparticle

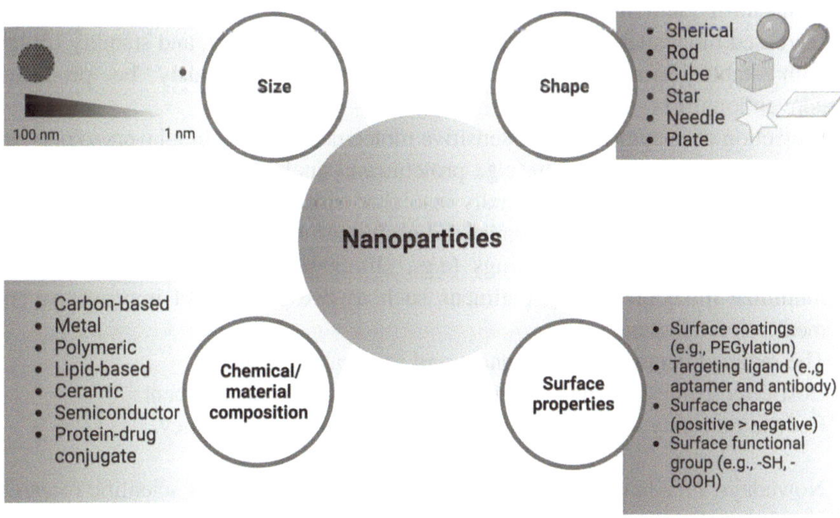

Fig. 11.2 General physicochemical properties of nanoparticles

molecular chaos in oncogenic pathways), which could be beneficial for the therapeutic field since human anatomy and physiology limit therapy (Gavas et al. 2021).

However, at nanometric size, materials are discovered to have divergent sets of characteristics than in bulk. For instance, the component in bulk may be non-toxic. However, when reduced to the nanoscale, it may become poisonous, which implies that different and size-dependent effects are exerted, which are strictly tied to the nanoparticulate nature (Recordati et al. 2015). The increased reactivity from the larger surface area may also cause the changes. Besides, different sizes of the same material of NPs also possess different absorption wavelengths, as smaller NPs can absorb photons of shorter wavelengths, directly affecting the absorption efficiency (Fagiolari and Bella 2019). Furthermore, the shape of the NPs influences their reactivity and other features (Ahmed and Ali 2020).

The multifunctional of NPs for diagnostic and therapeutic purposes removes major limitations in conventional medicine (Mitchell et al. 2021; Bharathala and Sharma 2019) including:

- Offer rapid onset of action, a small and highly sensitive drug carrier (drugs can be directed to particular intracellular compartments using intracellular targeting agent), and diagnostic tools that can provide direct interaction with target cells (e.g., immune cells) or tissues; non-immunogenicity; membrane-permeation moieties to improve cellular uptake.
- It targets and controls drug release that provides dose uniformity and a more diminutive dosage form that can reduce or prevent side effects (e.g., toxicity), which is ideal for long-term treatments, enhanced drug retention time, drug tissue permeability, and maximum tissue absorption.
- Increased bioavailability, biodegradability, biocompatibility, and stability (drugs with stable dose formulations that could not previously be produced conventionally).
- Protection of enzyme- and pH-sensitive molecules by providing layers to prevent biological cargo degradation (e.g., proteins and nucleic acids).
- Improved solubility of traditionally insoluble substances with a longer shelf life (e.g., lipophilic drugs are administered orally into the systemic circulation).
- Reduce adverse effects of drugs (e.g., elimination of abrasive properties to minimize medicine dosing regimens such as eye drops) and improve patient medication adherence.
- The reduced degree of invasiveness and surgical procedures.
- Cost-effectiveness of medical treatments and disease management.
- Reduce mortality and morbidity rates and higher life expectancy.

Nowadays, NPs have been regarded as ideal probes in various scientific research studies for diagnostics and imaging purposes due to their high stability, sensitivity, various shapes, and sizes that can be comparable to macromolecules such as receptors and polypeptides but small enough to permeate every single cell in the human body that conventional devices fail to do so (Baptista et al. 2011). By conjugating them to specific biomarkers, they can improve the biological sensing of cellular and molecular changes related to disease conditions as probes. For instance, detect NPs conjugated antibodies, which are being utilized as an early diagnostic tool to detect tumor accumulation or infections, can act as a tracker device to prevent harm and lead to specific targeted treatment in the early stage of diseases (Allison et al. 2008; Brigger et al. 2012; Colino et al. 2018).

NPs in medicine have been proven to be very useful from a therapeutic standpoint. The monotonous role of NPs as drug delivery has been transformed into multiple carriers for substances including drug and gene therapy (Wahane et al. 2020), tissue regeneration (Sahai et al. 2018), biomolecules and cell isolation (Meng et al. 2019), and tumor destructor (Yang et al. 2020). In particles, drugs or chemicals can be precisely delivered to the targeted body region using NPs, with improved bioavailability (Yetisgin et al. 2020). Other than improving drug delivery, NPs also have been exploited to analyze cellular DNA within minutes (Mathur and Medintz 2017) and effectively to overturn deposited plague in arteries (Flores et al. 2020), artificial intelligent-based biosensors for rapid diagnosis (Thapa et al. 2022), transdermal patches (Abu-Huwaij et al. 2022), and biocompatible bioprosthetic implants (Aminian et al. 2022). Of importance, most cancer medications are cleansed out of the body via the renal before reaching the tumor. The features of NPs can be customized in such a way as to be induced explicitly by endogenous stimuli such as magnetic field, pH, heat, and hypoxia-responsiveness to damage the cancer cells directly. For example, Kolosnjaj-Tabi and colleagues developed poly(ethylene glycol) (PEG)-coated magnetite NPs that showed excellent tumor regression in the epidermoid carcinoma xenograft mouse model following magnetic hyperthermia (Kolosnjaj-Tabi et al. 2014). By using engineered NPs, researchers have successfully delivered personalized cancer medicines directly to tumors, which lower the amount of chemotherapy required, maximize beneficial effects and limit adverse effects while usually sparing healthy cells and being less aggressive than the conventional therapies (Gavas et al. 2021). Additionally, NPs have improved blood compatibility against drug degradation (Guo et al. 2021). Since specific necessary yet sensitive drugs can be rapidly degraded by enzymes in the blood, they must be shielded to operate. Today, there are at least 14 clinically FDA-approved nanotechnologies for cancer therapy worldwide (Cheng et al. 2021), with the majority being liposomal-based nanocarriers and more than 100 NP-based therapies in clinical trials at various stages. These are only a fraction of the many applications of NPs in the biomedical field. This chapter briefly introduces how nanotechnology is employed in several parts of medicine, including diagnostics/imaging, drug delivery, and therapies.

11.3 Potential Risks and Concerns Associated with Nanoparticles

No doubt, nanotechnology has greatly benefited society by improving health and making life easier. Even with the benefits, the enhanced performance of NPs can be destructive from a safety viewpoint. It may only sometimes benefit the sustainable environment, especially for large-scale manufacturing (Zaytseva and Neumann 2016). In addition, nanomaterial research and development have significantly outpaced the discovery of toxicological data on these materials. Herein, the potential health, safety, and environmental issues concerning NP-based products are addressed.

The consequences of NPs on health and ecosystems still need to be better understood (Nel et al. 2015). The adoption of modern material technologies largely relies on NPs due to their superior surface activity compared to others and their versatility in particle fabrication for a wide range of purposes (Bharathala and Sharma 2019). There are many different engineered NPs; not all are the same or pose the same risks. Several factors should be considered when assessing the potential risks, such as the toxicological profiles of the particles through in vitro and in vivo studies, adverse biological implications from different levels of exposure, and the fate of the shed NPs in humans and the environment (Schulte et al. 2013). The capacity to evaluate and mitigate the newly developed NPs that may pose potential health hazards is necessary to guarantee that nanotechnology development is sustainable, safe to establish and use, and appropriate procedures for controlling the nanotechnology.

Although studies focusing on the profiles of nanoscale materials have only recently emerged. Studies on the behavior and toxicity mainly concern the insolubility of NPs and nanoformulations (Tirumala et al. 2021). Because of the physicochemical properties of NPs, including their ability to modify the cellular uptake, which eventually leads to the displacement from an entrance point to the target site, it is inevitable to suggest the potential of the particles to cause tissue injury. The following section is directed at the risk of nanoformulations in diagnostic, drug delivery, and therapeutic applications.

11.3.1 Toxicology Properties of Nanoparticles

Nanoparticles have been advanced in drug delivery; however, they are relatively new in drug formulation, and the lack of availability of detailed toxicology profiles can be hazardous and adverse to human health. The properties of NPs in toxicity testing medium, such as hydrodynamics of particle size, surface charge, temperature, surface functionalization, and colloidal stabilities, are known to affect in vitro toxicity evaluations significantly. Other than these features, the outcome of toxicity tests is also based on bacterial and eukaryotic cell viability, protein-containing media, and the state of organic matter of NPs, which may influence the interpretation of the results. To generate accurate information, selecting specific cell lines and executing a consistent testing approach are imperative. Thongkam et al. investigated the oxidant-generating properties of several nanomaterials on the epithelial cells of liver, kidney, and lung (Thongkam et al. 2016). The results illustrated that the effects of DNA damage and cytotoxicity differed greatly depending on the cell line, emphasizing that it is crucial to use appropriate types of cells when evaluating the organ or tissue-specific health risks of NPs. The data upholds the idea that the essential oxidant-generating activity of NPs is a poor indicator of their DNA-damaging potential. As such, they were also reported by Monteiro-Riviere, and the group determined the toxicity of several nanomaterials in different cell types (Monteiro-Riviere et al. 2009). They discovered that the physical interference of the particles on the dye staining for screening tests causes unpredictable findings that may not be appropriate for assessing NPs cytotoxicity, implying that more than one assay is necessary to establish the NPs toxicity for risk assessment.

The same may be extrapolated from protein-containing cell culture media, where the protein may be adsorbed on the surface of the NPs or in the presence of organic matter, hence altering the release of molecules and increasing their size that could lead to more significant aggregation, resulting in an underestimation of NP toxicity. Thus, toxicokinetic and biodistribution in organisms are urgently needed to assess the potential toxicity effects of NPs in humans (Buzea et al. 2007). Readers are directed to the cited reviews for further understanding (Cupi et al. 2015; Yang et al. 2013). Because of the complexity of employing experimental animal models that required months or years of monitoring, as well as the ethical problems involved with the high number of animals slaughtered, most studies were done in vitro (Wu and Tang 2018). Of note, toxicological testing on NPs has been reported to vary depending on the tested model organisms (e.g., bacteria, crustaceans, fish, plants, and algae). The physicochemical properties of NPs can be modified depending on the settings (e.g., methodologies, media, endpoints, and concentrations), leading to a variety of cellular responses, each of which has its own set of underlying mechanisms, such as algae and crustaceans, being the most vulnerable organisms to NPs. Several potential mechanisms of action have been indicated, including protein and nucleic acid damage, membrane modification, and reactive oxygen species (ROS) generation. Studies of genotoxicity are also necessary to determine the toxicity of NPs since the risk of genetic disorders, and cancer is common. Chinde and his group have investigated the biodistribution of tungsten oxide NPs (WO_3) in different tissues of rats after oral exposure in dose-dependent, time-dependent, and organ-dependent (Chinde et al. 2017). The outcomes of the study demonstrated that WO_3 NPs at 1000 mg/kg dose caused significant irregularities in DNA structure, micronuclei, and chromosomes.

A kidney is no longer a concern when adopting NPs to treat malignant tumors; however, the liver is because foreign entities like NPs are filtered and eliminated by the liver (Baboci et al. 2020). Therefore, the immune cells need to be deactivated or eradicated before the NPs can reach the tumor in the liver. While NPs are a highly successful cancer treatment, researchers assert that if an NP has poor solubility, structure, surface charge, encapsulation, or efficiency to aggregate, it may facilitate unwarranted chemical and biological events that cause cancer (Khan et al. 2019). Notably, the toxicity profiles depend on various factors of NPs properties under different conditions (e.g., type of disease and genetic predisposition), including size, shape, and chemical composition. Besides, currently, no regulations are available to assure the safety aspects of nanocarriers of drugs in terms of health and their impact on the environment (Nel et al. 2015).

11.3.1.1 Size

Nanoparticles range between 1 and 100 nm in size. Since their size is easily comparable to or smaller than some of the cellular components, including protein globules, cell membrane thickness, and DNA, these enable them to penetrate cells and their organelles through epithelial and endothelial barriers into cellular structures in several organs, including the bloodstream, lymphatic, and nervous systems (Sukhanova et al. 2018). They are delivered by the bloodstream, lymphatic, or nervous system into the targeted tissues or organs via transcytosis mechanisms or diffuse into them. When evaluated based on mass dosage, it is hypothesized that the NPs can cause a significant risk than bigger particles of similar insoluble substances (Sukhanova et al. 2018). An equivalent dose-dependent relationship between low toxicity and solubility, varied-sized particles, was obtained when a dose response was expressed based on particle size (Oberdörster et al. 2005). Although the minuscule size and large surface area to volume ratio are two notable features that typically determine the mechanism of interactions of NPs with living systems, these characteristics may have a substantial impact on generating intrinsic toxicity due to additional molecules on the surface (Schmid and Stoeger 2016). The extensive surface area also increases collision frequency, which dictates their high capacity and reactivity. These features may interfere with the normal functions of the body and develop tissue inflammation, indicating that the surface area of NPs as the driver for inflammogenicity for a particle can trigger aberrant cell functioning or potential cell death (Schmid and Stoeger 2016). Peters et al. (2004) examined the effects of five particles with different sizes; polyvinyl chloride (PVC), titanium oxide (TiO_2), silicon dioxide (SiO_2), cobalt (CO), and nickel (Ni), on endothelial viability and function (Peters et al. 2004). The toxicity in endothelial cells was only exhibited when exposed to CO NP, followed by the generation of the pro-inflammatory cytokine interleukin 8 (IL-8). TiO_2 and SiO_2 generated minimal and significant IL-8 production, respectively, compared to the other particles. The discrepancies in cytotoxicity were not addressed; however, given particle size ranged from 14 to 120 nm, with clusters of 420 nm, they might be attributed to both material variations and size differences at the nanoscale (Peters et al. 2004).

Furthermore, particles of titanium oxide (TiO_2 NP) caused a severe proinflammatory response compared to the same dose of fine TiO_2 (Monteiller et al. 2007), representing the dosage as assessed by particle surface area compared to exposed cell surface area. One of the most mass-produced NPs globally, silica NP (SiNP), is widely used in food and agriculture. Comparing SiNP to micro-sized crystalline silica, the higher surface area to mass ratio of the former appears to be more potent in pulmonary toxicity, which is potentially due to its translocation to the systemic circulation through the airway barrier (Skuland et al. 2014), hence may exert direct effects in secondary organs (Skuland et al. 2020).

Due to their relatively small size, metallic NPs such as silver (Ag) and gold (Au) can penetrate the cellular membrane via endocytosis or diffusion processes. They enter the cell and reach the mitochondria, where they can damage mitochondrial function by altering the electron transport chain, producing oxidative stress (Kang et al. 2015). The prolonged oxidative stress results can cause DNA damage by cross-linking or formation of DNA adducts, as well as protein oxidation and lipid peroxidation, which may ultimately affect cell growth. Readers are directed to a comprehensive review by Ferdous and Nemmar on the health impact of the various routes of exposure to silver NPs (AgNPs) (Ferdous and Nemmar 2020). For instance, 20 nm citrate-AgNPs exacerbated the cardiac ischemic reperfusion injury in rats more significantly than 110 nm particles, indicating particle size as an essential contributing factor associated with cellular toxicity (Holland et al. 2016). Studies on size-dependent toxicity of gold NPs (AuNPs) demonstrated that the larger AuNPs (15 nm) were found to have a lesser toxic effect on macrophages, fibroblasts, melanoma cells, and epithelial cells (Pan et al. 2007). It is also in correspondence to the smaller AuNPs (<6 nm) that can effectively penetrate the cell nucleus (Huo et al. 2014), suggesting size-dependent nucleus penetration of nanoparticles that can potentially exacerbate cell death.

11.3.1.2 Chemical Composition

The toxicity of NPs is predominantly affected by their size and surface area, but other factors, for instance, the chemical composition, which can influence their intrinsic toxicological properties, should also be addressed. The chemical composition of NPs, upon interactions with cells, may generate changes in pH, ionic strength, protein concentrations, and solubility that may generate toxicity in the intracellular environment. Consequently, NP degradation can occur, leading to a loss of stability and functionality, alleviating the toxicity level (Soenen et al. 2015). Given the apparent indicators of enhanced toxicity or loss of NP functioning, comprehensive characterization of NP with verifiable data to assess the degree of this degradation is crucial for sustainable development.

The toxicological behavior of inorganic NPs with varied chemical compositions can be generated from the same physical state. For example, a study of the effects of 20 nm silicon dioxide (SiO_2) and zinc oxide (ZnO) NPs on mouse fibroblast cells indicated that the toxicity mechanisms varied, whereby the former generated oxidative stress. Simultaneously, ZnO altered the DNA structure (Yang et al. 2009). Compared to TiO_2 NPs, Ag, ZnO, and CuO NPs often show more significant toxicity. Johnston et al. (2015) reported that Ag nanomaterials, but not TiO_2, induced neutrophil-like cells to undergo time- and concentration-dependent oxidative stress and an inflammatory response triggered by variations in intracellular calcium ion and ROS concentration (Johnston et al. 2015). The study implied that ion release has been primarily responsible for the toxicity, independent of the testing system.

Metal ions such as lead, cadmium, mercury, and silver have a toxic nature that can change through oxidation or reduction processes. Their leakage from the NP core is the most prevalent source of the toxic effect upon interacting with cells (Roane et al. 2015). Although NPs released in the air can be used to estimate the toxicity of the particles, it is crucial to highlight that particles in ambient air, owing to combustion pollution, are coated with various reactive chemicals and even biological substances (Carty et al. 2003). Consequently, data gathered from ambient air for NP toxicity should consider the potential impact of particle composition and pollution. While other metal ions, such as iron and zinc, can be beneficial from the biological standpoint, they are hazardous at high concentrations and trigger toxicity responses. Interestingly, the concerns can be overcome or reduced by coating the cores of the NPs in silica sheets, gold, or thick polymer shells rather than short ligands or by employing new NP synthesized with non-toxic compounds or other metals to modify the composition of the core (Soenen et al. 2015). This may strengthen the safety and chemical stability of metal ion NPs in the body against degradation and leakage.

Furthermore, molecules adsorbing on the surface may influence the reactivity of NPs. The detrimental effects, such as mitochondrial dysfunction and increased production of oxidative stress, of fractions separated from particulate air pollutants were demonstrated in vitro (Xia et al. 2004). The composition of NPs in ambient air may be complex, with numerous components such as organics and metals interacting. For instance, the efficacy of carbon black NPs can be improved when interacting with metallic iron; however, the combination can generate unnecessary oxidative stress (Wilson et al. 2002). On the other hand, surface modification of NPs with their derivatives can overturn the cytotoxicity-induced behavior or morphology of the NPs. Likewise, the cytotoxicity of superparamagnetic iron oxide NPs in vitro and iron oxide NPs in vitro can be avoided by coating the surface of the particles with hydrogel pullulan and dextran or albumin, respectively (Berry et al. 2003; Gupta and Gupta 2004).

Their crystal structure also determines the toxicity of NPs. A human bronchial epithelial cell line and TiO_2 NPs with various crystal lattices were used to investigate the association between crystal structure and toxicity (Gurr et al. 2005). TiO_2 NPs are known as biocompatible materials but are prone to generate potent cytotoxicity in the presence of photoactivation. Unlike this concept, a study done by Gurr et al. (2005) revealed otherwise, where the results suggested that in the absence of photoactivation, TiO_2 particles with rutile-like crystal forms (10 and 20 nm) stimulated cytotoxicity damage to the BEAS-2B cells, a human bronchial epithelial cell line, but not TiO_2 particles with the size of 200 nm and more. The smaller the NPs with rutile-like crystal structures, the easier it is to cause oxidative damage, but same-size NPs with an anatase-like crystal structure are non-toxic. In addition, depending on the environment, the formation of the crystal-like structure of NP might vary, such as when it encounters water, biological fluids, or other dispersion mediums. Zhang et al. (2003) also investigated that when zinc sulfate (ZnS) NPs are exposed to contact with water, their crystal lattice is restructured into a more organized shape (Zhang et al. 2003).

11.3.1.3 Shape

Most NPs have been developed into the following shapes: spheres, tubes, sheets/ films, discs, fibers, cubes, and rods. Although there is no definitive proof, the shape is also considered a pivotal component that allows NPs to penetrate biological barriers, which considerably can contribute to the toxicity of the NPs. This has been shown for several NPs of various forms. Spherical NPs, for example, are more vulnerable to endocytosis than nanotubes and nanofibers (Champion and Mitragotri 2006). Conversely, carbon nanotubes with a single wall have been discovered to block calcium channels more than spherical fullerenes (Park et al. 2003). NPs with the shapes of plates or needles were shown to inflict more significant necrosis proportions than spherical and rod-like NPs due to their higher tendency to impose severe damage to cells and living tissues via direct contact (Zhao et al. 2013). This is demonstrated by the different shapes of hydroxyapatite NPs, which are plate-, needle-, rod, and spherical shapes on BEAS-2B (human lung epithelial) cell cultured where the plate and needle-like NPs damaging a more significant proportion of cells through a direct contact than rod-like and spherical NPs (Zhao et al. 2013).

Of note, AuNPs are increasingly utilized in various areas of medicine, particularly cancer treatment. Steckiewicz and group investigated the cytotoxicity effect of AuNPs rods, stars, and spheres in human fetal osteoblastic cells (hFOB), osteosarcoma (143B, MG63), and pancreatic duct cell (hTERT-HPNE) lines, where AuNPs stars generated the most prominent cytotoxic effect with significant anticancer potential. In contrast, Au nanospheres were less cytotoxic (Steckiewicz et al. 2019). Also, the form of AuNPs substantially impacts accumulation kinetics and excretion, with only star-like shapes being able to be reserved in the lung. Moreover, in vitro studies have established that the variations in shape and geometrical form have no significant effects on their ability to cross the blood-brain barriers (Talamini et al. 2017).

Similarly, toxic effects with physical damage to the cell membrane were revealed by the graphene oxide (GO) nanosheets when directly interacting with the human cell (Hu et al. 2011). On the contrary, the cytotoxicity was lower with a higher concentration of fetal bovine serum in the culture medium. This protective effect is primarily contributed by the high protein adsorption ability of GO NPs, which covers the surface area with shape modification of the NPs, substantially preventing damage to the cell membranes (Hu et al. 2011).

11.4 Waste Management and Nanoparticle Environmental Impact

Following the increasing application of NPs with product diversity in various sectors, the management of nanowaste is a new challenge in waste disposal. Questions arise on the diversification of emission sources in the environment. Do they attach to them, or are there particles that cluster on their own or are repelled by the cellular membrane of an organism? (Mu et al. 2014). Little is understood about how NPs function, where they go, and how harmful they are in their natural ecosystems. Significant gaps in our understanding of the environment and toxicity of NPs after they have been utilized for a specific purpose have been alarming to researchers, policymakers, and consumers. Not to mention, although recent advancements in NPs are rapid, there still needs to be a consensus or standardized protocols for the monitoring of NPs ecotoxicity, limited knowledge of the actual behavior of NP, and no report on the long-term assessment of the impacts of NPs in the ecosystem. Although model ambient (air pollutant) concentration estimates for various nanomedicines have been documented, current data implies that the biological properties of NPs and their consequences to the ecosystems are poorly described. Indeed, sophisticated studies on the health impacts of incorporating nanomaterials in medical devices, diagnostics, and therapeutics have dramatically increased since the 2000s (Bharathala and Sharma 2019). Despite the fact that NPs are excellent DDS, they come with severe drawbacks, including high-dose toxicity, volatilization by various inorganics, and poor pH sensitivity. Overall, this section highlights the routes of emissions of NPs from the continuous and large production of pharmaceutical products; however, the release process is beyond the scope of the writing.

Continuous use of NPs as an alternative to enhance the functions of present healthcare products is now resulting in the emergence of a high number of novel and complex materials, with the subsequent high release of the particles and to their accumulation into the environment. The increment in production may eventually cause the release of these nanoscale materials into the environment and lead to their pile, potentially jeopardizing ecosystems apart from human health. NPs can be disseminated into the environment generally considered by planned or inadvertent releases:

- Production of raw material and NP-based products (e.g., drugs and medical devices).
- Use by consumers.
- After disposal or waste, handling of NP-based products, which are commonly associated with erosion and degradation of the materials.

The entry of NPs into specific environmental compartments depends on their life cycle and disposal method used, such as for the large production of textiles. Washing nanofunctionalized textiles will release NPs into the sewer systems, which will then be transported to natural water streams. Air, soil, and water are just some environmental compartments in which NPs can be detected, but there is presently no gold criterion for assessing how harmful they are to biodiversity (Baysal et al. 2017).

Initial attempts to evaluate NP emissions across the whole life cycle showed that the vast majority of NP is released during the utilization phase and after disposal, often in landfills (Keller et al. 2013). In contrast, only 2% of the manufacturing volume is released during production (Gottschalk and Nowack 2011). NPs are immediately released directly or indirectly into the environment through systems, such as wastewater treatment facilities or landfills. In comparison, indirect emissions likely occur via the discharge from wastewater treatment facilities, the implementation of organic wastes into the soil, or the byproducts from landfills. Taking into account that the fate of NPs released into the proper systems, such as wastewater treatment facilities, is predominantly reflected by their conditions or properties, such as whether they are uncoated or coated and modified physically or chemically (Kaegi et al. 2011; Zuin et al. 2014).

Nonetheless, lands, soils, and sediments were the ultimate reservoirs for NPs, whereas wastewater treatment facilities are temporary receivers of NPs (Westerhoff et al. 2013). Similarly, under various environmental circumstances, NP-based pharmaceutical products can be transformed depending on the features of the NPs and environmental matrices. For example, a study done by Hong et al. (2009) on the transport of magnetic NPs using several iron-based NPs in aquatic environments revealed that the magnetic NPs aggregate with high stability when the pH level is close to neutral, which offers more excellent retention, but they scatter at higher pH (Hong et al. 2009). Other than aggregation, adsorption, oxidation or reduction, and dissolution, are some primary transformation processes for NPs (Lowry et al. 2012). To avoid hazardous impacts, NPs must be promptly eliminated from the body after functioning. There are several entrance points into the environment for various NP-based medical applications. Taken orally or injectable NPs-based medicine may undergo metabolic activation in the system due to the activity of numerous enzymes and are eliminated mainly through urine or feces (Lipka et al. 2010). The primary route of NPs elimination from the human body that does not disintegrate in vivo into biologically benign components is through the liver, bile, feces, or renal filtration (Zhang et al. 2020). From these pathways, the particles were excreted to residential sewage, then to wastewater treatment facilities, receiving water bodies (e.g., rivers, lakes, and ponds), and land, with a considerable proportion of wastewater remaining untreated.

Several studies have been conducted to determine the clearance processes of NPs for medical applications. For example, studies on the biodistribution of AuNPs have discovered that the clearance pathway of the NPs depends on their size, type of surface coating, and charge. A study done by Lipka and the group revealed that 10 kDa PEG-coated AuNPs were excreted via the hepatobiliary system (Lipka et al. 2010). Unlike the coated AuNPs, the uncoated 20 nm AuNPs were excreted mainly in the urine (Balasubramanian et al. 2010). A 2011 study found a negative relationship between hepatobiliary clearance and negatively charged AuNPs of varying sizes. Hirn and colleagues reported that the minuscule size and negatively charged AuNPs was excreted via the hepatobiliary pathway due to the dynamic protein binding and exchange to regulate AuNP deposition in tissues and organs (Hirn et al. 2011). Additionally, the biodistribution of encapsulated and unbound drugs may differ. According to a study, administering the unchanged medication inside poly(epsilon-caprolactone) NPs increased urine excretion of the drug by 15% compared to administering the drug without encapsulation (Ferranti et al. 2001). Consequently, a more significant environmental release will occur unless doses are changed. When the biological fate of PEGylated model protein was investigated, it was discovered that 46.5% of the delivered volume of PEG was discharged through the urine during a short period. It was also revealed that the whole PEGylated model protein was eliminated in the first several days after delivery (Elliott et al. 2012). NPs could be released into the air through alternative pathways such as via inhalers, the disposal of unneeded pharmaceuticals at hospitals, clinical research, and R&D institutes by inexperienced and untrained workers, material discharge during manufacture, transport, unintentional spills, and incinerators. Because of their high specific surface areas, NPs may be adsorbed on waste materials by the combustion processes in an incinerator or detected in the off-gas (Walser et al. 2012). Even after cutting-edge discharge gas treatment procedures, the total elimination of NPs smaller than 100 nm from incinerator off-gas may not occur (Ounoughene et al. 2017).

11.4.1 Lands and Soil

In drug development, durable NPs like monodispersed iron oxide NPs that can be directed to specific cells are essential to completely alleviate their unique features, extend shelf life, and impact physiological responses more effectively (Ho et al. 2011; Nam et al. 2003). The same design may also determine the fate and danger of the NPs in the environment by not only impeding growth and reproduction but also with aerobic metabolism of the organisms, hence disrupting the balance of the ecosystem. However, organisms may be able to survive nanomedicine exposure to some degree, although this is rarely defined. Also, NPs, like organic polymer-coated NPs, can readily evade the mononuclear phagocytic system of the body or to directly target cells. Nonetheless, as documented by Măruţescu et al. (2019), continuous manufacturing and utilization of AgNPs result in increasing release into the natural environment, with vastly greater concentrations in soil (Măruţescu et al. 2019). Because of the ecotoxicity of discrete AgNPs, aggregated AgNPs, AgNPs converted to silver sulfide (Ag2S), or Ag ions generated by AgNPs, the accumulation may pose a threat to living terrestrial organisms.

In addition to generating contamination in the environment that affects many animal habitats. NP-based pesticides can endanger human health. More research is required to improve the composition of NPs for agricultural uses and subsequently limit their toxicity to non-target organisms. On the other hand, several terrestrial species have been utilized to evaluate NPs soil toxicity. Jacques et al. (2017) investigated the toxicological assessment of pesticide-loaded NPs on a soil organism, *Caenorhabditis elegans* (Jacques et al. 2017). The authors used three different NPs with loaded or unloaded pesticides; (1) solid lipid NPs loaded with atrazine and simazine, (2) polymeric NPs loaded with atrazine, and (3) chitosan/tripolyphosphate NPs loaded with paraquat. The results exhibited a dose-dependent increase in lethality with comparable LC50. Even at low concentrations, polymeric NPs, both loaded and unloaded NPs were the most hazardous to the growth and development of *Caenorhabditis elegans* causing them to get shorter. Chitosan/tripolyphosphate NPs, on the other hand, were the least toxic even at higher concentrations, which could be due to the biocompatibility of chitosan. These results show that additional research is needed to enhance the composition of NPs and thereby limit their toxicity to non-target organisms.

11.4.2 Aquatic Organisms

Limited aquatic organisms are employed for nanotoxicity studies, especially for biomedical applications. Algae are the most frequently studied and sensitive organisms, followed by crustaceans, fish, and bacteria (Freixa et al. 2018). Although there is no imminent risk from carbon-based NPs (e.g., graphene, fullerenes, and carbon nanotubes) in low concentrations, there have been reports of detrimental impacts at higher concentrations. However, other studies found considerable variations in these findings, which may be related to their physicochemical properties (Freixa et al. 2018). Since most NP-based drug delivery and therapy utilizes metallic materials, most of the nanotoxicity studies focused on the source materials, fate, and effects of metallic NPs, i.e., AuNPs and AgNPs. For a review of the uncertainty and implications of AuNPs and AgNPs in the environment, readers are directed to other in-depth literature (Ferdous and Nemmar 2020; Mahapatra et al. 2015).

The use of AgNPs, which are present in several common consumer goods other than pharmaceutical products, including cosmetics, paints, plastics, cosmetics, textiles, and paints, has significantly increased during the past 10 years. As a result, discharging such AgNPs into aquatic environments may disturb the ecology. Sendra et al. (2017) study the toxicity of AgNPs on freshwater microalgae, *Chlamydomonas reinhardtii*, and marine microalgae, *Phaeodactylum tricornutum*, suggesting the significance of intrinsic and extrinsic variables on NP behavior in various aquatic habitats and interactions with microalgae. Indirect effects, mainly dissolved Ag ions, appear to regulate toxicity to *Chlamydomonas reinhardtii*. In contrast, *Phaeodactylum tricornutum* toxicity seems to be controlled by direct effects, such as attaching to the cell surface and cellular internalization by Ag NPs (Sendra et al. 2017).

Overall, biowaste influences the water management system. It has been discovered that NP-based biomedical applications can be hazardous to aquatic animals and plants, implying that these ostensibly biocompatible materials may represent a substantial risk to plants and wildlife. Further studies are needed to confirm the notion. Studies on possible flow and concentrations of NPs of various materials in anthropogenic and biological systems are critically needed to aid frame risk assessments.

11.5 Green Synthesized Nanoparticles for Sustainability in Biomedical Application

The extent of the acceptance of nanotechnology attracted many investors and gave rise to growing marketing. For example, gold, silver, gold, aluminum, copper, zinc, and titanium metals and their oxide-based NPs and carbon-based nanostructures are often used in consumer products (Pandey et al. 2020). Increase in the frequency and intensity of climate changes have, however, increased awareness of the importance of nanomaterials for sustainable development and the necessity of taking action in this direction (Chee et al. 2022).

In 2017, the United Nations issued a call to action for all nations in the form of its Sustainable Development Goals, which recognize that reducing inequality in all areas requires simultaneous efforts to end poverty and other forms of deprivation and to improve health, the environment, and the economy (Pokrajac et al. 2021). Therefore, bolstering present nanotechnology to benefit humanity, the environment, and the planet, is crucial.

Sustainability is defined as direction and scope that fulfils the requirements of the present without jeopardizing the capacity of future generations to achieve their objectives (Subramanian et al. 2014). A sustainable healthcare system strives to provide high-quality care with little harm to the environment while taking the advantage of every opportunity to improve and restore the environment for the benefit of both current and future generations. Safer alternative approaches must be implemented to synthesize multifaceted nanocompositions or structures that can ensure sustainability. For example, ecofriendly methodologies and surfactants mediated by plants offer a green tool for nanomaterial fabrication. Of importance, recent developments in green-influenced nanomaterial fabrication technologies demonstrate their significant relevance in science and nanotechnology. Indeed, recent advancements in NPs using green functionalized materials to improve functionality while limiting adverse effects are employed in green chemistry and progress toward sustainable nanotechnology.

In previous decades, conventional NPs were primarily manufactured using non-green methods, most of which contained hazardous substances that caused significant concern due to their toxicity. Reducing and stabilizing agents were used to minimize metal ion toxicity and avoid unwanted agglomeration of generated NPs. Nonetheless, bio-incompatible materials endanger the environment. Multiple works of literature have reported other harmful reducing agents for synthesizing NPs, particularly metallic NPs (Chou and Ren 2000; Dhas et al. 1998; Valodkar et al. 2011). As a result, in recent years, the scientific community has come to terms with the development of simplistic and greener frameworks in which safe materials can replace substantial toxic and unstable precursors to minimize the various risks associated with existing chemicals or substances. Using NPs in combination with green nanotechnology is a considerably superior solution since it avoids the need for harmful and expensive chemicals, requires less energy, and is ecologically friendly. Green nanotechnology aims to adjust the size of NPs within the nanometer scale to develop a biologically active particle. Green NPs can be classified into several sources (see Table 11.1). However, this chapter will focus on plant-based green NPs in biomedical applications.

NPs made up of materials that can be divided into two categories: natural materials and synthetic materials, which are aligned with three criteria (1) easy to synthesize, (2) ecofriendly synthesis (safe humans and environment), and (3) cost-effective. A wide range of NPs have been synthesized by adopting a green method and effectively implemented in various fields, including medicine, wastewater treatment, agriculture, textile, and solar energy (Al-Khalaf and Hussein 2018). Indeed, the excellent properties of green NPs are fundamental for their improved performance, which reveals promising prospects for diagnostic and therapeutic applications. Some of the significant biomedical applications using green synthesized NPs with the potential for sustainability are highlighted below.

11.5.1 Diagnostics and Imaging

Early diagnosis for intervention is the key to treating almost all diseases successfully. Medical imaging technology is vital in diagnosing and assessing therapy

Table 11.1 Types of green NPs with examples

Types of Green NPs	NPs	Source	References
Bacteria-mediated	Copper	*Rhodotorula mucilaginosa*	Salvadori et al. (2014)
	Metallic	*Clostridium pasteurianum*	Nordmeier et al. (2018)
	Silver	*Shewanella oneidensis*	Suresh et al. (2010)
Fungi-mediated	Silver	*Sporothrix schenckii*	Wang et al. (2021)
	Silver	*Candida parapsilosis*	Wang et al. (2021)
	Silver	*Trichoderma harzianum*	Guilger-Casagrande and de Lima (2019); Rajeshkumar and Sivapriya (2020)
Actinomycetes-mediated	Gold	*Rhodococcus* sp.	Ahmad et al. (2003)
	Silver	*Streptomyces* sp. NH21	Składanowski et al. (2017)
	Zinc	*Streptomyces spp.*	Kumari et al. (2020)
Yeast-mediated	Gold	*Pichia jadinii (Candida utilis)*	Gericke and Pinches (2006)
	Gold	*Yarrowia lipolytica*	Pimprikar et al. (2009)
	Cadmium sulfide	*Candida glabrata*	Krumov et al. (2007)
Algae-mediated	Silver	*Turbinaria conoides*	Rajeshkumar et al. (2012)
	Gold	*Fucus vesiculosus*	Mata et al. (2009)
	Gold	*Rhizoclonium fontinales*	Dhanalakshmi et al. (2012)
Biomolecular-mediated	Gold	Terminalia catappa	Ankamwar (2010)
	Silver	Syzygium cumini	Kumar et al. (2010)
	Silver	Desmodium triflorum	Ahmad et al. (2011)
Plant-mediated	Silver	*Prunus yedoensis*	Sathishkumar et al. (2012)
	Silver	*Grewia flaviscences*	Sana et al. (2015)
	Copper	*Citrus medica*	Vijayaraghavan and Ashokkumar (2017)

responses to many medical conditions. X-ray radiography, ultrasound, magnetic resonance imaging, computed tomography, fluorescence imaging, and positron emission tomography are some of the mostly employed and successful bioimaging and diagnostic tools in healthcare. Unfortunately, in the case of cancer, for example, the typical diagnostic procedures cannot detect anomalies in the early stages of development. Besides, they are limited in distinguishing malignant, benign, and sometimes healthy cells. It is important to note that the spatial resolution generated by imaging technology, such as contrast-enhanced computed tomography, cannot identify a hypervascular hepatoma at 3 mm, let alone estimate a tumour during the early development of the disease (Cuccurullo et al. 2018). Moreover, the anatomic and functional medical imaging contrast agents are utilized to differentiate between healthy tissue and pathological lesions to attain more precise information. Nonetheless, current medical imaging contrast agents are small substances with rapid metabolism, non-specific distribution, and potentially harmful toxicity (Chronos et al. 1993; Petrik et al. 2005).

Due to their novel passive, active, and physical targeting capabilities, nanomaterials have sparked interest in enhancing diagnostics and bioimaging. Although the green synthesized NP is very new in this field, multiple studies in the imaging diagnosis of different cancers have been optimistic. In contrast to the previously mentioned, imaging agents using green NPs exhibit the potential to be more sensitive and selective, allowing for better cancer detection and other pathologies. Because of their diminutive size, NPs have increased permeability and retention effects in tumors, resulting in relative increases in local tumor contrast agent concentrations. Among all features of the NPs, size plays a vital role in tumor imaging. They strongly support their potential as an alternative to conventional diagnostic tests. The physicochemical properties of green synthesized NPs make them ideal gateways for imaging and diagnostics. Because several studies have demonstrated the immense potential of green NPs in the theranostic of various cancers, this chapter will focus on cancer. For example, He et al. (2013) incorporated green AgNPs in clinical ultrasound gel to enhance cancer therapy (He et al. 2013). Likewise, Gao and group investigated near-infrared fluorescence imaging of cancer cells, revealing the presence of Ag nanoclusters within tumor cells, where they protect normal cells despite prolonged exposure to Ag ions, suggesting the highly precise and selective green AgNPs for cancer cells imaging (Gao et al. 2014).

There is a growing demand for environmentally friendly AuNP biosynthesis approaches to assure their safety and prevent unintended implications in various biomedical applications. Green AuNPs are currently used in imaging diagnosis tools, including X-ray radiography, ultrasound, and fluorescence imaging. Notably, the extensive supply of biological precursors employed in plant-based AuNPs enables the development of greener large-scale manufacturing. Green synthesized AuNPs, in addition to AgNPs, have garnered attention due to their excellent properties in bioimaging, particularly for cancer detection, such as high stability for long-term imaging and monitoring target cells, nonimmunogenicity, low cytotoxicity, and high permeability (Gong et al. 2020). Compared to the contrast agents generally used in clinical settings, AuNPs offer long-term imaging and target cell monitoring. These superior features of AuNPs are precisely due to their longer circulation time in blood, enabling simple modification with higher uptake using ligands identified by cancer cell surface receptors (Arifin et al. 2011). Further supporting the notion, Cai and colleagues examined the pharmacokinetics and computed tomography imaging effectiveness of colloidal AuNPs as a blood pool agent for X-ray (Cai et al. 2007). The findings show that green AuNPs have a high potential as a blood pool contrast agent, extending drug retention duration within the blood vessels. Another green-based NP shown to label human cervical cancer cells in DNA binding tests is the green synthesis of AuNPs using polysorbate (De et al. 2014). Compared to the commercial dye Hoechst, the authors noticed staining capability even at extremely low AuNPs concentrations. This finding was corroborated by Emmanuel et al. (2014) study on a diagnostic application in a mouse model of cervical cancer, showing that AuNPs prepared from *Acacia nilotica* twig bark extract could detect low concentrations of a potentially lethal pollutant that causes methemoglobinemia (Emmanuel et al. 2014).

Since detecting tumors at an early stage is so critical, research into new tools to aid in this endeavor is vital to minimizing the severe consequences of this disease in the future and reducing the risk of one life.

11.5.2 Drug Delivery and Therapeutic

One of the outstanding innovations of nanotechnology is the function of NPs, which can efficiently and precisely deliver drugs or substances directly to the target sites. It is becoming vital when used as part of the treatment to prevent adverse effects and complications. The ability of NPs to deliver numerous substances at once is a massive benefit and a possible solution for various treatments (Sun et al. 2021). Besides being an attractive model for diagnostic and imaging applications in terms of being environmentally safe and cost-effective, green NPs also offer ample potential for drug delivery applications. The burgeoning need for green nanotechnology DDS has resulted in adoption of several delivery mechanisms. Specific targeting and high biodistribution, biocompatibility, and safety are indeed the positive aspects of NPs over conventional drug delivery methods. The relatively small size and beneficial surface properties facilitate NPs for the rapid uptake of the loaded substance by all the cellular compartments of the human body (Ferdous and Nemmar 2020). Green synthesized NPs using plants have been attracting much attention due to their low cost and ecofriendly properties. Additionally, they are beneficial as drug delivery agents for cancer therapy, antibacterial agents, and other diseases, which were highlighted in the following paragraphs.

Recently, using six medicinal plants, Khanzada et al. conjugate diclofenac to the AuNPs and FeNPs. *Syzygium cumini* and *Ficus microcarpa* proved to have good antibacterial properties of gram-positive and gram-negative bacteria (Khanzada et al. 2021). Also, FeNPs and AuNPs were discovered to have benign effects despite having a spherical shape and typical sizes of less than 50 nm. Although FeNPs showed potential antioxidant activity when synthesized with *Phyllanthus Emblica* and *Syzygium cumini*, AuNPs with the same materials were protective against DNA damage, with the latter showing potent anti-inflammatory effect compared to DFP and diclofenac-loaded FeNPs of *Ficus microcarpa*. The results suggest potential green synthesized AuNPs and FeNPs as promising approaches for targeted drug delivery (Khanzada et al. 2021). In another study, AgNPs were developed using *Juniperus procera* extract, which exhibited dual roles as a reducing and stabilizing agent (Khan et al. 2022). The synthesized AgNPs displayed significant antimicrobial activity compared to the plant extract alone and were almost comparable to the standard microbial drugs. Overall, AgNPs using *Juniperus procera* can be used as an efficient drug delivery agent due to their properties including uniform size (~23 nm) and spherical shape with aqueous colloidal stability of nanoparticles offer the scope for further development. Thirumurugan and his group demonstrated that the biological fabrication of AgNPs and platinum NPs (PtNPs) with curry leaves and neem extracts could be utilized as an anticancer DDS (Thirumurugan et al. 2018). AgNPs and PtNPs encapsulating doxorubicin (DOX), an anticancer drug against breast cancer cell line, revealed better cytotoxic activity of AgNPs with DOX than PtNPs. It also shows that secondary metabolites from plants played a crucial function as a reducing agent in synthesizing metallic complexes into NPs.

The biocompatible quality of green NPs and their affordable and simple fabrication may assist in anticancer research. Indeed, combining anticancer drugs with biosynthesized metallic NPs demonstrated outstanding anticancer efficacy. For example, using an in vivo lymphoma mouse model, research reported that AgNPs synthesized from *Melia azedarach* significantly reduced cytotoxic effects by inhibiting the proliferation of human epithelial. In contrast, the normal human breast lactating cell line remained secured, demonstrating the excellent biocompatible nature of the green synthesized NPs (Sukirtha et al. 2012). This study is also in line with work done by Barua and colleagues, where synthesized AgNPs using an aqueous extract of *Thuja occidentalis* leaves exhibited nonhazardous anticancer properties and were biocompatible with normal cell lines (Barua et al. 2013). AgNPs synthesized using *Datura inoxia* resulted in anticancer activity where the proliferation of human breast cancer cell line (MCF) was inhibited at low concentrations by inducing apoptosis (Gajendran et al. 2014). Various studies have revealed that growth suppression, cell cycle arrest, DNA synthesis inhibition, and apoptosis are potential mechanisms for the cytotoxic effect of the NPs, as mentioned earlier. Also, the successful delivery of imatinib with apoptosis-inducing ability using co-delivery NPs with AgNPs depicted the green AgNPs as a promising delivery of therapeutics (Azadpour et al. 2022). Undoubtedly, the drug-release nanocomposite of green synthesized AgNPs coated with targeted chitosan NPs for efficient anticancer drug delivery demonstrated excellent suppression against cancer cells. Another study of AuNPs biosynthesized with *Dysosma pleiantha* rhizome rich in lignans and flavonoid demonstrated anti-metastatic activity in human fibrosarcoma cell line HT-1080 (Karuppaiya et al. 2013). The effect of ZnO surface properties on anti-cancer activity was also observed using green ZnO NPs, which were revealed to provide higher toxicity effects on Hut-78 lymphoma T cells than chemically synthesized ZnO NPs (Punnoose et al. 2014). Other than ZnO NPs, copper NPs synthesized with *Euphorbia nivulia* exhibited a cytotoxic effect on human lung carcinoma cells (A549) where the NPs can easily penetrate the cells dose-dependently (Valodkar et al. 2011). The proposed scheme might be beneficial for delivering anticancer drugs and accurate and selective tumor imaging.

The emergence of increased resistance cases to traditional antibiotics has been a primary concern for treating bacterial infections. According to several pieces of evidence, multidrug-resistant strains of bacteria (MDRS) may be immune to third-generation drugs (Terreni et al. 2021). Besides, antibiotic resistance in bacterial strains has rapidly developed due to the unrestricted and unauthorized use of antibiotics (Singh et al. 2020). Hence, more effective antibacterial/antimicrobial agents or novel substances in the current pipeline are required to combat the increased microbial infections and are crucial for primary care. Interestingly, green NPs have been widely established as the alternate strategy to combat MDR microorganisms for having potent antimicrobial properties. Because NPs have a large surface area, they may quickly enter microorganisms, but the specific process is still uncertain (Terreni et al. 2021). In addition to serving as a reducing agent, which helps to minimize the cytotoxic effect of NPs, green synthesized NPs such as antibiotic NPs complex (nanoplex) have been revealed to overcome the poor antibiotics or substances loading. For example, ofloxacin and levofloxacin are used as antibiotics, with dextran sulfate nanoplex possessing very high loading and optimal size (400 nm) ideal for sputum penetration (Cheow and Hadinoto 2012). The antibacterial efficacy of the commercial drug ofloxacin was also significantly improved when administered in combination with these NPs.

Green synthesized metal and metal oxide NPs have also been proven to be highly toxic to gram-positive bacteria like *S. aureus*. However, they are safe, less harmful, and biocompatible compared to other NPs. Readers are directed to comprehensive reviews for further thought on this subject (Pillai et al. 2020; Sharma et al. 2009; Vanlalveni et al. 2021). Green metallic NPs, for instance, AgNPs synthesized using latex and leaf extracts of *Ficus sycomorus*, were recently found to have superior antibacterial properties against many gram-positive and gram-negative bacteria (Salem et al. 2014). Indeed, *Ficus sycomorous* is known to possess potent antioxidant and antimicrobial activities against *S. aureus* and fungal infections, further validating its potential against pathogenic microorganisms resistant to antibiotics (Özdenefe et al. 2020). Using polysaccharides obtained from marine algae, El-Rafie et al. synthesized AgNPs and showed the potent antibacterial properties of the treated textiles (El-Rafie et al. 2013). Biosynthesized Au and Ag NPs have the antimicrobial potential to prevent the growth of bacteria and other microorganisms. They exhibit immobilization in cotton, hyaluronan fibers, or other materials to form antiseptic dressings or bandages for sustained and improved antimicrobial effects.

11.6 Challenges and Future Perspective of Green Synthesis Nanoparticles

Sustainable nanomaterials are defined as safe nano-objects for humans and the environment, incorporating green chemistry and engineering. Sustainable NPs are not only defined by their inherent nature from natural precursors (plants, animals, human derivatives, agricultural waste, and small biomolecules) but also involve the process that would reduce the intrinsic hazard. The nanoscale processes and characteristics are fundamentally different from those of larger sizes, which drive their design and implementation; however, it is not reasonable to assume that these nanoscale features will always be beneficial or free of consequences when applied to humans. The importance of sustainability arises from the individual and collective effects of environmental, political, and economic challenges the world is currently facing.

The challenges of sustainable NPs that need to be addressed are their competency compared to the traditional toxic nanoparticles, the cost, and scalability. In the biomedical field, efficacy and bioavailability are inherently prioritized in the development of therapeutics, and the processes involved that pose environmental threats are usually overlooked. The biomedical field contributes through R&D, industrial production, and consumption to waste management of medicinal products. Since various high-throughput tests and strategies for evaluating NPs safety risks have progressed with tremendous success, standardized regulations that account for the possible toxicological behaviors of various nanoformulations are critically needed. The majority of the types of nanoparticles highlighted in this chapter are metallic and metallic oxides, as there is growing research employing this material and their application in recent years. Albeit favorable over other nanomaterials, just like other NPs, they are inevitably accompanied by the release of NPs residues to the environment. This demonstrates one of the key routes through which hazardous materials enter the ecosystem and the food chain and, as such, disturb the ecological equilibrium. It, therefore, remains the primary threat in the future.

Another issue is the biocompatibility of NPs, mainly metal and metallic oxide. This necessitates additional modification of these particles to introduce biocompatible groups. The NPs are also unstable in various matrix mixes, particularly in physiological conditions desirable for biomedical applications. Surface coating is thus required to increase the stability of the particles, which brings about additional synthesis and is time-consuming, cost-ineffective, and significantly further complicates the characteristics of the NPs, which frequently conflict with the expected features.

Conventionally, the advancement of nanotechnology equates to a high cost, hence low affordability in middle-/low-income countries. In the Sustainable Development Goals by the United Nations, growth in the economy, health, and education must go hand-in-hand with reducing inequality and other deprivations in all countries. Uskoković (2021) addressed this issue using COVID-19 vaccines as a current example and outlined two primary routes to address this challenge; contextual considerations from the sociocultural climate in nanomedical research and challenging the profit-seeking objectives (Uskoković 2021). However, these routes must be addressed promptly, and the outreach goes beyond the framework of scientific discovery.

The COVID-19 pandemic has awakened science to move faster, and bioinformatics has emerged as an efficient discipline to acknowledge this challenge. Hence, incorporating machine learning (ML) to address problems in nanotechnology is considered one of the strategies for sustainable NPs. ML can be used to predict the proteins in NPs protein corona and identify features predicting NPs toxicity (Findlay et al. 2018; Labouta et al. 2019). Daly Jr. and Hernandez (2020) have used ML consisting of artificial neural networks and uncovered features for designing NPs with increased viability, suggesting experiments only to extend to the next phase of nanoparticle development (Daly Jr and Hernandez 2020).

11.7 Conclusion

The conventional NPs possess distinctive properties mainly due to their diminutive size, which allows them to penetrate deep tissues while dispersing throughout the body, which is the primary reason for their toxicity. Besides producing very unhealthy products using harsh chemicals, they are costly and have been shown to significantly impact the environment by releasing hazardous wastes into the soil, water, and air. Therefore, the alternate approaches found in green nanotechnology are at the center of progress in the field to alleviate the concerns associated with synthesizing conventional NPs. This article focused on the green synthesis of metallic NPs using the plant, as it has shown tremendous success in various areas. However, reports pouring on them are limited to AuNPs and AgNPs on their anticancer and antimicrobial activities. Green synthesis is a relatively cost-effective approach for synthesizing NPs in large-scale production and regulation. The biological sources provide a natural compound that serves a versatile yet robust role in boosting the function and activity of NPs. They significantly influence the characteristics of the particles. Unique features in the matter of biocompatibility, stability, and therapeutic activity. Taking these factors into account, the green paradigm tackles several elements of NP risks. It has given rise to a new vision for progressing nanotechnology in a considerably more adaptive and environmentally responsible manner. The chapter highlighted the risks and concerns of the currently implemented NPs in biomedical applications and the impact of a green approach for sustainability in terms of synthesis materials and methods in curbing these potential malignant issues. This work has also highlighted recent trends in the green synthesis of NPs, particularly metallic NPs, and their prospects for diagnosis and drug delivery. With a growing interest in green approaches in recent years, healthy growth in this expanding field is expected, and more toxicity research is warranted; the mechanisms must yet be investigated to comprehend better and long-term sustainability in biomedical applications.

Acknowledgment Acknowledgment for the Fundamental Research Grant Scheme with Project Code: FRGS/1/2020/SKK06/USM/03/2 to the "Ministry of Higher Education Malaysia."

Conflict of Interest All authors declare that the writing was conducted in the absence of financial or any commercial ties that could be viewed as a potential conflict of interest.

References

Abu-Huwaij R et al (2022) Synthesis of transdermal patches loaded with greenly synthesized zinc oxide nanoparticles and their cytotoxic activity against triple negative breast cancer. Appl Nanosci 12(1):69–78

Ahmad A et al (2003) Extracellular biosynthesis of monodisperse gold nanoparticles by a novel extremophilic actinomycete, Thermomonospora sp. Langmuir 19(8):3550–3553

Ahmad N et al (2011) Biosynthesis of silver nanoparticles from Desmodium triflorum: a novel approach towards weed utilization. Biotechnol Res Int 2011:454090

Ahmed S, Ali W (2020) Green nanomaterials: processing, properties, and applications, vol 126. Springer Nature, Singapore

Al-Khalaf AKH, Hussein FH (2018) Green and sustainable advanced nanomaterials. In: Ahmed S, Hussain CM (eds) Green and sustainable advanced materials, pp 93–106

Allison R et al (2008) Bio-nanotechnology and photodynamic therapy—state of the art review. Photodiagn Photodyn Ther 5(1):19–28

Aminian A et al (2022) Nanoparticles to overcome bacterial resistance in orthopedic and dental implants. Nanomed Res J 7(2):107–123

Ankamwar B (2010) Biosynthesis of gold nanoparticles (green-gold) using leaf extract of Terminalia catappa. E-J Chem 7(4):1334–1339

Arifin DR et al (2011) Trimodal gadolinium-gold microcapsules containing pancreatic islet cells restore normoglycemia in diabetic mice and can be tracked by using US, CT, and positive-contrast MR imaging. Radiology 260(3):790–798

Azadpour A, Hajrasouliha S, Khaleghi S (2022) Green synthesized-silver nanoparticles coated with targeted chitosan nanoparticles for smart drug delivery. J Drug Deliv Sci Technol 74:103554

Baboci L et al (2020) The dual role of the liver in nanomedicine as an actor in the elimination of nanostructures or a therapeutic target. J Oncol 2020:4638192

Balasubramanian SK et al (2010) Biodistribution of gold nanoparticles and gene expression changes in the liver and spleen after intravenous administration in rats. Biomaterials 31(8): 2034–2042

Baptista PV et al (2011) Nanoparticles in molecular diagnostics. Prog Mol Biol Transl Sci 104:427–488

Barua S et al (2013) Non-hazardous anticancerous and antibacterial colloidal 'green' silver nanoparticles. Colloids Surf B Biointerfaces 105:37–42

Baysal A, Saygin H, Ustabasi GS (2017) Interaction of PM2.5 airborne particulates with ZnO and TiO2 nanoparticles and their effect on bacteria. Environ Monit Assess 190(1):34

Berry CC et al (2003) Dextran and albumin derivatised iron oxide nanoparticles: influence on fibroblasts in vitro. Biomaterials 24(25):4551–4557

Bharathala S, Sharma P (2019) Biomedical applications of nanoparticles. In: Nanotechnology in modern animal biotechnology. Elsevier, pp 113–132

Brigger I, Dubernet C, Couvreur P (2012) Nanoparticles in cancer therapy and diagnosis. Adv Drug Deliv Rev 64:24–36

Buzea C, Pacheco II, Robbie K (2007) Nanomaterials and nanoparticles: sources and toxicity. Biointerphases 2(4):MR17–MR71

Cai Q-Y et al (2007) Colloidal gold nanoparticles as a blood-pool contrast agent for X-ray computed tomography in mice. Investig Radiol 42(12):797–806

Carty CL et al (2003) Seasonal variability of endotoxin in ambient fine particulate matter. J Environ Monit 5(6):953–958

Champion JA, Mitragotri S (2006) Role of target geometry in phagocytosis. Proc Natl Acad Sci U S A 103(13):4930–4934

Chee P et al (2022) Introduction of nanotechnology and sustainability

Cheng S et al (2021) Soluble tumor necrosis factor-alpha-induced hyperexcitability contributes to retinal ganglion cell apoptosis by enhancing Nav1. 6 in experimental glaucoma. J Neuroinflammation 18(1):1–19

Cheow WS, Hadinoto K (2012) Green preparation of antibiotic nanoparticle complex as potential anti-biofilm therapeutics via self-assembly amphiphile–polyelectrolyte complexation with dextran sulfate. Colloids Surf B: Biointerfaces 92:55–63
Chinde S et al (2017) Toxicological assessment of tungsten oxide nanoparticles in rats after acute oral exposure. Environ Sci Pollut Res Int 24(15):13576–13593
Chou K-S, Ren C-Y (2000) Synthesis of nanosized silver particles by chemical reduction method. Mater Chem Phys 64(3):241–246
Chronos N et al (1993) Profound platelet degranulation is an important side effect of some types of contrast media used in interventional cardiology. Circulation 88(5):2035–2044
Colino CI, Millán CG, Lanao JM (2018) Nanoparticles for signaling in biodiagnosis and treatment of infectious diseases. Int J Mol Sci 19(6):1627
Cuccurullo V et al (2018) Microvascular invasion in HCC: the molecular imaging perspective. Contrast Media Mol Imaging 2018:9487938
Cupi D, Hartmann NB, Baun A (2015) The influence of natural organic matter and aging on suspension stability in guideline toxicity testing of silver, zinc oxide, and titanium dioxide nanoparticles with Daphnia magna. Environ Toxicol Chem 34(3):497–506
Daly CA Jr, Hernandez R (2020) Learning from the machine: uncovering sustainable nanoparticle design rules. J Phys Chem C 124(24):13409–13420
De S et al (2014) Green synthesis of gold nanoparticles for staining human cervical cancer cells and DNA binding assay. J Photochem Photobiol B Biol 140:130–139
Dhanalakshmi P et al (2012) Synthesis of silver nanoparticles using green and brown seaweeds. Phykos 42(2):39–45
Dhas NA, Raj CP, Gedanken A (1998) Synthesis, characterization, and properties of metallic copper nanoparticles. Chem Mater 10(5):1446–1452
Elliott VL et al (2012) Evidence for metabolic cleavage of a PEGylated protein in vivo using multiple analytical methodologies. Mol Pharm 9(5):1291–1301
El-Rafie H, El-Rafie M, Zahran M (2013) Green synthesis of silver nanoparticles using polysaccharides extracted from marine macro algae. Carbohydr Polym 96(2):403–410
Emmanuel R et al (2014) Green synthesis of gold nanoparticles for trace level detection of a hazardous pollutant (nitrobenzene) causing Methemoglobinaemia. J Hazard Mater 279:117–124
Fagiolari L, Bella F (2019) Carbon-based materials for stable, cheaper and large-scale processable perovskite solar cells. Energy Environ Sci 12(12):3437–3472
Ferdous Z, Nemmar A (2020) Health impact of silver nanoparticles: a review of the biodistribution and toxicity following various routes of exposure. Int J Mol Sci 21(7):2375
Ferranti V et al (2001) Effects of encapsulation of primidone on its oxidative metabolism in rats. Drug Metabol Drug Interact 18(3–4):191–208
Feynman R (2018) There's plenty of room at the bottom. In: Feynman and computation. CRC Press, pp 63–76
Findlay MR et al (2018) Machine learning provides predictive analysis into silver nanoparticle protein corona formation from physicochemical properties. Environ Sci Nano 5(1):64–71
Flores AM et al (2020) Pro-efferocytic nanoparticles are specifically taken up by lesional macrophages and prevent atherosclerosis. Nat Nanotechnol 15(2):154–161
Francia V et al (2020) The biomolecular corona of lipid nanoparticles for gene therapy. Bioconjug Chem 31(9):2046–2059
Freixa A et al (2018) Ecotoxicological effects of carbon based nanomaterials in aquatic organisms. Sci Total Environ 619-620:328–337
Gajendran B et al (2014) Biosynthesis and characterization of silver nanoparticles from Datura inoxia and its apoptotic effect on human breast cancer cell line MCF7. Mater Lett 122:98–102
Gao S et al (2014) Near-infrared fluorescence imaging of cancer cells and tumors through specific biosynthesis of silver nanoclusters. Sci Rep 4(1):4384
Gavas S, Quazi S, Karpiński TM (2021) Nanoparticles for cancer therapy: current progress and challenges. Nanoscale Res Lett 16(1):173

Gericke M, Pinches A (2006) Biological synthesis of metal nanoparticles. Hydrometallurgy 83(1–4):132–140
Gong F et al (2020) Tumor microenvironment-responsive intelligent nanoplatforms for cancer theranostics. Nano Today 32:100851
Gottschalk F, Nowack B (2011) The release of engineered nanomaterials to the environment. J Environ Monit 13(5):1145–1155
Guilger-Casagrande M, de Lima R (2019) Synthesis of silver nanoparticles mediated by fungi: a review. Front Bioeng Biotechnol 7:287
Guo S et al (2021) Relationship and improvement strategies between drug nanocarrier characteristics and hemocompatibility: what can we learn from the literature. Asian J Pharm Sci 16(5):551–576
Gupta M, Gupta AK (2004) In vitro cytotoxicity studies of hydrogel pullulan nanoparticles prepared by AOT/N-hexane micellar system. J Pharm Pharm Sci 7(1):38–46
Gurr J-R et al (2005) Ultrafine titanium dioxide particles in the absence of photoactivation can induce oxidative damage to human bronchial epithelial cells. Toxicology 213(1–2):66–73
He Y et al (2013) Green synthesis of silver nanoparticles by Chrysanthemum morifolium Ramat. Extract and their application in clinical ultrasound gel. Int J Nanomedicine 8:1809–1815
Hirn S et al (2011) Particle size-dependent and surface charge-dependent biodistribution of gold nanoparticles after intravenous administration. Eur J Pharm Biopharm 77(3):407–416
Ho D, Sun X, Sun S (2011) Monodisperse magnetic nanoparticles for theranostic applications. Acc Chem Res 44(10):875–882
Holland NA et al (2016) Impact of pulmonary exposure to gold core silver nanoparticles of different size and capping agents on cardiovascular injury. Part Fibre Toxicol 13(1):48
Hong Y et al (2009) Transport of iron-based nanoparticles: role of magnetic properties. Environ Sci Technol 43(23):8834–8839
Hu W et al (2011) Protein corona-mediated mitigation of cytotoxicity of graphene oxide. ACS Nano 5(5):3693–3700
Huo S et al (2014) Ultrasmall gold nanoparticles as carriers for nucleus-based gene therapy due to size-dependent nuclear entry. ACS Nano 8(6):5852–5862
Jacques MT et al (2017) Safety assessment of nanopesticides using the roundworm Caenorhabditis elegans. Ecotoxicol Environ Saf 139:245–253
Johnston H et al (2015) Mechanism of neutrophil activation and toxicity elicited by engineered nanomaterials. Toxicol In Vitro 29(5):1172–1184
Kaegi R et al (2011) Behavior of metallic silver nanoparticles in a pilot wastewater treatment plant. Environ Sci Technol 45(9):3902–3908
Kang H et al (2015) Pharmacokinetics, pharmacodynamics and toxicology of theranostic nanoparticles. Nanoscale 7(45):18848–18862
Karuppaiya P et al (2013) Anti-metastatic activity of biologically synthesized gold nanoparticles on human fibrosarcoma cell line HT-1080. Colloids Surf B: Biointerfaces 110:163–170
Keller AA et al (2013) Global life cycle releases of engineered nanomaterials. J Nanopart Res 15(6): 1–17
Khan I, Saeed K, Khan I (2019) Nanoparticles: properties, applications and toxicities. Arab J Chem 12(7):908–931
Khan M et al (2022) Green synthesis of silver nanoparticles using Juniperus procera extract: their characterization, and biological activity. Crystals 12(3):420
Khanzada B et al (2021) Green synthesis of gold and iron nanoparticles for targeted delivery: an in vitro and in vivo study. J Chem 2021:1581444
Kheirollahpour M et al (2020) Nanoparticles and vaccine development. Pharm Nanotechnol 8(1): 6–21
Kolosnjaj-Tabi J et al (2014) Heat-generating iron oxide nanocubes: subtle "destructurators" of the tumoral microenvironment. ACS Nano 8(5):4268–4283
Kralj S, Makovec D (2015) Magnetic assembly of superparamagnetic iron oxide nanoparticle clusters into nanochains and nanobundles. ACS Nano 9(10):9700–9707

Krumov N et al (2007) Accumulation of CdS nanoparticles by yeasts in a fed-batch bioprocess. J Biotechnol 132(4):481–486

Kumar V, Yadav SC, Yadav SK (2010) Syzygium cumini leaf and seed extract mediated biosynthesis of silver nanoparticles and their characterization. J Chem Technol Biotechnol 85(10): 1301–1309

Kumari S et al (2020) Actinomycetes mediated synthesis, characterization, and applications of metallic nanoparticles. Inorg Nano-Met Chem 51(10):1386–1395

Labouta HI et al (2019) Meta-analysis of nanoparticle cytotoxicity via data-mining the literature. ACS Nano 13(2):1583–1594

Lipka J et al (2010) Biodistribution of PEG-modified gold nanoparticles following intratracheal instillation and intravenous injection. Biomaterials 31(25):6574–6581

Lowry GV et al (2012) Transformations of nanomaterials in the environment. Environ Sci Technol 46(13):6893–6899

Mahapatra I et al (2015) Probabilistic modelling of prospective environmental concentrations of gold nanoparticles from medical applications as a basis for risk assessment. J Nanobiotechnol 13(1):93

Măruțescu L et al (2019) Chapter 17 - Nanoparticles' toxicity for humans and environment. In: Grumezescu AM (ed) Nanomaterials for drug delivery and therapy. William Andrew Publishing, pp 515–535

Mata Y et al (2009) Gold (III) biosorption and bioreduction with the brown alga Fucus vesiculosus. J Hazard Mater 166(2–3):612–618

Mathur D, Medintz IL (2017) Analyzing DNA nanotechnology: a call to arms for the analytical chemistry community. Anal Chem 89(5):2646–2663

Meng Q-F et al (2019) Biomimetic Immunomagnetic nanoparticles with minimal nonspecific biomolecule adsorption for enhanced isolation of circulating tumor cells. ACS Appl Mater Interfaces 11(32):28732–28739

Mitchell MJ et al (2021) Engineering precision nanoparticles for drug delivery. Nat Rev Drug Discov 20(2):101–124

Monteiller C et al (2007) The pro-inflammatory effects of low-toxicity low-solubility particles, nanoparticles and fine particles, on epithelial cells in vitro: the role of surface area. Occup Environ Med 64(9):609–615

Monteiro-Riviere NA, Inman AO, Zhang LW (2009) Limitations and relative utility of screening assays to assess engineered nanoparticle toxicity in a human cell line. Toxicol Appl Pharmacol 234(2):222–235

Mu Q et al (2014) Chemical basis of interactions between engineered nanoparticles and biological systems. Chem Rev 114(15):7740–7781

Nam JM, Thaxton CS, Mirkin CA (2003) Nanoparticle-based bio-bar codes for the ultrasensitive detection of proteins. Science 301(5641):1884–1886

Nel AE et al (2015) Where are we heading in nanotechnology environmental health and safety and materials characterization? vol 9. ACS Publications, pp 5627–5630

Nordmeier A et al (2018) Microbial synthesis of metallic molybdenum nanoparticles. Chemosphere 203:521–525

Oberdörster G, Oberdörster E, Oberdörster J (2005) Nanotoxicology: an emerging discipline evolving from studies of ultrafine particles. Environ Health Perspect 113(7):823–839

Ounoughene G et al (2017) Thermal disposal of waste containing nanomaterials: first investigations on a methodology for risk management. J Phys Conf Ser 838:012024

Ozdenefe MS et al (2020) Antioxidant and antimicrobial activity of Ficus sycomorus fruit and leaf extracts. Marmara Med J 33(3):133–140

Pan Y et al (2007) Size-dependent cytotoxicity of gold nanoparticles. Small 3(11):1941–1949

Pandey A et al (2020) Reductive metabolites of curcumin and their therapeutic effects. Heliyon 6(11):e05469

Park KH et al (2003) Single-walled carbon nanotubes are a new class of ion channel blockers. J Biol Chem 278(50):50212–50216

Peters K et al (2004) Effects of nano-scaled particles on endothelial cell function in vitro: studies on viability, proliferation and inflammation. J Mater Sci Mater Med 15(4):321–325

Petrik M et al (2005) No detectable nephrotoxic side effect using a dimer, non-ionic contrast media in cerebral perfusion computed tomography in case of suspected brain ischemia. Rofo 177(9): 1242–1249

Pillai AM et al (2020) Green synthesis and characterization of zinc oxide nanoparticles with antibacterial and antifungal activity. J Mol Struct 1211:128107

Pimprikar P et al (2009) Influence of biomass and gold salt concentration on nanoparticle synthesis by the tropical marine yeast Yarrowia lipolytica NCIM 3589. Colloids Surf B Biointerfaces 74(1):309–316

Pokrajac L et al (2021) Nanotechnology for a sustainable future: addressing global challenges with the international network4sustainable nanotechnology, vol 15. ACS Publications, p 18608

Punnoose A et al (2014) Cytotoxicity of ZnO nanoparticles can be tailored by modifying their surface structure: a green chemistry approach for safer nanomaterials. ACS Sustain Chem Eng 2(7):1666–1673

Rajeshkumar S, Sivapriya D (2020) Fungus-mediated nanoparticles: characterization and biomedical advances. In: Nanoparticles in Medicine. Springer, pp 185–199

Rajeshkumar S, Kannan C, Annadurai G (2012) Green synthesis of silver nanoparticles using marine brown algae Turbinaria conoides and its antibacterial activity. Int J Pharm Bio Sci 3(4): 502–510

Recordati C et al (2015) Tissue distribution and acute toxicity of silver after single intravenous administration in mice: nano-specific and size-dependent effects. Part Fibre Toxicol 13(1):1–17

Roane TM, Pepper IL, Gentry TJ (2015) Microorganisms and metal pollutants. In: Environmental microbiology. Elsevier, pp 415–439

Sahai N, Ahmad N, Gogoi M (2018) Nanoparticles based drug delivery for tissue regeneration using biodegradable scaffolds: a review. Curr Pathobiol Rep 6(4):219–224

Salem WM et al (2014) Antibacterial activity of silver nanoparticles synthesized from latex and leaf extract of Ficus sycomorus. Ind Crop Prod 62:228–234

Salvadori MR et al (2014) Intracellular biosynthesis and removal of copper nanoparticles by dead biomass of yeast isolated from the wastewater of a mine in the Brazilian Amazonia. PLoS One 9(1):e87968

Sana SS et al (2015) Eco-friendly synthesis of silver nanoparticles using leaf extract of Grewia flaviscences and study of their antimicrobial activity. Mater Lett 145:347–350

Sathishkumar G et al (2012) Phyto-synthesis of silver nanoscale particles using Morinda citrifolia L. and its inhibitory activity against human pathogens. Colloids Surf B Biointerfaces 95:235–240

Schmid O, Stoeger T (2016) Surface area is the biologically most effective dose metric for acute nanoparticle toxicity in the lung. J Aerosol Sci 99:133–143

Schulte PA et al (2013) Occupational safety and health criteria for responsible development of nanotechnology. J Nanopart Res 16(1):2153

Sendra M et al (2017) Direct and indirect effects of silver nanoparticles on freshwater and marine microalgae (Chlamydomonas reinhardtii and Phaeodactylum tricornutum). Chemosphere 179: 279–289

Sharma VK, Yngard RA, Lin Y (2009) Silver nanoparticles: green synthesis and their antimicrobial activities. Adv Colloid Interf Sci 145(1–2):83–96

Singh A, Amiji MM (2022) Application of nanotechnology in medical diagnosis and imaging. Curr Opin Biotechnol 74:241–246

Singh A et al (2020) Green synthesis of metallic nanoparticles as effective alternatives to treat antibiotics resistant bacterial infections: a review. Biotechnol Rep 25:e00427

Składanowski M et al (2017) Silver and gold nanoparticles synthesized from Streptomyces sp. isolated from acid forest soil with special reference to its antibacterial activity against pathogens. J Clust Sci 28(1):59–79

Skuland T et al (2014) Role of size and surface area for pro-inflammatory responses to silica nanoparticles in epithelial lung cells: importance of exposure conditions. Toxicol In Vitro 28(2): 146–155
Skuland T et al (2020) Pro-inflammatory effects of crystalline-and nano-sized non-crystalline silica particles in a 3D alveolar model. Part Fibre Toxicol 17(1):1–18
Soenen SJ et al (2015) (intra) cellular stability of inorganic nanoparticles: effects on cytotoxicity, particle functionality, and biomedical applications. Chem Rev 115(5):2109–2135
Steckiewicz KP et al (2019) Impact of gold nanoparticles shape on their cytotoxicity against human osteoblast and osteosarcoma in in vitro model. Evaluation of the safety of use and anti-cancer potential. J Mater Sci Mater Med 30(2):22
Subramanian V et al (2014) Sustainable nanotechnology: defining, measuring and teaching. Nano Today 9(1):6–9
Sukhanova A et al (2018) Dependence of nanoparticle toxicity on their physical and chemical properties. Nanoscale Res Lett 13(1):44
Sukirtha R et al (2012) Cytotoxic effect of green synthesized silver nanoparticles using Melia azedarach against in vitro HeLa cell lines and lymphoma mice model. Process Biochem 47(2): 273–279
Sun T et al (2021) Engineered nanoparticles for drug delivery in cancer therapy. Nanomater Neoplasms 31–142
Suresh AK et al (2010) Silver nanocrystallites: biofabrication using Shewanella oneidensis, and an evaluation of their comparative toxicity on gram-negative and gram-positive bacteria. Environ Sci Technol 44(13):5210–5215
Talamini L et al (2017) Influence of size and shape on the anatomical distribution of endotoxin-free gold nanoparticles. ACS Nano 11(6):5519–5529
Taniguchi N (1974) On the basic concept of nanotechnology. Proceeding of the ICPE
Terreni M, Taccani M, Pregnolato M (2021) New antibiotics for multidrug-resistant bacterial strains: latest research developments and future perspectives. Molecules 26(9)
Thapa S et al (2022) State-of-the-art smart and intelligent nanobiosensors for SARS-CoV-2 diagnosis. Biosensors 12(8):637
Thirumurugan A, Blessy V, Karthikeyan M (2018) Comparative study on doxorubicin loaded metallic nanoparticles in drug delivery against MCF-7 cell line. In: Applications of nanomaterials. Elsevier, pp 303–313
Thongkam W et al (2016) Oxidant generation, DNA damage and cytotoxicity by a panel of engineered nanomaterials in three different human epithelial cell lines. Mutagenesis 32(1): 105–115
Tirumala MG et al (2021) Novel methods and approaches for safety evaluation of nanoparticle formulations: a focus towards in vitro models and adverse outcome pathways. Front Pharmacol 12:612659
Tolfree D (2006) Commercialising nanotechnology concepts–products–markets. Int J Nanomanufacturing 1(1):117–133
Usković V (2021) Nanomedicine for the poor: a lost cause or an idea whose time has yet to come? Nanomedicine 16(14):1203–1218
Valodkar M et al (2011) Biocompatible synthesis of peptide capped copper nanoparticles and their biological effect on tumor cells. Mater Chem Phys 128(1):83–89
Vanlalveni C et al (2021) Green synthesis of silver nanoparticles using plant extracts and their antimicrobial activities: a review of recent literature. RSC Adv 11(5):2804–2837
Vijayaraghavan K, Ashokkumar T (2017) Plant-mediated biosynthesis of metallic nanoparticles: a review of literature, factors affecting synthesis, characterization techniques and applications. J Environ Chem Eng 5(5):4866–4883
Wahane A et al (2020) Role of lipid-based and polymer-based non-viral vectors in nucleic acid delivery for next-generation gene therapy. Molecules 25(12):2866
Walser T et al (2012) Persistence of engineered nanoparticles in a municipal solid-waste incineration plant. Nat Nanotechnol 7(8):520–524

Wang D et al (2021) Fungus-mediated green synthesis of nano-silver using Aspergillus sydowii and its antifungal/antiproliferative activities. Sci Rep 11(1):1–9
Westerhoff PK, Kiser MA, Hristovski K (2013) Nanomaterial removal and transformation during biological wastewater treatment. Environ Eng Sci 30(3):109–117
Wilson MR et al (2002) Interactions between ultrafine particles and transition metals in vivo and in vitro. Toxicol Appl Pharmacol 184(3):172–179
Wu T, Tang M (2018) Review of the effects of manufactured nanoparticles on mammalian target organs. J Appl Toxicol 38(1):25–40
Xia T et al (2004) Quinones and aromatic chemical compounds in particulate matter induce mitochondrial dysfunction: implications for ultrafine particle toxicity. Environ Health Perspect 112(14):1347–1358
Yang H et al (2009) Comparative study of cytotoxicity, oxidative stress and genotoxicity induced by four typical nanomaterials: the role of particle size, shape and composition. J Appl Toxicol 29(1):69–78
Yang Y, Zhang C, Hu Z (2013) Impact of metallic and metal oxide nanoparticles on wastewater treatment and anaerobic digestion. Environ Sci: Processes Impacts 15(1):39–48
Yang H et al (2020) Enhancement of tumour penetration by nanomedicines through strategies based on transport processes and barriers. J Control Release 328:28–44
Yao CG, Martins PN (2020) Nanotechnology applications in transplantation medicine. Transplantation 104(4):682–693
Yetisgin AA et al (2020) Therapeutic nanoparticles and their targeted delivery applications. Molecules 25(9):2193
Zaytseva O, Neumann G (2016) Carbon nanomaterials: production, impact on plant development, agricultural and environmental applications. Chem Biol Technol Agric 3(1):1–26
Zhang H et al (2003) Water-driven structure transformation in nanoparticles at room temperature. Nature 424(6952):1025–1029
Zhang A et al (2020) Absorption, distribution, metabolism, and excretion of nanocarriers in vivo and their influences. Adv Colloid Interf Sci 284:102261
Zhao X et al (2013) Cytotoxicity of hydroxyapatite nanoparticles is shape and cell dependent. Arch Toxicol 87(6):1037–1052
Zuin S et al (2014) Leaching of nanoparticles from experimental water-borne paints under laboratory test conditions. J Nanopart Res 16(1):1–17

Development of Nanomaterials from Natural Resources for Biosensing and Biomedical Technology

12

Juan Matmin, Mohamad Hamdi Zainal-Abidin,
Nik Ahmad Nizam Nik Malek, and Roswanira Abdul Wahab

Abstract

Advancement in nanotechnological sciences has considerably revolutionized nanostructured and nanocomposite materials in biosensor devices, disease diagnostics, and biomedical applications. The trend has since leaned toward sustainable manufacturing of such materials from renewable resources (polysaccharides, plant extracts, cellulose, etc.) as readily cost-effective technology. This chapter discusses the green synthesis of nanomaterials from natural resources and their improved functionalization techniques for biomedical usages, ranging from drug delivery to molecular probing. This innovative concept will expand the functionality of sustainable nanomaterials in biomedical engineering, especially in biosensing and biomedical technology.

Keywords

Nanomaterials · Natural resources · Green synthesis · Biosensing · Biomedical technology · Healthcare

J. Matmin (✉) · M. H. Zainal-Abidin · R. A. Wahab
Department of Chemistry, Faculty of Science, Universiti Teknologi Malaysia, Johor Bahru, Malaysia
e-mail: juanmatmin@utm.my

N. A. N. N. Malek
Department of Biosciences, Faculty of Science, Universiti Teknologi Malaysia, Johor Bahru, Malaysia

W. S. Wan Kamarul Zaman, A. A. Nurul (eds.), *Sustainable Material for Biomedical Engineering Application*, https://doi.org/10.1007/978-981-99-2267-3_12

Abbreviations

°C	Degree Celsius
3D	Three-dimensional
Ag	Silver
AgNP	Silver nanoparticle
AIV H7	Avian influenza virus
CNT	Carbon nanotube
COVID-19	Coronavirus disease 2019
DNA	Deoxyribonucleic acid
ECM	Extracellular matrix
GO	Graphene oxide
LOD	Limit of detection
LOQ	Limit of quantification
MERS-CoV	Middle east respiratory syndrome coronavirus
MRSA	Methicillin-resistant *Staphylococcus aureus*
MWCNT	Multiwall carbon nanotubes
$NaBH_4$	Sodium borohydride
nm	Nanometer
NP	Nanoparticle
OPB	Oil palm biomass
OPL	Oil palm leaves
pH	Potential hydrogen
PoC	Point-of-care
POME	Palm oil effluent
RNA	Ribonucleic acid
ROS	Reactive oxygen species
SARS-CoV-2	Severe Acute Respiratory Syndrome Coronavirus
SDS	Sodium dodecyl sulfate
TiO_2	Titanium nanoparticles
XDR	Extremely drug-resistant

12.1 Introduction to Nanomaterials

Nanomaterials incorporated in electronic devices, foods, cosmetics, pharmaceuticals, and biomedical devices are currently in high demand for various applications and emerging technologies. From those nanotechnology applications, medicine and pharmaceuticals have significant roles, especially during the pandemic (Adhikari et al. 2020; Naghdi et al. 2018). This is because nanoparticles of sizes between 1 and 100 nm allow smaller samples to be used for various biochemical analyses and point-of-care diagnostics. At the nanoscale, the ability of nanomaterials to control macroscopic physicochemical qualities and the engagement of individual molecules with respect to bulk macroscopic properties are quite distinct (Damodharan et al. 2021; Huang et al. 2018).

For medical applications, the topic of nanomaterials has gained interest in recent years due to the enormous documented potential of these materials. Such a change as seen here is driven by advancements in nanotechnology and the growing need for high-quality medical treatment. Nanomaterials development may pave the way for the emergence of new systems such as prosthetic devices, implants, and artificial organs. This is exemplified in the rapid changes in nanomedicine, which see the broad-ranging applications of nanomaterials, viz., for diagnosis, prevention, and treatment of diseases. The larger surface area to mass ratio of nanoparticles, along with their reduced size and increased reactivity, is an exciting field to explore, especially in biological, optical, and electronic applications (Damodharan et al. 2021; Sebastian et al. 2014). The increasing demand and use of nanomaterials demonstrate the research progress and innovative inventions for different applications by increasing the product's potency through groundbreaking solutions (Matmin et al. 2018; Zhao et al. 2020). The scope of medical nanotechnology has substantially increased over the span of a decade due to the nanoparticles/materials' unique features that trump large-scale particles, thus enhancing the efficiency of the former in many treatment processes, which hinge on the materials' size, shape, and surface area (Bobo et al. 2016; Nardecchia et al. 2019).

12.1.1 Conventional Preparation of Nanomaterials

The synthesis of nanomaterials can be classified into two primary approaches, viz. the top-down method and the bottom-up method, followed by specific classifications of synthesis methods such as physical, chemical, and biological (Patra et al. 2018; Chen and Ostrom 2015). In particular, the top-down method applies the concept of disincorporation, where a large particle is broken down into tiny particles (Chugh et al. 2018; Chakrabartty et al. 2022). A bulk solid material can be transformed into smaller fragments with the aid of external energy from physical, chemical, or thermal sources (Nabavifard et al. 2020; Zhang et al. 2019). Contrariwise, the bottom-up method refers to the concept of combination, where tiny single atoms or molecules are brought together into a larger nanostructure (Chugh et al. 2018; Zhu et al. 2017).

Both approaches have their own advantages and limitations. For example, the top-down approach is more expensive as extra energy sources are required. Moreover, the sizes and morphologies of the synthesized nanoparticles (e.g., Ag: silver, Zn: zinc oxide, Mn: manganese, and Mg: magnesium) cannot be controlled perfectly, and the yield of NPs is relatively low (Bouafia et al. 2021; Ghiuță et al. 2017; Gregorczyk and Knez 2016). However, the top-down synthesis approach uses water as a liquid medium, thus averting the need for an organic or inorganic solvent. The resultant synthesized NPs are contamination-free and do not require further purification (Gregorczyk and Knez 2016; Iravani 2011; Hoseinpour and Ghaemi 2018; Sirelkhatim et al. 2015). On the other hand, the bottom-up approach is cheaper than the top-down approach, as the sizes and morphologies of the synthesized nanomaterials can be controlled through several factors, such as concentration, temperature, and pH, to afford a more satisfying outcome and product yield (Jeyaraj et al. 2019; Ali and Makhlouf 2021; Sanchez et al. 2005).

Popular examples of the bottom-up approach are the chemical reducing agents used to reduce Ag ions (Ag^+) into Ag^0 (AgNP) and also to reduce Mn ions (Mn^+) into M^0 (MnNP) in the chemically-assisted synthesis technique, with sodium borohydride ($NaBH_4$) being the commonplace chemical reducing agent (Hoseinpour and Ghaemi 2018; Mavani and Shah 2013; Pradhan et al. 2013). Although $NaBH_4$ is undoubtedly an excellent reducing agent for expediting the reduction process and producing AgNPs of uniform size and shape, the risks that the compound poses limit its use in large-scale synthetic reactions. Irritation, flammability, corrosiveness, and acute toxic effects are among the risks related to the handling of $NaBH_4$. Due to such circumstances, extra precautions are needed in the large-scale manufacturing of AgNPs, such as increased vigilance when handling raw materials and a wastewater treatment section to manage excess $NaBH_4$ in water. In addition, a stabilizer such as sodium dodecyl sulfate (SDS) must be used during the synthesis of AgNPs by chemical reduction to prevent their agglomeration or aggregate, resulting in larger particles. Likewise, physical methods such as laser ablation can be used to create (Aadim and Jasim 2022; Yang et al. 2013). The laser ablation synthesis method is a top-down strategy in which a bulk silver precursor, i.e., a highly pure silver sheet, is bombarded with laser energy to yield tiny Ag particles. Also, AgNPs could be synthesized using electrochemical techniques that use an electrical potential to collect at an electrode (Rodriguez-Sanchez et al. 2000; Abid et al. 2022). While the physical approach averts the use of toxic chemicals, it has several shortcomings, such as high energy consumption, an extensive experimental setup, and the need for an extra stabilizer to keep the AgNPs stable.

12.2 Green Synthesis of Nanomaterials

Recently, the era of "green synthesis" ideas has been attracting much attention in modern materials science and technology development. Recent research has focused on biological synthesis, particularly in bacteria, fungi, and plants, because of their potential environmental benefits. The advantages of green synthesis over physical and chemical processes are their non-toxicity, cost-effectiveness, quick synthesis, eco-friendliness, monodispersity, low waste output, and ability to allow large-scale production. One of the suggested solutions is to use a unique bottom-up process for making nanoparticles, known as "green synthesis," which is recognized as a key instrument and receiving much attention from the scientific community (Messaoudi and Bendahou 2020; Hussain et al. 2016; Kashani et al. 2019). Developing dependable, sustainable, and environmentally friendly synthesis processes is necessary to prevent the generation of undesirable or dangerous by-products. Natural resources, including plants, fungi, bacteria, actinomycetes, yeast, and algae, could reduce metal ions into metallic nanoparticles.

It is crucial to employ different solvents and natural resources (including organic systems) to accomplish this. At high metal concentrations, organisms have evolved to survive by changing the chemical composition of the harmful metals to reduce or eliminate their toxicity. The "consequence" of an organism's defense mechanism against a particular metal is the production of nanoparticles. Green synthesis methods based on biological precursors require the careful selection of several process conditions, such as chemical, heat, pressure, and pH settings (acidic, basic, or neutral). Similarly, an increasingly popular greener synthesis technique is the biosynthesis of AgNPs by the diverse, powerful phytochemicals found in leaves, which contain compounds such as ketones, aldehydes, flavonoids, amides, terpenes, carboxylic acids, phenolics, and ascorbic acids (Nguyen et al. 2021; Khandel et al. 2018). This bio-based synthesis approach is preferable due to the absence of toxic solvents or hazardous by-products, as the plants' extracts function as reducing agents, which reduce silver (I) ions (Ab Razak et al. 2021; Sharma et al. 2019, 2021). The plant extracts also act as capping agents for the produced AgNPs, preventing their agglomeration, reducing toxicity, and enhancing their antibacterial action. Therefore, more efforts must be focused on developing greener, more sustainable, and low-cost synthesis techniques, which do not compromise the performance of the produced AgNPs (Fig. 12.1).

As a general rule, green synthesis techniques have many advantages, for example, more energy efficiency, lower emissions of toxic and dangerous substances, simplicity of equipment, and high output. More importantly, there are wide-ranging eco-friendly products that could simultaneously function as solvents and end-capping compounds, in addition to being energy-saving and not needing any toxic chemicals. The major agents of green synthesis are microorganisms (fungi, bacteria, and algae) or leaf extracts (Pal et al. 2019; Salem et al. 2022; Salem and Fouda 2021; Abdelghany et al. 2018). Likewise, metal nanoparticles can be produced when green materials are present and under optimized parameters.

12.2.1 Phytochemical Synthesis of Silver Nanoparticles (AgNPs)

In recent years, AgNPs have been utilized in an array of applications, viz., as the active ingredient in hand sanitizers, handwash, cosmetics, home items, paint, and other consumer goods. The potential of AgNPs to kill microorganisms or prevent the growth of pathogenic bacteria is the fundamental reason for their widespread use. Notably, AgNPs reportedly exhibit good antibacterial activity against various microbes, and unlike other nanomaterials, AgNPs are simple to manufacture. As a result, the AgNPs market is enormous. This rapid increase is due to its antibacterial ability and the recent coronavirus disease 2019 (COVID-19) pandemic. The latter was the key driving factor for consumers seeking specialized, antimicrobial products to prevent the spread of pathogenic microbes. Furthermore, according to Allied Market Research, the increasing demand for AgNPs is driven mainly by the rapid consumer demand for pharmaceutical products. The increased demands include biosensors, tissue engineering, nanorobots, nanomedicine, and biomarkers. Because

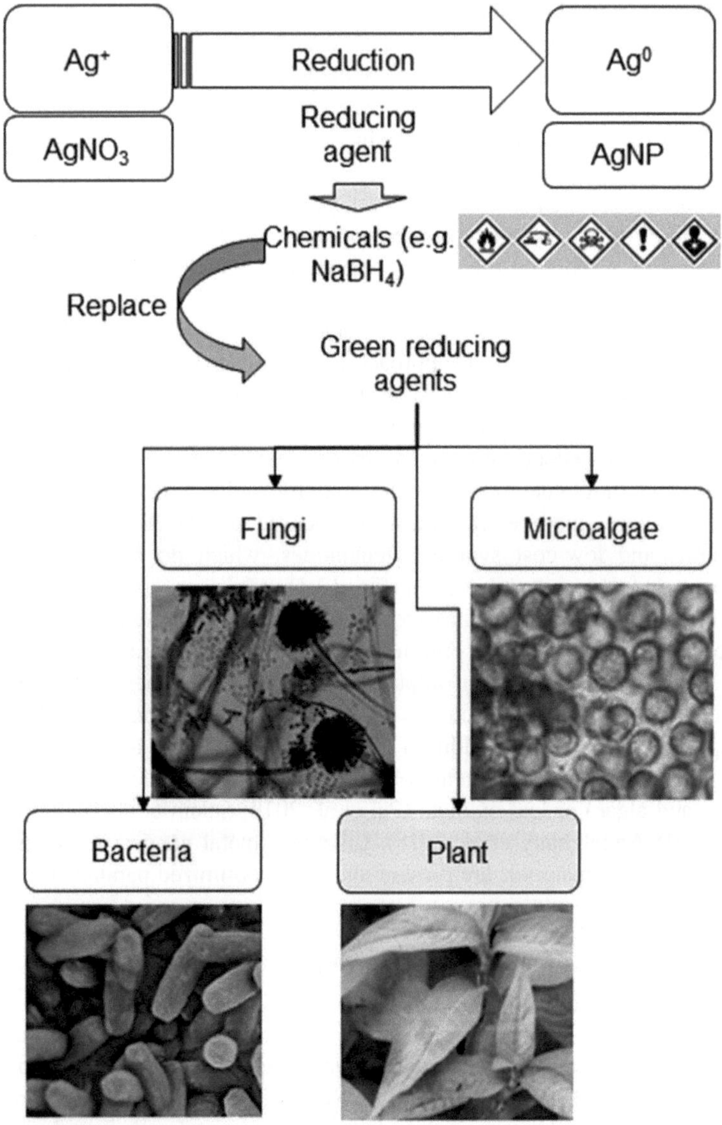

Fig. 12.1 The reduction process of Ag ions to AgNP using green reducing agents and the various green reducing agents

of the large market for AgNPs, the demand for AgNPs requires a high production volume at a lower cost. Thus, the AgNP market is expected to benefit from biological synthesis techniques.

The biological synthesis of nanomaterials, particularly AgNPs, is an alternative approach to addressing difficulties that arise from chemical and physical synthesis methods. This is because various organic chemicals and biomolecules found in

Table 12.1 Biosynthesis of AgNP using microorganisms

Microorganism	Characteristics of AgNPs	Application/activity	References
Bacteria *Pseudomonas indica*	Spherical (2.4–53.5 nm)	• Antifungal and antioxidant • No cytotoxicity on the Vero normal cell line	Salem et al. (2022)
Freshwater green alga, *Chlorella vulgaris*	Spherical (Average size of 55.06 ± 9.67 nm)	Photocatalytic decolorization activity	Rajkumar et al. (2021)
Fungus *Aspergillus brunneoviolaceus*	Spherical (0.72–15.21 nm)	• Anti-bacterial activity • Antioxidant activity	Mistry et al. (2021)
Endophytic fungus *Penicillium oxalicum*	Spherical (Average size of 52.26 nm)	• Anti-bacterial activity • Cytotoxic effect against cell lines derived from human adenocarcinoma breast cancer • Larvicidal activity	Seetharaman et al. (2021)
Fungus *Fusarium oxysporum*	Spherical (40 nm)	• Anti-bacterial activity • Antifungal activity • Photocatalytic activity • Enhancing seed germination	Gupta and Chundawat (2020)

microorganisms and natural products can act as reducing agents and potentially replace chemical-based reducing agents. Table 12.1 enlists the recent investigations on AgNP production using different microorganisms, including bacteria, fungi, and microalgae (Ghosh and Webster 2021). As can be seen, the biosynthesized AgNPs exhibit antibacterial, antioxidant, anticancer, antifungal, and antilarval properties and the ability to photo-catalytically decolorize pollutants in water. These investigations proved the feasibility of microorganisms as reducing agents to produce AgNPs as a greener strategy with many possible applications. However, the complexity of preparing and handling microbes is a rather expensive affair, limiting their large-scale utilization. Among the challenges to overcome are the susceptibility to contamination by foreign elements, the need for a sterile environment and additional precautions to be taken, specific storage requirements, and high-end facility and infrastructure costs. The strict aseptic environment is a factor that increases production costs when scaling up the manufacturing process. Furthermore, extreme caution must be taken to avoid the microorganisms contaminating the handlers, facility, infrastructure, and the environment. That said, biosynthesis of AgNPs using aqueous plant extract appears to be a better option than microorganisms-assisted ones (Ghosh and Webster 2021; Hemlata et al. 2020; Das et al. 2017; Aslam et al. 2021).

In general, the term "phyto" refers to a plant that contains various chemical substances that are beneficial to humans. Herbs and spices are examples of key plant classifications with strong antioxidant activity and are commonly used as herbal supplements for human health and wellness. These substances interact with

Table 12.2 Biosynthesis of AgNP using plant extract

Plant	Extraction	AgNPs characteristics	Application/activity	References
Clinacanthus nutans	Aqueous	Spherical (20–30 nm)	• Anti-bacterial	Waiezi et al. (2022)
Orthosiphon aristatus	Aqueous	Spherical (mean size of 20.01 nm)	• Anti-bacterial • In vitro cell viability and migration	Asraf et al. (2022)
Persicaria odorata	Aqueous	Spherical (mean diameter of 11 ± 3 nm)	• Anti-bacterial • In vitro cytocompatibility and cell migration	Lubis et al. (2022)
Diplazium esculentum	Aqueous	Polyhedra shapes (mean size of 23.385 ± 8.349 nm)	• Catalytic activity • Anti-bacterial	Hadi et al. (2022)
Plantago major L.	Ethanol	Spherical (10–20 nm)	• Anti-bacterial	Sukweenadhi et al. (2021)
Kedrostis foetidissima	Ethanol	Spherical (8.38–33.20 nm)	• Anticancer against human lung cancer cell lines	Muthusamy and Pottail (2021)
Allium sativum L.	Acetone	Spherical (22.3–34.4 nm).	• Larvicidal activity against the dengue vector *Aedes aegypti* L.	Nasir et al. (2022)

free radicals, preventing them from causing harm to our bodies. Specifically, the high antioxidant activity and high reducing ability of AgNPs allow them to reduce Ag^+ to Ag^0 and produce AgNPs. Alkaloids, saponins, flavonoids, triterpenoids, diterpenes, steroids, and other phytochemical substances are found in most plants. Table 12.2 lists several recent articles on the biosynthesis of AgNPs using plant extracts. As can be seen, the extraction methods varied depending on the study. The use of aqueous and solvent extractions both yielded AgNPs in spherical forms and in the nanometer range (1–100 nm). When utilizing a green technique, organic solvents must be avoided because they are detrimental to humans and the environment. Moreover, an additional step to evaporate or remove the organic solvent is required to obtain AgNPs. As a result, the aqueous extract is preferred and is more environmentally friendly. There is also an increasing interest in plants with medicinal characteristics and high antioxidant activity. For example, *Clinacanthus nutans*, also known as "Belalai Gajah," is a medicinal plant, rich in phytochemical substances, which reportedly exhibit an array of biological actions, viz. antiproliferative, antibacterial, anti-inflammatory, and antitumorigenic (Zulkipli et al. 2017). Likewise, *Diplazium esculentum* is another medicinal herb showing a wide range of biological activity due to a variety of phytochemical components (Semwal et al. 2021). Ipso facto, the successful synthesis of AgNPs very much hinges on choosing the right plant extract.

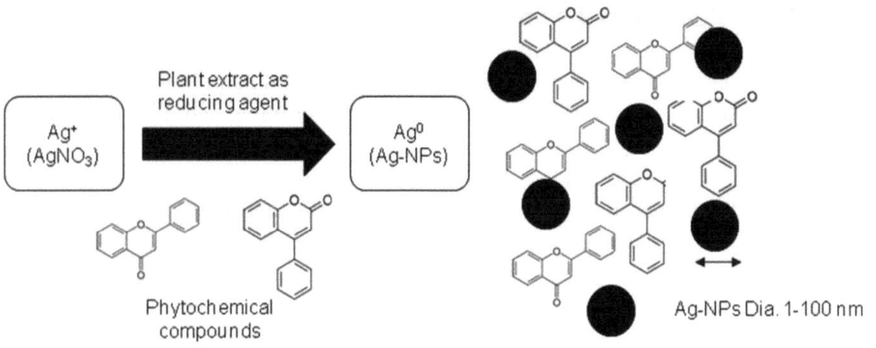

Fig. 12.2 Formation of AgNPs through reduction of Ag ions and stabilizing the nanoparticles by the phytochemical compounds

The past decade has seen numerous studies worldwide on the biosynthesis of AgNPs using a variety of phytochemicals extracted from different parts of plants. For instance, Malaysia's tropical climate supports the growth of an abundance of plants, including medicinal herbs, spices, ginseng, fruits, and many others. Furthermore, different plant organs will contain various phytochemical substances, as different beneficial phytochemical substances can be found in the leaf or root. However, not all plants can be employed as bio-reducing agents. The suitable ones are phytochemical components in plant extracts with high antioxidant activity, especially those with high total phenolic and flavonoid contents.

The synthesis of AgNPs employing plant extract as a reducing agent is depicted in Fig. 12.2. In this process, the inorganic, water-insoluble AgNPs will be distributed or suspended in another substance, namely, a solution comprising organic phytochemical components from the plant extract. However, it is worth mentioning here that phytochemical compounds are both effective reducing and stabilizing agents in the AgNPs colloid. These phytocompounds play a key role in preventing the AGNPs particles from agglomerating into big particles. Most importantly, phytochemicals derived from aqueous extracts are safe for both humans and the environment. As a result, no extra steps are required in the biosynthesis process, such as particle separation or adding polymers to the colloidal solution. The plant extract-biosynthesized AgNPs are also termed a hybrid organic-inorganic substance, with the inorganic component being AgNPs and the organic part being phytochemical compounds. Then again, the hybrid organic-inorganic material is a difficult-to-understand material, and more research is needed for each plant employed in the biosynthesis process. Compared to a single substance, the hybrid material has a synergistic effect between the two chemicals. To summarize, AgNP synthesis is possible using a variety of plant extracts. However, the chemical properties of the plant extracts must first be understood before furthering into the synthesis part.

12.2.2 Oil Palm-Based Nanocomposite

Sustainable nanomaterials are currently synthesized by oil palm biomass (OPB), the agricultural by-products left behind from replanting, pruning, and milling activities. Such biomass is often underutilized, used as animal fodder, and/or left to decompose in plantations (Onoja and Wahab 2020; Jacob et al. 2021). Notably, the oil palm tree comprising the kernel, leaves, empty fruit bunch trunk, and shell, ranks among the sources with an abundance of natural carbon- and silica-based components, whose chemical compositions can be physically and/or chemically converted into an array of advanced materials. As a matter of fact, the oil palm leaves (OPL) and other parts of the tree are prospectively greener and sustainable biopolymers sources given their all-year-round availability (Onoja and Wahab 2020; Kurnia et al. 2016). In addition, the OPB contains carbon-based components and other atoms, such as alkaline earth and heavy metals, though in minute amounts (Jacob et al. 2021). Modern technology has made it possible to create sustainable paths using different means to convert OPB into high-quality products, thus adding value to such wastes (Onoja and Wahab 2020; Abdul Khalil et al. 2010).

The expansion of nanotechnology-based industries has further intensified the demand for nanocomposites in the manufacturing industry. This changing trend puts the oil-palm-derived nanocomponents, viz. nanocellulose and cellulose nanofibers, as well as silica, into the spotlight, where these components are increasingly finding their way into products and utensils that we use daily. For instance, silica, an integral structural component in palm oil trees, imparts stiffness to the cell wall, enhances mechanical strength, and safeguards from both abiotic and biotic stresses (Jacob et al. 2021; Lim et al. 2022; Wasim et al. 2021; Faruk et al. 2012; Boussetta et al. 2022). Silica is prevalently employed as the precursor for fabricating an array of innovative materials for different applications, viz. catalyst production and thin film or coating for electronic and optic materials for chromatography and dendrimer synthesis, and as a key material to fabricate ceramics and corrosion inhibitors (Onoja and Wahab 2020; Madlangbayan et al. 2021). Pertinently, the amorphic biogenic silica extraction process from agricultural wastes is less energy-intensive compared to the high-purity inorganic silica manufacturing process, which requires heating as high as 1700°C (Sirelkhatim et al. 2015; Omar et al. 2016; Barlianti et al. 2015). Likewise, the lignocellulosic component of the oil palm tree holds a particular interest in the manufacturing of smart biopolymers and composites. This is because the abundance of hydroxyl groups could form a solid network of intermolecular hydrogen bonds between the same and different polymers and/or nanoparticles. The compact crystalline regions of the cellulose fibrils are tough, water-insoluble, and unyielding to most organic solvents.

To actualize a sustainable community, the exploration of material sciences for pioneering nanocomposite development using oil palm biomass remains at the forefront of green initiatives that capitalize on nano-based technologies. However, greener means of processing and utilization of OPB are challenging due to the natural recalcitrance of lignocellulose to microbial and enzymatic degradation by fermentation technology (Saratale and Oh 2012; Thapa et al. 2020). That said,

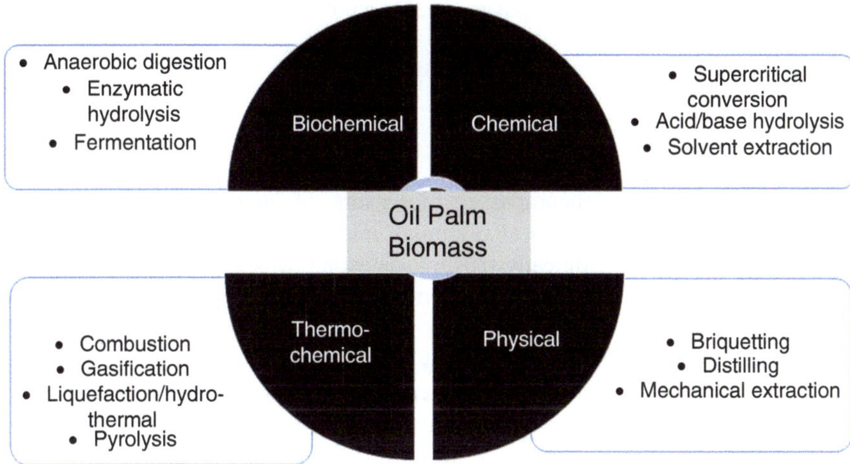

Fig. 12.3 The technologies used in the pre-treatment of OPB to obtain nanomaterials from lignocellulose and silica for nanocomposite development

physicochemical treatments are the most popular methods to process biomass (Teo and Wahab 2020; Yu et al. 2017). To enable the utilization of the oil palm trunks, leaves, empty fruit bunches, fronds, kernel shell, palm oil effluent (POME), and boiler ash, each type of biomass must undergo physical, chemical, thermochemical, or biological processes. The different technologies for treating OPB are illustrated in Fig. 12.3. The treatments are necessary for obtaining the desired level of physicochemical characteristics of the extracted nanomaterials to match their intended applications as nanocomposites (Teo and Wahab 2020). Then again, the series of treatments also represents a major hurdle to the industrial valorization of OPB into innovative nanocomposites and/or biomaterials.

Since palm oil by-products have high fixed carbon and volatile contents, they make particularly good carbon precursors. The lignocellulosic OPB comprising lignin, hemicellulose, and cellulose must first be fractionated into monomeric substrates before further chemical modifications can be done, followed by integration with other nanomaterials to form nanocomposites. Notably, the treatment protocol for each type of biomass tends to differ from one biomass to another, due to differences in the ratio of lignin:hemicellulose:cellulose. In fact, researchers have had varying success in treating oil palm lignocellulosic biomass. It has been consistently shown that these technical challenges must be overcome, including issues related to inhibitory effects from carryover components of the pretreatment and economic limitations. That being said, enzymatic hydrolysis and fermentation technology are being explored to connect the physicochemical treatment steps, hoping to boost each pretreatment step's efficiency. An alternative pretreatment tactic is to remove some of the individual physicochemical pretreatment steps via a consolidated bioprocessing approach. This would ensure that the OPB valorization process into nanocomposites leaves a smaller carbon footprint and is eco-friendly, and future OPB-based nanocomposites could be produced using clean carbon-based processes.

12.3 Biosensing

Biosensing technology encompasses biosensors of highly specific analytical devices, which convert biological responses into electrical signals. The system's components include a biologically sensitive detecting receptor, a biotransducer, and a detector element that detects an analyte. In addition, such sensors should be independent of physical parameters, viz. pH and temperature, and reusable for cost-effectiveness. However, the assembly of biosensors warrants multidisciplinary knowledge in biology, chemistry, and engineering. Biosensors must be highly selective, accurate, and show good diversity and sensitivity. Detection by biosensors is based on conductometric, potentiometric, and voltammetric methods, with suitable biomaterials chosen from DNA and RNA, antibodies, and aptamers. Biosensors can be classified as electrochemical devices, mass-sensitive devices, magnetic devices (oxygen monitors), electrical devices, optical devices, and thermometric devices (catalytic sensors) (Arunadevi 2022; Thévenot et al. 2001; Topkaya et al. 2016; Zhu et al. 2015). However, the growing popularity of metal and bio-based nanocomposites and their favorable physical and chemical properties, these composites are now prevalent in optical, electrochemical, and piezoelectric biosensors for biomolecular recognition (Arunadevi 2022; Mehrotra 2016).

12.3.1 Nanomaterials-Based Biosensors

This group of biosensors detects biological components, for instance, pathogens and antibodies, and then binds to the analyte and bioreceptor. This interaction consequently alters the physicochemical signal into an electrical signal via a transducer. Notably, nanomaterials are exceptional materials in biosensors thanks to their large surface area-to-volume ratios, excellent operational stability, and sensitivity. These features play an essential role in augmenting the immobilization of receptor molecules and biorecognizers, resulting in improved electrochemical properties and good biocompatibility (Arunadevi 2022). There have been severe impacts on the global economy and the physical and mental wellness of mankind due to the outbreak of COVID-19. In fact, the nonexistence of an effective vaccine against COVID-19 exacerbated the demand for point-of-care (PoC) diagnostic methods to rapidly screen individuals and the general public.

Immunoassay platforms using antibodies, enzymes, aptamers, proteins, cells, and nucleic acids as receptor probes are preferred over molecular diagnostics. The former are typically immobilized by physisorption or chemisorption forces on the transducer surface. The biochemical interaction events on the transducer surface are then gauged using a variety of transduction modalities, such as electrochemical, optical, piezoelectric, thermal, magnetic, and mechanical. Currently, nanoparticles are utilized in optical PoC and electrochemical biosensors for the ultrasensitive detection of biomarkers in biofluidic samples (Gowri et al. 2021). For example, pyrrolidinyl peptide nucleic acid-activated AgNPs and graphene/gold nanorode/polythionine biosensors were fabricated for detecting the Middle East respiratory syndrome coronavirus (MERS-CoV), human papillomavirus (Huang et al. 2015), and *Mycobacterium tuberculosis* (Arunadevi 2022). Biodegradable poly(D,L-glycolide), poly(lactic acid), and chitosan nanoparticles conjugated to biorecognition drugs were employed as therapeutic cancer vaccine formulations. Notably, graphene oxide (GO)-based capacitive biosensors were developed for SARS-CoV-2 detection. The highly sensitive and selective AgNPs electrode coated with AuNP-graphene nanocomposites identified and quantified the avian influenza virus (AIV H7). The AgNPs-based nano biosensor that detects the antibodies displayed an exceptional operating scope of 1.6×10^{-3} ng/mL and a limit of detection (LOD) of 1.5 pg/mL (Sharma et al. 2021; Verma et al. 2021). Notably, innovative smart textile-based wearable biosensors aid in the early diagnosis of an individual's health by providing real-time information on the current health. The nanocomposite-treated textile detects certain analytes and metabolites from human biofluids like sweat, saliva, and urine, which showed LOD of 0.4 mM and a limit of quantification (LOQ) of 1.3 mM, respectively (Khan et al. 2022).

Adding to that, carbon-based materials (carbon nanotubes (CNTs) and carbon nanoparticles, graphene-based materials), polymers, and organic and inorganic nanomaterials, viz., magnetic and metal NPs, have found applicability in biomedical engineering for therapeutic and monitoring purposes (Banigo et al. 2020). Meanwhile, CNTs are a popular choice of nanomaterials due to their extraordinary properties related to flexible physical geometric features and electronic conductivity. Plus, their dynamic physicomechanical properties are due to exceptional functionalization abilities, high aspect ratios, high folding abilities, and mechanical strength (Malik et al. 2013). For instance, the ligation of CNTs with synthetic polymers produced a biocompatible material with high mechanical properties for tissue regeneration (MacDonald et al. 2005). In short, the complexity of biochemical production strategies following enhanced nano-based synthetic biology technology warrants better streamlined and standardized engineering techniques. While biosensor development undoubtedly remains cumbersome and costly due to procedural complexities, the technology is friendly and provides result-oriented experimental support.

12.3.2 Antibody Immobilization for Immuno-Biosensor

Immuno-biosensor has been a viable device owing to its high sensitivity and selectivity performances. Immuno-biosensors are originally based on immunological principles that utilize receptors (abdominal antibodies, proteins, etc.) to identify antigens, giving rise to an immune complex (Lim and Ahmed 2016). The formation of an immunocomplex enables high-throughput quantification of the antigens of interest. The recent development of immuno-biosensors facilitates their miniaturization with better detection performances using nanotechnology. Numerous types of micro-biomarkers, such as glycated hemoglobin, carcinoembryonic antigen, prostate-specific antigen, urinary lactoferrin, and carbohydrate antigen, can be quantitatively analyzed by immuno-nanobiosensors with high accuracy, sensitivity, and specificity (Zhang and Zhao 2019). The use of proteins or antibodies in biosensor applications was reported as early as the 1950s (Morgan et al. 1996). In an immunosensor for antigen detection, the antibody is immobilized on a nanomaterial, and the type of antibody applied depends on the analyte being detected (Arris et al. 2018). Three major strategies are typically employed for antibody immobilization; namely physical adsorption-, affinity-, and covalent-based immobilization techniques (Shohibuddin et al. 2020; Abd-Wahab et al. 2019).

Physical adsorption is a conventional immobilization technique for immunosensors. The antibody adsorption typically occurs through the drop-casting technique directly onto the nanosensor platform. The interaction between antibodies and nanosensor is generally weak, i.e., via hydrophobic and electrostatic interactions (Cao 2005). One of the most significant limitations of the technique is that any changes in temperature and pH can easily lead to the leaching out of the antibody from the sensor (Shohibuddin et al. 2020). The antibody binds to the platform in a random orientation, resulting in the low selectivity, and sensitivity of the immunosensors. Therefore, the physical adsorption technique is not highly recommended for applications involving a low detection limit. Several immunosensors using physical adsorption have been previously reported, such as the immobilization of ovalbumin conjugate on a surface plasmon resonance thin-film gold chip biosensor (Gobi et al. 2007), physically adsorption of immunoglobulin G onto the mixed hydrophobic self-assembled monolayers (Wang et al. 2006), immobilization of *Shigella flexneri* on the multiwalled carbon nanotube-based biosensor (Zhao et al. 2011), and immobilization of anti-Ochratoxin A on the green synthesized graphene oxide for label-free immunosensor (Sharifuzzaman et al. 2019).

An affinity-based immobilization strategy, also known as immunoaffinity immobilization, aims to regulate the orientation of antibodies, allowing for high chances of antigen binding as well as excellent enzyme stabilization (Karav et al. 2017). In addition, the technique is typically reversible, and the sensor platform with the antibody can be reused (Saleemuddin 1999; Demir and Önal 2018; Asal et al. 2019). Affinity interaction between biotin and avidin has been widely explored for the immobilization of antibodies onto immunosensors (Chinnadayyala and Cho 2021; Wang et al. 2020a; Dong et al. 2017; Ma et al. 2019). A glycoprotein called avidin, consists of four subunits, and each unit is compatible for biotin-binding (Livnah et al. 1993; Wilchek and Bayer 1984; Izrailev et al. 1997). The sensor surface (e.g., nano-electrode) is functionalized with avidin molecules, providing a spot for avidin-biotin complex formation (Pan et al. 2018; Wang et al. 2020b; Trexler et al. 2021).

A covalent attachment-based immobilization strategy is an approach that requires a chemical reaction between nanosensor and antibodies (Hartmann and Kostrov 2013; Hosseini et al. 2018; Bilek et al. 2020). One of the most common methods for covalent attachment is carbodiimide-assisted amidation. The carbodiimide acts as a cross-linker that connects amines with carboxylated platforms (Shohibuddin et al. 2020; Jal et al. 2004; Schuster et al. 2001; Karyakin et al. 2000). The formation of an ester is the outcome of the reaction between amines and carboxylates, which may react with the free amide group of the antibody or protein of interest (Jal et al. 2004; Becke et al. 2018).

12.4 Biomedical Applications

Nanotechnology applications for improving human health are of special interest in medical biotechnology. Nanomedicine is one of the newest and most promising aspects of biomedical technology. Nanomaterials have wide applications in biomedical engineering, such as scaffold material for bone tissue engineering, wound-healing, and antibacterial agents (Bharathala and Sharma 2019; Wang et al. 2021; Zhao et al. 2016). Despite numerous advances in nanomaterial devices for biomedical applications, there are still no rules and regulations regarding pharmaceutical products and biomedicines developed from nanomaterials. With that in mind, there is no way to regulate the interest area effectively and comprehensively without proper legislation and regulations. Nevertheless, the ethical discussion regarding biomedicine has advanced in recent years, leading to an increasing number of focused ethical issues in the context of nanomedicine, including tissue engineering, wound-healing, and antibacterial agents (Santhini and Kirupha 2020; Arts et al. 2015; Zhang et al. 2018; Birhanu et al. 2018).

12.4.1 Scaffold Material for Bone Tissue Engineering

An emerging field, tissue engineering, may create organs and tissues that can restore function to injured and diseased tissues. The first step in tissue engineering is to develop a biomaterial scaffold made from biodegradable materials such as collagen, alginate, or calcium (Guo and Ma 2018; Griffith and Naughton 2002). Scaffolds are essentially created to imitate the extracellular matrix (ECM) function of bone, which can offer a three-dimensional (3D) environment for cell adhesion, proliferation, and differentiation that is physically suited for bone regeneration (Qu et al. 2019). The type of scaffold material and its surface properties are vital in bone tissue engineering.

It has been demonstrated that bone healing can be facilitated by magnetic bone scaffolds, particularly when external magnetic stimulation is used (Salahinejad et al. 2013a, b). Metallic implants have several disadvantages, including persistent inflammation induced by corrosion, stress shielding during post-healing, fatigue, and implant loosening (Salahinejad et al. 2013c; Razavi et al. 2014). In addition, following bone healing, a second surgery is essential to remove the metallic implants, which is inconvenient and costly for patients (Shahini et al. 2014; Yazdimamaghani et al. 2014a). Biodegradable materials, such as chitosan and hydroxyapatite, have gained massive attention as a potential replacement for metallic implants in bone stimulation (Zhao et al. 2002; Yazdimamaghani et al. 2014b). Such materials can either be used alone or in combination with other polymers or ceramics in bone tissue engineering applications (Chocholata et al. 2019; Saravanan et al. 2016). For instance, Heidari et al. (Heidari et al. 2015) introduced a novel method to synthesize iron oxide nanomaterials in chitosan and hydroxyapatite bone scaffolds via in situ synthesis. The study discovered that the nanoparticles were superparamagnetic, which may be advantageous in stimulating the bone healing process.

Several nanocomposites effectively stimulated cell proliferation, differentiation, and bone tissue regeneration (Bharathala and Sharma 2019; Salahinejad et al. 2013a, b; Peng et al. 2020; Bhattacharya et al. 2011). Scaffolds comprised of nanocomposite of multiwall carbon nanotubes (MWCNTs) and chitosan demonstrated excellent biocompatibility in terms of preosteoblast cell survival, adhesion, spreading, and proliferation (Abarrategi et al. 2008). In several studies, hydroxyapatite-CNT nanocomposite strengthened the scaffold's mechanical properties, promoting cell adhesion and proliferation (Xu et al. 2009; Gomes et al. 2019; Li et al. 2016). Likewise, the hydrophilicity of chitosan-MWCNT and hydroxyapatite-CNT scaffolds was improved due to hydroxyl and carboxyl functional groups in the nanocomposite (Zancanela et al. 2016; Shokri et al. 2015).

12.4.2 Wound Healing

Biosynthesized nanoparticles, bionanocellulose, and nanofiber are examples of nanomaterials derived from natural resources that can be employed as wound-healing agents (Fig. 12.4). A wound is defined as an injury to living tissue produced under various circumstances, and the healing process can be hastened by using appropriate materials to treat the wound. The major goal of designing wound healing materials is to help the skin proliferate cells and tissue at the wound site while being safe for human skin. Bacteria and other microbes can infect the wound region, causing the lesion to worsen, and become more serious. Killing the bacteria or preventing it from growing in the initial stage of the wound is one way to solve this problem. Ag-based products, such as silver sulfadiazine or AgNPs, can be used to do this because the antibacterial action of Ag is well-known for bacteria, and at low concentrations, it is safe for humans (Paladini and Pollini 2019). The phytochemical compounds are bonded together on the surface of AgNPs as a stabilizer and capping agent. The AgNPs can be manufactured using plant extracts as a bioreducing agent. Plant extracts can be used as a bioreducing agent to make AgNPs, and phytochemical compounds can be bonded to the surface of AgNPs as a stabilizer and capping agent. As a result, AgNPs and plant extract have synergistic effects in killing germs and boosting human cells, resulting in faster wound healing (Abdellatif et al. 2022; Sheng et al. 2022). A study used plant extracts known for wound-healing properties, such as curcumin, with AgNPs, and the compounds were combined in a bacterial cellulose-based hydrogel (Gupta et al. 2020; Loh et al. 2018).

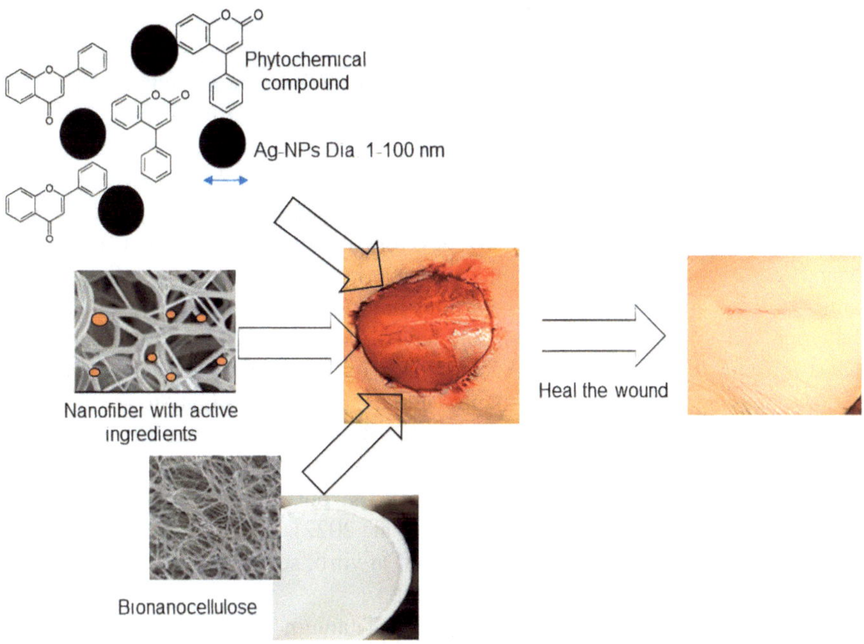

Fig. 12.4 Types of nanomaterials from natural resources suitable for wound healing

Some nanoparticles are employed to deliver other active substances that have been shown to speed wound healing. The major feature of these nanomaterials is that they are biocompatible, meaning they do not impair human cell viability. Bionanocellulose is an example of a biocompatible nanomaterial since it is made from the biomolecule cellulose, which is safe for humans and inert against certain active substances. Sultana and her colleague (2021) employed bionanocellulose to encapsulate chitosan and lawsone, resulting in a hemostatic-antibacterial substance (Sultana et al. 2021). The hemostatic feature is crucial to stop bleeding at the wound site, but the antibacterial ingredient can kill pathogenic germs and expedite healing. The nanofibers created by the electrospinning technology, on the other hand, could be employed to deliver active chemicals for wound treatment (Li et al. 2016). The origin of the nanofiber is an important aspect in evaluating its suitability for topical cutaneous use. Because of its biocompatibility, biodegradability, and high exudate-absorbing capacity, gelatin is a promising option for this function (Li et al. 2022). Another critical feature of nanofiber is its mechanical stability, which can be used to support and transfer other active substances like curcumin (Nguyen et al. 2021, 2013; Bello et al. 2022; Ranjbar-Mohammadi and Bahrami 2016) or essential oil (Zare et al. 2021; Cui et al. 2019; Ardekani et al. 2019) to the wound site. These active substances will then perform the wound-healing function.

12.4.3 Antibacterial Activity

Antibiotics such as penicillin and streptomycin are commonly employed to kill pathogenic germs. However, because these antibiotics were used for such a long time, microorganisms developed resistance to them, as exemplified by the increased cases of methicillin-resistant *Staphylococcus aureus* (MRSA). To overcome this problem, alternative and effective antibacterial agents are needed, particularly to counteract bacteria that have already developed resistance to current antibiotics. Many studies have looked into the antibacterial activity of nanoparticles as an antibacterial agent. Table 12.3 lists various natural-resource nanomaterials that can be employed as antibacterial agents and have toxicity tests on human cells in vitro. Most investigations combine antibacterial activity with cytotoxicity against normal human cells to ensure that the nanomaterials under investigation are not only antibacterial but also biocompatible, allowing their use in humans. Biosynthesized AgNPs using *Persicaria odorata* (Kesum leaf) were reported to be bactericidal against *Staphylococcus epidermidis* and MRSA, but nontoxic to normal human fibroblast cells (Lubis et al. 2022; Hadi et al. 2022). Furthermore, AgNPs have been demonstrated to increase cell migration in vitro, suggesting that they could be used as an antibacterial wound healing agent.

Apart from AgNPs, biosynthesized TiO_2 (Titanium nanoparticles) (Arya et al. 2021; Raliya et al. 2015; Rajakumar et al. 2012), magnetic Fe_3O_4 nanoparticles (Sandhya and Kalaiselvam 2020; Mahdavi et al. 2013; Jacinto et al. 2021; Sulaiman et al. 2018), and zinc oxide nanoparticles (Abdelmigid et al. 2022; Basnet et al. 2018; Bhuyan et al. 2015; Vanathi et al. 2014) from bioresources are some other

Table 12.3 Examples of nanomaterials from natural resources as an antibacterial agent

Nanomaterials	Natural resources	Antibacterial and cytotoxicity activity	References
Silver nanoparticles	Biosynthesis using *Persicaria odorata*	• Bactericidal against Gram-positive bacteria • Non-toxic to human cells • Enhancing cell migration in vitro	Lubis et al. (2022)
Zinc oxide nanoparticles	Biosynthesis using Pomegranate Fruit Peel and Solid Coffee Grounds	• Antibacterial activity • Lower cytotoxic effect on Vero cells	Abdelmigid et al. (2022)
Titanium nanoparticles	Biosynthesis using *Trichoderma citrinoviridae*	• Antibacterial activity against extremely drug-resistant (XDR) *Pseudomonas aeruginosa* clinical isolates	Arya et al. (2021)
Magnetic iron oxide nanoparticles	Biosynthesis using *borassus flabellifer* seed	• Antimicrobial activity against bacteria and fungi • Cytocompatibility in vitro	Sandhya and Kalaiselvam (2020)
Nanocellulose	Chitosan/nanocellulose electrospun fibers	• Antibacterial activity • Absence of cytotoxicity in non-tumor cells	Ribeiro et al. (2021)

nanomaterials that can be used as antibacterial agents. In contrast to antibiotics, which are mostly made up of organic molecules, they are considered inorganic antibacterial agents. As a result, these inorganic antibacterial agents differ from organic compounds in terms of stability at high pH, temperature, or other hostile environments, their inability to dissolve in any solvent, and a simple structure. It gives these nanomaterials their distinctive qualities, namely their high surface area and reactivity, when combined with their nanosized structure. Unlike antibiotics, which have limited antibacterial activity against a specific bacterium, these inorganic compounds show a broad antibacterial effect and is effective against both gram-positive and gram-negative bacteria (Gaidhani et al. 2013). In addition, these inorganic nanoparticles could only be applied topically or as a coating on surfaces, not eaten. Despite this, many clinical studies have shown that these inorganic nanoparticles are safe and biocompatible for humans (Abdelmigid et al. 2022; Arya et al. 2021; Sandhya and Kalaiselvam 2020; Ribeiro et al. 2021). Foulkes and his colleagues discussed the regulation of nanomaterials for clinical use and discovered that there is still much work to be done regarding regulatory guidance for employing nanoparticles for human purposes (Foulkes et al. 2020). The lack of widely accepted vocabulary and definitions and the inconsistency of experimental outcomes were contributing causes. As a result, improving the antibacterial effect of these nanomaterials and their biocompatibility is critical before they can be marketed.

The specific anti-bacterial mechanism of nanoparticles is yet unknown, as they can kill both gram-positive and gram-negative bacteria. Moreover, due to these nanomaterials' unique properties, the mechanism of their antimicrobial action remains an active research area. Specifically, the bacterial death caused by AgNPs, the most widely used antibacterial nanomaterial, is thought to be caused by damage to the bacteria's membrane or the generation of reactive oxygen species (ROS) (Fig. 12.5). When the cell membrane ruptures, the cell is no longer intact, causing damage to the entire system and, eventually, cell damage. Additionally, ROS can damage DNA and proteins inside bacterial cells and cause the cells to stop working, killing the bacteria. Later, the remaining AgNPs disrupt the action on sulfur and phosphorus in proteins (Rahman et al. 2019; Soshnikova et al. 2018; Nisar et al. 2019; Hamouda et al. 2019).

12.5 Conclusion and Future Perspectives

In the future, the rapid development of nanomaterials in nanotechnology will impact many areas, especially in medicine and pharmaceuticals, such as cancer therapy and medical diagnosis, optical imaging and tissue engineering, and drug delivery technology. The rapidly expanding discoveries in nanotechnology are much needed to design, characterize, synthesize, and control the shape and size of materials at the nanoscale. By achieving these, cutting-edge materials could be created for biomedical applications and even better pharmaceutical products.

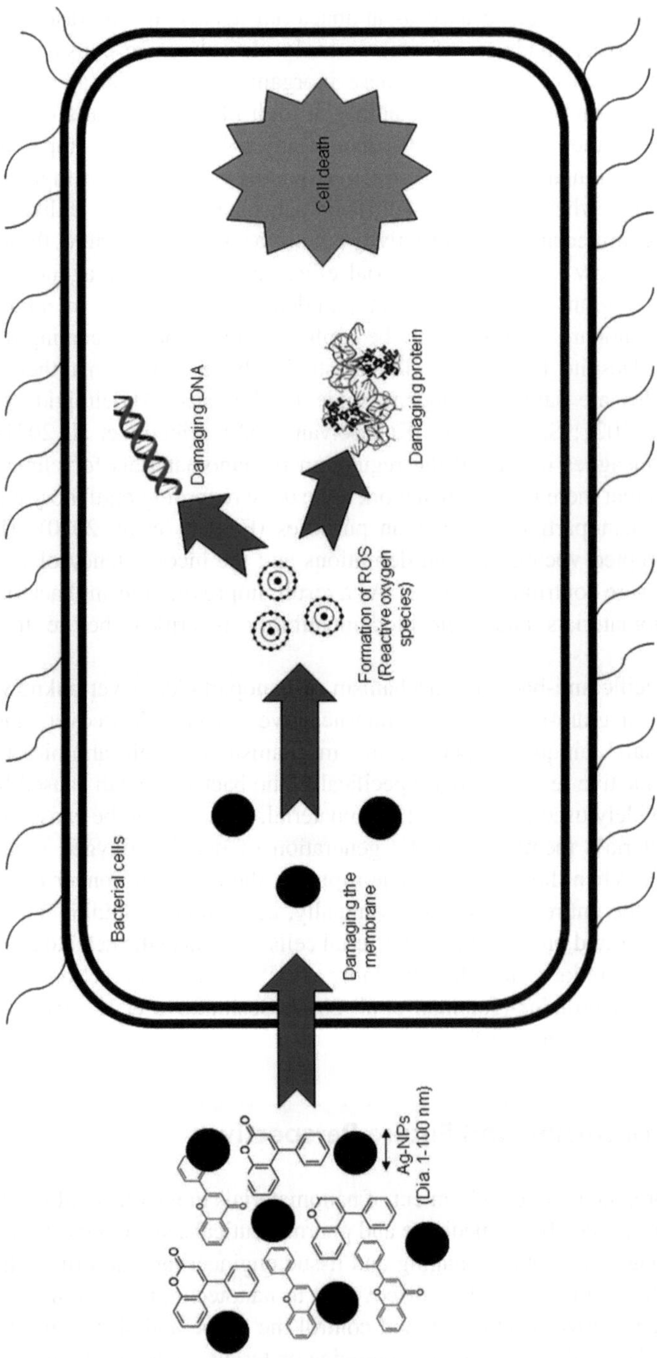

Fig. 12.5 Possible antibacterial mechanism by AgNPs

In the medical field, nanoparticles are attractive due to their surface-to-mass ratio, which is more significant than any other material, and their function as carriers of other particles. The large surface area of nanoparticles enables drugs, probes, and proteins to travel in the body, thus carrying the correct doses of drugs to specific targets. While nanomaterials have proven their effectiveness in medicine and pharmaceuticals, their toxicity in nanomedicine cannot be denied. However, nanomedicine should not pose any risks to the human body when used appropriately. Thus, concerted efforts are required to tackle this issue and improve the effectiveness of nanomedicine. Consequently, nanomaterials developed from natural resources are of great importance and can be explored in greater depth to further extend their applicability for other uses.

Acknowledgment The authors highly acknowledged the financial support from Universiti Teknologi Malaysia (UTM).

Conflict of Interest The authors declare no conflict of interest.

References

Aadim KA, Jasim AS (2022) Silver nanoparticles synthesized by Nd:YAG laser ablation technique: characterization and antibacterial activity. Karbala Int J Mod Sci 8(1):71–82

Ab Razak NH, Nizam NA, Matmin J, Dagang WRZW, Zawawi NA, Chundawat TS (2021) Brief review on bioresources green synthesis of silver nanoparticles. J Adv Res Mater Sci 79(1):1–10

Abarrategi A, Gutiérrez MC, Moreno-Vicente C, Hortigüela MJ, Ramos V, López-Lacomba JL, Ferrer ML, del Monte F (2008) Multiwall carbon nanotube scaffolds for tissue engineering purposes. Biomaterials 29(1):94–102. https://doi.org/10.1016/j.biomaterials.2007.09.021

Abdelghany TM, Al-Rajhi AMH, Al Abboud MA, Alawlaqi MM, Ganash Magdah A, Helmy EAM, Mabrouk AS (2018) Recent advances in green synthesis of silver nanoparticles and their applications: about future directions. A review. BioNanoScience 8(1):5–16. https://doi.org/10.1007/s12668-017-0413-3

Abdellatif AAH, Mahmood A, Alsharidah M, Mohammed HA, Alenize SK, Bouazzaoui A, Al Rugaie O, Alnuqaydan AM, Ahmad R, Vaali-Mohammad M-A, Alfayez M, Traiki TB, Al-Regaiey KA, Ali AT, Hassan YAH, Abdulla M-H (2022) Bioactivities of the green synthesized silver nanoparticles reduced using Allium cepa L Aqueous extracts induced apoptosis in colorectal cancer cell lines. J Nanomater 2022:1746817. https://doi.org/10.1155/2022/1746817

Abdelmigid HM, Hussien NA, Alyamani AA, Morsi MM, Al Sufyani NM, Kadi HA (2022) Green synthesis of zinc oxide nanoparticles using pomegranate fruit peel and solid coffee grounds vs. chemical method of synthesis, with their biocompatibility and antibacterial properties investigation. Molecules 27(4):1236

Abdul Khalil H, Bhat AH, Jawaid M, Amouzgar P, Ridzuan R, Said M (2010) Agro-wastes: mechanical and physical properties of resin impregnated oil palm trunk core lumber. Polym Compos 31(4):638–644

Abd-Wahab F, Abdul Guthoos HF, Wan Salim WWA (2019) Solid-state rGO-PEDOT: PSS transducing material for cost-effective enzymatic sensing. Biosensors 9(1):36

Abid N, Khan AM, Shujait S, Chaudhary K, Ikram M, Imran M, Haider J, Khan M, Khan Q, Maqbool M (2022) Synthesis of nanomaterials using various top-down and bottom-up approaches, influencing factors, advantages, and disadvantages: a review. Adv Colloid Interf Sci 300:102597. https://doi.org/10.1016/j.cis.2021.102597

Adhikari SP, Meng S, Wu YJ, Mao YP, Ye RX, Wang QZ, Sun C, Sylvia S, Rozelle S, Raat H, Zhou H (2020) Epidemiology, causes, clinical manifestation and diagnosis, prevention and control of coronavirus disease (COVID-19) during the early outbreak period: a scoping review. Infect Dis Poverty 9(1):29. https://doi.org/10.1186/s40249-020-00646-x

Ali GA, Makhlouf ASH (2021) Fundamentals of waste recycling for nanomaterial manufacturing. In: Waste recycling technologies for nanomaterials manufacturing. Springer, New York, NY, pp 3–24

Ardekani NT, Khorram M, Zomorodian K, Yazdanpanah S, Veisi H, Veisi H (2019) Evaluation of electrospun poly (vinyl alcohol)-based nanofiber mats incorporated with Zataria multiflora essential oil as potential wound dressing. Int J Biol Macromol 125:743–750. https://doi.org/10.1016/j.ijbiomac.2018.12.085

Arris FA, Benoudjit AM, Sanober F, Wan Salim WWA (2018) Characterization of electrochemical transducers for biosensor applications. In: Multifaceted protocol in biotechnology. Springer, New York, NY, pp 119–137

Arts JHE, Hadi M, Irfan M-A, Keene AM, Kreiling R, Lyon D, Maier M, Michel K, Petry T, Sauer UG, Warheit D, Wiench K, Wohlleben W, Landsiedel R (2015) A decision-making framework for the grouping and testing of nanomaterials (DF4nanoGrouping). Regul Toxicol Pharmacol 71(2, Suppl):S1–S27. https://doi.org/10.1016/j.yrtph.2015.03.007

Arunadevi N (2022) Metal nanocomposites for advanced futuristic biosensing applications. Mater Lett 309:131320

Arya S, Sonawane H, Math S, Tambade P, Chaskar M, Shinde D (2021) Biogenic titanium nanoparticles (TiO2NPs) from Tricoderma citrinoviride extract: synthesis, characterization and antibacterial activity against extremely drug-resistant Pseudomonas aeruginosa. Int Nano Lett 11(1):35–42

Asal M, Özen Ö, Şahinler M, Baysal HT, Polatoğlu İ (2019) An overview of biomolecules, immobilization methods and support materials of biosensors. Sens Rev 39(3):377–386. https://doi.org/10.1108/SR-04-2018-0084

Aslam M, Fozia F, Gul A, Ahmad I, Ullah R, Bari A, Mothana RA, Hussain H (2021) Phyto-extract-mediated synthesis of silver nanoparticles using aqueous extract of Sanvitalia procumbens, and characterization, optimization and photocatalytic degradation of azo dyes orange G and direct blue-15. Molecules 26(20):6144. https://doi.org/10.3390/molecules26206144

Asraf MH, Sani NS, Williams CD, Jemon K, Malek NANN (2022) In situ biosynthesized silver nanoparticle-incorporated synthesized zeolite A using Orthosiphon aristatus extract for in vitro antibacterial wound healing. Particuology 67:27–34

Banigo A, Azeez T, Ejeta K, Lateef A, Ajuogu E (2020) Nanobiosensors: applications in biomedical technology. In: IOP Conference Series: Materials science and engineering. IOP Publishing, Bristol, p 012028

Barlianti V, Dahnum D, Hendarsyah H, Abimanyu H (2015) Effect of alkaline pretreatment on properties of lignocellulosic oil palm waste. Proc Chem 16:195–201. https://doi.org/10.1016/j.proche.2015.12.036

Basnet P, Inakhunbi Chanu T, Samanta D, Chatterjee S (2018) A review on bio-synthesized zinc oxide nanoparticles using plant extracts as reductants and stabilizing agents. J Photochem Photobiol B Biol 183:201–221. https://doi.org/10.1016/j.jphotobiol.2018.04.036

Becke TD, Ness S, Sudhop S, Gaub HE, Hilleringmann M, Schilling AF, Clausen-Schaumann H (2018) Covalent immobilization of proteins for the single molecule force spectroscopy. J Vis Exp (138):58167. https://doi.org/10.3791/58167

Bello M, Abdullah F, Yusoff ARM, Mahmood WMAW, Chee TY, Malek NANN, Jemon KB, Sathishkumar P (2022) Curcumin-loaded electrospun poly (ε-caprolactone) nanofibrous membrane: an efficient and biocompatible wound-dressing material. Mater Lett 315:131910

Bharathala S, Sharma P (2019) Chapter 8 - Biomedical applications of nanoparticles. In: Maurya PK, Singh S (eds) Nanotechnology in modern animal biotechnology. Elsevier, Amsterdam, pp 113–132

Bhattacharya M, Wutticharoenmongkol-Thitiwongsawet P, Hamamoto DT, Lee D, Cui T, Prasad HS, Ahmad M (2011) Bone formation on carbon nanotube composite. J Biomed Mater Res A 96A(1):75–82. https://doi.org/10.1002/jbm.a.32958

Bhuyan T, Mishra K, Khanuja M, Prasad R, Varma A (2015) Biosynthesis of zinc oxide nanoparticles from Azadirachta indica for antibacterial and photocatalytic applications. Mater Sci Semicond Process 32:55–61. https://doi.org/10.1016/j.mssp.2014.12.053

Bilek MMM, Vandrovcová M, Shelemin A, Kuzminova A, Kylián O, Biederman H, Bačáková L, Weiss AS (2020) Plasma treatment in air at atmospheric pressure that enables reagent-free covalent immobilization of biomolecules on polytetrafluoroethylene (PTFE). Appl Surf Sci 518: 146128. https://doi.org/10.1016/j.apsusc.2020.146128

Birhanu G, Akbari Javar H, Seyedjafari E, Zandi-Karimi A, Dusti Telgerd M (2018) An improved surface for enhanced stem cell proliferation and osteogenic differentiation using electrospun composite PLLA/P123 scaffold. Artif Cells Nanomed Biotechnol 46(6):1274–1281. https://doi.org/10.1080/21691401.2017.1367928

Bobo D, Robinson KJ, Islam J, Thurecht KJ, Corrie SR (2016) Nanoparticle-based medicines: a review of FDA-approved materials and clinical trials to date. Pharm Res 33(10):2373–2387

Bouafia A, Laouini SE, Ahmed AS, Soldatov AV, Algarni H, Feng Chong K, Ali GA (2021) The recent progress on silver nanoparticles: synthesis and electronic applications. Nanomaterials 11(9):2318

Boussetta A, Benhamou AA, Barba FJ, Grimi N, Simirgiotis MJ, Moubarik A (2022) Effect of cellulose microfibers from sugar beet pulp by-product on the reinforcement of HDPE composites prepared by twin-screw extrusion and injection molding. J Bionic Eng 20:349. https://doi.org/10.1007/s42235-022-00260-7

Cao L (2005) Adsorption-based Immobilization. In: Carrier-bound immobilized enzymes. Wiley, New York, NY, pp 53–168

Chakrabartty I, Hakeem KR, Mohanta YK, Varma RS (2022) Greener nanomaterials and their diverse applications in the energy sector. Clean Techn Environ Policy 24:3237. https://doi.org/10.1007/s10098-022-02368-0

Chen A, Ostrom C (2015) Palladium-based nanomaterials: synthesis and electrochemical applications. Chem Rev 115(21):11999–12044. https://doi.org/10.1021/acs.chemrev.5b00324

Chinnadayyala SR, Cho S (2021) Electrochemical immunosensor for the early detection of rheumatoid arthritis biomarker: anti-cyclic citrullinated peptide antibody in human serum based on avidin-biotin system. Sensors 21(1):124

Chocholata P, Kulda V, Babuska V (2019) Fabrication of scaffolds for bone-tissue regeneration. Materials 12(4):568

Chugh H, Sood D, Chandra I, Tomar V, Dhawan G, Chandra R (2018) Role of gold and silver nanoparticles in cancer nano-medicine. Artif Cells Nanomed Biotechnol 46(Suppl 1): 1210–1220

Cui H, Zhang C, Li C, Lin L (2019) Preparation and antibacterial activity of Litsea cubeba essential oil/dandelion polysaccharide nanofiber. Ind Crop Prod 140:111739. https://doi.org/10.1016/j.indcrop.2019.111739

Damodharan D, Gopal K, Sathiyagnanam A, Rajesh Kumar B, Depoures MV, Mukilarasan N (2021) Performance and emission study of a single cylinder diesel engine fuelled with n-octanol/WPO with some modifications. Int J Amb Energy 42(7):779–788

Das M, Mondal A, Patowary K, Malipeddi H (2017) Biosynthesis of AgNPs using aqueous leaf extract of Ipomoea eriocarpa and their anti-inflammatory effect on carrageenan-induced paw edema in male Wistar rats. IET Nanobiotechnol 11(3):225–229. https://doi.org/10.1049/iet-nbt.2016.0034

Demir T, Önal S (2018) Bioaffinity immobilization and characterization of α-galactosidase on aminophenylboronic acid derivatized chitosan and Sepabeads EC-EA. LWT 90:547–555. https://doi.org/10.1016/j.lwt.2017.12.073

Dong X-X, Yuan L-P, Liu Y-X, Wu M-F, Liu B, Sun Y-M, Shen Y-D, Xu Z-L (2017) Development of a progesterone immunosensor based on thionine-graphene oxide composites platforms:

improvement by biotin-streptavidin-amplified system. Talanta 170:502–508. https://doi.org/10.1016/j.talanta.2017.04.054

Faruk O, Bledzki AK, Fink H-P, Sain M (2012) Biocomposites reinforced with natural fibers: 2000–2010. Prog Polym Sci 37(11):1552–1596. https://doi.org/10.1016/j.progpolymsci.2012.04.003

Foulkes R, Man E, Thind J, Yeung S, Joy A, Hoskins C (2020) The regulation of nanomaterials and nanomedicines for clinical application: current and future perspectives. Biomater Sci 8(17): 4653–4664

Gaidhani S, Singh R, Singh D, Patel U, Shevade K, Yeshvekar R, Ananda Chopade B (2013) Biofilm disruption activity of silver nanoparticles synthesized by Acinetobacter calcoaceticus PUCM 1005. Mater Lett 108:324–327. https://doi.org/10.1016/j.matlet.2013.07.023

Ghiuță I, Cristea D, Munteanu D (2017) Synthesis methods of metallic nanoparticles-an overview. Bull Transil Univ Brasov Eng Sci Ser I 10(2):133–140

Ghosh S, Webster TJ (2021) Nanobiotechnology: microbes and plant assisted synthesis of nanoparticles, mechanisms and applications. Elsevier, Philadelphia, PA

Gobi KV, Kim SJ, Tanaka H, Shoyama Y, Miura N (2007) Novel surface plasmon resonance (SPR) immunosensor based on monomolecular layer of physically-adsorbed ovalbumin conjugate for detection of 2,4-dichlorophenoxyacetic acid and atomic force microscopy study. Sensors Actuators B Chem 123(1):583–593. https://doi.org/10.1016/j.snb.2006.09.056

Gomes D, Santos A, Neves G, Menezes R (2019) A brief review on hydroxyapatite production and use in biomedicine. Cerâmica 65:282–302

Gowri A, Kumar NA, Anand BS (2021) Recent advances in nanomaterials based biosensors for point of care (PoC) diagnosis of COVID-19–a minireview. TrAC Trends Anal Chem 137: 116205

Gregorczyk K, Knez M (2016) Hybrid nanomaterials through molecular and atomic layer deposition: top down, bottom up, and in-between approaches to new materials. Prog Mater Sci 75:1–37. https://doi.org/10.1016/j.pmatsci.2015.06.004

Griffith LG, Naughton G (2002) Tissue engineering--current challenges and expanding opportunities. Science 295(5557):1009–1014. https://doi.org/10.1126/science.1069210

Guo B, Ma PX (2018) Conducting polymers for tissue engineering. Biomacromolecules 19(6): 1764–1782. https://doi.org/10.1021/acs.biomac.8b00276

Gupta K, Chundawat TS (2020) Time and size-dependent biogenically synthesized nanoparticles using fungus fusarium oxysporum: a review on their preparation, characterization and biological activities. Nanosci Nanotechnol Asia 10(2):95–108

Gupta S, Hemlata H, Tejavath K (2020) Synthesis, characterization and comparative anticancer potential of phytosynthesized mono and bimetallic nanoparticles using Moringa oleifera aqueous leaf extract. Beilstein Arch 1:95

Hadi AA, Ng JY, Shamsuddin M, Matmin J, Malek NANN (2022) Green synthesis of silver nanoparticles using Diplazium esculentum extract: catalytic reduction of methylene blue and antibacterial activities. Chem Pap 76(1):65–77

Hamouda RA, Hussein MH, Abo-elmagd RA, Bawazir SS (2019) Synthesis and biological characterization of silver nanoparticles derived from the cyanobacterium Oscillatoria limnetica. Sci Rep 9(1):13071. https://doi.org/10.1038/s41598-019-49444-y

Hartmann M, Kostrov X (2013) Immobilization of enzymes on porous silicas – benefits and challenges. Chem Soc Rev 42(15):6277–6289. https://doi.org/10.1039/C3CS60021A

Heidari F, Bahrololoom ME, Vashaee D, Tayebi L (2015) In situ preparation of iron oxide nanoparticles in natural hydroxyapatite/chitosan matrix for bone tissue engineering application. Ceram Int 41(2, Part B):3094–3100. https://doi.org/10.1016/j.ceramint.2014.10.153

Hemlata, Meena PR, Singh AP, Tejavath KK (2020) Biosynthesis of Silver Nanoparticles Using Cucumis prophetarum Aqueous Leaf Extract and Their Antibacterial and Antiproliferative Activity Against Cancer Cell Lines. ACS Omega 5(10):5520–5528. https://doi.org/10.1021/acsomega.0c00155

Hoseinpour V, Ghaemi N (2018) Green synthesis of manganese nanoparticles: applications and future perspective–a review. J Photochem Photobiol B Biol 189:234–243. https://doi.org/10.1016/j.jphotobiol.2018.10.022
Hosseini SH, Hosseini SA, Zohreh N, Yaghoubi M, Pourjavadi A (2018) Covalent immobilization of cellulase using magnetic poly(ionic liquid) support: improvement of the enzyme activity and stability. J Agric Food Chem 66(4):789–798. https://doi.org/10.1021/acs.jafc.7b03922
Huang H, Bai W, Dong C, Guo R, Liu Z (2015) An ultrasensitive electrochemical DNA biosensor based on graphene/Au nanorod/polythionine for human papillomavirus DNA detection. Biosens Bioelectron 68:442–446
Huang Y, Mei L, Chen X, Wang Q (2018) Recent developments in food packaging based on nanomaterials. Nanomaterials 8(10):830
Hussain I, Singh NB, Singh A, Singh H, Singh SC (2016) Green synthesis of nanoparticles and its potential application. Biotechnol Lett 38(4):545–560. https://doi.org/10.1007/s10529-015-2026-7
Iravani S (2011) Green synthesis of metal nanoparticles using plants. Green Chem 13(10): 2638–2650
Izrailev S, Stepaniants S, Balsera M, Oono Y, Schulten K (1997) Molecular dynamics study of unbinding of the avidin-biotin complex. Biophys J 72(4):1568–1581. https://doi.org/10.1016/S0006-3495(97)78804-0
Jacinto MJ, Silva VC, Valladão DMS, Souto RS (2021) Biosynthesis of magnetic iron oxide nanoparticles: a review. Biotechnol Lett 43(1):1–12. https://doi.org/10.1007/s10529-020-03047-0
Jacob AG, Wahab RA, Mahat NA (2021) Ternary biogenic silica/magnetite/graphene oxide composite for the hyperactivation of Candida rugosa lipase in the esterification production of ethyl valerate. Enzym Microb Technol 148:109807
Jal PK, Patel S, Mishra BK (2004) Chemical modification of silica surface by immobilization of functional groups for extractive concentration of metal ions. Talanta 62(5):1005–1028. https://doi.org/10.1016/j.talanta.2003.10.028
Jeyaraj M, Gurunathan S, Qasim M, Kang M-H, Kim J-H (2019) A comprehensive review on the synthesis, characterization, and biomedical application of platinum nanoparticles. Nanomaterials 9(12):1719
Karav S, Cohen JL, Barile D, de Moura Bell JMLN (2017) Recent advances in immobilization strategies for glycosidases. Biotechnol Prog 33(1):104–112. https://doi.org/10.1002/btpr.2385
Karyakin AA, Presnova GV, Rubtsova MY, Egorov AM (2000) Oriented immobilization of antibodies onto the gold surfaces via their native thiol groups. Anal Chem 72(16):3805–3811. https://doi.org/10.1021/ac9907890
Kashani HM, Madrakian T, Afkhami A, Mahjoubi F, Moosavi MA (2019) Bottom-up and green-synthesis route of amino functionalized graphene quantum dot as a novel biocompatible and label-free fluorescence probe for in vitro cellular imaging of human ACHN cell lines. Mater Sci Eng B 251:114452. https://doi.org/10.1016/j.mseb.2019.114452
Khan A, Winder M, Hossain G (2022) Modified graphene-based nanocomposite material for smart textile biosensor to detect lactate from human sweat. Biosens Bioelectron 10:100103
Khandel P, Yadaw RK, Soni DK, Kanwar L, Shahi SK (2018) Biogenesis of metal nanoparticles and their pharmacological applications: present status and application prospects. J Nanostruct Chem 8(3):217–254
Kurnia JC, Jangam SV, Akhtar S, Sasmito AP, Mujumdar AS (2016) Advances in biofuel production from oil palm and palm oil processing wastes: a review. Biofuel Res J 3(1):332
Li H, Zhao Q, Li B, Kang J, Yu Z, Li Y, Song X, Liang C, Wang H (2016) Fabrication and properties of carbon nanotube-reinforced hydroxyapatite composites by a double in situ synthesis process. Carbon 101:159–167. https://doi.org/10.1016/j.carbon.2016.01.086
Li T, Sun M, Wu S (2022) State-of-the-art review of electrospun gelatin-based nanofiber dressings for wound healing applications. Nanomaterials 12(5):784

Lim SA, Ahmed MU (2016) Electrochemical immunosensors and their recent nanomaterial-based signal amplification strategies: a review. RSC Adv 6(30):24995–25014

Lim HJ, Cheng WK, Tan KW, Yu LJ (2022) Oil palm-based nanocellulose for a sustainable future: where are we now? J Environ Chem Eng 10(2):107271. https://doi.org/10.1016/j.jece.2022.107271

Livnah O, Bayer EA, Wilchek M, Sussman JL (1993) Three-dimensional structures of avidin and the avidin-biotin complex. Proc Natl Acad Sci 90(11):5076–5080. https://doi.org/10.1073/pnas.90.11.5076

Loh EYX, Mohamad N, Fauzi MB, Ng MH, Ng SF, Mohd Amin MCI (2018) Development of a bacterial cellulose-based hydrogel cell carrier containing keratinocytes and fibroblasts for full-thickness wound healing. Sci Rep 8(1):2875. https://doi.org/10.1038/s41598-018-21174-7

Lubis FA, Malek NANN, Sani NS, Jemon K (2022) Biogenic synthesis of silver nanoparticles using Persicaria odorata leaf extract: antibacterial, cytocompatibility, and in vitro wound healing evaluation. Particuology 70:10–19

Ma X, Wang Z, He S, Chen C, Luo F, Guo L, Qiu B, Lin Z, Chen G, Hong G (2019) Development of an immunosensor based on the exothermic reaction between H2O and CaO using a common thermometer as readout. ACS Sens 4(9):2375–2380. https://doi.org/10.1021/acssensors.9b00968

MacDonald RA, Laurenzi BF, Viswanathan G, Ajayan PM, Stegemann JP (2005) Collagen–carbon nanotube composite materials as scaffolds in tissue engineering. J Biomed Mater Res A 74(3):489–496

Madlangbayan MS, Diola CNB, Tapia AKG, Peralta MM, Peralta EK, Almeda RA, Bayhon MAL, Sundo MB (2021) Corrosion inhibition of sodium silicate with nanosilica as coating in pre-corroded steel. Civ Eng J 7(11):1806–1816

Mahdavi M, Namvar F, Ahmad MB, Mohamad R (2013) Green biosynthesis and characterization of magnetic iron oxide (Fe3O4) nanoparticles using seaweed (Sargassum muticum) aqueous extract. Molecules 18(5):5954–5964

Malik P, Katyal V, Malik V, Asatkar A, Inwati G, Mukherjee TK (2013) Nanobiosensors: concepts and variations. Int Scholar Res Notic 2013:327435

Matmin J, Affendi I, Ibrahim SI, Endud S (2018) Additive-free rice starch-assisted synthesis of spherical nanostructured hematite for degradation of dye contaminant. Nanomaterials 8(9):702

Mavani K, Shah M (2013) Synthesis of silver nanoparticles by using sodium borohydride as a reducing agent. Int J Eng Res Technol 2(3):1–5

Mehrotra P (2016) Biosensors and their applications–a review. J Oral Biol Craniofac Res 6(2):153–159

Messaoudi O, Bendahou M (2020) Biological synthesis of nanoparticles using endophytic microorganisms: current development. In: Nanotechnology and the environment. IntechOpen, Rijeka, pp 37–48

Mistry H, Thakor R, Patil C, Trivedi J, Bariya H (2021) Biogenically proficient synthesis and characterization of silver nanoparticles employing marine procured fungi Aspergillus brunneoviolaceus along with their antibacterial and antioxidative potency. Biotechnol Lett 43(1):307–316

Morgan CL, Newman DJ, Price C (1996) Immunosensors: technology and opportunities in laboratory medicine. Clin Chem 42(2):193–209

Muthusamy A, Pottail L (2021) Rapid green synthesis of gold and silver nanoparticles using ethanol extract of Kedrostis foetidissima (Jacq.) Cogn. and its anticancer efficacy against A549 human lung cancer cell lines. Indian Drugs 58:30–40

Nabavifard S, Jalili S, Rahmati F, Vasseghian Y, Ali GA, Agarwal S, Gupta VK (2020) Application of dendrimer/gold nanoparticles in cancer therapy: a review. J Inorg Organomet Polym Mater 30(11):4231–4244

Naghdi S, Rhee KY, Hui D, Park SJ (2018) A review of conductive metal nanomaterials as conductive, transparent, and flexible coatings, thin films, and conductive fillers: different deposition methods and applications. Coatings 8(8):278

Nardecchia S, Sánchez-Moreno P, de Vicente J, Marchal JA, Boulaiz H (2019) Clinical trials of thermosensitive nanomaterials: an overview. Nanomaterials 9(2):191
Nasir S, Walters KF, Pereira RM, Waris M, Chatha AA, Hayat M, Batool M (2022) Larvicidal activity of acetone extract and green synthesized silver nanoparticles from Allium sativum L. (Amaryllidaceae) against the dengue vector Aedes aegypti L. (Diptera: Culicidae). J Asia Pac Entomol 25:101937
Nguyen TTT, Ghosh C, Hwang S-G, Tran LD, Park JS (2013) Characteristics of curcumin-loaded poly (lactic acid) nanofibers for wound healing. J Mater Sci 48(20):7125–7133. https://doi.org/10.1007/s10853-013-7527-y
Nguyen TM-T, Nguyen TA-T, Tuong-Van Pham N, Ly Q-V, Tran TT-Q, Thach T-D, Nguyen C-L, Banh K-S, Le V-D, Nguyen L-P (2021) Biosynthesis of metallic nanoparticles from waste Passiflora edulis peels for their antibacterial effect and catalytic activity. Arab J Chem 14(4): 103096
Nisar P, Ali N, Rahman L, Ali M, Shinwari ZK (2019) Antimicrobial activities of biologically synthesized metal nanoparticles: an insight into the mechanism of action. J Biol Inorg Chem 24(7):929–941. https://doi.org/10.1007/s00775-019-01717-7
Omar FN, Hanipah SH, Xiang LY, Mohammed MAP, Baharuddin AS, Abdullah J (2016) Micromechanical modelling of oil palm empty fruit bunch fibres containing silica bodies. J Mech Behav Biomed Mater 62:106–118
Onoja E, Wahab RA (2020) Robust magnetized oil palm leaves ash nanosilica composite as lipase support: immobilization protocol and efficacy study. Appl Biochem Biotechnol 192(2):585–599
Pal G, Rai P, Pandey A (2019) Chapter 1 - Green synthesis of nanoparticles: a greener approach for a cleaner future. In: Shukla AK, Iravani S (eds) Green synthesis, characterization and applications of nanoparticles. Elsevier, Amsterdam, pp 1–26
Paladini F, Pollini M (2019) Antimicrobial silver nanoparticles for wound healing application: progress and future trends. Materials 12(16):2540
Pan Y, Wei X, Liang T, Zhou J, Wan H, Hu N, Wang P (2018) A magnetic beads-based portable flow cytometry immunosensor for in-situ detection of marine biotoxin. Biomed Microdevices 20(3):60. https://doi.org/10.1007/s10544-018-0304-6
Patra JK, Das G, Fraceto LF, Campos EVR, Rodriguez-Torres MP, Acosta-Torres LS, Diaz-Torres LA, Grillo R, Swamy MK, Sharma S, Habtemariam S, Shin H-S (2018) Nano based drug delivery systems: recent developments and future prospects. J Nanobiotechnol 16(1):71
Peng Z, Zhao T, Zhou Y, Li S, Li J, Leblanc RM (2020) Bone tissue engineering via carbon-based nanomaterials. Adv Healthc Mater 9(5):1901495. https://doi.org/10.1002/adhm.201901495
Pradhan S, Patra P, Das S, Chandra S, Mitra S, Dey KK, Akbar S, Palit P, Goswami A (2013) Photochemical modulation of biosafe manganese nanoparticles on Vigna radiata: a detailed molecular, biochemical, and biophysical study. Environ Sci Technol 47(22):13122–13131. https://doi.org/10.1021/es402659t
Qu H, Fu H, Han Z, Sun Y (2019) Biomaterials for bone tissue engineering scaffolds: a review. RSC Adv 9(45):26252–26262. https://doi.org/10.1039/C9RA05214C
Rahman S, Rahman L, Khalil AT, Ali N, Zia D, Ali M, Shinwari ZK (2019) Endophyte-mediated synthesis of silver nanoparticles and their biological applications. Appl Microbiol Biotechnol 103(6):2551–2569
Rajakumar G, Rahuman AA, Roopan SM, Khanna VG, Elango G, Kamaraj C, Zahir AA, Velayutham K (2012) Fungus-mediated biosynthesis and characterization of TiO2 nanoparticles and their activity against pathogenic bacteria. Spectrochim Acta A Mol Biomol Spectrosc 91: 23–29. https://doi.org/10.1016/j.saa.2012.01.011
Rajkumar R, Ezhumalai G, Gnanadesigan M (2021) A green approach for the synthesis of silver nanoparticles by Chlorella vulgaris and its application in photocatalytic dye degradation activity. Environ Technol Innov 21:101282
Raliya R, Biswas P, Tarafdar JC (2015) TiO2 nanoparticle biosynthesis and its physiological effect on mung bean (Vigna radiata L.). Biotechnol Rep 5:22–26. https://doi.org/10.1016/j.btre.2014.10.009

Ranjbar-Mohammadi M, Bahrami SH (2016) Electrospun curcumin loaded poly(ε-caprolactone)/gum tragacanth nanofibers for biomedical application. Int J Biol Macromol 84:448–456. https://doi.org/10.1016/j.ijbiomac.2015.12.024

Razavi M, Fathi M, Savabi O, Beni BH, Razavi SM, Vashaee D, Tayebi L (2014) Coating of biodegradable magnesium alloy bone implants using nanostructured diopside (CaMgSi2O6). Appl Surf Sci 288:130–137. https://doi.org/10.1016/j.apsusc.2013.09.160

Ribeiro AS, Costa SM, Ferreira DP, Calhelha RC, Barros L, Stojković D, Soković M, Ferreira IC, Fangueiro R (2021) Chitosan/nanocellulose electrospun fibers with enhanced antibacterial and antifungal activity for wound dressing applications. React Funct Polym 159:104808

Rodriguez-Sanchez L, Blanco MC, López-Quintela MA (2000) Electrochemical synthesis of silver nanoparticles. J Phys Chem B 104(41):9683–9688

Salahinejad E, Hadianfard M, Macdonald D, Mozafari M, Walker K, Rad AT, Madihally S, Vashaee D, Tayebi L (2013a) Surface modification of stainless steel orthopedic implants by sol–gel ZrTiO4 and ZrTiO4–PMMA coatings. J Biomed Nanotechnol 9(8):1327–1335

Salahinejad E, Hadianfard MJ, Macdonald DD, Sharifi-Asl S, Mozafari M, Walker KJ, Rad AT, Madihally SV, Tayebi L (2013b) In vitro electrochemical corrosion and cell viability studies on nickel-free stainless steel orthopedic implants. PLoS One 8(4):e61633. https://doi.org/10.1371/journal.pone.0061633

Salahinejad E, Hadianfard MJ, Macdonald DD, Mozafari M, Vashaee D, Tayebi L (2013c) A new double-layer sol–gel coating to improve the corrosion resistance of a medical-grade stainless steel in a simulated body fluid. Mater Lett 97:162–165. https://doi.org/10.1016/j.matlet.2013.01.111

Saleemuddin M (1999) Bioaffinity based immobilization of enzymes. In: Bhatia PK, Danielsson B, Gemeiner P, Grabley S, Lammers F, Mukhopadhyay A, Ramanathan K, Saleemuddin M, Scheper T, Stefuca V, Thiericke R, Xie B (eds) Thermal biosensors, bioactivity, bioaffinitty. Springer, Berlin, pp 203–226

Salem SS, Fouda A (2021) Green synthesis of metallic nanoparticles and their prospective biotechnological applications: an overview. Biol Trace Elem Res 199(1):344–370. https://doi.org/10.1007/s12011-020-02138-3

Salem SS, Ali OM, Reyad AM, Abd-Elsalam KA, Hashem AH (2022) Pseudomonas indica-mediated silver nanoparticles: antifungal and antioxidant biogenic tool for suppressing mucormycosis fungi. J Fungi 8(2):126

Sanchez C, Julián B, Belleville P, Popall M (2005) Applications of hybrid organic–inorganic nanocomposites. J Mater Chem 15(35–36):3559–3592. https://doi.org/10.1039/B509097K

Sandhya J, Kalaiselvam S (2020) Biogenic synthesis of magnetic iron oxide nanoparticles using inedible Borassus flabellifer seed coat: characterization, antimicrobial, antioxidant activity and in vitro cytotoxicity analysis. Mater Res Express 7(1):015045

Santhini OE, Kirupha SD (2020) Nanomaterials and ethical issues. In: Khan FA (ed) Applications of nanomaterials in human health. Springer, Singapore, pp 275–286

Saratale GD, Oh SE (2012) Lignocellulosics to ethanol: the future of the chemical and energy industry. Afr J Biotechnol 11(5):1002–1013

Saravanan S, Leena RS, Selvamurugan N (2016) Chitosan based biocomposite scaffolds for bone tissue engineering. Int J Biol Macromol 93:1354–1365. https://doi.org/10.1016/j.ijbiomac.2016.01.112

Schuster M, Meyer WH, Wegner G, Herz HG, Ise M, Schuster M, Kreuer KD, Maier J (2001) Proton mobility in oligomer-bound proton solvents: imidazole immobilization via flexible spacers. Solid State Ionics 145(1):85–92. https://doi.org/10.1016/S0167-2738(01)00917-1

Sebastian V, Arruebo M, Santamaria J (2014) Reaction engineering strategies for the production of inorganic nanomaterials. Small 10(5):835–853. https://doi.org/10.1002/smll.201301641

Seetharaman PK, Chandrasekaran R, Periakaruppan R, Gnanasekar S, Sivaperumal S, Abd-Elsalam KA, Valis M, Kuca K (2021) Functional attributes of myco-synthesized silver nanoparticles from endophytic fungi: a new implication in biomedical applications. Biology 10(6):473

Semwal P, Painuli S, Painuli KM, Antika G, Tumer TB, Thapliyal A, Setzer WN, Martorell M, Alshehri MM, Taheri Y, Daştan SD, Ayatollahi SA, Petkoska AT, Sharifi-Rad J, Cho WC (2021) Diplazium esculentum (Retz.) Sw.: ethnomedicinal, phytochemical, and pharmacological overview of the Himalayan ferns. Oxidative Med Cell Longev 2021:1917890. https://doi.org/10.1155/2021/1917890
Shahini A, Yazdimamaghani M, Walker KJ, Eastman MA, Hatami-Marbini H, Smith BJ, Ricci JL, Madihally SV, Vashaee D, Tayebi L (2014) 3D conductive nanocomposite scaffold for bone tissue engineering. Int J Nanomedicine 9:167
Sharifuzzaman M, Barman SC, Zahed MA, San NJ, Park JY (2019) Green synthesis of reduced graphene oxide decorated with few-layered MoS2-nanoroses and Au/Pd/Ag trimetallic nanoparticles for ultrasensitive label-free immunosensing platforms. J Electrochem Soc 166(4):B249
Sharma D, Kanchi S, Bisetty K (2019) Biogenic synthesis of nanoparticles: a review. Arab J Chem 12(8):3576–3600. https://doi.org/10.1016/j.arabjc.2015.11.002
Sharma PK, Dorlikar S, Rawat P, Malik V, Vats N, Sharma M, Rhyee JS, Kaushik AK (2021) Nanotechnology and its application: a review. In: Nanotechnology in cancer management. Elsevier, Amsterdam, pp 1–33
Sheng Y, Narayanan M, Basha S, Elfasakhany A, Brindhadevi K, Xia C, Pugazhendhi A (2022) In vitro and in vivo efficacy of green synthesized AgNPs against Gram negative and Gram positive bacterial pathogens. Process Biochem 112:241–247. https://doi.org/10.1016/j.procbio.2021.12.012
Shohibuddin IUS, Barthasarathy PR, Wan Salim WWA (2020) Graphene synthesis and antibody immobilization techniques for immunosensors. In: Jameel AT, Yaser AZ (eds) Advances in nanotechnology and its applications. Springer, Singapore, pp 21–34
Shokri S, Movahedi B, Rafieinia M, Salehi H (2015) A new approach to fabrication of Cs/BG/CNT nanocomposite scaffold towards bone tissue engineering and evaluation of its properties. Appl Surf Sci 357:1758–1764. https://doi.org/10.1016/j.apsusc.2015.10.048
Sirelkhatim A, Mahmud S, Seeni A, Kaus NHM, Ann LC, Bakhori SKM, Hasan H, Mohamad D (2015) Review on zinc oxide nanoparticles: antibacterial activity and toxicity mechanism. Nano-Micro Lett 7(3):219–242. https://doi.org/10.1007/s40820-015-0040-x
Soshnikova V, Kim YJ, Singh P, Huo Y, Markus J, Ahn S, Castro-Aceituno V, Kang J, Chokkalingam M, Mathiyalagan R, Yang DC (2018) Cardamom fruits as a green resource for facile synthesis of gold and silver nanoparticles and their biological applications. Artif Cells Nanomed Biotechnol 46(1):108–117. https://doi.org/10.1080/21691401.2017.1296849
Sukweenadhi J, Setiawan KI, Avanti C, Kartini K, Rupa EJ, Yang D-C (2021) Scale-up of green synthesis and characterization of silver nanoparticles using ethanol extract of Plantago major L. leaf and its antibacterial potential. S Afr J Chem Eng 38(1):1–8
Sulaiman GM, Tawfeeq AT, Naji AS (2018) Biosynthesis, characterization of magnetic iron oxide nanoparticles and evaluations of the cytotoxicity and DNA damage of human breast carcinoma cell lines. Artif Cells Nanomed Biotechnol 46(6):1215–1229. https://doi.org/10.1080/21691401.2017.1366335
Sultana T, Hossain M, Rahaman S, Kim YS, Gwon J-G, Lee B-T (2021) Multi-functional nanocellulose-chitosan dressing loaded with antibacterial lawsone for rapid hemostasis and cutaneous wound healing. Carbohydr Polym 272:118482
Teo HL, Wahab RA (2020) Towards an eco-friendly deconstruction of agro-industrial biomass and preparation of renewable cellulose nanomaterials: a review. Int J Biol Macromol 161:1414–1430
Thapa S, Mishra J, Arora N, Mishra P, Li H, O'Hair J, Bhatti S, Zhou S (2020) Microbial cellulolytic enzymes: diversity and biotechnology with reference to lignocellulosic biomass degradation. Rev Environ Sci Biotechnol 19(3):621–648. https://doi.org/10.1007/s11157-020-09536-y

Thévenot DR, Toth K, Durst RA, Wilson GS (2001) Electrochemical biosensors: recommended definitions and classification. Biosens Bioelectron 16(1):121–131. https://doi.org/10.1016/S0956-5663(01)00115-4

Topkaya SN, Azimzadeh M, Ozsoz M (2016) Electrochemical biosensors for cancer biomarkers detection: recent advances and challenges. Electroanalysis 28(7):1402–1419. https://doi.org/10.1002/elan.201501174

Trexler M, Brusatori M, Auner G (2021) Avidin-biotin complex-based capture coating platform for universal Influenza virus immobilization and characterization. PLoS One 16(2):e0247429. https://doi.org/10.1371/journal.pone.0247429

Vanathi P, Rajiv P, Narendhran S, Rajeshwari S, Rahman PKSM, Venckatesh R (2014) Biosynthesis and characterization of phyto mediated zinc oxide nanoparticles: a green chemistry approach. Mater Lett 134:13–15. https://doi.org/10.1016/j.matlet.2014.07.029

Verma MK, Sharma PK, Verma HK, Singh AN, Singh DD, Verma P, Siddiqui AH (2021) Rapid diagnostic methods for SARS-CoV-2 (COVID-19) detection: an evidence-based report. J Med Life 14(4):431

Waiezi S, Malek NANN, Asraf MH, Sani NS (2022) Preparation, characterization, and antibacterial activity of green-biosynthesised silver nanoparticles using Clinacanthus nutans extract. Biointerf Res Appl Chem 13(2):171

Wang Z-H, Viana AS, Jin G, Abrantes LM (2006) Immunosensor interface based on physical and chemical immunoglobulin G adsorption onto mixed self-assembled monolayers. Bioelectrochemistry 69(2):180–186. https://doi.org/10.1016/j.bioelechem.2006.02.001

Wang Y, Zhao G, Wang H, Zhang Y, Zhang N, Wei D, Feng R, Wei Q (2020a) Label-free electrochemical immunosensor based on biocompatible nanoporous Fe3O4 and biotin–streptavidin system for sensitive detection of zearalenone. Analyst 145(4):1368–1375. https://doi.org/10.1039/C9AN02543J

Wang S, Hossain MZ, Han T, Shinozuka K, Suzuki T, Kuwana A, Kobayashi H (2020b) Avidin–Biotin technology in gold nanoparticle-decorated graphene field effect transistors for detection of biotinylated macromolecules with ultrahigh sensitivity and specificity. ACS Omega 5(46):30037–30046. https://doi.org/10.1021/acsomega.0c04429

Wang N, Thameem Dheen S, Fuh JYH, Senthil Kumar A (2021) A review of multi-functional ceramic nanoparticles in 3D printed bone tissue engineering. Bioprinting 23:e00146. https://doi.org/10.1016/j.bprint.2021.e00146

Wasim M, Shi F, Liu J, Khan MR, Farooq A, Sanbhal N, Alfred M, Xin L, Yajun C, Zhao X (2021) Extraction of cellulose to progress in cellulosic nanocomposites for their potential applications in supercapacitors and energy storage devices. J Mater Sci 56(26):14448–14486. https://doi.org/10.1007/s10853-021-06215-3

Wilchek M, Bayer EA (1984) The avidin-biotin complex in immunology. Immunol Today 5(2):39–43. https://doi.org/10.1016/0167-5699(84)90027-6

Xu JL, Khor KA, Sui JJ, Chen WN (2009) Preparation and characterization of a novel hydroxyapatite/carbon nanotubes composite and its interaction with osteoblast-like cells. Mater Sci Eng C 29(1):44–49. https://doi.org/10.1016/j.msec.2008.05.009

Yang J, Ling T, Wu W-T, Liu H, Gao M-R, Ling C, Li L, Du X-W (2013) A top–down strategy towards monodisperse colloidal lead sulphide quantum dots. Nat Commun 4(1):1695. https://doi.org/10.1038/ncomms2637

Yazdimamaghani M, Razavi M, Vashaee D, Tayebi L (2014a) Microstructural and mechanical study of PCL coated Mg scaffolds. Surf Eng 30(12):920–926. https://doi.org/10.1179/1743294414Y.0000000307

Yazdimamaghani M, Vashaee D, Assefa S, Shabrangharehdasht M, Rad AT, Eastman MA, Walker KJ, Madihally SV, Köhler GA, Tayebi L (2014b) Green synthesis of a new gelatin-based antimicrobial scaffold for tissue engineering. Mater Sci Eng C 39:235–244. https://doi.org/10.1016/j.msec.2014.03.007

Yu C, Harrold DR, Claypool JT, Simmons BA, Singer SW, Simmons CW, Van der Gheynst JS (2017) Nitrogen amendment of green waste impacts microbial community, enzyme secretion and potential for lignocellulose decomposition. Process Biochem 52:214–222. https://doi.org/10.1016/j.procbio.2016.11.002

Zancanela DC, de Faria AN, Simão AMS, Gonçalves RR, Ramos AP, Ciancaglini P (2016) Multi and single walled carbon nanotubes: effects on cell responses and biomineralization of osteoblasts cultures. J Mater Sci Mater Med 27(3):62. https://doi.org/10.1007/s10856-016-5673-x

Zare MR, Khorram M, Barzegar S, Asadian F, Zareshahrabadi Z, Saharkhiz MJ, Ahadian S, Zomorodian K (2021) Antimicrobial core–shell electrospun nanofibers containing ajwain essential oil for accelerating infected wound healing. Int J Pharm 603:120698

Zhang J, Zhao J (2019) Chapter 5 - Immuno-biosensor. In: Li G (ed) Nano-inspired biosensors for protein assay with clinical applications. Elsevier, Amsterdam, pp 115–137

Zhang B, Yan W, Zhu Y, Yang W, Le W, Chen B, Zhu R, Cheng L (2018) Nanomaterials in neural-stem-cell-mediated regenerative medicine: imaging and treatment of neurological diseases. Adv Mater 30(17):1705694. https://doi.org/10.1002/adma.201705694

Zhang L, Doyle-Davis K, Sun X (2019) Pt-based electrocatalysts with high atom utilization efficiency: from nanostructures to single atoms. Energy Environ Sci 12(2):492–517. https://doi.org/10.1039/C8EE02939C

Zhao F, Yin Y, Lu WW, Leong JC, Zhang W, Zhang J, Zhang M, Yao K (2002) Preparation and histological evaluation of biomimetic three-dimensional hydroxyapatite/chitosan-gelatin network composite scaffolds. Biomaterials 23(15):3227–3234. https://doi.org/10.1016/S0142-9612(02)00077-7

Zhao G, Zhan X, Dou W (2011) A disposable immunosensor for Shigella flexneri based on multiwalled carbon nanotube/sodium alginate composite electrode. Anal Biochem 408(1):53–58. https://doi.org/10.1016/j.ab.2010.08.039

Zhao W, Li J, Jin K, Liu W, Qiu X, Li C (2016) Fabrication of functional PLGA-based electrospun scaffolds and their applications in biomedical engineering. Mater Sci Eng C 59:1181–1194. https://doi.org/10.1016/j.msec.2015.11.026

Zhao L, Lu L, Wang A, Zhang H, Huang M, Wu H, Xing B, Wang Z, Ji R (2020) Nano-biotechnology in agriculture: use of nanomaterials to promote plant growth and stress tolerance. J Agric Food Chem 68(7):1935–1947. https://doi.org/10.1021/acs.jafc.9b06615

Zhu C, Yang G, Li H, Du D, Lin Y (2015) Electrochemical sensors and biosensors based on nanomaterials and nanostructures. Anal Chem 87(1):230–249. https://doi.org/10.1021/ac5039863

Zhu C, Fu S, Song J, Shi Q, Su D, Engelhard MH, Li X, Xiao D, Li D, Estevez L, Du D, Lin Y (2017) Self-assembled Fe–N-doped carbon nanotube aerogels with single-atom catalyst feature as high-efficiency oxygen reduction electrocatalysts. Small 13(15):1603407. https://doi.org/10.1002/smll.201603407

Zulkipli IN, Rajabalaya R, Idris A, Sulaiman NA, David SR (2017) Clinacanthus nutans: a review on ethnomedicinal uses, chemical constituents and pharmacological properties. Pharm Biol 55(1):1093–1113

3D Bioprinted Scaffolds from Sustainable Materials for Tissue Engineering: Evolution and Current Challenges

13

Anusha Wei Asohan, Nurshafiqah Jasme, Nik Nur Farisha Nik Md Noordin Kahar, Yazmin Bustami, and Zuratul Ain Abdul Hamid

Abstract

Three-dimensional (3D) bioprinting techniques have become an important research area in the past two decades. These techniques are used to fabricate different tissue and organ constructs of varying functional and structural complexity. Bioinks are one of the main elements in bioprinting, which include multiple biomaterials, either natural or synthetic, as well as different types of cells and soluble factors. To date, there is a considerable demand for using sustainable materials in 3D-printed bioinks because they offer immense advantages such as reducing processing waste, encouraging customisation and thus lowering production costs. Recent bioprinting techniques are reviewed in this chapter, discussing the evolution of sustainable materials for bioinks, including their properties, structure requirements and preparation of bioinks. In the end, we highlighted the current challenges in terms of materials evolution, the bioprinting process and their biocompatibility.

Keywords

Bioinks · Sustainable materials · Tissue engineering · Hydrogel · 3D bioprinting · Polymer scaffolds

A. W. Asohan · N. Jasme · Y. Bustami
School of Biological Science, Universiti Sains Malaysia, Gelugor, Penang, Malaysia
e-mail: nurshafiqahjasme@student.usm.my; ybustami@usm.my

N. N. F. N. M. N. Kahar · Z. A. A. Hamid (✉)
School of Materials & Mineral Resources Engineering, Universiti Sains Malaysia, Nibong Tebal, Penang, Malaysia
e-mail: srzuratulain@usm.my

W. S. Wan Kamarul Zaman, A. A. Nurul (eds.), *Sustainable Material for Biomedical Engineering Application*, https://doi.org/10.1007/978-981-99-2267-3_13

Abbreviations

AM	Additive manufacturing
DFSAM	Design for sustainable additive manufacturing
ECM	Extracellular matrix
GelMA	Gelatin-methacryloyl
LbL	Layer-by-layer
PBAT	Polybutylene adipate terephthalate
PBS	Poly(butylene succinate)
PBSA	Poly(butylene succinate-co-butylene adipate)
PCL	Polycaprolactone
PEGDA	Poly (ethylene glycol) diacrylate
PHA	Poly(hydroxyalkanoate)
PHB	Poly(hydroxybutyrate)
PHBV	Polyhydroxybutyrate-valerate
PLA	Polylactic acid
RGD	Arginylglycylaspartic acid
SEC	Specific energy consumption
SHS	Selective heat sintering
SIS	Selective inhibition sintering
SMS	Selective mask sintering

13.1 Introduction

Tissue engineering is a rapidly evolving field with the purpose of restoring and regenerating damaged or injured tissues and organs. The most important part that contributes to the feasibility of tissue engineering is the fabrication of comparable tissues and organs that are mainly used for replacement and transplantation (Agarwal et al. 2020). Due to this reason, several requirements need to be carefully considered, such as scaffold biocompatibility, biodegradability, formation of a variety of shapes and sizes, as well as mechanical properties that are consistent with the target tissues (Gu et al. 2018). At present, conventional approaches such as gas foaming, phase separation, electrospinning, melt moulding and freeze-drying techniques are lacking in certain elements to satisfy the aforementioned requirements (Gu et al. 2018; Tan et al. 2021).

These methods are inadequate in achieving an accurate representation of the native tissue's architecture due to the lack of control over the scaffold design (Gu et al. 2018). In general, the microstructures and interconnectivity of the pores are the cornerstones in encouraging cell proliferation and interaction, which determine the regeneration possibility of the fabricated tissue (Rider et al. 2018). While these conventional methods allow the formation of pores, their lack of control over pore size formation and distribution could hinder cell migration and nutrient delivery

(Thavornyutikarn et al. 2014). Additionally, the release of toxic by-products and the use of toxic solvents during the production process are harmful to the cells and can elicit an inflammatory response (Tan et al. 2021; Thavornyutikarn et al. 2014).

The emergence of 3D bioprinting paved the way for the fabrication of artificial tissue or organs by allowing rapid, automated and precise manufacturing, which is considerably fascinating. 3D bioprinting is a layer-by-layer (LbL) additive manufacturing (AM) technique that allows the recreation of native tissues with anatomically accurate structures since it has full control over the scaffold design (Agarwal et al. 2020). The use of bioinks in 3D bioprinting efficiently delivers cells to specific locations of the constructed tissue and simultaneously promotes homogenous cell distribution within the scaffold (Gu et al. 2018). Interestingly, the bioprinted scaffolds are evidently superior in sustaining their shape and dimensions and display much higher viability than the cast scaffold (Rider et al. 2018). Since tissue engineering is a long-term pursuit for improving patient care and quality of life, the sustainability aspect should be non-negotiable. Thus, the methods and processes should be reflected in its development (Williams 2019).

13.2 Principle of 3D Bioprinting

Over the last few decades, advances in 3D printing and the growing demand for organ and tissue replacement surgeries have allowed for the development of the bioprinting industry, which involves the production of intricate living cell structures, such as functional tissues using bioink (Ozbolat and Yu 2013). The development of 3D bioprinting is associated with AM techniques, which were used earlier to manufacture nonbiological products and have been applied in a variety of industries (Wohlers and Gornet 2014).

The use of three-dimensional (3D) digital models, which can be converted into physical models with defined materials and controlled process parameters, is an example of AM manufacturing. For that reason, 3D bioprinting is commonly known as a subset of the AM technique, which is used to create artificial 3D tissue constructs from biomaterials, either incorporated with living cells (cell-laden constructs) or without the incorporation of living cells (functional constructs) (Whenish et al. 2022) Therefore, it is advantageous and can cater to urgent needs in tissue replacement and organ transplantation. Interestingly, it can be further used to imitate the tissue matrix or organ using patient-derived cells that could minimise the risk of rejection after transplantation.

Several considerations, such as material selection, formulation of the printable ink and production of sufficient quantities of bioink for printing, need to be identified before conducting 3D bioprinting (Maan et al. 2022). On top of that, it requires additional considerations such as the selection of the type of cells, adequate nutrient medium for cells and differentiation factors to provide an optimum cell growth condition (Murphy and Atala 2014). These considerations are mostly related to bioink, which is essential in bioprinting. Bioink can be referred to as biomaterials consisting of biological entities or cells that can be utilised in bioprinting (Gillispie et al. 2020). It is important for the bioink to exhibit distinctive characteristics relating to rheological, mechanical and biocompatibility to ensure it can support cell survival and functional 3D living structure (Kačarević et al. 2018).

Rheological assessments such as shear rate, shear stress, viscosity, and critical shear point of the bioink are the most important parameters, with the speculation that bioink exhibits shear thinning behaviour and could improve print quality without impacting the bioink's internal bonding structure (Pössl et al. 2021). Moreover, the information on its mechanical characteristics, such as elastic modulus, shear elastic modulus, Young's modulus or degradation rates, is beneficial for controlling the stiffness during extrusion-based biofabrication using soft gel.

Most bioinks are hydrogels, a highly hydrated polymeric network used to homogenously encapsulate cells by imitating the natural extracellular matrix found in vivo (Maan et al. 2022). A hydrogel that shows biocompatibility is described as a substance's ability to incorporate the host biological environment without triggering any adverse reactions, retain high cell viability, promote cell growth and proliferation of osteocytes and preserve healthy cell populations. Albeit their appealing properties, hydrogels commonly need integration with other materials to increase their efficiency and capability to be crosslinked (Wan Jusoh et al. 2022).

13.3 Types of 3D Bioprinting

The most widely used methods for 3D bioprinting are extrusion-based bioprinting, inkjet-based bioprinting, and laser-assisted bioprinting, as shown in Fig. 13.1. This technology has been used to create artificial blood vessels (Papaioannou et al. 2019), cartilage structures (Barcelo et al. 2022), bones (Tan et al. 2022), wound healing (Antezana et al. 2022), and artificial skin tissue (Joseph et al. 2019).

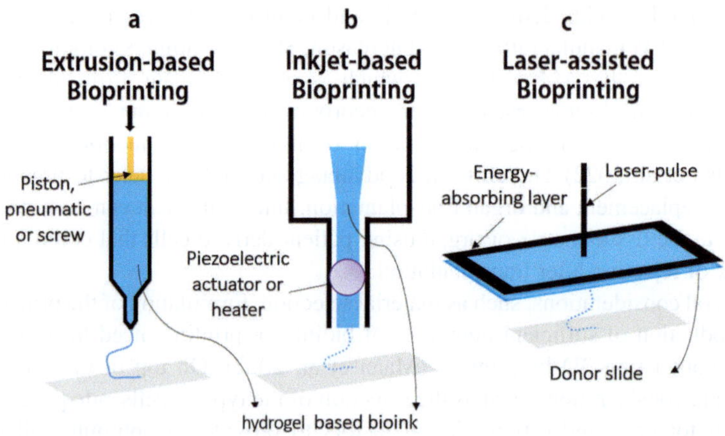

Fig. 13.1 3D bioprinting techniques for tissue and organ, (**a**) Extrusion-based bioprinting, (**b**) Inkjet-based bioprinting, (**c**) Laser-assisted bioprinting. (Reproduced from Lee et al. 2021 under the Creative Commons CC-BY Licence from Frontiers)

13.3.1 Extrusion-Based Bioprinting

Among all 3D bioprinting methods, extrusion-based bioprinting is the most widely used due to its capability of accommodating a broad spectrum of fluid viscosities ranging from 30 mPa/s to just under 6×10^7 mPa/s (Li et al. 2021; Khoeini et al. 2021). The viscoelasticity will ensure sufficient printability and shape to support the long-term mechanical demands of 3D structures. Extrusion-based bioprinting is based on the extrusion principle, which uses the continuous pressure-driven system, enabling the dispersion of bioink filaments through a small or even a micro-sized nozzle. The bioink is extruded using pneumatic, piston or screw syringes and deposited layer by layer into a predesigned structure (Fatimi et al. 2022). Importantly, it shows versatility, practicality, utilises different concentrations of cells, allows high cell density and has the ability to generate a large-scale 3D structure (Ozbolat and Hospodiuk 2016). However, it can also cause low resolution, nozzle clogging, and an insufficient survival rate of cells since cells may be subjected to increased shear stress and mechanical pressure during the dispensing process.

13.3.2 Inkjet-Based Bioprinting

The principle of inkjet bioprinting is simple and similar to home printing techniques. It is the first technology used for organ printing, which is noncontact printing. The biomaterials are printed separately, layer-by-layer, to form 3D object using thermal or acoustic methods (Lei and Wang 2016). The heat generated at the printer head forced the cells and biomaterial out of the nozzle via the pressure pulses (Kačarević et al. 2018). However, this approach is sometimes disabled by its bioprinting head, preventing continuous flow during the bioprinting process (Khoeini et al. 2021).

13.3.3 Laser-Assisted Bioprinting

Laser-assisted bioprinting uses a laser to deposit the bioink onto the substrate surfaces (Ji and Guvendiren 2017). It is composed of three primary components: a pulsed laser source, a target and a receiver plate. The target layer consists of a ribbon coated with a metallic absorbing layer (gold or titanium), a layer of bioink, and a transparent substrate (Li et al. 2021), while the receiving layer refers to an ordinary glass slide serving as a buffer layer. During the bioprinting process, a pulsed laser beam scans over the ribbon, which triggers the vaporisation of the sacrificial absorbing layer. Bubbles formed through vapour pressure initially collapse on the absorbent layer surface (Lei and Wang 2016). This pressure subsequently shoots a small drop of bioink, including the cells, onto the receiving substrate until a tissue-like structure is produced (Kačarević et al. 2018). The technology could accommodate various viscosities, achieving high resolution, high density of encapsulated cells and a high cell survival rate (Fatimi et al. 2022). However, the complex preparation of the ribbon, the trace amount of metallic residue in the scaffolds, and the high production cost limit its application.

13.4 Bioprinting: A Path to a More Sustainable Economy

The integration of sustainability and 3D bioprinting facilitates the realisation of the concept of productive coexistence between humans and nature. Bioprinting meets the needs of current and future generations by taking environmental and socioeconomic factors into account in product design. According to a survey of the literature, sustainable manufacturing is an important tool for environmental protection; 3D bioprinting, in particular, increases the feasibility of sustainable manufacturing (Whenish et al. 2022). This viewpoint seeks to assess the long-term solutions created by 3D bioprinting technology by incorporating nature-derived materials and structures into industrial design. It discusses how factors such as carbon management strategies, optimal recycling rates of surplus raw materials, energy conservation rates, production times, and other factors can affect human society and the global environment.

13.4.1 3D Bioprinting: Sustainable Materials

The biomaterials chosen for bioink development are essential to the research. As a result, researchers prefer to use biomaterials that have previously been reported to be compatible with cells in commercial or noncommercial products or devices, which are mostly comprised of sustainable biomaterials. The biomaterials chosen have also been approved by the Food and Drug Administration (FDA) for use in the medical field. Water-absorbing polymers, commonly known as hydrogels, are the gold standard used in bioprinting applications. Hydrogel-based bioinks are arguably regarded as the most suitable for bioprinting applications due to their cross-linking properties and ability to absorb and retain large quantities of water (Hospodiuk et al. 2017). The high-water content of hydrogels makes them ideal for mimicking the extracellular environment of the native tissues of the human body. The composition of a hydrogel network also allows for the permeability of oxygen, nutrients and other water-soluble compounds, which is important to encourage uniform cell growth (Teixeira et al. 2022). Hydrogel-based bioinks can be derived from natural or synthetic polymers. Naturally derived bioinks are popularly used in bioprinting applications due to their excellent biocompatibility, low cytotoxicity and biodegradable nature (Teixeira et al. 2022; Tamay and Hasirci 2021). Some commonly used naturally derived bioinks include alginate, gelatine, collagen and hyaluronic acid.

Alginate, for example, is one of the most widely used biomaterials in bioprinting applications due to its unique characteristics that closely resemble native tissue. Its fluid nature and ability to retain large amounts of water make it highly desirable as a bioink (Axpe and Oyen 2016). The high water content along with the fast gelation property of alginate is highly desirable in bioprinting applications as it is able to maintain the viability of encapsulated cells and protect them from shear stress during printing (Tamay and Hasirci 2021; Datta et al. 2019). A study by Christensen and colleagues reported using alginate bioink with mouse fibroblast cells to develop printed vascular-like structures using calcium chloride as a crosslinker (Christensen et al. 2015). Williams and colleagues also reported positive results in achieving homogenously dispersed viable adipose-derived stromal fraction cells in alginate spheroids using a direct write 3D bioprinting system (Williams et al. 2013). Another study by Lee et al. (2016) provided an account of the study of phage as a versatile nanoink for printing 3D cell-laden scaffolds. In their study, the arginylglycylaspartic acid (RGD) phage solution was used to create a versatile bioink with cell printing capability and M13 phages, in particular, were found to have good blending properties with alginate. Also, the proliferation of MC3T3-E1 cells was shown to improve proportionally with phage concentration.

However, using a single biomaterial alone is often inadequate to satisfy all the requirements of an ideal bioink. For example, although alginate shows favourable outcomes as a bioink, it lacks mechanical stability when printing larger tissue and organ structures (Weng et al. 2021). Additionally, alginate lacks cell binding sites, resulting in poor cell attachment and proliferation (Weng et al. 2021; Gungor-Ozkerim et al. 2018). To address this limitation, researchers have looked into combining alginate with other natural-based polymers. Kim and colleagues reported that the incorporation of carrageenan into alginate-based bioink improved the rheological properties and printability of the bioink without compromising the viability of the cells (Kim et al. 2019). Saljo and colleagues have also demonstrated that the 3D bioprinted construct printed using a combination of nanocellulose and alginate demonstrated evidence of vascularisation and good viability (Säljö et al. 2020). Similarly, collagen, although a reputable choice as a biomaterial, is not suitable to be used on its own. Collagen has low mechanical stability and a slow gelation rate at a temperature below 37 C, making it difficult to hold its shape when printing at physiological temperatures. Yang and colleagues found that adding Type 1 collagen to alginate-based bioink not only enhances its mechanical strength but also promotes cell adhesion and proliferation owing to its bioactive properties (Yang et al. 2018). This combination emphasises the advantage of mixing two or more biomaterials, as they complement each other's limitations. While combining natural polymers does help improve the limitations, they are still far from superior when compared to native tissues, especially in terms of mechanical properties (Weng et al. 2021). The soft nature of naturally derived polymers makes them more susceptible to losing their shape and deforming over extended periods of time. Thus, recent works have studied the incorporation of synthetic polymers to be used in combination with natural polymers (Gungor-Ozkerim et al. 2018). Synthetic polymers have the advantages of mechanical strength, tunable properties and controllable degradability and thus can be tailored to suit the desired requirements (Mao et al. 2020).

Koch and colleagues reported the use of hybrid bioprinting to achieve high mechanical stability along with high iMSC cell viability when using a combination of polycaprolactone (PCL) and alginate/gelatine hydrogel bioink (Koch et al. 2022). Decoupling mechanical and cellular requirements allows a compromise between the desired mechanical strength and maintaining cell viability. Another commonly used hybrid bioink is gelatine-methacryloyl (GelMA). The desirable properties of GelMA include a favourable environment for the differentiation and proliferation of cells (Ashammakhi et al. 2019). GelMA has been combined with other synthetic polymers, such as polyacrylamide and poly(ethylene glycol) diacrylate, to match the properties required for cartilage and bone bioprinting (Han et al. 2017; Wang et al. 2018a). Asohan and colleagues have also reported a preliminary study using alginate, nanocellulose and poly (ethylene glycol) diacrylate (PEGDA) as a bioinks for cartilage bioprinting applications. The double network mechanism employed in the study utilises the alginate sol/gel transition to print the initial structure of the construct and later secures the structure post-printing using photo crosslinking to form a stable structure (Asohan et al. 2022). These recent studies and reported work over the years have demonstrated that using just a single biomaterial is insufficient to fulfill the properties needed for an ideal bioink. Incorporating two or more biomaterials, both naturally and synthetically derived, also known as hybrid bioinks, is necessary to achieve the proper mechanical and biological properties of the target tissues and organs (Gungor-Ozkerim et al. 2018).

13.4.2 3D Bioprinting: Energy Sustainability

The life-cycle inventory of various bioprinting techniques demonstrates the utility of precise machine tools as an ecofriendly material that requires less energy to manufacture. However, compared to other bioprinting techniques, the laser-assisted bioprinting technique consumes more energy, as it consumes 40% of the system's energy for pre-heating and power feed. Moreover, about 25% is spent on building piston stepper motors and feeding them. Also, about 16% is spent on laser system operation (Jackson et al. 2016). As a result, considerable energy and carbon emission were consumed and released throughout this process. Therefore, in order to eliminate these shortcomings, 3D bioprinting techniques should be incorporated into a better thermal management system that can replace lasers, namely selective mask sintering (SMS), selective inhibition sintering (SIS), and selective heat sintering (SHS) (Whenish et al. 2022; Jackson et al. 2016).

Researchers have introduced a new concept to reduce electricity and material consumption using the Design for Sustainable Additive Manufacturing (DFSAM) approach, which is the predictive model for sustainable manufacturing to minimise adverse environmental impacts. In order to reduce the environmental impact, one AM machine manufacturer, OS, has followed the "cut material consumption" principle (Liu et al. 2018). This approach can further reduce the materials consumption rate by 75%, subsequently resulting in 40% reduction in CO_2 emissions. Additionally, the interaction between specific energy consumption (SEC) of bioprinting materials and machine tools should be thoroughly investigated while developing sustainable bioprinting practices, as it may have environmental consequences such as carbon emissions (Yang et al. 2017). The SEC of bioprinting techniques is influenced not only by the processing materials but also by the specific process parameters used.

13.5 Development of 3D Bioprinting Associated with Sustainability Aspects

13.5.1 Evolution of Bioprinting Techniques

Sustainability is essential in 3D bioprinting and is the root of its development. Using biodegradable and recyclable polymers can assist in creating a more sustainable environment. The infographic in Fig. 13.2 illustrates the evolution of the sustainable AM technique and bioprinting techniques. After the world's conservative strategy introduced 'sustainable development' in the 1980s, the AM technique was initiated and witnessed rapid progress with several patented 3D printers. The concept of 3D printing using a UV-based prototyping system was first described by Hideo Kodama (Whenish et al. 2022; Klebe 1998), followed by Chuck Hull using stereolithography (Klebe 1998; Dababneh and Ozbolat 2014). After the successful attempt at 3D printing, an inkjet-based bioprinter was introduced in 1988 and demonstrated the bioprinting process for the first time (Klebe 1998).

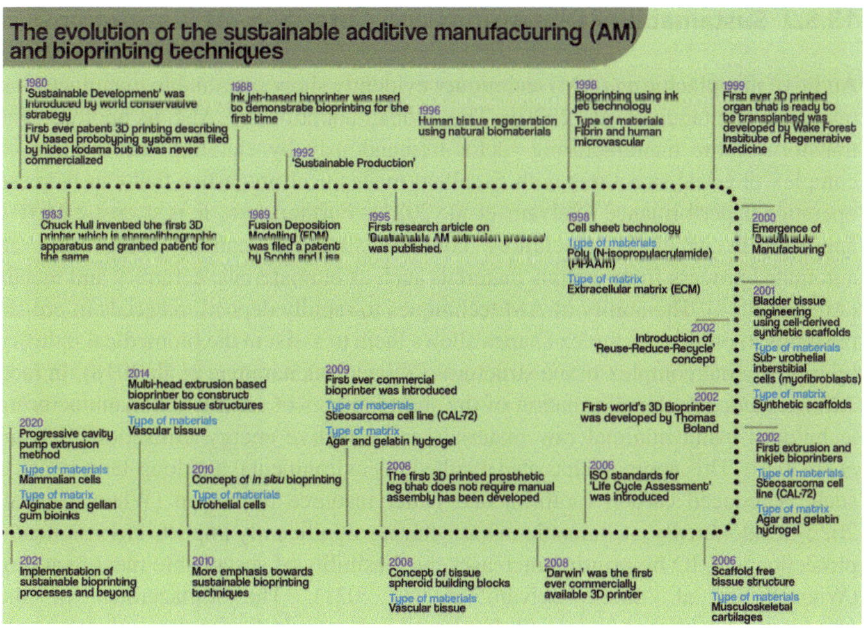

Fig. 13.2 Timeline of the evolution of sustainable additive manufacturing (AM) and bioprinting techniques

In the 1990s, the focus shifted towards sustainable production and described several types of matrix and biomaterials used in AM technology. The first category includes natural polymers such as lipids, polysaccharides and proteins derived from plants or animals (Yaashikaa et al. 2022; Wen et al. 2021; Liu et al. 2019; Díez-Pascual 2022; Bilal and Iqbal 2019). The second group of polymers includes polymers such as poly(hydroxyalkanoate) (PHA), polyhydroxybutyrate-valerate (PHBV) and poly(hydroxybutyrate) (PHB), which are produced by genetically modified microorganisms or bacteria (Morales et al. 2021; Yashavanth et al. 2021; Ibrahim et al. 2016). In the third group, synthetic polymers are created using bio-based monomers via various condensation methods or ring-opening polymerisation (Meereboer et al. 2020; Zoungranan et al. 2020; Pagno et al. 2020). Poly(butylene succinate) (PBS), polylactic acid (PLA), poly(butylene succinate-co-butylene adipate) (PBSA), polycaprolactone (PCL) and polybutylene adipate terephthalate (PBAT) are among these polymers (Karmanov et al. 2021; Meiabadi et al. 2021; Askari et al. 2021).

With the established study of AM techniques and the focus on the bioink study area, the first extrusion-based bioprinter was successfully developed in 2002 using agar and gelatin hydrogel as the matrix (Landers et al. 2002). Later, emphasis was given to the implementation of sustainable bioprinting processes (Whenish et al. 2022), and it later moved towards more sophisticated developments, such as multi-head extrusion and in-situ bioprinting (Chadha et al. 2022).

13.5.2 Sustainability Dimensions and Criteria in 3D Bioprinting

Additive manufacturing (AM) techniques evidently show sustainable manufacturing practices (Beltagui et al. 2020). The global adoption of AM technology has transformed the manufacturing sector tremendously by enabling the creation of complex physical structures with excellent mechanical properties that can increase operational performance (Selvam et al. 2021). Furthermore, it is a cost-effective approach in which architecturally complex structures can be built using an automated process from various materials such as biomaterials, ceramics and metals (Attaran 2017). The ability of AM techniques to rapidly deposit materials in pre-set positions to produce complex shapes allows them to assist in the biomedical industry by reproducing complex tissue structures (Vijayavenkataraman et al. 2018). In fact, AM also aided in the elimination of the shortcomings of subtractive manufacturing techniques using minimal raw materials and effective energy utilisation (Selvam et al. 2020). This approach has also resulted in environmental and long-term benefits such as reduced carbon footprint and better resource utilisation (Whenish et al. 2022). Undoubtedly, sustainable manufacturing significantly impacts environmental protection and 3D bioprinting increases the feasibility of sustainable manufacturing (Whenish et al. 2022; Selvam et al. 2021). The replacement of the non-biodegradable polymers with bio-based polymers indicates that polymers possess unique characteristics such as sustainability, renewability, biodegradability and non-toxic (Abdou et al. 2021; Khalid et al. 2021) are highly preferred. As a result, a

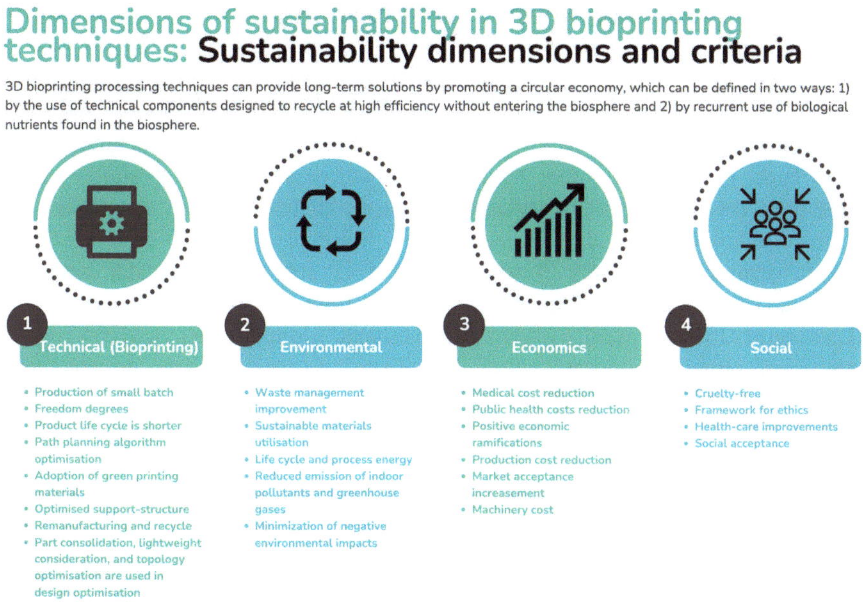

Fig. 13.3 Dimensions of sustainability in 3D bioprinting techniques in terms of social, economic, environmental and technical

wide range of biopolymers are replacing synthetic polymers in various commercial sectors, including agricultural, packaging and biomedical fields. Moreover, the ever-increasing demand for sustainable next-generation materials has also increased the number of opportunities for AM techniques.

For that reason, four sustainability dimensions and criteria were identified and categorised according to the technical aspect of bioprinting, environmental impact, circular economy and social impacts (Sauerwein et al. 2019), as shown in Fig. 13.3. Several factors must be considered when deciding on a particular manufacturing method, such as energy consumption rates, process costs and material waste (Beltagui et al. 2020). A circular economy aims to recreate capital while also improving the flow of goods and services (Sauerwein et al. 2019). In the technical aspect, 3D bioprinting offers several advantages, including reduced post-printing procedures, optimised design and short supply-production chains, optimised energy, and raw material utilisation, optimised multi-material processing time and shape consolidation in a single step (Selvam et al. 2020). As a result, bioprinting has the potential to significantly reduce material costs and increase the realisation of a circular economy in terms of product and goods maintenance, recycling and remanufacturing. Additionally, shorter supply chains and material reductions can further reduce production and transportation costs (Bahraminasab 2020).

13.6 Current Challenges in 3D Bioprinting

While bioprinting is a very promising area of research, with numerous progresses made over the years, many challenges need to be addressed. One of the main challenges of bioprinting is printing complex tissue and organ structures. There have been many studies that have reported success in printing simple tissue structures, for instance, the cartilage (Sun et al. 2020; Levato et al. 2017), skin (Cubo et al. 2016; Lee et al. 2014) and trachea (Bae et al. 2018; Huo et al. 2022), but the fabrication of a fully functional organ is still heavily under research. Vascularisation is the primary concern when printing complex tissues and organs (Bajaj et al. 2014). Vascular organs like the heart, kidney, lungs and liver require continuous circulation that will allow the delivery of gases, nutrients and metabolic products to grow and develop into functional structures (Bajaj et al. 2014; Wragg et al. 2019).

The kidney is perhaps one of the most intricate organs to recreate due to the complex nature of the internal structure and the multiple cell types present (Wragg et al. 2019; Fransen et al. 2021). With more than 20 different renal cell types, the fabrication process becomes slightly more complicated. This is because they require more considerations that would need to be factored in, including how to print different cell populations for the different regions of the kidney (Fransen et al. 2021). Recently, Mirdamadi and colleagues reported success in printing cardiac tissue that exhibited partial functionality with synchronous contraction (Wang et al. 2018b). The researchers have managed to print a full-sized adult human heart, including its internal anatomic features that mimic the mechanical properties of cardiac tissue. While this is definitely significant progress, given that there was once a time when printing full-sized organs were not yet possible; however, there are still challenges that need to be overcome.

These challenges include the duration of the printing process. For example, the full-sized adult heart model takes 4 days to complete printing, and this time span is not ideal for the cell-laden bioinks as the cells would die during the printing process (Mirdamadi et al. 2020). This will pose a problem for most full-sized organ structures with intricate architecture that requires a longer time to print. Efforts to speed up the printing process while also maintaining cell viability are crucial areas of improvement to look into. The current bioprinting process is time-consuming, compromising cell viability (Bishop et al. 2017). However, care should be taken to balance the printing process's speed with the print's precision (Bishop et al. 2017). The change in speed will have an effect on the printing resolution of the tissue and organ structures. While the higher resolution will result in a more accurate and precise microenvironment of the printed structure, the speed to achieve such accuracy will also be much slower (Miri et al. 2019). Therefore, future research will need to look into ensuring the cells are in their physiological environment and remain viable throughout the prolonged printing process.

Another major challenge is cellularising a full organ-size model because it requires culturing billions of cells for each cubic centimetre of volume (Mirdamadi et al. 2020). A strategy to generate consistent and reproducible large-scale culture will be necessary for the biofabrication of full-sized tissue constructs (Shao et al. 2020). One of the concerns of large-scale culture is also the high costs incurred. Additionally, when considering scaling and commercialising the bioprinted tissues, the costs in fabricating personalised organs for each patient must be considered, as individualised materials and tissue design will be required (Murphy et al. 2020). Not to mention the logistics factor, whereby the patient-derived cells and materials need to be transported from the initial biofabrication facility to the patient. Proper standard operating procedures will need to be standardised to ensure quality control throughout the fabrication process right up to the transplantation stage of the tissues or organs (Murphy et al. 2020).

13.7 Conclusion

The advancement of 3D bioprinting technology has opened up possibilities and opportunities in the fabrication of artificial tissues or organs and potentially revolutionised tissue engineering. It is arguable that the sustainability aspect is essential in 3D bioprinting to confer long-term application with sufficient functionality. Interestingly, sustainability is associated with 3D bioprinting, specifically with the bioink's biomaterials, which are at the heart of bioprinting. This involvement starts even at the initial stage of development and provides aid towards more sophisticated development. Bioink biomaterial selection is crucial to ensure biocompatibility properties with sustainability elements. Hence, four main sustainability dimensions and criteria have been identified to support the successful and continuous application of 3D bioprinting, including the technical aspect of bioprinting, environmental impact, circular economy and social impact. Despite the great achievement and successful implementation of 3D bioprinting in tissue engineering, several issues remain challenging, and they are mostly related to the fabrication of complex-tissue structures, which involve the complexity of the targeted tissues, printing process, large-scale culture and logistic issues.

Acknowledgement The authors would like to acknowledge the Ministry of Higher Education Malaysia for the Fundamental Research Grant Scheme with project Code: FRGS/1/2018/STG05/USM/02/6 for financial funding of study in the 3D bioprinting area, which provides an overview of the importance of the sustainability aspect in 3D bioprinting.

Conflict of Interest The authors declare no conflict of interest.

References

Abdou TR, Botelho Junior AB, Espinosa DCR, Tenório JAS (2021) Recycling of polymeric composites from industrial waste by pyrolysis: deep evaluation for carbon fibers reuse. Waste Manag 120:1–9. https://doi.org/10.1016/j.wasman.2020.11.010

Agarwal S, Saha S, Balla VK, Pal A, Barui A, Bodhak S (2020) Current developments in 3D bioprinting for tissue and organ regeneration – a review. Front Mech Eng 6(10):589171. https://doi.org/10.3389/fmech.2020.589171

Antezana PE, Municoy S, Álvarez-Echazú MI, Santo-Orihuela PL, Catalano PN, Al-Teh TH, Kadumudi FB, Dolatshahi-Pirouz A, Orive G, Federico M (2022) The 3D bioprinted scaffolds for wound healing. Pharmaceutics 14(2):1–46. https://doi.org/10.3390/pharmaceutics14020464

Ashammakhi N, Ahadian S, Xu C, Montazerian H, Ko H, Nasiri R, Barros N, Khademhosseini A (2019) Bioinks and bioprinting technologies to make heterogeneous and biomimetic tissue constructs. Mater Today Bio 1:100008. https://doi.org/10.1016/j.mtbio.2019.100008

Askari M, Afzali Naniz M, Kouhi M, Saberi A, Zolfagharian A, Bodaghi M (2021) Recent progress in extrusion 3D bioprinting of hydrogel biomaterials for tissue regeneration: a comprehensive review with focus on advanced fabrication techniques. Biomater Sci 9(3):535–573. https://doi.org/10.1039/d0bm00973c

Asohan AW, Hashim R, Ishak KMK, Hamid ZAA, Jasme N, Bustami Y (2022) Preparation and characterisation of cellulose nanocrystal/alginate/polyethylene glycol diacrylate (CNC/Alg/PEGDA) hydrogel using double network crosslinking technique for bioprinting application. Appl Sci 12(12):771. https://doi.org/10.3390/APP12020771

Attaran M (2017) The rise of 3-D printing: the advantages of additive manufacturing over traditional manufacturing. Bus Horiz 60(5):677–688. https://doi.org/10.1016/j.bushor.2017.05.011

Axpe E, Oyen ML (2016) Applications of alginate-based bioinks in 3D bioprinting. Int J Mol Sci 17(12):1976. https://doi.org/10.3390/ijms17121976

Bae SW, Lee KW, Park JH, Lee J, Jung CR, Yu J, Kim HY, Kim DH (2018) 3D bioprinted artificial trachea with epithelial cells and chondrogenic-differentiated bone marrow-derived mesenchymal stem cells. Int J Mol Sci 19(6):1624. https://doi.org/10.3390/IJMS19061624

Bahraminasab M (2020) Challenges on optimization of 3D-printed bone scaffolds. Biomed Eng 19(1):1–33. https://doi.org/10.1186/s12938-020-00810-2

Bajaj P, Schweller RM, Khademhosseini A, West JL, Bashir R (2014) 3D Biofabrication strategies for tissue engineering and regenerative medicine. Annu Rev Biomed Eng 16:247. https://doi.org/10.1146/annurev-bioeng-071813-105155

Barcelo X, Eichholz KF, Garcia O, Kelly DJ (2022) Tuning the degradation rate of alginate-based bioinks for bioprinting functional cartilage tissue. Biomedicine 10(7):1621

Beltagui A, Kunz N, Gold S (2020) The role of 3D printing and open design on adoption of socially sustainable supply chain innovation. Int J Prod Econ 221:107462. https://doi.org/10.1016/j.ijpe.2019.07.035

Bilal M, Iqbal HMN (2019) Naturally-derived biopolymers: potential platforms for enzyme immobilization. Int J Biol Macromol 130:462–482. https://doi.org/10.1016/J.IJBIOMAC.2019.02.152

Bishop ES, Mostafa S, Pakvasa M, Luu HH, Lee MJ, Wolf JM, Ameer GA, He TC, Reid RR (2017) 3-D bioprinting technologies in tissue engineering and regenerative medicine: current and future trends. Genes Dis 4(4):185–195. https://doi.org/10.1016/j.gendis.2017.10.002

Chadha U, Abrol A, Vora NP, Tiwari A, Shanker SK, Selvaraj SK (2022) Performance evaluation of 3D printing technologies: a review, recent advances, current challenges, and future directions. Prog Addit Manuf 7:1–34. https://doi.org/10.1007/s40964-021-00257-4

Christensen K, Xu C, Chai W, Zhang Z, Fu J, Huang Y (2015) Freeform inkjet printing of cellular structures with bifurcations. Biotechnol Bioeng 112(5):1047–1055. https://doi.org/10.1002/BIT.25501

Cubo N, Garcia M, Del Cañizo JF, Velasco D, Jorcano JL (2016) 3D bioprinting of functional human skin: production and in vivo analysis. Biofabrication 9(1):015006. https://doi.org/10.1088/1758-5090/9/1/015006

Dababneh AB, Ozbolat IT (2014) Bioprinting technology: a current state-of-the-art review. J Manuf Sci Eng 136(6):061016

Datta S, Barua R, Das J (2019) Importance of alginate bioink for 3D bioprinting in tissue engineering and regenerative medicine. In: Alginates - recent uses of this natural polymer. IntechOpen, Rijeka. https://doi.org/10.5772/INTECHOPEN.90426

Díez-Pascual AM (2022) Biopolymer composites: synthesis, properties, and applications. Int J Mol Sci 23(4):2257. https://doi.org/10.3390/ijms23042257

Fatimi A, Okoro OV, Podstawczyk D, Siminska-Stanny J, Shavandi A (2022) Natural hydrogel-based bio-inks for 3D bioprinting in tissue engineering: a review. Gels 8(3):179. https://doi.org/10.3390/gels8030179

Fransen MFJ, Addario G, Bouten CVC, Halary F, Moroni L, Mota C (2021) Bioprinting of kidney in vitro models: cells, biomaterials, and manufacturing techniques. Essays Biochem 65(3):587. https://doi.org/10.1042/EBC20200158

Gillispie G et al (2020) Assessment methodologies for extrusion-based bioink printability. Biofabrication 19,12(2):022003

Gu BK, Choi DJ, Park SJ, Kim YJ, Kim CH (2018) 3D bioprinting technologies for tissue engineering applications. In: Advances in experimental medicine and biology, vol 1078. Springer, New York, NY, pp 15–28

Gungor-Ozkerim PS, Inci I, Zhang YS, Khademhosseini A, Dokmeci MR (2018) Bioinks for 3D bioprinting: an overview. Biomater Sci 6(5):915–946. https://doi.org/10.1039/c7bm00765e

Han L, Xu J, Lu X, Gan D, Wang Z, Wang K, Zhang H, Yuan H, Weng J (2017) Biohybrid methacrylated gelatin/polyacrylamide hydrogels for cartilage repair. J Mater Chem 5(4): 731–741. https://doi.org/10.1039/c6tb02348g

Hospodiuk M, Dey M, Sosnoski D, Ozbolat IT (2017) The bioink: a comprehensive review on bioprintable materials. Biotechnol Adv 35(2):217–239. https://doi.org/10.1016/j.biotechadv.2016.12.006

Huo Y, Xu Y, Wu X, Gao E, Zhan A, Chen Y, Zhang Y, Hua Y, Swieszkowski W, Zhang YS, Zhou G (2022) Functional trachea reconstruction using 3D-bioprinted native-like tissue architecture based on designable tissue-specific bioinks. Adv Sci 9:e2202181. https://doi.org/10.1002/ADVS.202202181

Ibrahim S, Nawwar GAM, Sultan M (2016) Development of bio-based polymeric hydrogel: green, sustainable and low cost plant fertilizer packaging material. J Environ Chem Eng 4(1):203–210. https://doi.org/10.1016/j.jece.2015.10.028

Jackson MA, Van Asten A, Morrow JD, Min S, Pfefferkorn FE (2016) A comparison of energy consumption in wire based and powder-based additive-subtractive manufacturing. Procedia Manuf 5(1):989–1005

Ji S, Guvendiren M (2017) Recent advances in bioink design for 3D bioprinting of tissues and organs. Front Bioeng Biotechnol 5(4):1–8. https://doi.org/10.3389/fbioe.2017.00023

Joseph J, Deshmukh K, Tung T, Chidambaram K, Khadheer Pasha SK (2019) 3D printing technology of polymer composites and hydrogels for artificial skin tissue implementations. Sadasivuni K et al Polymer nanocomposites in biomedical engineering. Lecture notes in bioengineering. Springer, Cham. https://doi.org/10.1007/978-3-030-04741-2_7

Kačarević ŽP, Patrick MR, Alkildani S, Retnasingh S, Smeets R, Jung O, Ivanišević Z, Barbeck M (2018) An introduction to 3D bioprinting: possibilities, challenges and future aspects. Materials 11(11):2199. https://doi.org/10.3390/ma11112199

Karmanov AP, Kanarsky AV, Kocheva LS, Belyy VA, Semenov EI, Rachkova NG, Bogdanovich NI, Pokryshkin SA (2021) Chemical structure and polymer properties of wheat and cabbage lignins–valuable biopolymers for biomedical applications. Polymer 220:123571. https://doi.org/10.1016/j.polymer.2021.123571

Khalid MY, Arif ZU, Al Rashid A, Shahid MI, Ahmed W, Tariq AF, Abbas Z (2021) Interlaminar shear strength (ILSS) characterization of fiber metal laminates (FMLs) manufactured through VARTM process. Forces Mech 4:100038. https://doi.org/10.1016/j.finmec.2021.100038

Khoeini R, Nosrati H, Akbarzadeh A, Eftekhari A, Kavetskyy T, Khalilov R, Ozbolat IT (2021) Natural and synthetic bioinks for 3D bioprinting. Adv NanoBiomed Res 1(8):2000097. https://doi.org/10.1002/anbr.202000097

Kim MH, Lee YW, Jung WK, Oh J, Nam SY (2019) Enhanced rheological behaviors of alginate hydrogels with carrageenan for extrusion-based bioprinting. J Mech Behav Biomed Mater 98(10):187–194. https://doi.org/10.1016/J.JMBBM.2019.06.014

Klebe RJ (1998) Cytoscribing: a method for micropositioning cells and the construction of two- and three-dimensional synthetic tissues. Exp Cell Res 179(2):362–373. https://doi.org/10.1016/0014-4827(88)90275-3

Koch F, Thaden O, Conrad S, Tröndle K, Finkenzeller G, Zengerle R, Kartmann S, Zimmermann S, Koltay P (2022) Mechanical properties of polycaprolactone (PCL) scaffolds for hybrid 3D-bioprinting with alginate-gelatin hydrogel. J Mech Behav Biomed Mater 130:105219. https://doi.org/10.1016/J.JMBBM.2022.105219

Landers R, Hübner U, Schmelzeisen R, Mülhaupt R (2002) Rapid prototyping of scaffolds derived from thermoreversible hydrogels and tailored for applications in tissue engineering. Biomaterials 23(23):4437–4447. https://doi.org/10.1016/S0142-9612(02)00139-4

Lee V, Singh G, Trasatti JP, Bjornsson C, Xu X, Tran TN, Yoo SS, Dai G, Karande P (2014) Design and fabrication of human skin by three-dimensional bioprinting. Tissue Eng Part C Methods 20(6):473–484. https://doi.org/10.1089/TEN.TEC.2013.0335

Lee DY, Lee H, Kim Y, Yoo SY, Chung WJ, Kim G (2016) Phage as versatile nanoink for printing 3-D cell-laden scaffolds. Acta Biomater 29:112–124

Lee SY, Ma J, Khoo TS, Abdullah N, Noordin NM, Kahar NNF, Abdul Hamid ZA, Mustapha M (2021) Polysaccharide-based hydrogels for microencapsulation of stem cells in regenerative medicine. Front Bioeng Biotechnol 9(10):947. https://doi.org/10.3389/fbioe.2021.735090

Lei M, Wang X (2016) Biodegradable polymers and stem cells for bioprinting. Molecules 21(5): 1–14. https://doi.org/10.3390/molecules21050539

Levato R, Webb WR, Otto IA, Mensinga A, Zhang Y, van Rijen M, van Weeren R, Khan IM, Malda J (2017) The bio in the ink: cartilage regeneration with bioprintable hydrogels and articular cartilage-derived progenitor cells. Acta Biomater 61:41–53. https://doi.org/10.1016/J.ACTBIO.2017.08.005

Li N, Guo R, Zhang ZJ (2021) Bioink formulations for bone tissue regeneration. Front Bioeng Biotechnol 9(2):1–17. https://doi.org/10.3389/fbioe.2021.630488

Liu Z, Jiang Q, Ning F, Kim H, Cong W, Xu C et al (2018) Investigation of energy requirements and environmental performance for additive manufacturing processes. Sustainability 10(10): 3606

Liu J, Sun L, Xu W, Wang Q, Yu S, Sun J (2019) Current advances and future perspectives of 3D printing natural-derived biopolymers. Carbohydr Polym 207:297–316. https://doi.org/10.1016/j.carbpol.2018.11.077

Maan Z, Masri ZN, Willerth SM (2022) Smart bioinks for the printing of human tissue models. Biomolecules 12(1):141. https://doi.org/10.3390/biom12010141

Mao H, Yang L, Zhu H, Wu L, Ji P, Yang J, Gu Z (2020) Recent advances and challenges in materials for 3D bioprinting. Prog Nat Sci Mater Int 30(5):618–634. https://doi.org/10.1016/J.PNSC.2020.09.015

Meereboer KW, Misra M, Mohanty AK (2020) Review of recent advances in the biodegradability of polyhydroxyalkanoate (PHA) bioplastics and their composites. Green Chem 22(17): 5519–5558. https://doi.org/10.1039/D0GC01647K

Meiabadi MS, Moradi M, Karamimoghadam M, Ardabili S, Bodaghi M, Shokri M, Mosavi AH (2021) Modeling the producibility of 3D printing in polylactic acid using artificial neural networks and fused filament fabrication. Polymers 13(19):219. https://doi.org/10.3390/polym13193219

Mirdamadi E, Tashman JW, Shiwarski DJ, Palchesko RN, Feinberg AW (2020) Fresh 3D bioprinting a full-size model of the human heart. ACS Biomater Sci Eng 6(11):6453–6459. https://doi.org/10.1021/acsbiomaterials.0c01133

Miri AK, Mirzaee I, Hassan S, Oskui SM, Nieto D, Khademhosseini A, Zhang YS (2019) Effective bioprinting resolution in tissue model fabrication. Lab Chip 19(11):2019. https://doi.org/10.1039/C8LC01037D

Morales A, Labidi J, Gullón P, Astray G (2021) Synthesis of advanced biobased green materials from renewable biopolymers. Curr Opin Green Sustain Chem 29:100436. https://doi.org/10.1016/J.COGSC.2020.100436
Murphy SV, Atala A (2014) 3D bioprinting of tissues and organs. Nat Biotechnol 32(8):773–785. https://doi.org/10.1038/nbt.2958
Murphy SV, De Coppi P, Atala A (2020) Opportunities and challenges of translational 3D bioprinting. Nat Biomed Eng 4(4):370–380. https://doi.org/10.1038/s41551-019-0471-7
Ozbolat IT, Hospodiuk M (2016) Current advances and future perspectives in extrusion-based bioprintinG. Biomaterials 76:321–343. https://doi.org/10.1016/j.biomaterials.2015.10.076
Ozbolat IT, Yu Y (2013) Bioprinting toward organ fabrication: challenges and future trends. IEEE Trans Biomed Eng 60(3):691–699. https://doi.org/10.1109/TBME.2013.2243912
Pagno V, Modenes AN, Dragunski DC, Fiorentin-Ferrari LD, Caetano J, Guellis C, Goncalves BC, dos Anjos EV, Pagno F, Martinelli V (2020) Heat treatment of polymeric PBAT/PCL membranes containing activated carbon from Brazil nutshell biomass obtained by electrospinning and applied in drug removal. J Environ Chem Eng 8(5):104159. https://doi.org/10.1016/j.jece.2020.104159
Papaioannou TG, Manolesou D, Dimakakos E, Tsoucalas G, Vavuranakis M, Tousoulis D (2019) 3D bioprinting methods and techniques: applications on artificial blood vessel fabrication. Acta Cardiol Sin 35(3):284–289. https://doi.org/10.6515/ACS.201905_35(3).20181115A
Pössl A, Hartzke D, Schmidts TM, Runkel FE, Schlupp P (2021) A targeted rheological bioink development guideline and its systematic correlation with printing behavior. Biofabrication 13(3):035021. https://doi.org/10.1088/1758-5090/abde1e
Rider P, Kačarević ŽP, Alkildani S, Retnasingh S, Barbeck M (2018) Bioprinting of tissue engineering scaffolds. J Tissue Eng 9:2041731418802090. https://doi.org/10.1177/2041731418802090
Säljö K, Orrhult LS, Apelgren P, Markstedt K, Kölby L, Gatenholm P (2020) Successful engraftment, vascularization, and in vivo survival of 3D-bioprinted human lipoaspirate derived adipose tissue. Bioprinting 17(3):e00065. https://doi.org/10.1016/J.BPRINT.2019.E00065
Sauerwein M, Doubrovski E, Balkenende R, Bakker C (2019) Exploring the potential of additive manufacturing for product design in a circular economy. J Clean Prod 226:1138–1149. https://doi.org/10.1016/j.jclepro.2019.04.108
Selvam A, Mayilswamy S, Whenish R (2020) Strength improvement of additive manufacturing components by reinforcing carbon fiber and by employing bioinspired interlock sutures. J Vinyl Addit Technol 26(4):511–523. https://doi.org/10.1002/vnl.21766
Selvam A, Mayilswamy S, Whenish R, Velu R, Subramanian B (2021) Preparation and evaluation of the tensile characteristics of carbon fiber rod reinforced 3d printed thermoplastic composites. J Compos Sci 5(1):8. https://doi.org/10.3390/jcs5010008
Shao L, Gao Q, Xie C, Fu J, Xiang M, He Y (2020) Synchronous 3D bioprinting of large-scale cell-laden constructs with nutrient networks. Adv Healthc Mater 9(15):1901142. https://doi.org/10.1002/ADHM.201901142
Sun Y, You Y, Jiang W, Wang B, Wu Q, Dai K (2020) 3D bioprinting dual-factor releasing and gradient-structured constructs ready to implant for anisotropic cartilage regeneration. Sci Adv 6(37):eaay1422. https://doi.org/10.1126/SCIADV.AAY1422/SUPPL_FILE/AAY1422_SM.PDF
Tamay DG, Hasirci N (2021) Bioinks—materials used in printing cells in designed 3D forms. J Biomater Sci Polym Ed 32(8):1072–1106. https://doi.org/10.1080/09205063.2021.1892470
Tan T, Gan S, Wang X, Liu W, Li X (2021) Applications of 3D bioprinting in tissue engineering: advantages, deficiencies, improvements, and future perspectives. J Mater Chem B 9(27):5385–5413. https://doi.org/10.1039/d1tb00172h
Tan G, Xu J, Yu Q, Zhang J, Hu X, Sun C, Zhang H (2022) Photo-crosslinkable hydrogels for 3D bioprinting in the repair of osteochondral defects: a review of present applications and future perspectives. Micromachines 13(7):1038

Teixeira MC, Lameirinhas NS, Carvalho JPF, Silvestre AJD, Vilela C, Freire CSR (2022) A guide to polysaccharide-based hydrogel bioinks for 3D bioprinting applications. Int J Mol Sci 23(12): 6564. https://doi.org/10.3390/ijms23126564

Thavornyutikarn B, Chantarapanich N, Sitthiseripratip K, Thouas GA, Chen Q (2014) Bone tissue engineering scaffolding: computer-aided scaffolding techniques. Prog Biomater 3(2–4):61–102. https://doi.org/10.1007/S40204-014-0026-7

Vijayavenkataraman S, Yan WC, Lu WF, Wang CH, Fuh JFY (2018) 3D bioprinting of tissues and organs for regenerative medicine. Adv Drug Deliv Rev 132:296–332. https://doi.org/10.1016/j.addr.2018.07.004

Wan Jusoh WNL, Sajab MS, Mohamed Abdul P, Kaco H (2022) Recent advances in 3D bioprinting: a review of cellulose-based biomaterials ink. Polymers 14(11):2260. https://doi.org/10.3390/polym14112260

Wang Y, Ma M, Wang J, Zhang W, Lu W, Gao Y, Zhang B, Guo Y (2018a) Development of a photo-crosslinking, biodegradable GelMA/PEGDA hydrogel for guided bone regeneration materials. Materials 11(8):1345. https://doi.org/10.3390/ma11081345

Wang Z, Lee SJ, Cheng HJ, Yoo JJ, Atala A (2018b) 3D bioprinted functional and contractile cardiac tissue construct. Acta Biomater 70:48. https://doi.org/10.1016/J.ACTBIO.2018.02.007

Wen K, Wei L, Ren Z, Wang B, Lu J (2021) Enhanced photocatalytic degradation of cationic and anionic dyes by Ag-modified g-C_3N_4 composite: insights on different mechanisms under visible light. J Mater Res 36(7):1549–1560. https://doi.org/10.1557/s43578-021-00203-8

Weng T, Zhang W, Xia Y, Wu P, Yang M, Jin R, Xia S, Wang J, You C, Han C, Wang X (2021) 3D bioprinting for skin tissue engineering: current status and perspectives. J Tissue Eng 12: 20417314211028574. https://doi.org/10.1177/20417314211028574

Whenish R, Ramakrishna S, Jaiswal AK, Manivasagam G (2022) A framework for the sustainability implications of 3D bioprinting through nature-inspired materials and structures. Bio-Design Manuf 5(2):412–423. https://doi.org/10.1007/s42242-021-00168-x

Williams DF (2019) Challenges with the development of biomaterials for sustainable tissue engineering. Front Bioeng Biotechnol 7(5):127. https://doi.org/10.3389/FBIOE.2019.00127/BIBTEX

Williams SK, Touroo JS, Church KH, Hoying JB (2013) Encapsulation of adipose stromal vascular fraction cells in alginate hydrogel spheroids using a direct-write three-dimensional printing system. Bioresources 2(6):448–454. https://doi.org/10.1089/BIORES.2013.0046/ASSET/IMAGES/LARGE/FIGURE8.JPEG

Wohlers T, Gornet T (2014) History of additive manufacturing. Wohlers Rep 24(2014):118. https://doi.org/10.4018/978-1-5225-2289-8.ch001

Wragg NM, Burke L, Wilson SL (2019) A critical review of current progress in 3D kidney biomanufacturing: advances, challenges, and recommendations. Ren Replace Ther 5(1):1–16. https://doi.org/10.1186/S41100-019-0218-7

Yaashikaa PR, Senthil PK, Karishma S (2022) Review on biopolymers and composites – evolving material as adsorbents in removal of environmental pollutants. Environ Res 212:113114. https://doi.org/10.1016/j.envres.2022.113114

Yang S, Talekar T, Sulthan MA, Zhao YF (2017) A generic sustainability assessment model towards consolidated parts fabricated by additive manufacturing process. Procedia Manuf 10: 831–844

Yang X, Lu Z, Wu H, Li W, Zheng L, Zhao J (2018) Collagen-alginate as bioink for three-dimensional (3D) cell printing based cartilage tissue engineering. Mater Sci Eng C 83:195–201. https://doi.org/10.1016/J.MSEC.2017.09.002

Yashavanth PR, Das M, Maiti SK (2021) Recent progress and challenges in cyanobacterial autotrophic production of polyhydroxybutyrate (PHB), a bioplastic. J Environ Chem Eng 9(4):105379. https://doi.org/10.1016/J.JECE.2021.105379

Zoungranan Y, Lynda E, Dobi-Brice KK, Tchirioua E, Bakary C, Yannick DD (2020) Influence of natural factors on the biodegradation of simple and composite bioplastics based on cassava starch and corn starch. J Environ Chem Eng 8(5):104396. https://doi.org/10.1016/j.jece.2020.104396

Sustainable Biomaterials for 3D Printing 14

Dasmawati Mohamad, Abdul Manaf Abdullah, and Marzuki Omar

Abstract

Suitable implant materials and surgery time are some of the factors that surgeons must consider prior to surgery. With the development of current technology, applying a patient's specific implant in craniofacial reconstruction may reduce the surgery time. However, not all alloplast implants are suitable for 3D printing. This chapter covers the utilisation of selected polymeric biomaterials such as alloplastic material that are suitable for 3D printing with affordable 3D printers due to its sustainability, eco-friendly and excellent processability.

Keywords

Alloplastic material · Polymeric biomaterials · 3D printing · Patient specific implant

Abbreviations

3D-CPACC	3D Orbital polyamide customised composite
ABS	Acrylonitrile butadiene styrene
AM	Additive manufacturing
BAGs	Bioactive glasses
CAD	Computer-aided design

D. Mohamad (✉) · M. Omar
School of Dental Sciences, Universiti Sains Malaysia, Kubang Kerian, Kelantan, Malaysia
e-mail: dasmawati@usm.my

A. M. Abdullah
School of Mechanical Engineering, College of Engineering, Universiti Teknologi MARA, Shah Alam, Selangor, Malaysia

W. S. Wan Kamarul Zaman, A. A. Nurul (eds.), *Sustainable Material for Biomedical Engineering Application*, https://doi.org/10.1007/978-981-99-2267-3_14

CAM	Computer-aided manufacturing
CAS	Computer-assisted surgery
EBM	Electron beam melting
FDA	Food and Drug Administration
FDM	Fused deposition modelling
HDPE	High-density polyethylene
MRT	Magnetic resonance tomography
PA	Polyamide
PCL	Polycaprolactone
PE	Polyethylene
PEEK	Polyethether ketone
PGA	Polyglycolide acid
PLA	Polylactic acid
PLGA	Polylactic glycolic acid
PLLA	Poly(L-lactide)
PMMA	Polymethyl methacrylate
PSI	Patient specific implant
SLS	Selective laser sintering
SSI	Surgical site infection
UHMWPE	Ultra-high-molecular-weight polyethylene
β-TCP	β-Tricalcium Phosphate

14.1 Biomaterials as Implant Materials

Sustainable implant material is one of the crucial factors in determining the future development of biomaterials. Nowadays, designing and fabricating implant materials for a specific application are crucial as they will permanently be in the patient's body. The ideal implant should possess adequate mechanical characteristics, a surface texture that supports cell adhesion, biocompatible and reproducible (Ramakrishna et al. 2001). However, selecting the optimal material continues to be challenging. There are three groups of implant materials derived from human and animal donors, namely, autograft, allograft and xenograft. Autogenous bone grafts or autografts are obtained from the same individual, and they are well known for their biocompatibility as they are incorporated as living tissue and do not elicit an immune reaction to self-antigens that minimise foreign body reactions such as infection and extrusion. Other advantages of autografts include inherent strength and rigidity, lack of sharp edges, fixable to the adjacent bone with vascularisation potential and radiopaque. However, autografts display different degrees of resorption that make them unpredictable. They also lack pliability that poses difficulty for adequate contouring into complex shapes as they are breakable (Mok et al. 2004). The main issue with this particular choice of implant material is harvesting; it thus increases the operating time and time under general anaesthesia.

Other major drawbacks include donor site morbidity, which include risk of infection, hematoma, seroma, neurovascular injury, use of drains, increased time of recovery, increased postoperative pain, a bony defect and an additional surface scar at the donor site as well as site-specific complications of autografts (Mok et al. 2004).

Allografts are transplanted tissues from a donor of the same species in the form of fresh/frozen, freeze-dried bone or demineralised freeze-dried bone (Eppley et al. 2005). Several advantages of allografts in comparison to autografts include lack of donor site morbidity, decreased operating time, preoperative customisability and abundant availability of tissue (Mok et al. 2004). However, some of the disadvantages of allografts include a higher resorption rate in comparison to autografts and the requirement of immunosuppressive pharmacotherapy in post treatment (Mok et al. 2004; Boyette et al. 2015).

Allografts also impose a bigger concern with the possibility of disease transmissions such as HIV and Hepatitis C virus. Transmission of Creutzfeldt-Jakob disease had been reported in cadaveric dura transplantation. Furthermore, demineralised bone sheets of 100–300 μm thickness had provided insufficient support in cases with enophthalmos for orbital implant (Dubois et al. 2016).

Xenografts are tissues transplanted from a donor of a different species from the recipient and mainly involve collagen membrane, porcine sclera, porcine skin gelatin, bovine bone or sclera. Despite the advantages of reduced operative time and lack of donor site morbidity, xenografts are rarely utilised due to the association with disease transmission, immunological transplant rejection, unpredictability and high resorption rates.

Another option for implant material is an alloplastic material that is synthetically produced. A recent trend has shown an increase in the popularity of alloplastic material in a number of maxillofacial and craniofacial reconstruction. This may be attributed to several advantages of alloplastic material such as availability, ease of use, technological advancements, absence of donor site morbidity and increasing level of evidence of safety and efficacy of synthetic materials for craniomaxillofacial reconstruction (Mok et al. 2004; Boyette et al. 2015; Dubois et al. 2016). Alloplastic implants are made from a variety of materials including metals, ceramics, polymers and composites. They are either permanent implants, both metallic and non-metallic, or resorbable.

Gold and silver had been used by the Incas of Peru since 3000 BC to repair skull defects following the trephination procedure (Thorne 2007). It then evolved to a metal-based implant of aluminium and titanium. Due to several complications with aluminium-based implants obtained from findings of studies conducted which had shown an association with epilepsy (Aydin et al. 2011) and Alzheimer's disease (Walton 2007), the usage of aluminium implant was stopped. On the other hand, titanium was deemed as the material of choice since its first utilisation as reported in 1965 by Simpson (Simpson 1965) for the application in craniofacial reconstruction, due to its high mechanical strength. The mechanical strength of a titanium implant is reported to be comparable to the cortical bone. However, the elastic modulus of titanium is about 5× higher than human bone, ranging from 11–20 GPa. Mismatch of this property would hinder the growth of surrounding tissue post implantation.

Moreover, heterogeneity of Young's modulus between titanium implant and surrounding bone tissue would lead to a stress shielding effect phenomenon. This phenomenon occurred when the loaded stress is not well distributed and shared between the implant and existing bone. Hence, the result would be an increase in bone porosity, possibility of bone refracture and implant loosening (Niinomi and Nakai 2011). The application of metal implants often leads to a dispute as they easily corrode. The lifespan of the metal implant, especially in orthopaedic application, will also be affected from the exposure to high stress and load that can accelerate the formation of wear debris leading to fatigue wear. Therefore, a surface modification process such as hydroxyapatite (HA) coating on metal-based implants had been introduced to improve their chemical, biological and mechanical properties (Liu et al. 2004).

Ceramics are also an option for implant materials, and they can be classified into bioinert and bioactive ceramics. Bioinert ceramic refers to the ceramic material that does not produce any reaction either to the host body or the body of the material. The host body also does not reject the implant material as it only acts as the replacement of the body parts. As for the bioactive ceramic material, it will facilitate bone growth. Common example of the bioactive material is HA. It is chemically and crystallographically similar to the bone mineral and has been utilised in craniofacial surgery since the 1990s. Another bioactive material that is commonly used is β-Tricalcium Phosphate (β-TCP). It is considered as ideal bone graft material due to its excellent properties such as biocompatible, osteoconductive, osteoinductive, biodegradable and unlimited supply quantity (Kondo et al. 2005). It is extensively used as void filler in orthopaedic and dental field for bone fractures repair and replacement post tumour removal as well as coating material for metal implants (Barrere et al. 2006). Another bioactive ceramic materials are bioactive glasses (BAGs). They are synthetic blocks or granules that can bond chemically to the bone. They have osteoinductive and osteoconductive properties with minimal foreign body reaction, infection, extrusion, displacement and resorption. However, these ceramic implant materials are brittle in nature with poor mechanical properties and they are difficult to mould, shape and fix (Aitasalo et al. 2001).

Composite is a mixture of two or more different classes of materials. These combined materials would possess unique physicomechanical properties different from their original individual materials. Composite is popular in restorative dentistry but yet to be explored as implant material.

Synthetic polymers as implant material have been revolutionised. One such example is the use of the hybridisation technique that utilises the advantages of a selected material while reducing the disadvantages (Kim et al. 2012). Titanium-reinforced polyethylene (PE) used as the orbital floor implant utilises the high strength, ease of contour and radiopacity of titanium, while the use of porous PE avoids sharp edges and allows infiltration of host tissue. Details of the polymeric biomaterials group are discussed further in a subtopic below.

14.2 Alloplastic Polymeric Materials

Polymeric biomaterials have become a material of choice due to their versatility. In fact, the properties of the polymers can be altered to suit certain applications. Although there are a wide range of polymeric biomaterials, the choice of polymers for craniofacial reconstruction is relatively limited. The major setback that hinders the use of polymeric biomaterials for that particular purpose is their mechanical integrity. In general, clinicians utilise resorbable as well as non-resorbable polymers for craniofacial reconstruction, where the choice of the materials typically depends on the age of the patients and the required properties for the intended application.

The classification of polymeric biomaterials and the examples are summarised in Fig. 14.1.

14.3 Resorbable Polymer and Non-resorbable Polymer

Resorbable polymer indicates a polymer that undergoes the degradation process in biological environments. The mechanism of degradation generally results in the elimination of by-product. In most cases, the polymers will be completely resorbed by the host tissue (Liu et al. 2017). It should be noted that resorbable polymers range from Poly(L-lactide) (PLLA), polylactic acid (PLA), polylactic glycolic acid (PLGA), polyglycolide acid (PGA) and polycaprolactone (PCL). Due to the nature of the degradation, resorbable polymers are generally used in paediatric patients. The respective polymers serve as a temporary fixation, where the timely degradation of the implant will be replaced by a growing tissue. Among the listed resorbable polymers, PGA and PLA are the earliest polymers to be used in surgery. Both

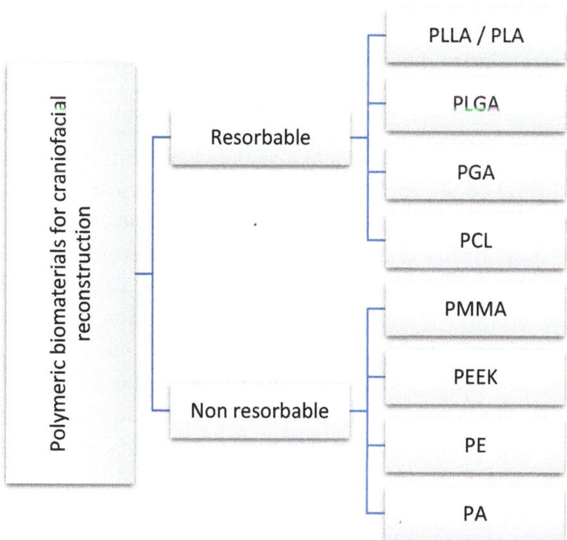

Fig. 14.1 Classification of polymeric biomaterial and the examples

polymers were assessed and introduced as suture materials (Herrmann et al. 1970; Cutright et al. 1971).

Besides PGA and PLA, the attempt to use polymeric materials for craniofacial reconstruction expands with the introduction of PLLA. PLLA is a homopolymer of the PLA family and was used as screws to immobilise the cranioplasty plate (Bos et al. 1987). However, the patients returned back 3 years after the operations due to adverse events following the operation. The patients suffer intermittent swelling surrounding the implantation site due to body reaction during the occurrence of the degradation process, and the finding was confirmed with the discovery of PLLA in the cytoplasm of the cell (Bergsma et al. 1993).

The rising need for polymeric materials that were able to be recontoured to suit the defects' part resulted in the debut of PLA and a combination of PGA (PLA/PGA copolymer) in 1996 (Moe and Weisman 2001). The polymer is malleable and responds to the surrounding heat, which allows the surgeon to re-contour. The process was made possible by immersion of the implant in the saline bath at a controlled temperature of 56 °C. This type of implant proved to be effective in securing a temporary fixation in paediatric patients with craniosynostosis (Eppley 2002). The resorption of this temporary implant in the human body, which at the same time permits bone growth, avoids secondary operation and eventually reduces the cost as well as the risk of infection (He and Shi 2017).

Non-resorbable polymer refers to polymers that do not substantially resorb, dissolve or degrade after implantation in the biological environment under typical physiological conditions. The polymers exhibit a strong molecular bonding derived from a repetition of a long hydrocarbon chain. The commonest non-resorbable polymers used in craniofacial reconstruction are polymethyl methacrylate (PMMA), polyethylene (PE), polyether ether ketone (PEEK) and polyamide (PA).

PMMA has been used since 1943 and become the earliest material for craniofacial reconstruction. In the early years during the evaluation stage, PMMA was also used to investigate the intracranial phenomena in macacus monkeys while repairing head injuries due to its transparent nature (Shelden et al. 1944). PMMA has been a preferred material after autologous bone due to its excellent functional and cosmetic results for long-term use (Marchac and Greensmith 2008). It also possesses adequate mechanical properties and is widely available at affordable cost. However, the limits of PMMA are its brittleness and high shrinkage (Hamad et al. 2016). Furthermore, an exothermic reaction during polymerisation of PMMA may harm the surrounding tissues. In the cranioplasty procedure, preoperative PMMA implant fabrication is recommended due to the high temperature of more than 50 °C was observed inside the PMMA implant during polymerisation (Golz et al. 2010). The effect of sterilisation on mechanical properties of the PMMA implant also should not be compromised. Different sterilisation methods such as autoclave, ethylene oxide, hydrogen oxide gas plasma and γ-irradiation effects on the mechanical properties need to be elucidated to ensure the survival of implants for long-term usage (Münker et al. 2018).

Polyethylene (PE) is another polymeric material that is commonly used (Ridwan-Pramana et al. 2015). It was approved by the Food and Drug Administration (FDA) for commercial usage in 1984 (Ellis and Messo 2004). However, the utilisation of PE for craniofacial reconstruction had been documented in 1954, which was to augment the congenitally missing condyles of a 12-year-old girl in England (Prowler and Glassman 1954). Currently, the PE-derived implant based on high-density polyethylene (HDPE) and ultra-high-molecular-weight polyethylene (UHMWPE), commercially known as of MedPor and Synpor, respectively, are widely used. Commonly PE is engineered into a porous structure with a pore size of 100–200 μm to achieve osteoconduction. Being flexible, yet strong makes it popular for the reconstruction of the craniofacial region. While PMMA and PE are established materials for the purposes, the usage of other variations of polymeric materials for craniofacial reconstruction is evolving.

Alternatively, polyether ether ketone (PEEK) is a remarkable high-performance polymeric material. Victrex, a company based in the United Kingdom, has started commercialising PEEK since April 1998 (Green and Schelegel 2001). PEEK is preferred to titanium owing to its lightweight and its mechanical properties that resemble the properties of cortical bone (Petrovic et al. 2006). Moreover, it is rarely associated with artifacts in magnetic resonance tomography (MRT) images compared to titanium (Maier 2009). PEEK is considered an inert polymer since it did not show the osteoinductive property when it was implanted in rat (Li et al. 2005). In 2007, PEEK had been used as an alternative material to correct a complex orbito-fronto-temporal defect of a previously failed treatment. The initial reconstruction was conducted using titanium with PMMA, which resulted in infection and wound dehiscence (Scolozzi et al. 2007). On follow-up a year after the second operation, the patient had regained satisfactory facial cosmesis.

Another type of non-resorbable polymer that is widely used is polyamide (PA) or commonly known as nylon. Polyamide 6,6 is one type of PA available which was invented by Carothers in 1935 at DuPont, U.S.A. Later, Paul Schlack at IG Farben, Germany, discovered polyamide 6 and similarly, Toray Japan also announced success in synthesising Polyamide 6. Thence, in the early years, the production of Polyamide 6,6 was dominated by the United States, while polyamide 6 was mainly produced by Europe and Japan (Sastri 2014). Early literature in the German language referred PA for biomedical application as supramid. Although polyamide was first established as a suture material, it was then expanded to the craniofacial region reconstruction where one of the initial attempts included a flat saddle nose correction using a supramid splint (Ulrich 1957). Later on, polyamide had also been successfully utilised as an orbital floor implant (Breitbart and Ablaza 2007) and as a condylar implant (Li et al. 2011). The move towards using polyamide is due to the presence of a polar molecular structure (CO-NH), which imitates the structure of collagen, a crucial factor that can induce osteoblast. Despite the presence of CO-NH seemingly advantageous in terms of biocompatibility, it also contributes to the hygroscopic nature of the material, whereby the storage condition must be well defined.

14.4 Clinical Aspects of Ideal Properties in Alloplastic Materials

Certain features for the purpose of orbital implant stability are essential in the craniofacial implant. The orbital implants should be able to reproduce and reposition the supporting tissues without significant resorption, as well as retain their shape once it is manipulated and remain stable over time upon supporting orbital content and related forces. Fixation is necessary to prevent migration, which in turn can cause infection, fibrosis and scarring, which may incidentally result in diplopia and blindness (Dubois et al. 2016).

Restoring a complex anatomy in the orbital implant also requires a great effort in contouring the implant; therefore, it is crucial for the implants to be easily mouldable to fit the orbital defect and regional anatomy. The alloplastic also should have a smooth surface without sharp edges. Till date, the easiest alloplastic material to shape anatomically is titanium mesh, especially by using an intraoperative skull model for adaptation of the implant. Implant contouring has become an increasingly important factor in large defects for correct globe reposition as large fractures have more variability in defect shape. Failure of correct repositioning may lead to atrophy, contractions and tissue herniation. Ideal orbital implants must allow drainage of orbital fluids to prevent retrobulbar hematoma, as it is one of the most serious complications. However, no study has been published on the advantages of a permeable material that can drain a possible intra-orbital haemorrhage into the maxillary and ethmoidal sinuses. In this case, titanium mesh has the best implant design due to its perforation, making it suitable for large orbital wall defects. Radiopacity is a desirable property of an implant to assess and monitor implant position. A good implant in a bad position would compromise the function or aesthetics of it.

An ideal implant also should not increase surgical complication rate or cause donor site morbidity such as pain and swelling. One of the reasons for the selection of alloplastic material is the lack of requirement for extra surgeries. Craniomaxillofacial implants should display good biocompatibility without complications of infection, extrusion, migration or foreign body reaction. Other than being chemically inert which include non-allergenic and non-carcinogenic, it should also be durable with minimal resorption, ideally osteoinductive and osteoconductive, which can facilitate the growth of bone formation. Moreover, it should exhibit high tissue incorporation yet able to dissect readily in events of implant removal if secondary reconstruction is necessary.

It should be noted that the risk of surgical site infection (SSI) following craniectomy, craniotomy, cranioplasty and burrhole procedures are approximately around 7.7%, and it is mainly caused by *S. aureus*, methicillin-resistant *S. aureus* (MRSA) and *P. aeruginosa* (Buang and Haspani 2012). However, the SSI rate following cranioplasty using PMMA and autologous bone are 8.2% and 6.7%, respectively (Kim et al. 2017). While infection post craniectomy, craniotomy and burrhole procedure can be resolved via intravenous or oral administration of drugs, drawbacks such as poor drug distribution and side effects in non-related tissue are issues that should not be overlooked. On the other hand, surgical site infection post

Table 14.1 Type of implant material and its infection rate. Reproduced from Kwarcinski et al. (2017)

Material	Number of implants	Number of infections	Average infection rate (%)
Autologous Bone	686	72	10.50
PMMA (hand-formed)	282	31	10.98
Titanium (plate)	638	53	8.31
PEEK	152	12	7.89
Titanium (mesh)	415	32	7.71
PMMA (prefabricated)	274	19	6.99
PMMA (templated)	102	7	6.86
Calcium-phosphate	179	12	6.70
PE	118	7	5.93
HA	1765	36	2.04

cranioplasty procedure can be reduced by introducing an implant with antibacterial properties, in which a local approach is projected to permit effective distribution, focused and targeted administration of antimicrobial agents. A broader study conducted has revealed that the SSI rates of various craniofacial implants are in the range of 5.93–10.98% (Kwarcinski et al. 2017). The details of the implants and their associated infection rates are summarised in Table 14.1.

The average infection rates are higher in autologous bone and hand formed PMMA as compared to other types of materials. The SSI rate is often linked to the duration of surgery (Cheng et al. 2017). By using an autologous bone, an additional time is needed to remove the bone from its original part prior to the implantation procedure which may expose it to bacterial infection. While similar additional time is required to hand contour the PMMA intraoperatively, a prefabricated implant could be opted to reduce the time and minimise the contact with the implant to prevent microbial infection.

The cost and availability of an implant are other factors that must be considered. The supply of alloplastic materials is limited in some parts of the world, which makes bone grafts the preferable choice. However, alloplast materials are advantageous in terms of donor site morbidity as well as reducing operation time and pre-bent or pre-formed alloplastic materials would further reduce operating time.

14.5 Patient-Specific Implants (PSI) in Craniomaxillofacial Implant Reconstruction

In craniofacial surgery, the restoration of function and cosmesis are both important to ensure good quality of life. Restoration of hard tissue defects due to trauma or pathologies using the conventional method requires the surgeons to shape the implant manually. Moulding done intraoperatively would result in lengthy surgery associated with a higher risk of blood loss and infection. The surgery outcome also depends on surgeon factor in terms of skills and experience. Hence, it is more prone to produce ill-fitting implants with poor aesthetic. Furthermore, inaccurate prosthesis will have a higher tendency to migrate and become displaced. Hence, a patient-specific implant (PSI) with accurate morphology in the craniomaxillofacial defect is highly desirable. Patient-specific implants (PSIs) offer several distinct advantages as the implants are engineered to precisely match the margins of the bony defect, which improves the stability of fixation and offers superior restoration of cosmesis. This is crucial in craniofacial surgery as PSIs are manufactured pre-operatively with reference to the 1:1 scale model of the patient's craniofacial defect and avoid intraoperative fabrication methods, which may improve surgical workflow efficiencies and patient outcomes (Thayaparan et al. 2021).

PSI can also reduce the infection rate. Low et al. (2019) had utilised PSI hybrid cranioplasty in which the original bone flap was reconstructed with PMMA among 12 craniectomy patients. PMMA was constructed in the gypsum mould intra-operatively, and the polymerisation had occurred prior to re-implantation of bone flap. Hence, the exothermic and toxic effect of polymerisation on the brain and surrounding structures could be avoided. The post-operative follow-up at 6 weeks and 3 months were unremarkable with no reported adverse effects. Satisfactory cosmesis were achieved in all the subjects, which was also supported by improvement in their quality of life as measured by the SF-36 scale.

The patient-specific implant (PSI) is a custom-made implant that is manufactured to each patient's individual anatomical specifications utilising additive manufacturing technology. The process begins with the image acquisition of the patient from CT and MRI scan datasets in the 2D model. Patient-specific implant constructs are specifically designed to mirror the non-affected craniomaxillofacial area by utilising preoperative CT data. 3D virtual reconstruction is derived and converted to an STL file. 2D and 3D modelling is constructed based on Computer-Aided Design (CAD) software. Then, the PSI implant of the required anatomical part is printed via Computer Aided-Manufacturing (CAM) or 3D printing. This advent of new manufacturing techniques and materials opens the opportunity for closer integration into the clinical routine. These anatomically ideal models are intended to reduce the need for intraoperative manipulation, thus reducing the operative time with more accurate reconstruction. Initial studies have demonstrated accurate fitting on all implants with no persistent postoperative visual impairments with zero patient reported sensation of foreign body (Boyette et al. 2015). Vignesh et al. (2017) have further supported the concept of a customised implant with 3D virtual treatment planning. 3D stereolithographic models and CAD-CAM have greatly improved the correction of extremely difficult late post-traumatic orbital defects. Dubois et al. (2016), on the other hand, have agreed on the benefits of computer-assisted surgery (CAS) in orbital reconstruction by facilitating complex preoperative planning and intraoperative navigation. 3D reconstruction in CT imaging and virtual planning offers accurate and individualised assessment and planning for the restoration of orbital walls and orbital volume. In lieu of this, the stereolithographic files of pre-shaped or individually designed implants have provided the best possible anatomical fit for orbital repair.

Common alloplastic materials that have been utilised for PSIs are manufactured from titanium, PEEK and glass-bioceramics. 3D printing for high melting temperature materials such as titanium and PEEK requires a higher cost 3D printer such as selective laser sintering (SLS) compared to a lower melting temperature material that is derived from polymer based. Metal PSI also possesses several disadvantages such as stress shielding, bone wear and easily conducts heat and cold (Niinomi and Nakai 2011). Metal alloplasts in patients who are undergoing CT scans also would cause artifacts and scattering of the image (Maier 2009), which can affect the accuracy of the reconstruction.

14.6 3D Printing

3D printing is a process of developing a three-dimensional solid object virtually of any shape from a digital computer model. 3D printing has been revolutionised in many sectors including manufacturing, food industries and health sectors due to its advantages. There are various 3D printing techniques available in the market such as jetting photopolymer, electron beam melting (EBM), selective laser sintering (SLS) and fused deposition modelling (FDM). Among all of the 3D printers, the one option that can accommodate the needs is FDM as it offers affordable processing techniques. Benchtop Fused Deposition Modelling (FDM) is the most common 3D printer, and it is affordable, especially for developing countries. The price had decreased dramatically in 2012, and currently, it costs around below RM 6000. Figure 14.2 illustrates examples of benchtop FDM 3D printers that are available in the market. However, the limitation of this type of 3D printer is a small area for printing with a dimension of 20 cm × 20 cm × 20 cm. Nevertheless, it produces less wastage than other traditional manufacturing throughout the printing process in which a large number of materials are trimmed away from the end product. Therefore, this type of 3D printer offers more sustainable manufacturing, and the wastage from the manufacturing can be reduced.

The emergence of 3D printing technology can be manipulated to produce a complex patient-specific craniofacial implant. However, many advantages of 3D printing technology in the medical field are restricted due to the limited availability of biocompatible feed stock.

This 3D printer utilised a thermoplastic polymer material as its feedstock material as it has to be melted and extruded via a nozzle. The feedstock of polymeric is in the filament form with a constant diameter ranging from 1.65 to 1.85 mm. The feedstock also requires winding. The common feedstock for FDM 3D printers is a thermoplastic polymer material in which it can be softened at high temperature and easily reshaped or moulded. Then, the material is solidified upon cooling. FDM 3D printer usually has temperatures ranging from 100 to 300 °C, and the melting feedstock must be below 250 °C. Acrylonitrile butadiene styrene (ABS) is commonly used as the feedstock with a melting point of 230 °C. However, it is not a biocompatible

Fig. 14.2 Desktop FDM 3D printer available in the market, MakerBot Replicator

material. ABS is made up from polymerisation of acrylonitrile and styrene with the presence of butadiene. ABS is said to be a carcinogenic material that is not suitable to be used as biomaterial, as exposure to butadiene and styrene is reported to have an association with leukaemia in humans (Sathiakumar et al. 2015). There are a few biocompatible feedstock materials or alloplastic materials in the market, which are polylactic acid (PLA) and polyether ether ketone (PEEK). PLA is a bioresorbable material, as stated in Sect. 14.3, hence it is suitable for an implant that does not require a permanent reconstruction as it will degrade after a certain time. On the other hand, PEEK is a new alloplastic material that has shown high mechanical properties due to its molecular arrangement that is reflected on a high melting point around 343 °C. It can only be processed via a high-end polymer processing machine that is equipped with higher heating capabilities such as SLS; thus it can be costly for the 3D printer. Although exclusive and expensive in nature, the introduction of PEEK as PSI that is processed via current additive manufacturing (AM) techniques has helped patients with craniofacial anomalies regain their regular cosmesis (Lethaus et al. 2011). This material, however, lacks osteointegrative properties, which have led to various complications such as infection and implant loss (Khonsari et al. 2014).

Alasseri and Alasraj (2020) had reported a series of case studies from six patients who utilised PSI by using PEEK and titanium (Alasseri and Alasraj 2020). They concluded that the use of PSI for maxillofacial reconstructions had advantages such as predictable outcomes, eliminated the usual complications seen in non-custom-made implants and boasted excellent patient satisfaction. However, the high cost was deemed as a major drawback of this technique. For example, a craniofacial reconstruction in Malaysia would cost up to RM 10,000–15,000 for a customised cranial prosthesis, depending on the size of cranial defect and types of material used. Furthermore, a new material of PEEK can cost up to RM 20,000, while a customised titanium implant would cost around RM 15,000. A much lower price with PMMA is available, but the technique used is not customised and quite conventional, whereby surgeons have to mould the implant based on their experience to fit the size of the cranium, which would take a longer time in the operation room. However, the cost of production of customised hybrid PMMA-autologous bone implant using a prefabricated cranial implant and gypsum mould is reported around RM 3000–4000, which is far more cost-effective than other types of implants (Low et al. 2019).

Cranio-maxillofacial trauma involving the face and skull may involve craniectomy as part of the surgical procedure. This surgical reconstruction is a challenge to the surgeons due to limited visualisation and complex computer-aided intervention. Apart from the risks that can affect the patients, the cost of the surgery is high with increased operating time.

To date, research on costs of injuries is scarce, specifically on craniofacial trauma. According to a study by Geraerds et al. (2019), it is believed that both medical costs and productivity costs would generally increase with the severity of the injury. Hence, all factors must be considered in developing a sustainable alloplastic material through the utilisation of 3D printing.

14.7 Innovations on Sustainable Alloplastic for 3D Printing

Alloplastic feedstock offers many sustainable parameters including reducing the operation time, affordable cost and reducing waste implant material. Recently, there have been several studies done on the polyamide 12 composite that is carried out as a potential alloplastic feedstock 3D printer (Abdullah et al. 2018; Rahim et al. 2015). Firstly, polyamide 12, β-Tricalcium Phosphate (β-TCP) and zirconia powders were compounded using twin screws before they were pelletized. The pelletized form of polyamide composite 12 was processed in a single screw extruder (Fig. 14.3) and wound into a filament with a diameter of 1.75–1.80 mm, as shown in Fig. 14.4.

The size of the filament diameter is crucial as it had to be fixed to the affordable FDM 3D printer for the PSI printing. This hybrid ceramics filled with PA 12 had shown a sufficient flow ability for FDM 3D printing and fulfilled the standards requirement of bone replacement with adequate mechanical properties.

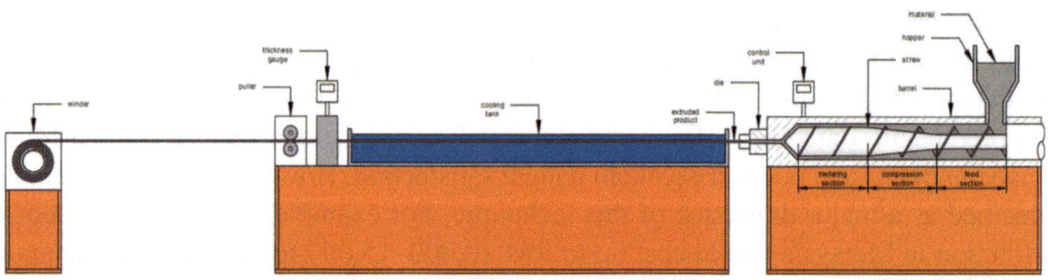

Fig. 14.3 Single screw extruder

Fig. 14.4 Polyamide 12 composite filament

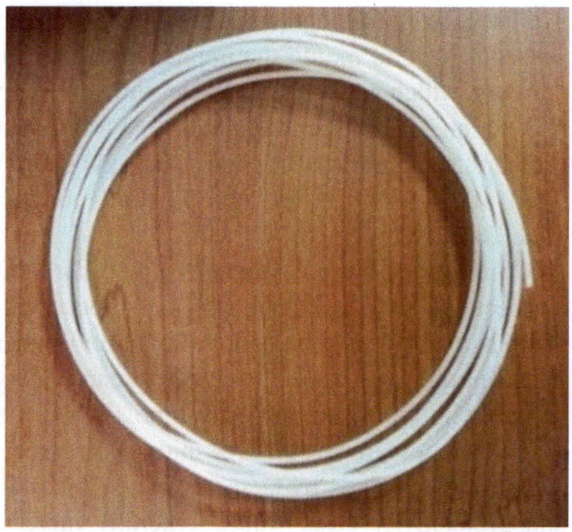

Biocompatibility evaluation of the polyamide 12 composite, namely, 3D Orbital Polyamide Customised Composite (3D-OPACC) has been studied on direct cell viability and attachment on material surfaces by live-dead staining fluorescence microscopic images (Sheng et al. 2020). The results in Fig. 14.5 had shown that the viable cells on top of 3D-OPACC were clearly visible with more cells observed as compared to Synpor and Medpor. This is due to the (–CO–NH–) structure of the PA chain, which is a polar molecule similar to that of collagen that may possibly stimulate osteoblast. Thus, the mimicked structure of PA could be of advantage in the development of biomaterial for craniofacial reconstruction.

This future application of polyamide composite is summarised in Fig. 14.6 as an alternative alloplast material with direct printing which can fit in each patient as PSI would reduce the operating time without preparation time of cutting and bending. This would result in the cosmesis of the patients as it fits the curvature and their anatomy.

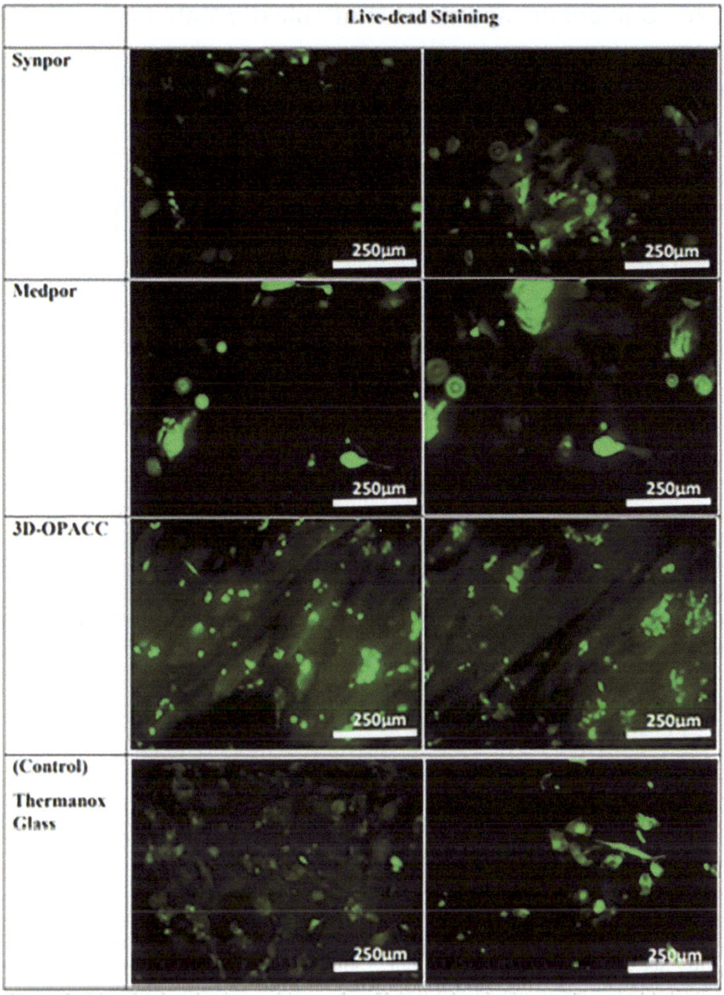

Fig. 14.5 Evaluation of cell viability on 3D-OPACC via fluorescence microscope. (Adapted with permission from Sheng et al. 2020)

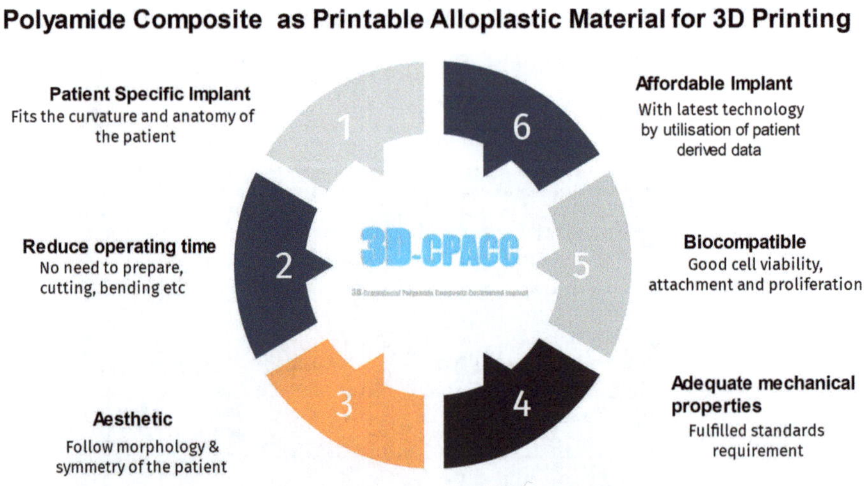

Fig. 14.6 Infographic summary of alloplastic PA 12 composite or 3D-CPACC

Acknowledgement The authors would like to acknowledge the Fundamental Research Grant Scheme (FRGS/1/2021/SKK0/USM/02/13) from the Ministry of Higher Education (MOHE) Malaysia.

Conflict of Interest None.

References

Abdullah AM, Rahim TNAT, Hamad WNFW, Mohamad D, Akil HM, Rajion ZA (2018) Mechanical and cytotoxicity properties of hybrid ceramics filled polyamide 12 filament feedstock for craniofacial bone reconstruction via fused deposition modelling. Dent Mater 34(11):e309–e316

Aitasalo K, Kinnunen I, Palmgren J, Varpula M (2001) Repair of orbital floor fractures with bioactive glass implants. J Oral Maxillofac Surg 59(12):1390–1395. https://doi.org/10.1053/joms.2001.27524; discussion 1395-1396.

Alasseri N, Alasraj A (2020) Patient-specific implants for maxillofacial defects: challenges and solutions. Maxillofac Plast Reconstr Surg 42(1):15. https://doi.org/10.1186/s40902-020-00262-7

Aydin S, Kucukyuruk B, Abuzayed B, Aydin S, Sanus GZ (2011) Cranioplasty: review of materials and techniques. J Neurosci Rural Pract 2(2):162–167

Barrere F, van Blitterswijk CA, de Groot K (2006) Bone regeneration: molecular and cellular interactions with calcium phosphate ceramics. Int J Nanomedicine 1(3):317–332

Bergsma EJ, Rozema FR, Bos RRM, Bruijn WCD (1993) Foreign body reactions to resorbable poly (l-lactide) bone plates and screws used for the fixation of unstable zygomatic fractures. J Oral Maxillofac Surg 51:666–670. https://doi.org/10.1016/S0278-2391(10)80267-8

Bos RRM, Boering G, Rozema FR, Leenslag JW (1987) Resorbable poly(l-lactide) plates and screws for the fixation of zygomatic fractures. J Oral Maxillofac Surg 45:751–753. https://doi.org/10.1016/0278-2391(87)90194-7

Boyette JR, Pemberton JD, Bonilla-Velez J (2015) Management of orbital fractures: challenges and solutions. Clin Ophthalmol 9:2127–2137. https://doi.org/10.2147/opth.S80463

Breitbart AS, Ablaza VJ (2007) Implant materials. In: Thorne CH (ed) Grabb and Smith's plastic surgery, 6th edn. Lippincott Williams & Wilkins, Philadelphia, PA, pp 58–67

Buang SS, Haspani MS (2012) Risk factors for neurosurgical site infections after a neurosurgical procedure: a prospective observational study at hospital Kuala Lumpur. Med J Malaysia 67(4): 393–398

Cheng H, Chen BP-H, Soleas IM, Forke NC, Cameron CG, Hinoul P (2017) Prolonged operative duration increases risk of surgical site infections: a systematic review. Surg Infect 18(6): 722–735

Cutright DE, Beasley J, Perez B (1971) Histologic comparison of polylactic and polyglycolic acid sutures. Oral Surg Oral Med Oral Pathol 32(1):165–173. https://doi.org/10.1016/0030-4220(71)90265-9

Dubois L, Steenen SA, Gooris PJJ, Bos RRM, Becking AG (2016) Controversies in orbital reconstruction—III. Biomaterials for orbital reconstruction: a review with clinical recommendations. Int J Oral Maxillofac Surg 45(1):41–50. https://doi.org/10.1016/j.ijom.2015.06.024

Ellis E, Messo E (2004) Use of nonresorbable alloplastic implants for internal orbital reconstruction. J Oral Maxillofac Surg 62(7):873–881. https://doi.org/10.1016/j.joms.2003.12.025

Eppley BL (2002) Use of resorbable and screw fixation in pediatric craniofacial surgery. Oper Tech Plast Reconstr Surg 9(1):36–45. https://doi.org/10.1016/S1071-0949(03)90007-2

Eppley BL, Pietrzak WS, Blanton MW (2005) Allograft and alloplastic bone substitutes: a review of science and technology for the craniomaxillofacial surgeon. J Craniofac Surg 16:981–989. https://doi.org/10.1097/01.scs.0000179662.38172.dd

Geraerds AJLM, Haagsma JA, de Munter L, Kruithof N, de Jongh M, Polinder S (2019) Medical and productivity costs after trauma. PLoS One 14(12):e0227131. https://doi.org/10.1371/journal.pone.0227131

Golz T, Graham CR, Busch LC, Wulf J, Winder RJ (2010) Temperature elevation during simulated polymethylmethacrylate (PMMA) cranioplasty in a cadaver model. J Clin Neurosci 17(5): 617–622. https://doi.org/10.1016/j.jocn.2009.09.005

Green S, Schelegel J (2001) A polyaryletherketone biomaterial for use in medical implant applications polymers for the medical industry. Rapra Technology Limited, Brussels, pp 1–8

Hamad WNFW, Abdullah AM, Mohamad D (2016) Effect of zinc oxide on flexural and physical properties of PMMA composites. AIP Conf Proc 1791:020014-1–020014-4. https://doi.org/10.1063/1.4968869

He J, Shi W (2017) Inflammatory foreign body reaction caused by resorbable materials used for orbital fractures repair. Medicine 96(49):1–3. https://doi.org/10.1097/MD.0000000000008822

Herrmann JB, Kelly RJ, Higgins GA (1970) Polyglycolic acid sutures: laboratory and clinical evaluation of a new absorbable suture material. JAMA Surg 100(4):486–490. https://doi.org/10.1001/archsurg.1970.01340220162027

Khonsari RH, Berthier P, Rouillon T, Perrin J-P, Corre P (2014) Severe infectious complications after PEEK-derived implant placement: report of three cases. J Oral Maxillofac Surg Med Pathol 26(4):477–482

Kim CY, Jeong BJ, Lee SY, Yoon JS (2012) Comparison of surgical outcomes of large orbital fractures reconstructed with porous polyethylene channel and porous polyethylene titan barrier implants. Ophthalmic Plast Reconstr Surg 28(3):176–180. https://doi.org/10.1097/IOP.0b013e3182467c4a

Kim SH, Kang DS, Cheong JH, Kim JH, Song KY, Kang MH (2017) Comparison of complications following cranioplasty using a sterilized autologous bone flap or polymethyl methacrylate. Kr J Neurotr 13(1):15–23

Kondo N, Ogose A, Tokunaga K, Ito T, Arai K, Kudo N, Inoue H, Irie H, Endo N (2005) Bone formation and resorption of highly purified beta-tricalcium phosphate in the rat femoral condyle. Biomaterials 26(28):5600–5608

Kwarcinski J, Boughton P, Ruys A, Doolan A, Gelder JV (2017) Cranioplasty and craniofacial reconstruction: a review of implant material, manufacturing method and infection risk. Appl Sci 7:1–17

Lethaus B, Laak MPT, Beerens M, Koper D, Poukens J, Kessler P (2011) A treatment algorithm for patients with large skull bone defects and first results. J Craniomaxillofac Surg 39:435–440

Li H, Springer M, Zou X, Briest A, Bünger C (2005) Ectopic bone induction by equine bone protein extract. In: Fisher JP (ed) Advances in experimental medicine and biology. Springer, Cham, pp 393–402

Li J, Hsu Y, Luo E, Khadka A, Hu J (2011) Computer-aided design and manufacturing of rapid prototyped nanoscale hydroxyapatite/polyamide (n-HA/PA) construction for condylar defect caused by mandibular angle ostectomy. Aesthet Plast Surg 35:636–640. https://doi.org/10.1007/s00266-010-9602-y

Liu X, Chu PK, Ding C (2004) Surface modification of titanium, titanium alloys, and related materials for biomedical applications. Mater Sci Eng R Rep 47(3–4):49–121

Liu Y, Zheng Y, Hayes B (2017) Degradable, absorbable or resorbable—what is the best grammatical modifier for an implant that is eventually absorbed by the body? Sci China Mater 60(5): 377–391. https://doi.org/10.1007/s40843-017-9023-9

Low PH, Mohamad D, Abdullah AM, Abdullah JY, Yahya S, Idris Z (2019) Patient-specific reconstruction utilizing computer assisted 3D modelling for partial bone flap defect in hybrid cranioplasty. J Craniofac Surg 30(8):e720. https://doi.org/10.1097/SCS.0000000000005713

Maier W (2009) Biomaterials in skull base surgery. GMS Curr Top Otorhinolaryngol Head Neck Surg 8:1–14. https://doi.org/10.3205/cto000059

Marchac D, Greensmith A (2008) Long-term experience with methylmethacrylate cranioplasty in craniofacial surgery. J Plast Reconstr Aesthet Surg 61(7):744–752. https://doi.org/10.1016/j.bjps.2007.10.055

Moe KS, Weisman RA (2001) Resorbable fixation in facial plastic and head and neck reconstructive surgery: an initial report on polylactic acid implants. Laryngoscope 111:1697–1701. https://doi.org/10.1097/00005537-200110000-00005

Mok D, Lessard L, Cordoba C, Harris PG, Nikolis A (2004) A review of materials currently used in orbital floor reconstruction. Can J Plast Surg 12(3):134–140. https://doi.org/10.1177/229255030401200302

Münker TJAG, van de Vijfeijken SECM, Mulder CS, Vespasiano V, Becking AG, Kleverlaan CJ (2018) Effects of sterilization on the mechanical properties of poly(methyl methacrylate) based personalized medical devices. J Mech Behav Biomed Mater 81:168–172. https://doi.org/10.1016/j.jmbbm.2018.01.033

Niinomi M, Nakai M (2011) Titanium-based biomaterials for preventing stress shielding between implant devices and bone. Int J Biomaterials 2011:836587

Petrovic L, Pohle D, Munstedt H, Rechtenwald T, Schlegel KA, Rupprecht S (2006) Effect of betaTCP filled polyetheretherketone on osteoblast cell proliferation in vitro. J Biomed Sci 13(1): 41–46. https://doi.org/10.1007/s11373-005-9032-z

Prowler RJ, Glassman S (1954) Agenesis of the mandibular condyles: diagnostic findings and treatment of deformity by polyethylene implant. Oral Surg Oral Med Oral Pathol 7(2):133–139. https://doi.org/10.1016/0030-4220(54)90043-2

Rahim TNAT, Abdullah AM, Akil HM, Mohamad D, Rajion ZA (2015) Preparation and characterization of a newly developed polyamide composite utilising an affordable 3D printer. J Reinf Plast Compos 34:1628–1638

Ramakrishna S, Mayer J, Wintermantel E, Leong KW (2001) Biomedical applications of polymer-composite materials: a review. Compos Sci Technol 61:1189–1224

Ridwan-Pramana A, Wolff J, Raziei A, Ashton-James CE, Forouzanfar T (2015) Porous polyethylene implants in facial reconstruction: outcome and complications. J Craniomaxillofac Surg 43(8):1330–1334. https://doi.org/10.1016/j.jcms.2015.06.022

Sastri VR (2014) Engineering thermoplastics: acrylics, polycarbonates, polyurethanes, polyacetals, polyesters, and polyamides. In: Plastics in medical devices properties, requirements and applications. William Andrew, Norwich, NY, pp 121–172

Sathiakumar N, Brill I, Leader M, Delzell E (2015) 1,3-Butadiene, styrene and lymphohematopoietic cancer among male synthetic rubber industry workers - preliminary exposure-response analyses. Chem Biol Interact 241:40–49

Scolozzi P, Martinez A, Jaques B (2007) Complex orbito-fronto-temporal reconstruction using computer designed PEEK implant. J Craniofac Surg 18(1):224–228. https://doi.org/10.1097/01.scs.0000249359.56417.7e

Shelden H, Pudenz RH, Restarski JS, Craig WM (1944) The lucite calvarium - a method for direct observation of the brain. J Neurosurg 1:67–75. https://doi.org/10.3171/jns.1944.1.1.0067

Sheng SB, Abdullah AM, Omar M, Mohamad D (2020) Cytotoxicity and cell adherence evaluation of 3D orbital polyamide composite customised implant. Malay J Microsc 16(1):17–27

Simpson D (1965) Titanium in cranioplasty. J Neurosurg 22(3):292–293

Thayaparan GK, Lewis PM, Thompson RG, D'Urso PS (2021) Patient-specific implants for craniomaxillofacial surgery: a manufacturer's experience. Ann Med Surg 66:102420

Thorne CH (2007) Grabb and Smith's plastic surgery, 6th edn. Lippincott Williams & Wilkins, Philadelphia, PA

Ulrich L (1957) Technic of corrective surgery using supramid angular splints for flat saddle nose. Zeitschrift für Laryngologie, Rhinologie, Otologie und ihre Grenzgebiete 36(4):215–221

Vignesh U, Mehrotra D, Anand V, Howlader D (2017) Three dimensional reconstruction of late post traumatic orbital wall defects by customized implants using CAD-CAM, 3D stereolithographic models: a case report. J Oral Biol Craniofac Res 7(3):212–218. https://doi.org/10.1016/j.jobcr.2017.09.004. ISSN 2212-4268

Walton JR (2007) An aluminium-based rat model for Alzheimer's disease exhibits oxidative damage, inhibition of PP2A activity, phosphorylated tau, and granulovacuolar degeneration. J Inorg Biochem 101(2007):1275–1284

Biowaste as Candidates for Future Bone Materials

15

Yanny Marliana Baba Ismail, Nur Liyana Mohd Rosli, and Myat Myat-Htun

Abstract

Calcium magnesium silicate (CMS) bioceramic has been widely investigated as an alternative bone substitute material due to its good bioactivity and biocompatibility properties. CMS also provides good mechanical strength and controllable resorbability compared to calcium phosphate-based bioceramics. This bioceramic can be easily formulated and synthesised via various chemical routes. Interestingly, many biowastes such as eggshells, cockleshells, rice husks, dolomite, marble sludge, dental mould and plaster-of-Paris (POP) mould wastes have been recently explored as candidate raw materials for future bone materials. In addition, biowastes such as coconut choir, sugarcane, banana stem, corn hobs and kenaf powders have also been incorporated into the ceramics blend in making three-dimensional porous ceramic components by sintering at high temperatures. This chapter provides insight into the making of CMS bioceramics by transforming biowaste into a wealth of functional bone materials for a sustainable tomorrow.

Keywords

Biowaste · Calcium · Bone materials · Calcium phosphate · Calcium silicate

Y. M. Baba Ismail (✉) · N. L. Mohd Rosli
Biomaterials Niche Group, School of Materials and Mineral Resources Engineering, Universiti Sains Malaysia, Penang, Malaysia
e-mail: yannymarliana@usm.my

M. Myat-Htun
Department of Physics, Bago University, Payathonzu, Bago Region, Myanmar

W. S. Wan Kamarul Zaman, A. A. Nurul (eds.), *Sustainable Material for Biomedical Engineering Application*, https://doi.org/10.1007/978-981-99-2267-3_15

Abbreviations

Al_2O_3	Aluminium oxide
$CaCO_3$	Calcium carbonate
CaO	Calcium oxide
CaP	Calcium phosphate
CaSi	Calcium silicate
CHA	Carbonated hydroxyapatite
CMS	Calcium magnesium silicate
Fe_2O_3	Iron (III) oxide
HA	Hydroxyapatite
K_2O	Potassium oxide
$MgCO_3$	Magnesium carbonate
MgO	Magnesium oxide
Na_2O	Sodium oxide
PFA	Pore forming agent
POP	Plaster-of-Paris
RHA	Rice husk ash
SiO_2	Silicon dioxide
TCP	Tricalcium phosphate

15.1 Bioceramics in Biomedical Applications

Bioceramics are the class of ceramic materials that are compatible with the biological system. These subsets of biomaterials are very diverse, and their biocompatibility properties vary from inert, which remains inactive at the application site, to the other extreme of resorbable materials, which are eventually replaced by the body after they have assisted repair (Roy et al. 2017; Pina et al. 2018).

Bioceramics comprise inorganic and non-metallic elements bonded ionically and covalently that are being heat-treated at a high temperature (800–1300 °C), depending on their composition and intended applications. These materials can be in the form of ceramic powders or bulk (dense or porous). The powders are usually used as fillers and coatings in biomedical and dental applications. While bulk bioceramics have commonly been used in the form of dense implants or porous scaffolds (Ramesh et al. 2013).

Bioceramics have been established as a group of materials for medical application, mainly for orthopaedic surgery, maxillofacial surgery and dental implants since the twentieth century. Bioceramics are also used as endodontics material (Fopase and Pandey 2022). Bioceramics include calcium phosphate, CaPs (such as hydroxyapatite and tricalcium phosphate), and calcium silicate groups, CaSi (such as bioglass and Ca-Mg-Si-containing ceramics such as akermanite) (Mohammadi et al. 2020; Myat et al. 2020). The rationale for using these CaP-based materials is the fact that CaP is the major component of biological apatite and that it shows promises of biocompatibility, osteoconductivity and biodegradability. However, the selection of bioceramics is determined by the objective of applications and the desired characteristics, as each group has its advantages and disadvantages.

15.1.1 Relevancies of Using Ca-Based Bioceramics in Biomedical Applications

Among various bioceramics, the calcium phosphate (CaP) and calcium silicate (CaSi) families are the most used as bone and dental materials. This is due to the nature of these bioceramics, in particular in having chemical similarity to the inorganic elements of human native bone and teeth, which mainly consists of calcium and phosphate, with minor elements such as carbonate, magnesium, silicate, chloride and fluoride (LeGeros 1994; Landi et al. 2010; Eliaz and Metoki 2017) (Table 15.1). The weight proportions of the major component of human bone are 60–70% inorganic mineral substances with 20–30% organic collagenous and non-collagenous components, while the remaining is water. This has become the main benchmark directing most researchers and clinicians in formulating Ca-based bioceramics in biomedical applications.

Over the years, various chemical syntheses, including dry and wet methods, have been developed to produce Ca-based bioceramics. These include chemical precipitation, hydrothermal, sol–gel, nanoemulsion and mechanochemical methods. Generally, Ca-based bioceramics, either CaP or CaSi bioceramics, can be fabricated in crystalline or amorphous form from macro- to nanoscales depending on their intended applications.

Today, Ca-based ceramics can be easily formulated via chemical methods using expensive reagent-graded chemicals. Among the various commercial calcium sources used are calcium nitrate tetrahydrate, $Ca(NO_3)_2 \cdot 4H_2O$, calcium oxide, CaO, and calcium hydroxide, $Ca(OH)_2$ (Ezekiel et al. 2018; Myat et al. 2022). While di-ammonium hydrogen phosphate ($(NH_4)_2HPO_4$) or orthophosphoric acid (H_3PO_4) have been used as the phosphate precursor in making CaP-based bioceramics, particularly hydroxyapatite ($Ca_{10}(PO_4)_6(OH)_2$) (Ezekiel et al. 2018; Yanny Marliana et al. 2017). In synthesising CaSi-based materials such as wollastonite ($CaSiO_3$),

Table 15.1 The inorganic mineral in human bone and teeth (enamel and dentin) (wt%)

Constituent	Bone	Teeth	
		Enamel	Dentin
Calcium	34.80	36.60	35.10
Phosphorus	15.20	17.70	16.90
Carbonate	2.00–8.00	3.20	5.60
Magnesium	0.55–0.72	0.44	1.23
Silicate	0.03–0.50	0.04	0.05
Chloride	0.13	0.30	0.01
Fluoride	0.03	0.01	0.06

akermanite ($Ca_2MgSi_2O_7$), diopside ($CaMgSi_2O_6$) and bredigite ($Ca_7MgSi_4O_{16}$), silicon dioxide (SiO_2) and magnesium dioxide (MgO) have typically been used as Si^{4+} and Mg^{2+} precursors (Mohammadi et al. 2021; Myat et al. 2022). Interestingly, researchers have then realised that these bioceramics can also be derived from natural biowaste materials.

In recent years, various biowastes have been extensively investigated for their potential uses as alternative sources to form functional Ca-based bioceramics. These biowastes have been generated by the agricultural and food industries, including bovine bone, chicken bone, fish bone, eggshell, cockleshell, sugarcane bagasse and risk husks. This sustainable approach is considered a smart way to manage biowaste and reducing the environmental pollution. The success in transforming these biowastes into functional materials also reduces the production cost and provides a more affordable choice of bone materials to patients in need, particularly in helping the bottom billion nations worldwide (Bee and Abdul Hamid 2020).

15.2 Bioceramics as Bone Materials

Bioceramics designed as bone materials are ceramic materials that are biocompatible with bone cells as well as their neighbouring cells. These bioceramics, mimicking the mineral phase of bone tissue, make it one of the most promising choices (Liu et al. 2008). The response of tissue to the implant surface affects the mechanism of tissue interaction. This is due to their unique tissue responses, which are three types, i.e. bioinert ceramics, bioactive ceramics and bioresorbable ceramics. The classification of implant–tissue interactions is summarised below (Heness and Ben-Nissan 2004):

1. Bioinert ceramics: Refers to the minimal interaction of the material with its surrounding tissue once placed in the human body. Examples are stainless steel, titanium, carbon, alumina and partially stabilised zirconia.
2. Bioactive ceramics: Form direct chemical bonds with bone or even with the soft tissue of a living organism. It slowly dissolves when in contact with living tissue, forming a layer of biological apatite before reaching the bone and forming a direct bond with the bone. Examples of these materials are synthetic hydroxyapatite (HA), glass ceramic, apatite-wollastonite (A-W) and bioglass.
3. Bioresorbable ceramics: Materials that can be replaced by new tissue formation over time. This group of materials also actively participates in the metabolic processes of an organism. Examples of bioresorbable materials are carbonated HA (CHA) and β-tricalcium phosphate (β-TCP).

15.2.1 Calcium Phosphate-Based Bioceramics

Hydroxyapatite (HA), with a chemical formula of $Ca_{10}(PO_4)_6(OH)_2$, is one of the CaP ceramics, which has been widely used as bone material in the form of dense/

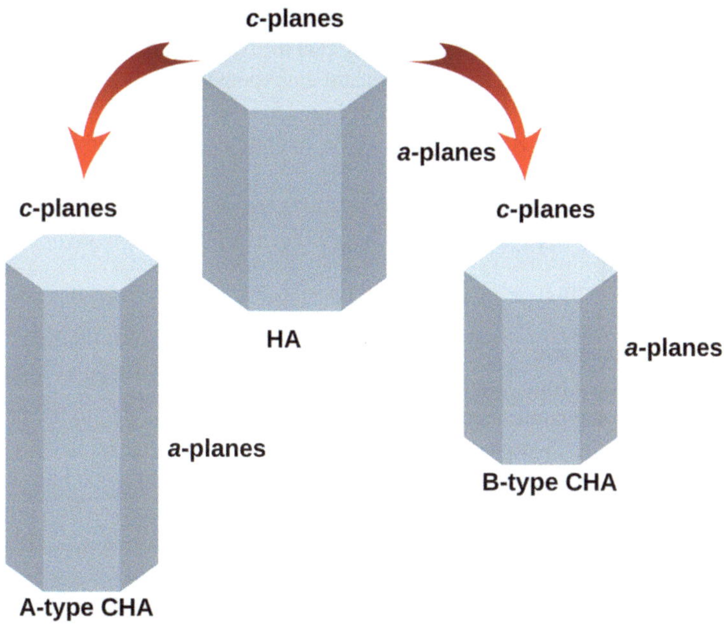

Fig. 15.1 Hexagonal crystal structure of stoichiometric HA. A-type CHA showed the contraction of the *c*-planes and an expansion in the *a*-planes and vice versa for the B-type CHA

porous bone substitute, implant coating and injectable cement in the clinical treatment. Stoichiometric HA belongs to the hexagonal crystal system with the crystal lattice $a = b \neq c$, as depicted in Fig. 15.1. The positively charged calcium, Ca^{2+}, normally positions at the *a*-plane, whereas the *c*-plane is rich in phosphate ions (negatively charged). In the last two decades, HA has attracted worldwide interest as an important substitute biomaterial in orthopaedic and dentistry as a result of its biocompatibility, bioactivity and osteoconductivity and the important is its similarity (chemically and structurally) to the mineral portion of natural human bone. However, the chemical composition of our native bone is slightly different from the stoichiometric HA in that they contain several additional trace elements substituted into the HA lattice, which play important roles in bone repair and regeneration. Besides, synthetic HA has serious drawbacks compared to biological apatites, in which this material shows negligible resorption even years after implantation.

One of the major trace elements is carbonate which is present about 2–8 wt% in bone minerals, and it depends on the individual age, sex and health of the patients. This makes carbonated hydroxyapatite (CHA) remains the most preferred material for bone repair, as the biological apatite is more accurately described as CHA rather than HA alone (Yanny Marliana et al. 2017; Awonusi et al. 2007). The carbonate ions may occupy either the hydroxyl, phosphate or both hydroxyl and phosphate ion positions; these result in A-, B- or AB-type CHA formation, respectively. B-type CHA is the favoured carbonate substitution and is typically found in the young bone.

However, the carbonate content varies depending on the age of the individual, with an increase of A-type in the old tissue (Ezekiel et al. 2018). The A-type CHA usually resulted in the contraction of the *c*-planes and an expansion in the *a*-planes and vice versa for the B-type CHA, as illustrated in Fig. 15.1.

In general, the A- and B-type CHA can be written using the following empirical formulae:

$$\text{A-type} : \mathrm{Ca_{10}(PO_4)_6(CO_3)_{\mathit{x}}(OH)_{2-\mathit{x}}} \tag{1}$$

$$\text{B-type} : \mathrm{Ca_{10-\mathit{x}/2}(PO_4)_{6-\mathit{x}}(CO_3)_{\mathit{x}}(OH)_2} \tag{2}$$

where $x =$ carbonate ions

Besides carbonate ions, various trace elements such as Mg^{2+}, Co^{2+}, Sr^{2+}, Zn^{2+}, Cu^{2+}, K^+ and Na^+ have attracted great interest from researchers due to the beneficial effects in making multi-doped HA for bone repair and regeneration (Bose et al. 2013; Yanny Marliana et al. 2021). The effect of these dopants had shown a profound effect on the biological response, particularly in improving bioresorbability and osteogenic capability; however, multi-doped HA still lacked mechanical strength.

Apart from HA-based materials, akermanite ceramic has been recently discovered as a potential candidate for bone material (Myat et al. 2022; Nur Liyana and Yanny Marliana 2022). Compared to phosphate-based ceramics, silicate-based ceramics have shown better performance in terms of mechanical properties owing to the basic Si–O–Si covalent bonding in their structures (Khandan and Ozada 2017). For example, diopside and akermanite have shown higher tensile strength and fracture toughness than HA and β-TCP (Hesaraki et al. 2009). Previously, hydroxyapatite/diopside ceramic composites were fabricated by introducing diopside into hydroxyapatite through uniaxial hot-pressing (Zhang et al. 2011). The fracture toughness was increased from 0.9 to 1.2 MPa $m^{1/2}$ after 10 wt% of diopside was composited into HA. Moreover, silicate-based ceramics have wider chemical compositions compared to phosphate-based ceramics, which might contribute to regulating the mechanical properties, bioactivity and biodegradability of these ceramics (Wu et al. 2013).

15.2.2 Calcium Silicate-Based Bioceramics

Akermanite ($Ca_2MgSi_2O_7$) is one of the members of the calcium silicate family. It contains calcium (Ca), magnesium (Mg) and silicon (Si) ions that have shown excellent bioactivity with good mechanical strength as compared to HA (Myat et al. 2022; Mohammadi et al. 2021). According to the literature, the structure of akermanite was first determined by Warren and Trautz in 1930 using the X-ray diffraction (XRD) technique (Ardit et al. 2012). Akermanite belongs to the melilite family of minerals, $X_2T^{(1)}T^{(2)}{}_2O_7$. It has a tetragonal crystal structure and the space group P-$\bar{4}$21m with $Z = 2$. The main advantage of Ca–Si-based bioceramic systems over CaP-based bioceramic is the rapid formation rate of the bioactive apatite layer on the surface because Si ions in Ca–Si are actively involved in the metabolism of bone growth (Mallick and Winnett 2014).

Previous researchers reported that alkaline-based silicates could be suitable hosts with chemical and thermal stabilities (Fei et al. 2005; Jiang et al. 2003). Despite the high degree of flexibility of the crystal structure, akermanite-based ceramics substituting with trace elements have not been significantly developed for bone repair thus far. Like the calcium phosphate family, various trace elements could be introduced into the akermanite crystal structure to further improve the biological and mechanical properties. Previously, several metallic cations (i.e. Mg^{2+}, Ba^{2+}, Cu^{2+}, $Fe^{2+/3+}$, Sr^{2+}, Zn^{2+}, etc.) and anions (i.e. F^{-}, CO2–3, SiO4–4, etc.) have been documented to participate in the bone life cycle involving bone formation (osteoblastic process) and bone resorption (osteoclastic process) (Bose et al. 2013). Through ion substitution, the lattice parameters, crystallinity, phase stability, sinterability, mechanical properties and solubility of the host ceramic could be chemically modified. Although the amount of these trace elements is relatively small, they play essential roles in bone regeneration.

Both akermanite and hydroxyapatite-based ceramics require a large amount of calcium as the main precursor. This has driven many researchers to seek alternative materials as the source of calcium besides silica, magnesium and phosphate sources.

15.3 From Biowaste to Bioceramics

In the last centuries, rapid growth in the animal agriculture and food industries due to the expansion of the world population has resulted in a decline in earth reserves, indirectly leading to the massive production of biowaste. About 1/3 of the food produced for human consumption is either lost or wasted. These inedible biowastes or their derivative products, such as eggshells, cockleshells and animal bones, are categorised as having no economic value, therefore, have been discarded directly into the landfill without fully utilising them (Bee and Abdul Hamid 2020). Irresponsible disposal of these wastes could cause environmental pollution and destroy the greens for the younger generation to enjoy in the future.

Besides the biowastes generated from animal agriculture and food industries, marble industries also produced a large amount of waste. The marble industry is one of the most environmentally destructive sectors, as the processes involve cutting stones from mother nature and producing a huge amount of heat, slurry and dust. Although marble waste generally includes non-radioactive by-products, it destroys plant life by changing soil alkalinity and disrupting photosynthesis and thus affecting the health of many herbivores. Yearly, about 200 million tons of marble waste are produced by industries in the United States, Turkey, Italy, Spain, Greece, India and Southeast Asia, particularly Indonesia and Malaysia. According to Sri Martek Marble, one of the biggest marble manufacturers in Malaysia, about 3000 tons of marble were processed monthly in 2018. About 40% of the processed marble turns into waste powder or dust during sawing, shaping and polishing. This natural biowaste is massive. Marble waste has been used as the base materials for building and construction, particularly in cement-based mortar and conventional aggregates in concrete. Researchers have attempted to blend this waste as fillers in making advanced polymers for different construction applications as it is believed to improve the physical and mechanical properties of the current application. Another form of marble waste is sludge. It is a slurry waste as water is used as a cooling medium during the cutting and polishing of the marble slabs. This marble sludge is seen as a great alternative material for high-grade calcium sources. The availability of the natural biowaste generated from the marble processing in high volumes would be the perfect source of raw material to consolidate the research needs.

Calcium sources derived from biowaste can also be obtained from clinical waste, for instance, dental mould waste (DMW). Dental mould or also known as a dental stone is a gypsum-based material. It is a model or working cast specifically designed for each individual patient. This model is then used in making dentures for dental restoration. In clinical, the used dental mould will be discarded into the landfill once the patients obtained their dentures fixed. This mould is calcium sulphate dihydrate, $CaSO_4 \cdot 2H_2O$ (also known as plaster-of-Paris, POP), with some minor elements (e.g. Cr^{3+}, Sr^{2+}, Cu^{2+}, etc.) added to provide certain properties to the mould. Therefore, dumping these wastes into landfills can produce toxic and odorous hydrogen sulphide gas and leaching of toxic ions into the soil. Driven by these environmental and economic impacts, massive research attention has been focused on transforming the biowastes into functional wealth materials in providing effective approaches for waste management at the same time providing more affordable biomedical products to patients in need (Nur Liyana and Yanny Marliana 2022).

Apart from seeking alternative materials as the main precursors in making bioceramic products, material engineers and scientists have also investigated the potential used as a natural pore-forming agent (PFA). Biowastes such as coconut coir and sugarcane serve as potential choices for natural PFA. These biowastes have been incorporated into the ceramics blend to make three-dimensional porous ceramic components followed by subsequent sintering at high temperatures. This process is known as the organic phase burning-out method.

15.4 Understanding the Potential Use of Biowaste

One smart approach that adds great value to animal bones, seafood shells and eggshells waste is to use them as precursors or raw materials in making new biomaterial products. For instance, these biowastes have a big potential to be transformed into a cheap and safe source of calcium, which is the main precursor in making artificial bone material or as a natural pore forming agent. But in order to do so, one should first understand the properties of these biowaste materials such as the composition, structure, morphology, thermal stability, etc. These properties could be identified using proper materials characterisation techniques by understanding their specific aims (Table 15.2).

For instance, in recent years, considerable efforts have been performed to transform eggshell wastes into valuable products, as eggshell contains mainly calcium

Table 15.2 Materials characterisation techniques and their aims

Characterisation techniques	Aim
X-Ray Fluorescence (XRF)	To quantify the elemental composition
X-Ray Diffraction (XRD)	To determine phase purity and crystallinity
Scanning Electron Microscopy (SEM)	To observe the morphological shape and size of the particle
Transmission Electron Microscopy (TEM)	
Thermogravimetry (TG)	To determine the mass change (e.g. decomposition, reduction and oxidation) with the change of temperature
Differential Thermal Analysis (DTA)	To show either decomposition is endothermic or exothermic reactions

carbonate ($CaCO_3$). The presence of calcium carbonate in the eggshell was confirmed by the results of XRF and FTIR. Owing to its composition, eggshells have been used as the starting material to prepare calcium phosphate bioceramics such as hydroxyapatite (HA) (Balázsi et al. 2007) and calcium silicate bioceramics (Palakurthy et al. 2020) as a bone substitute (Dupoirieux et al. 1995).

Interestingly, ostrich eggshell waste has also become an alternative source of calcium. An Ostrich egg is about 20–25 times bigger than a chicken egg. For a group of good-breeding ostriches, the females were observed to lay between 10 and 30 eggs during the first laying season and increased to between 40 and 70 eggs for the following laying season (Caliman et al. 2017). For commercial hybrid laying hens, the production of eggs lies in an average of 180–200 eggs/year (Zaheer 2015).

Besides eggshells, cockleshells have been widely investigated as a potential source of calcium. Cockles are an important source of protein. With the advancement in food technology, we can now enjoy cockles either fresh or as produced in canned food, but this also means an abundance of cockleshell waste has been produced. It was reported that the cockleshell contains 97.68% $CaCO_3$ (Sangsawang and Kaewdook 2018). For decades, calcium sources such as CaO and $Ca(OH)_2$ have been produced from eggshells and cockleshell wastes.

About 1 ton of husk is produced from 5 tons of rice paddy, and it has been identified as having the greatest potential to be reused or recycled because it is widely available (Nagrale et al. 2012). The rice husk is the outer layer of the rice paddy, and it covers about 20–25% of the rice weight. Rice husk ash (RHA) is obtained from burning the rice husk by combustion method to remove volatile organic carbon such as cellulose and lignin, which can be easily decomposed at a low temperature. RHA, in turn, contains around 85–96% amorphous silica (Mohammadi et al. 2020; Nagrale et al. 2012). Rice husk ash is considered a major source of silicon dioxide, SiO_2.

In our previous studies, the compositions of the chicken and ostrich eggshells and rice husks have been quantified using XRF analysis (Table 15.3). It is also important to take note that these biowastes may contain other trace elements. For instance, RHA also contains a small number of other oxides, including aluminum oxide (Al_2O_3), calcium oxide (CaO), magnesium oxide (MgO), potassium oxide (K_2O)

Table 15.3 The compositions of the chicken and ostrich eggshells and rice husks ash

	$CaCO_3$	$MgCO_3$	CaO	SiO_2	Al_2O_3	Fe_2O_3	MgO	Na_2O	K_2O	Others
Chicken eggshell	94.00	1.00	1.55	–	–	–	–	–	–	3.45
Ostrich eggshell	95.12	1.22	1.09	–	–	–	–	–	–	2.57
Rice husk ash	–	–	0.66	96.11	–	–	–	–	1.42	1.21

and other traces such as phosphorus pentoxide (P_2O_5), chromium (III) oxide (Cr_2O_3) and sulphur trioxide (SO_3) (Mohammadi et al. 2020). In designing materials for biomedical applications, one must ensure that these elements have no toxic effect either present in the final product or the by-products.

15.5 Transforming Biowastes to Sustainable Materials

The concept of transforming biowastes into sustainable materials has been used to produce calcium-based ceramics. This concept is considered environmentally friendly and innovative waste management. Eggshells, cockleshells, dolomite, rice husks, sugarcane bagasse and coconut coir have been used in making either dense or porous calcium silicate ceramics and in powder form.

Several synthesis methods can be applied in producing calcium silicate and calcium phosphate-based ceramics. Despite the synthesis method chosen, the wastes collected must be properly washed, cleaned and dried. This is then followed by crushing and milling to reduce the particle size. It is important to process these wastes into finer particles before synthesising HA or akermanite. Theoretically, the finer the particle size, the larger the surface area, resulting in a better reaction between the precursors used.

15.6 Processing Methods

15.6.1 Extraction of Calcium from Eggshells Waste

Eggshells make up approximately 8–12% of the total weight of the egg. It consists of three layers of structures, where the innermost shell is the membrane layer, the middle is the calcite layer (spongy and mammillary layer) and the outermost is called the cuticle. Both spongy and mammillary layers form a matrix of calcium carbonate crystals embedded in the protein fibre (Laca et al. 2017; Carvalho et al. 2011). While the eggshell membrane constitutes mostly protein fibres besides glucosamine, hyaluronic acid and chondroitin, which can also be commonly found in the structural component of human cartilage (Mittal et al. 2016). First, the thin protein membrane adhered to the inside of the eggshells must be carefully detached. The protein layer should be removed totally prior to the calcium extraction. The protein-free eggshells were then cleaned and crushed into powders. It is possible to extract about 40% calcium present in each individual eggshell. Besides calcium, eggshells contain some trace elements, such as boron and strontium, that can help prevent osteoporosis.

Ostrich eggshells and chicken eggshells mostly contain calcium carbonate ($CaCO_3$). Therefore, they must undergo calcination at 900 °C for 3 h to convert $CaCO_3$ to calcium oxide (CaO). Calcium oxide is commonly used as the starting powder to synthesise akermanite or hydroxyapatite powders.

15.6.2 Extraction of Calcium from Cockleshells Waste

The cockleshells collected from the food industries, markets and restaurants were first washed with water to remove dirt and impurities, followed by boiling in water for 30 min to remove unwanted protein coatings and microorganisms from the shells. This is then followed by drying in the oven at 100 °C overnight. In our previous study, the dried shells were calcined in a chamber furnace (Lenton, UK) at 900 °C with a heating rate of 10 °C/min for 4 h. The calcined solid was crushed using an agate mortar and pestle into powder form to be used as a calcium precursor. The obtained powders are calcium carbonate ($CaCO_3$) powders with a minor phase of calcium oxide (CaO) (Mohammadi et al. 2020).

15.6.3 Extraction of Calcium from Animal Bone

Mammalian bone can also be an alternative source of calcium, particularly from bovine, due to its size, abundantly available and closely resembling the chemical and architecture of human bone. However, the extraction of calcium from animal bones is more complex than the eggshell and cockleshell wastes. First, all the collected bones were subjected to preliminary cleaning procedures, boiling and chemical washing. The aims are as follows: (1) to ensure the adhered soft tissues and bone marrow are detached from the bone, (2) to remove proteins and fats and (3) to soften bone to ease the cutting of the bone in the next process. The cleaned bones were then calcined at about 600 °C to fully eliminate the collagenous and non-collagenous residues from the bone samples, leaving inorganic product, which is mainly calcium phosphate minerals (Bee and Abdul Hamid 2020).

Besides bovine bones, fish bones have also been used as a raw material in producing calcium phosphate (CaP). In one method, CaP powders have been produced by heat-treating the fish bones placed in a salt solution. The solution was then stirred at 75 °C and reacted with $CaCl_2 \cdot 2H_2O$. The mixture was then filtered, dried and ground into powders, yielding about 42% of CaP powders (Da Silva 2016).

15.6.4 Extraction of Calcium from Dental Mould Waste

A dental mould, a component used in making the model structures of dentures for patients with tooth defects, has great potential to be converted as a calcium source. The mould is calcium sulphate dihydrate ($CaSO_4 \cdot 2H_2O$), with additives such as chromium to provide certain properties to the mould. Dental mould is patient-specific and will be disposed of once the patient obtains his dentures. Dumping these wastes into landfills can produce toxic and odorous hydrogen sulphide gas and chromium ions leached into the soil. Annually, large quantities of $CaSO_4 \cdot 2H_2O$ are consumed for routine dental work in creating a model of this mould. As a result, a massive amount of waste has been produced.

Recently, our group has successfully innovated a smart way of transforming dental mould waste into calcium hydroxide (CaOH) via alkaline roasting followed by caustic leaching. The produced CaOH powders were then used as a calcium precursor in making akermanite or hydroxyapatite-based powders. Briefly, the dental mould waste (DMW) collected from the dentists was initially cleaned, crushed and ground into finer powders using a high-energy planetary ball mill. The milled powders were then reacted with sodium bicarbonate (Na_2CO_3) and roasted at 1000 °C. These powders were then mixed with sodium hydroxide (NaOH), filtered and washed with deionised water to completely remove the chromium ions in the DMW until the filtered solution changed from greenish to whitish. The filtered cake was dried in the oven and ground into powders, forming CaOH powders (Nur Liyana and Yanny Marliana 2022).

15.6.5 Extraction of Silica from Rice Husks Ash

The rice husk obtained is heated in a furnace at 800–850 °C to eliminate the organic compounds, forming white ash. This ash then undergoes digestion with caustic soda (NaOH), giving a sodium silicate solution. The solution is then filtered, where the filtered cake forms the silica, while sodium sulphate is the by-product collected at the bottom of the filter (Todkar et al. 2016).

In our previous study, akermanite ceramics have been successfully produced by reacting calcite (extracted from cockleshell waste) with silica (derived from rice husks ash) and magnesium oxide (obtained from dolomite) according to the following chemical reaction:

$$CaCO_3 \rightarrow CaO + CO_2$$

$$2CaO + MgO + 2SiO_2 \rightarrow Ca_2MgSi_2O_7$$

A highly densified Akermanite, $Ca_2MgSi_2O_7$ (about 94% dense), was obtained by sintering at a temperature above 1200 °C. This sintered Akermanite also possesses tensile strength (~9 MPa) and flexural strength (~22 MPa) values that fall in the range of human cancellous bone. This highlighted that the biowastes could be employed to be used as precursors in making Akermanite ceramics, particularly for Bone Tissue Engineering (BTE) applications (i.e. non-load-bearing bone substitute) (Mohammadi et al. 2020).

The extracted calcium has also been widely used in making hydroxyapatite (HA, $Ca_{10}(PO_4)_6(OH)_2$). One of the most common wet synthesis methods is the hydrothermal reaction. This process involves a chemical reaction between calcium (extracted from biowaste) and phosphate (phosphoric acid) at elevated pressure and temperature. The reactions involved are as shown below:

$$CaCO_3 \rightarrow CaO + CO_2$$

$$2CaO + H_3PO_4 \rightarrow CaHPO_4 + Ca(OH)_2$$

$$5CaHPO_4 + 5Ca(OH)_2 + H_3PO_4 \rightarrow Ca_{10}(PO_4)_6(OH)_2$$

15.7 Conclusion and Future Trends

Eggshells, cockleshells, animal bones and rice husks are often considered zero-value wastes and discarded, although they have potential multidisciplinary applications. Transforming these biowastes into functional materials is considered a sustainable approach to preserving the earth from environmental pollution and a smart way of managing waste. In biomaterial research, these biowastes can be easily converted into calcium and silica precursors, which could also reduce production costs. Instead of chemically derived pore-forming agents (PFA), sugarcane bagasse and coconut coir biowastes could be used as alternative materials. Apart from these biowastes, dental mould and marble sludge wastes generated from the medical and ceramic industries, which are calcium-rich wastes, could also potentially be the source of cheaper high-purity calcium precursors in bulk. "Reuse, recycle and repurpose of these waste materials" in developing functional products could help avoid dumping wastes and potentially reduce the impact of climate change.

Acknowledgements The authors sincerely acknowledge the Ministry of Higher Education Malaysia for Fundamental Research Grant Scheme with Project Code: FRGS/1/2021/TKO/USM/03/4 and Universiti Sains Malaysia for Research University Grant (RUI) with Project Code: 8014143.

Conflict of Interest The authors declare no competing interests.

References

Ardit M, Dondi M, Merlini M, Cruciani G (2012) Melilite-type and melilite- related compounds: structural variations along the join Sr2-xBaxMgSi2O7 (0≤x≤2) and high-pressure behaviour of the two end-members. Phys Chem Miner 39:199–211

Awonusi A, Morris MD, Tecklenburg MMJ (2007) Carbonate assignment and calibration in the Raman spectrum of apatite. Calcif Tissue Int 81:46–52

Balázsi C, Wéber F, Kövér Z, Horváth E, Németh C (2007) Preparation of calcium– phosphate bioceramics from natural resources. J Eur Ceram Soc 27:1601–1606

Bee S-L, Abdul Hamid ZA (2020) Hydroxyapatite derived from food industry bio- wastes: syntheses, properties and its potential multifunctional applications. Ceram Int 46:17149–17175

Bose S, Fielding G, Tarafder S, Bandyopadhyay A (2013) Understanding of dopant-induced osteogenesis and angiogenesis in calcium phosphate ceramics. Trends Biotechnol 31(10): 594–605

Caliman LB, Silva SND, Junkes JA, Sagrillo VPD (2017) Ostrich eggshell as an alternative source of calcium ions for biomaterials synthesis. Mater Res 20:413–417

Carvalho J, Araujo J, Castro F (2011) Alternative low-cost adsorbent for water and wastewater decontamination derived from eggshell waste: an overview. Waste Biomass Valoriz 2:157–167
Da Silva E (2016) Water remediation using calcium phosphate derived from marine residues. Water Air Soil Pollut 223(3):989–1003
Dupoirieux L, Pourquier D, Souyris F (1995) Powdered eggshell: a pilot study on a new bone substitute for use in maxillofacial surgery. J Cranio-Maxillofac Surg 23:187–194
Eliaz N, Metoki N (2017) Calcium phosphate bioceramics: a review of their history, structure, properties, coating technologies and biomedical applications. Materials 10:334
Ezekiel I, Shah Rizal K, Yanny Marliana BI, Ahmad-Fauzi MN (2018) Nanoemulsion synthesis of carbonated hydroxyapatite nanopowders: effect of variant CO3 2−/PO4 3− molar ratios on phase, morphology, and bioactivity. Ceram Int 44:13082–13089
Fei Q, Chang C, Mao D (2005) Luminescent properties of Sr2MgSi2O7 and Ca2MgSi2O7 long lasting phosphors activated by Eu^{2+}, Dy^{3+}. J Alloys Compound 390:133–137
Fopase R, Pandey LM (2022) Bioceramics for biomedical applications. In: Pandey LM, Hasan A (eds) Nanoscale engineering of biomaterials: properties and applications. Springer, Singapore
Heness G, Ben-Nissan B (2004) Innovative bioceramics. Mater Forum 27:104–114
Hesaraki S, Safari M, Shokrgozar MA (2009) Development of β-tricalcium phosphate/sol-gel derived bioactive glass composites: physical, mechanical, and *in vitro* biological evaluations. J Biomed Mater Res Part B 91:459–469
Jiang L, Chang C, Mao D (2003) Luminescent properties of CaMgSi2O6 and Ca2MgSi2O7 phosphors activated by Eu^{2+}, Dy^{3+} and Nd^{3+}. J Alloys Compound 360:193–197
Khandan A, Ozada N (2017) Bredigite-magnetite ($Ca_7MgSi_4O_{16}$-Fe_3O_4) nanoparticles: a study on their magnetic properties. J Alloys Compound 726:729–736
Laca A, Laca A, Díaz M (2017) Eggshell waste as a catalyst: a review. J Environ Manag 197:351–359
Landi E, Jacopo U, Sprio S, Tampieri A, Guizzzardi S (2010) Human osteoblast behavior on as-synthesized SiO(4) and B-CO(3) co-substituted apatite. J Biomed Mater Res A 94(1):59–70
LeGeros RZ (1994) Biological and synthetic apatites. In: Brown PW, Constantz B (eds) Hydroxy-apatite and related materials. CRC Press, Boca Raton, FL, pp 3–28
Liu Q, Cen L, Yin S, Chen L, Liu G, Chang J, Cui L (2008) A comparative study of proliferation and osteogenic differentiation of adipose-derived stem cells on akermanite and beta-TCP ceramics. Biomaterials 29:4792–4799
Mallick KK, Winnett J (2014) 3D bioceramic foams for bone tissue engineering. In: Mallick K (ed) Bone substitute biomaterials. Woodhead Publishing, New Delhi
Mittal A, Teotia M, Soni RK, Mittal J (2016) Applications of egg shell and eggshell membrane as adsorbents: a review. J Mol Liq 223:376–387
Mohammadi H, Boey ZH, Yanny Marliana BI, Khairul Anuar S, Ahmad-Fauzi MN (2020) Green synthesis of calcium magnesium silicate (CMS-Akermanite) using natural biowastes by solid-state sintering route. Malay J Microsc 16:66–76
Mohammadi H, Yanny Marliana BI, Khairul Anuar S, Ahmad-Fauzi MN (2021) Effect of strontium substitution on structural, sinterability, physicomechanical and biological properties of akermanite ceramic. J Mech Behav Biomed Mater 116:104379
Myat M-H, Ahmad-Fauzi MN, Masakazu K, Yanny Marliana BI (2020) Enhanced sinterability and in vitro bioactivity of barium-doped akermanite ceramic. Ceram Int 46:19062–19068
Myat M-H, Ahmad-Fauzi MN, Masakazu K, Yanny Marliana BI (2022) Tailoring mechanical and in vitro biological properties of calcium–silicate based bioceramic through iron doping in developing future material. Ceram Int 46:17149–17175
Nagrale SD, Hajare H, Modak PR (2012) Utilization of rice husk ash. Int J Eng Res Appl 2:1–5
Nur Liyana MR, Yanny Marliana BI (2022) Different concentrations of sodium hydroxide caustic leaching of Cr^{6+} for refinement of calcium source from dental mould waste. Malay J Microsc 18(1):46–56

Palakurthy S, Abdul Azeem P, Venugopal Reddy K, Vasudevarao P, Manavathi B (2020) A comparative study on in vitro behaviour of calcium silicate ceramics synthesized from biowaste resources. J Am Ceram Soc 103:933–943

Pina S, Reis RL, Oliveira JM (2018) Chapter 4 Ceramic biomaterials for tissue engineering. In: Thomas S, Balakrishnan P, Sreekala MS (eds) Fundamental biomaterials: ceramics, 1st edn. Woodhead Publishing, Cambridge, New Delhi, pp 95–116

Ramesh S, Aw KL, Tolouei R, Amiriyan M, Tan CY, Hamdi M, Purbolaksono J, Hassam MA, Teng WD (2013) Sintering properties of hydroxyapatite powders prepared using different methods. Ceram Int 39:111–119

Roy M, Bandyopadhyay A, Bose S (2017) Chapter 6 - Ceramics in bone grafts and coated implants. In: Bose S, Bandyopadhyay A (eds) Materials for bone disorders. Academic Press, London

Sangsawang M, Kaewdook D (2018) 5th International Conference on Business and Industrial Research (ICBIR). IEEE, Washington, DC, pp 230–235

Todkar BS, Deorukhkar OA, Deshmukh SM (2016) Extraction of silica from rice husk. Int J Eng Res Dev 12(3):69–74

Wu C, Chang J, Xiao Y (2013) Advanced bioactive inorganic materials for bone regeneration and drug delivery, 1st edn. Taylor & Francis, Boca Raton, FL

Yanny Marliana BI, Ian W, Oana B, Kenneth WD, Alicia JEH (2017) Development of multi-substituted hydroxyapatite nanopowders as biomedical materials for bone tissue engineering applications. J Biomed Mater Res A 105(6):1775–1785

Yanny Marliana BI, Muhammad Syazwan MN, Ahmad-Fauzi MN, Balestri W, Reinwald Y (2021) Influence of ternary divalent cations (Mg^{2+}, Co^{2+}, Sr^{2+}) substitution on the physicochemical, mechanical and biological properties of carbonated hydroxyapatite scaffolds. J Aust Ceram Soc 57:1499–1510

Zaheer K (2015) An updated review on chicken eggs: production, consumption, management aspects and nutritional benefits to human health. Food Nutr Sci 06:1208–1220

Zhang M, Liu C, Sun J, Zhang X (2011) Hydroxyapatite/diopside ceramic composites and their behaviour in simulated body fluid. Ceram Int 37:2025–2029

Hybrid Bioscaffolds Formation Using Natural and Synthetic Materials for Cartilage Tissue Engineering: The Case of Fibrin, Atelocollagen and Poly(Lactic-co-Glycolic Acid)

16

Munirah Sha'ban and Muhammad Aa'zamuddin Ahmad Radzi

Abstract

Approved by the US Food and Drug Administration (FDA), poly(lactic-co-glycolic acid) (PLGA) is used in many medical device productions. Degradation flexibility has made this synthetic copolymer convenient to use. Since the human body can metabolise lactic and glycolic acid monomers, minimal systemic toxicity has been reported associated with PLGA's acidic degradation. PLGA by-products reduce local pH low enough to create an autocatalytic environment. Nonetheless, fibrin and atelocollagen can be combined with PLGA to reduce inflammatory reactions caused by its acidic by-product. The former two scaffolds are naturally occurring biomaterials and are broadly used in biomedical and cartilage tissue engineering applications. Fibrin is a protein network produced following bodily injury. Atelocollagen is collagen without telopeptide regions. Like PLGA, these two natural materials can be engineered to scaffold tissue substitutes to aid biological healing with no significant side effects. This chapter discusses the research progress of selected natural and biomaterials, namely, fibrin, atelocollagen and PLGA copolymer, respectively, as potential scaffold materials for articular cartilage tissue engineering. Depending on research purposes, these materials can be used individually or in relevant combinations also known as hybrid bioscaffolds. The hybrid bioscaffolds based on natural and

M. Sha'ban (✉)
Department of Physical Rehabilitation Sciences, Kulliyyah of Allied Health Sciences, International Islamic University Malaysia (IIUM), Kuantan, Pahang, Malaysia
e-mail: munirahshaban@iium.edu.my

M. A. A. Radzi
Department of Obstetrics and Gynaecology, Sultan Ahmad Shah Medical Centre @IIUM, Kuantan, Pahang, Malaysia
e-mail: azam4z4m@gmail.my

W. S. Wan Kamarul Zaman, A. A. Nurul (eds.), *Sustainable Material for Biomedical Engineering Application*, https://doi.org/10.1007/978-981-99-2267-3_16

biocompatible synthetic materials may hold the key to sustainable materials for biomedical engineering applications.

Keywords

Hybrid bioscaffolds · Natural material · Synthetic material · Cartilage · Tissue engineering · Fibrin · Atelocollagen · PLGA

Abbreviations

3D	Three-dimensional
45SBG	45S Bioactive Glass/calcium sodium phosphosilicate
ACIC	Autologous collagen-induced chondrogenesis
AD-MSCs	Human adipose tissue-derived mesenchymal stem cells
Agili-C	Aragonite biphasic osteochondral scaffolds
AMIC	Autologous matrix-induced chondrogenesis
ASEAN	Association of Southeast Asian Nations
ATMPs	Advanced therapy medicinal products
BMAC	Bone marrow aspirate concentrate implantation
BMP-7	Bone morphogenetic protein 7
BMSCs	Bone marrow mesenchymal stem cells
$CaCl_2$	Calcium chloride
CAIS	Cartilage Autograft Implantation System
CartiMax	Viable cartilage allograft putty
CARTISEM	Human umbilical cord blood-derived mesenchymal stem cells
CGTPs	Cell and gene therapy products
COVID-19	Coronavirus Disease 2019
CVOCA	Cryopreserved viable osteochondral allograft
DMEM	Dulbecco's modified Eagle medium
ECM	Extracellular matrix
EMA	European Medicines Agency
FDA	Food and Drug Administration
hADSCs	Human adipose-derived stem cells
IIUM	International Islamic University Malaysia
KGN	Kartogenin
MACI	Matrix-induced autologous chondrocyte implantation
MCIC	Mesenchymal cell-induced chondrogenesis
MOH	Ministry of Health Malaysia
MPEG	Methoxy poly(ethylene glycol)
MSCs	Mesenchymal stem cells
MST	Bone marrow-stimulation techniques
NPCB	National Pharmaceutical Control Bureau
OAT	Osteochondral autograft transplant
OCA	Allograft transplant

PACI	Particulated autologous cartilage implantation
PEG	Poly(ethylene glycol)
pH	Potential hydrogen
PJAC	Particulated juvenile allograft cartilage
PLGA	Poly(lactic-co-glycolic acid)
PLL	Poly(L-lysine)
PLLA	Poly-L-lactic acid
PRP	Platelet-rich plasma
sGAG	Sulphated glycosaminoglycans
SOX-9	SRY-Box Transcription Factor 9
synMSCs	Synovium-resident mesenchymal stem cells
TEMPs	Tissue-engineered medical products
TERT	Telomerase reverse transcriptase
TGF-β1	Transforming growth factor beta 1
TWG	Technical Working Group
WOS	Web of Science
β-TCP	Tricalcium phosphate

16.1 Introduction

16.1.1 What Is Tissue Engineering?

With growing interest in precision medicine, tissue engineering approaches for regenerative medicine have become even more relevant in current healthcare settings post-COVID-19 pandemic. Also known as personalised medicine, precision medicine aims to customise health care with decisions and treatments tailored to each patient in every feasible way. The focus is on identifying practical approaches for individuals based on genetic, environmental and lifestyle factors. Tissue engineering is a multidisciplinary field often referred to as a subset of regenerative medicine. The field enables the regeneration of a patient's tissue (an autologous approach). It thus brings a glimpse of hope to lessen the dependency on conventional tissue/organ donation and transplantation and overcome many well-documented shortcomings following the procedures.

The tissue engineering term was officially coined at a National Science Foundation workshop in 1988 (Angel 2010). It represents a new scientific field focusing on functional tissue regeneration using cells, biomaterial scaffolds and growth factors. Langer and Vacanti spread tissue engineering awareness through their publication. They defined it as "an interdisciplinary field which *applies the principles of engineering and life sciences toward the development of biological substitutes that restore, maintain, or improve tissue function*" (Langer and Vacanti 1993). Regardless of current advances in the field and the fast pace of disruptive technologies, the fundamental guiding principles of tissue engineering are none-other than *cell sources* (committed and/or non-committed cells), *signalling factors* (chemical and/or mechanical) and *biomaterials scaffold* (natural and/or synthetics) (Sha'ban and Ahmad Radzi 2020; Sha'ban et al. 2014). These three working principles, typically known as the tissue engineering triad, can be studied and used individually or in combination to engineer tissues of interest.

Although initially intended for regenerating tissue-engineered medical products (TEMPs) for clinical applications, the multicellular laboratory-grown engineered tissues (also referred to as tissue-based products) have been used widely for non-clinical purposes. These include the in vitro model systems for human biology and medical research. This trend keeps growing and can be seen in the investigation into communicable diseases, especially during the recent COVID-19 pandemic (Kim et al. 2020). Besides being used to study cells' role and to know how the cells talk to each other (crosstalk) (Barbuzano 2017), cultured cells and three-dimensional (3D) engineered tissues or organoid technologies can be used to minimise the use of animal models in research (Cookson et al. 2022; Kawasaki et al. 2022). This step may help various, if not entirely, bypass animal ethical issues. Indeed, it is good news for people advocating for animal welfare or the animal rights movement.

16.1.2 Cartilage Tissue Engineering

Cartilage is among the most researched tissues following the tissue engineering field's inception in the 1980s (Sha'ban and Ahmad Radzi 2020). Cartilage research proponents targeted treatment development for focal cartilage lesions in joints (Abdul et al. 2015b; Munirah et al. 2007), osteoarthritis (Ude et al. 2018; Mohamad Sukri et al. 2022), intervertebral disc diseases (Mohamad et al. 2017, 2020), trachea defects (Dang et al. 2021), microtia (Yin et al. 2020), nasal reconstructions (Ren et al. 2021) and many more. Although doable, it is easier said than done. Laborious systematic research activities and highly skilled personnel are needed to materialise and achieve the targets. That includes isolating cartilage cells from viable tissue sources, growing the harvested cells in a proper microenvironment and engineering the desired 3D neocartilage tissue with or without biomaterial scaffold support. At the same time, sterility is maintained throughout the process by practising good aseptic techniques. Moreover, safety and efficacy aspects are necessary and must be carefully addressed when dealing with or introducing any new treatment modality leading to clinical applications (Md Nazir 2018). Ethical, legal and social implications shall be considered wisely and must not be ignored by those who aim to translate their research from bench to bedside (from lab-based work to clinical) (Radzi et al. 2020; Aa'zamuddin Ahmad Radzi et al. 2020; Syamimi et al. 2022).

After nearly four decades of earnest attempts, engineering functional articular cartilage tissue graft remains unresolved. The production of multizonal neocartilage that mimics the native tissue has not yet materialised fully. It is due to the knowledge gap and the lack of research into understanding specific zonal chondrocytes' attributes (Gomoll et al. 2014; Fu et al. 2020) and deciding on proper biomaterials. The intricacy of articular cartilage histoarchitecture may have been undermined throughout these years of research. It is so because cartilage is structurally avascular and has only one cell type, namely, chondrocyte (*chondro* = cartilage; *cyte* = cell). Moreover, these cells were found homogeneously distributed in immature cartilage, leading to the thought that the cells could be easily manipulated in vitro. Researchers may not realise a few details later that chondrocytes' arrangement or composition in the mature (adult) cartilage is somehow multilayered and complex compared to its immature counterpart (Fu et al. 2020; Gao et al. 2019). One practical idea is to mimic adult articular cartilage organisation by seeding zonal chondrocyte subpopulations in multilayer scaffolds. It leads to engineered neocartilage construct formation to enhance graft functionality (Fu et al. 2020; Yin et al. 2018). It needs concerted efforts from various perspectives to reach this goal.

Cartilage maintenance and biomechanics largely depend on its extracellular matrix. Although not sole, biomechanics plays an essential role in maintaining healthy articular cartilage (Eschweiler et al. 2021). Abnormal changes in cartilage biomechanics are also associated with the progression of osteoarthritis (Martínez-Moreno et al. 2019). Biomechanical behaviours of articular cartilage may be best understood when viewed as biphasic tissue comprising fluid and solid phases (Eschweiler et al. 2021). The interaction and relationship between proteoglycan and interstitial fluid within the extracellular matrix provide compressive resilience to articular cartilage through negative electrostatic repulsion forces (Sophia Fox et al. 2009). With multiple ongoing modelling being researched and developed, the complex structural organisation and cartilage biomechanics make cartilage repair using a tissue engineering approach challenging from a biomechanical point of view (Barbon et al. 2021).

As indicated earlier in this chapter, the tissue engineering triad, namely, cell sources, signalling factors and biomaterials scaffold, can be used independently or in a combination of two or three elements to form tissue aggregate or to engineer tissue. However, when it comes to weight-bearing tissues necessitating mechanical strength, like in the case of articular cartilage, many researchers will have to focus more on finding the best biomaterial scaffolds that serve as a reliable guiding matrix. A highly specialised connective tissue like cartilage has a high matrix-to-cell ratio, whereas its maintenance depends much on extracellular matrix homeostasis. Although good cell sources and signalling factors are essential equally to grow the cells and form a 3D tissue in vitro, the engineered neocartilage may not survive well in vivo without a proper biomaterial scaffold or extracellular matrix support.

16.1.3 Biomaterial Scaffold for Cartilage Tissue Engineering

Many biomaterials have been studied for cartilage tissue engineering applications (Sha'ban and Ahmad Radzi 2020; Yusof et al. 2021). The materials are constructed or engineered into bioscaffolds that help promote cell proliferation and provide cues for cell differentiation. Biomaterials can be either natural or synthetic. Depending on research objectives, the materials can be used individually or in a combination of natural–natural, natural–synthetic, synthetic–synthetic, or other feasible mixtures.

Natural materials are typically preferred because they have biochemical compositions comparable to that of tissue of interest. They are biocompatible and may cause a minimal immune reaction (if any) when used in an implant. However, natural materials are not easy to control, unlike synthetic materials, which can be tailored to fit into the desired shape and size of the intended scaffolds. In terms of processability and handling, synthetic materials could win over natural materials. However, biocompatibility and biodegradability concerns upon synthetic materials warrant thorough analyses and should be optimised for clinical use. In response to safety and efficacy issues, research into a decellularised extracellular matrix has emerged as an alternative biomaterial for tissue engineering. As another form of natural biomaterial, decellularised extracellular matrix minimises in vitro manipulation of the scaffold and preserves the native tissue environment suitable for cell growth (Yusof et al. 2019).

This chapter aims to summarise research findings on cartilage tissue engineering with a particular focus on the hybrid bioscaffold formation using selected biomaterials: fibrin and atelocollagen (natural materials) and PLGA copolymer (synthetic materials). However, the scope is confined only to the hyaline articular cartilage found mainly in the synovial joint. Unlike nasal septal cartilage, which is also hyaline, articular cartilage lines the end of the bones and holds a crucial weight-bearing function. Appears glassy (hyaline = *hýalos*), from which it has been characterised; articular cartilage looks simple macroscopically. However, when examined microscopically, articular cartilage exhibited multizonal layers: superficial, transitional, radial and calcified layers overlying the cancellous bone layer. Each zone contributes distinctively to cartilage biomechanics. Macromolecule roles, such as proteoglycans, and their interaction with interstitial fluids are crucial to understanding the joint's friction. Apart from being avascular and thus lacking natural healing properties, the attempt to repair and restore articular cartilage is understandably challenging simply by looking at its structural and functional aspects. This chapter also shows how much progress has been made in the field through updates in relevant clinical trials and applications.

The US FDA-approved PLGA is used in many medical device productions and nanomedicine (Rocha et al. 2022; Operti et al. 2021). Degradation flexibility has made this synthetic copolymer convenient to utilise. However, the PLGA by-product reduces local pH low enough to create an autocatalytic environment. Since the human body can metabolise lactic and glycolic acid monomers, minimal systemic toxicity has been reported to be associated with acidic degradation (Xu et al. 2021; Lim et al. 2022). A few efforts have been made to overcome this issue. For instance, fibrin and atelocollagen can be combined with PLGA to reduce inflammatory reactions caused by its acidic by-products. The former two scaffolds are naturally occurring biomaterials used in biomedical and cartilage tissue engineering applications. Like PLGA, these two natural materials can be engineered individually or in combination to scaffold tissue substitutes aiding natural or biological healing with no significant side effects.

Fibrin is a tough insoluble protein used extensively in orthopaedic procedures (Hinckel et al. 2021) and as cell delivery vehicles (Mohamad Sukri et al. 2022; Ali et al. 2019; Md Ali Tahir et al. 2019). It forms a major component in blood coagulation following injury or bleeding. Fibrin formation is preceded by fibrinogen found in blood plasma. Unlike fibrin, fibrinogen is a soluble protein produced by the liver. Fibrinogen is transformed into fibrin at the wound by a clotting enzyme thrombin. Fibrin is arranged in long fibrous chains. Fibrin molecules combine to form long fibrin threads, entrap thrombocytes or platelet cells and build up a spongy substance that gradually coagulates to form a blood clot to stop or prevent bleeding. This hardening process is stabilised by a fibrin-stabilising factor known as factor XIII (Britannica The Editors of Encyclopaedia 2018).

Atelocollagen is soluble collagen produced from tropocollagen, a collagen molecule that makes collagen fibrils. Atelocollagen, a firm gel-like material, is produced by eliminating or removing N- and C-terminal telopeptide regions (Gomoll et al. 2014). The telopeptide regions represent most of the collagen's antigenicity (Kanamoto et al. 2021). Hence, atelocollagen remains a type of collagen with no telopeptide regions. It has low antigenicity and is thus biocompatible and suitable for clinical use. It is widely used for medical implants, surgical products and as a drug delivery carrier (Matsubara et al. 2021; Gilson et al. 2017). Atelocollagen is considered a safe biomaterial due to the little immunogenicity it has. Figure 16.1 summarises the usage of PLGA, atelocollagen and fibrin for cartilage tissue engineering.

16.2 Methodology

16.2.1 To Summarise Cartilage Tissue Engineering Research Using Fibrin, Atelocollagen and PLGA

Literature was identified from the Scopus database on 9 August 2022. The searched keywords are "cartilage", "tissue engineering", "fibrin", "PLGA" and "atelocollagen", in different combinations using the relevant Boolean operators. No year limit was imposed. Fifty-two articles obtained from each search set were screened and checked for any duplication. Articles relevant to the objective and research question were included and summarised in a table.

16.2.2 To Summarise Clinical Trials and Applications Using Fibrin, Atelocollagen and PLGA For Cartilage Repair

All relevant studies were identified from the ClinicalTrials.gov database on 25 August 2022. The searched keywords are "cartilage" (based on condition or disease), "fibrin", "PLGA" and "atelocollagen" as individual keywords and in different combinations, e.g., cartilage–fibrin, cartilage–PLGA and cartilage–atelocollagen. Studies relevant to the objective and research question were included and summarised in a table.

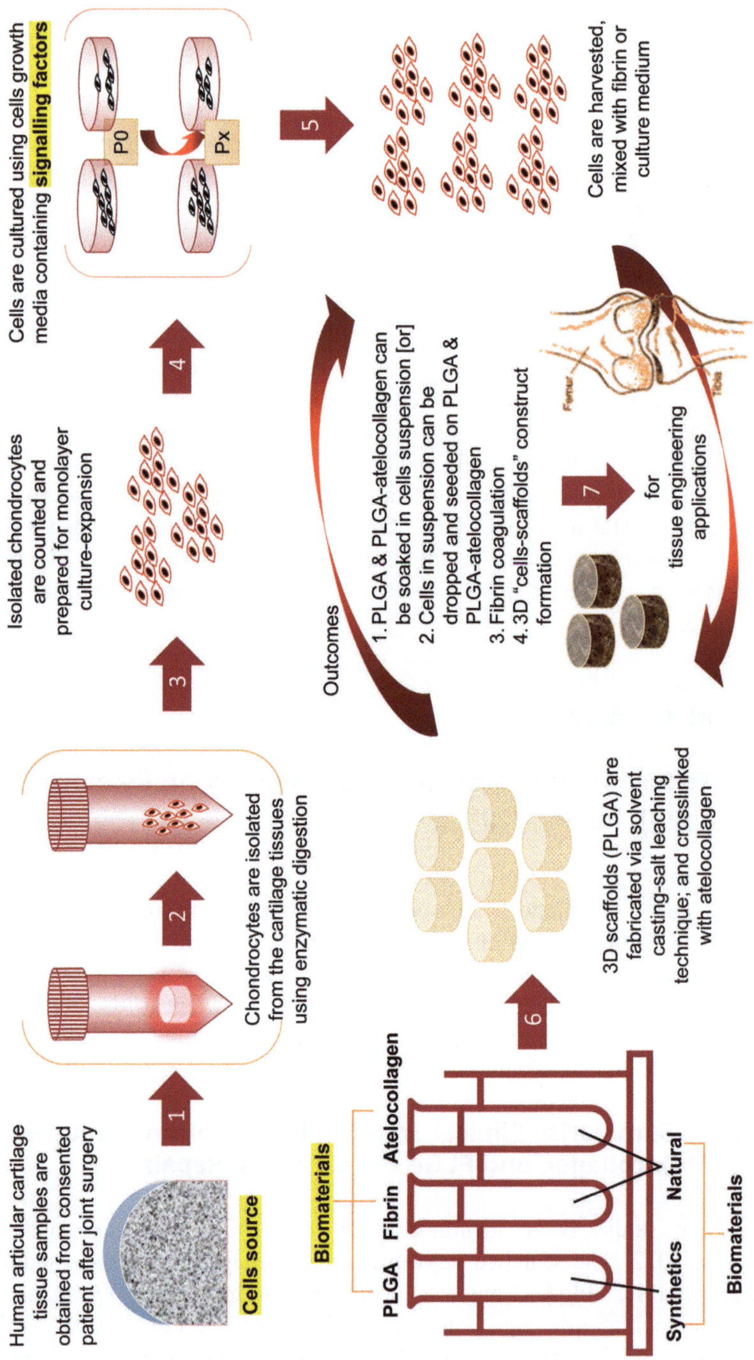

Fig. 16.1 The use of synthetic biomaterials, PLGA, and natural biomaterials atelocollagen and fibrin for cartilage tissue engineering. Besides biomaterials, the success of 3D "cells-scaffolds" construct formation depends on quality cell source and suitable signalling factors. The combination

16.3 Results

16.3.1 Updates on Cartilage Tissue Engineering Research Using Fibrin, Atelocollagen and PLGA

After duplicate removal, the search yielded 45 articles related to the keywords. However, 23 articles were excluded. There are ten review papers, one published in a non-English language and 14 articles related to other than the "articular cartilage" (i.e., the intervertebral disc, tracheal cartilage, meniscal, costal cartilage, tissue adhesives and general cartilage engineering applications). The remaining 21 original articles (2022–2008) related to the "articular cartilage" were included. Nonetheless, other identified articles were also considered to enrich the chapter. Table 16.1 summarises the 21 original articles presenting articular cartilage formation via tissue engineering technique using PLGA, fibrin and atelocollagen scaffolds. These scaffolds were seeded with various cell sources, including committed (primary chondrocytes) and non-committed cells (bone marrow- and adipose-derived stem cells). Most articles favour using PLGA-based scaffolds to regenerate cartilage, except one paper prefers biodegradable poly-L-lactic acid (PLLA) over PLGA (Asawa et al. 2012) due to PLGA's fast degradation rate. Based on the articles, the trend indicated that PLGA is used as the primary scaffold supported by fibrin and atelocollagen. Both materials function as cell delivery vehicles in many cases and are used to immobilise cells within the PLGA, thus strengthening the whole scaffold composites. Fibrin and atelocollagen are incorporated into PLGA because they are assumed to minimise the systemic effects caused by the acidic by-products following PLGA degradation. Another interesting observation is that fibrin and atelocollagen have been indicated to help slow-release activity if encapsulated by relevant cartilage-related growth factors. Relevant summaries or findings in each article are recorded in Table 16.1.

16.3.2 Updates on Clinical Trials and Applications Using Fibrin, Atelocollagen and PLGA for Cartilage Repair

A direct search using ClinicalTrials.gov data indicated that 461 clinical trials involving "cartilage", 37 "PLGA", 3998 "fibrin" and six "atelocollagen" were registered at ClinicalTrials.gov. Upon further inspection, no study was found addressing articular cartilage repair using PLGA and atelocollagen. Although unrelated to articular cartilage repair, two trials involving PLGA are linked to knee and spinal joint interventions (Table 16.2).

Fig. 16.1 (continued) of the three elements or working principles (namely biomaterials, cell source and signalling factors) is known as the tissue engineering triad

Table 16.1 Summary of research articles on articular cartilage tissue engineering using PLGA, atelocollagen and fibrin scaffolds

#	Scaffold(s)	Fabrication method(s)	Cells source(s)	Findings	References
1	PLGA, PLGA/fibrin	3D PLGA disc-shaped scaffold was constructed using solvent casting/salt leaching techniques, freeze-dried, and soaked in the fibrin to make a hybrid scaffold. The fibrin was coagulated using $CaCl_2$ (calcium chloride).	Articular cartilage cells from osteoarthritic patients transfected with SOX-9	PLGA/fibrin served as a better scaffold for articular cartilage formation using human chondrocytes than the PLGA alone.	Mohamad Sukri et al. (2022)
2	Atelocollagen gel, fibrin glue	Atelocollagen gel and fibrin glue were ready-to-use items, commercially available.	Articular cartilage cells from osteoarthritic patients	Using minced cartilage embedded in a 3-dimensional culture scaffold and fixed with fibrin glue is better than using isolated chondrocytes in atelocollagen to induce cell migration, proliferation, and matrix production cartilage from patients with knee joint osteoarthritis.	Tsuyuguchi et al. (2021)
3	PLGA/fibrin, Fibrin	PLGA scaffold was developed using the solvent casting/salt leaching technique. The hybrid PLGA/fibrin scaffold was made by incorporating fibrin.	Human adipose-derived stem cells (hADSCs)	Incorporating PLGA with fibrin enhances articular cartilage regeneration of hADSCs effectively in the presence of icariin.	Batool et al. (2020)
4	Fibrous PLGA (BMP-7-loaded), Fibrin	Fibrous PLGA scaffolds were prepared by wet spinning. Fibrin was a ready-to-use item, commercially available.	Synovium-resident mesenchymal stem cells (synMSCs)	Intra-articular co-delivery of synMSCs with BMP-7 via fibrous PLGA scaffolds may be a promising therapeutic modality for articular cartilage repair.	Kim et al. (2019)

5	PLGA, PLGA/fibrin	3D PLGA disc-shaped scaffold was constructed using solvent casting/salt leaching techniques, freeze-dried, and soaked in the fibrin to make a hybrid scaffold. The fibrin was coagulated using $CaCl_2$.	Articular cartilage cells from rabbits transfected with SOX9, SOX-9/TERT and TERT	Incorporation of SOX9/TERT-transfected chondrocytes into 3D PLGA-based scaffolds facilitates cartilage regeneration.	Md Nazir et al. (2019)
6	PLGA, PLGA/fibrin, PLGA/atelocollagen, PLGA/fibrin/atelocollagen	3D PLGA disc-shaped scaffold was developed using solvent casting/salt leaching, freeze-dried, and soaked in fibrin to make a hybrid scaffold. Atelocollagen was crosslinked with PLGA and mixed with fibrin to make a hybrid scaffold. The fibrin was coagulated using $CaCl_2$.	Articular cartilage cells from rabbits	PLGA/fibrin/atelocollagen served as a better scaffold for articular cartilage formation than the other scaffold groups.	Ali et al. (2019)
7	PLGA, PLGA/fibrin, PLGA/atelocollagen, PLGA/fibrin/atelocollagen	3D PLGA disc-shaped scaffold was constructed using solvent casting/salt leaching, freeze-dried, and soaked in fibrin to make a hybrid scaffold. Atelocollagen was crosslinked with PLGA and mixed with fibrin to make a hybrid scaffold. The fibrin was coagulated using $CaCl_2$.	Articular cartilage cells from rabbits	All scaffolds except PLGA/atelocollagen indicated positive cytocompatibility. PLGA is hydrophobic but has the highest swelling ratio compared to other groups though the difference is insignificant. PLGA/fibrin + PLGA/fibrin/atelocollagen groups were a better microenvironment to facilitate cellular growth.	Bin Mohamed Amin et al. (2018)

(continued)

Table 16.1 (continued)

#	Scaffold(s)	Fabrication method(s)	Cells source(s)	Findings	References
8	Poly (lactic-co-glycolic acid) (PLGA)/Hyaluronic Acid (Ha)/ Fibrin/45S Bioactive Glass (45SBG) nanocomposite	PLGA scaffold was formed using solvent casting/salt leaching. 45SBG nanoparticles were synthesised via the sol-gel technique. PLGA scaffolds were coated with suspensions of Ha/Fibrin/45SBG containing different amounts of the components.	Human adipose-derived mesenchymal stem cells (hADMSCs)	Increase bioactivity in nano-composite scaffolds, as compared to the pure PLGA scaffold. Adding Ha, fibrin, and 45SBG nanoparticles could modify the degradation rate of PLGA. The nano-composite scaffolds did not exhibit cytotoxicity, and the hADMSCs attached and proliferated properly in the scaffolds.	Mehdikhani-Nahrkhalaji et al. (2017)
9	Fibrin gel- and PLGA-poly (ethylene glycol) (PEG)-PLGA (PLGA-PEG-PLGA) hydrogel	A triblock PLGA-PEG-PLGA copolymer hydrogel was fabricated using a ring-opening polymerisation of lactic and glycolic acids. PEG acts as a macroinitiator. Stannous octoate was used as the catalyst.	Bone marrow mesenchymal stem cells (BMSCs)	The PLGA-PEG-PLGA group had better therapeutic efficacy than the fibrin gel group.	Liu et al. (2016)
10	PLGA-PEG-PLGA thermogel	PLGA-PEG-PLGA thermogel was fabricated through the ring-opening polymerisation of lactic and glycolic acids. PEG acts as a macroinitiator. Stannous octoate was used as the catalyst.	BMSCs	A thermogel PLGA-PEG-PLGA + Kartogenin (KGN, chondroinductive non-protein small molecule) make an appropriate microenvironment for BMSCs for effective cartilage regeneration.	Li et al. (2016)

11	PLGA, PLGA/fibrin	3D PLGA disc-shaped scaffold was constructed using solvent casting/salt leaching, freeze-dried, and soaked in fibrin to make a hybrid scaffold. The fibrin was coagulated using $CaCl_2$	BMSCs from rabbits	PLGA/fibrin constructs exhibited better reparation, more cartilaginous matrix, higher sGAG production, better compressive strength, and expression of chondrogenic markers than the PLGA-only group.	Abdul Rahman et al. (2015a)
12	PLGA, PLGA/fibrin	3D PLGA disc-shaped scaffold was constructed using solvent casting/salt leaching, freeze-dried, and soaked in fibrin to make a hybrid scaffold. The fibrin was coagulated using $CaCl_2$	Articular cartilage cells from osteoarthritic patients transfected with SOX-9	All PLGA/fibrin constructs showed higher cell viability, higher sGAG content, better histological features, and distribution of extracellular matrix with positive glycosaminoglycan (GAG) accumulation than the PLGA groups. The use of transfected cells was not necessarily better than the non-transfected cells.	Mohamad Sukri et al. (2015)
13	PLGA, PLGA/fibrin	3D PLGA disc-shaped scaffold was constructed using solvent casting/salt leaching, freeze-dried, and soaked in fibrin to make a hybrid scaffold. The fibrin was coagulated using $CaCl_2$	BMSCs from rabbits	PLGA/fibrin served as a better scaffold for articular cartilage formation using BMSCs than the PLGA alone.	Abdul et al. (2015b)
14	Fibrin glue, PLGA, alginate	PLGA scaffold was formed using solvent casting/salt leaching techniques.	Human adipose tissue-derived mesenchymal stem cells (AD-MSCs)	The proliferation of AD-MSCs differentiated-chondrogenic cells, expression of chondrogenic genes and structural appearance of constructs in fibrin glue were higher significantly than in other scaffold groups. Fibrin glue scaffolds were filled with	Sheykhhasan et al. (2015)

(continued)

Table 16.1 (continued)

#	Scaffold(s)	Fabrication method(s)	Cells source(s)	Findings	References
				uniformly distributed cells suggesting it as an appropriate environment for cartilage differentiation.	
15	PLGA, PLGA/fibrin gel	PLGA sponges were fabricated by a gelatine porogen-leaching method. The PLGA was incorporated with fibrin to make a hybrid scaffold. The fibrin was solidified using a thrombin solution.	Mesenchymal stem cells (MSCs) loaded lipofectamine/pDNA-TGF-β1 complexes	PLGA/fibrin group exhibited better cartilage tissue formation. The defects implanted with the PLGA/fibrin constructs demonstrated significantly higher TGF-β1 expression than those of the fibrin-free group.	Li et al. (2014)
16	Atelocollagen, biodegradable poly-L-lactic acid (PLLA) or PLGA polymer	Atelocollagen was incorporated with PLLA or PLGA. PLGA was prepared using solvent casting and particulate (sucrose) leaching methods.	Canine autologous chondrocytes	Tissue-engineered constructs using PLLA contained abundant cartilage compared to the PLGA constructs. The presence of proteoglycan and type II collagen also favour PLLA over PLGA. The PLGA biodegradation progressed much faster than PLLA. The PLGA degraded by-products may evoke a more severe tissue reaction at an early stage of transplantation than PLLA.	Asawa et al. (2012)
17	PLGA sponge, fibrin gel	PLGA sponges were produced by a gelatine porogen-leaching method. The fibrin was incorporated into PLGA to make a hybrid scaffold—fibrin was solidified using a thrombin solution.	Rabbits' BMSCs, transfected with transforming growth factor-β1 (pDNA-TGF-β1)	A composite construct comprising BMSCs, plasmid DNA encoding TGF-β1, fibrin gel and PLGA sponge was successfully designed. It can effectively restore cartilage defects when used to repair articular cartilage defects in New Zealand white rabbit joints.	Wang et al. (2010a)

18	PLGA sponge, fibrin gel	PLGA sponges were fabricated by a gelatine porogen-leaching method. The PLGA was incorporated with fibrin to make a hybrid scaffold. The fibrin was solidified using a thrombin solution.	Rabbits' BMSCs	BMSCs and recombinant protein TGF-β1 are encapsulated into the fibrin gel and then filled into the PLGA sponges to obtain a composite construct for articular cartilage repair in vivo. The defects implanted with the PLGA/fibrin constructs were packed with cartilage-like tissue expressing collagen type II and glycosaminoglycans (GAGs). The PLGA group exhibited fibrous-like tissues.	Wang et al. (2010b)
19	Fibrin hydrogel, methoxy poly (ethylene glycol) (MPEG)-PLGA/fibrin	MPEG-PLGA was dissolved in 1,4 dioxane at 50 °C, poured into a precooled aluminium and freeze-dried. Fibrinogen solution was dropped onto MPEG-PLGA. The fibrin was solidified using a thrombin/$CaCl_2$ solution.	Autologous chondrocytes from goats	The MPEG-PLGA/chondrocyte constructs exhibited positive results based on the macroscopic surface, histological scores, and defect filling. Fibrin was used to immobilise cells in the MPEG-PLGA scaffold. Although there was no difference between treatment groups mechanically, the MPEG-PLGA/chondrocyte constructs showed significantly better cartilage repair response than the osteochondral drilling fibrin hydrogel with chondrocytes and empty defects.	Lind et al. (2008)

(continued)

Table 16.1 (continued)

#	Scaffold(s)	Fabrication method(s)	Cells source(s)	Findings	References
20	PLGA, PLGA/fibrin	3D PLGA disc-shaped scaffold was constructed using solvent casting/salt leaching, freeze-dried, and soaked in fibrin to make a hybrid scaffold. Fibrin was a ready-to-use item, commercially available.	Rabbit articular chondrocytes	Fibrin/PLGA hybrid scaffold promotes early in vitro chondrogenesis of rabbit articular chondrocytes. It is confirmed by the presence of chondrocytes in lacunae embedded within the basophilic extracellular matrix. Cartilaginous tissue formation was observed as early as 2-week in the fibrin/PLGA construct.	Sha'ban et al. (2008)
21	PLGA, PLGA/fibrin	3D PLGA disc-shaped scaffold was constructed using solvent casting/salt leaching, freeze-dried, and soaked in fibrin to make a hybrid scaffold. Fibrin was a ready-to-use item, commercially available.	Rabbit articular chondrocytes	In vivo cartilaginous tissue formation in the fibrin/PLGA hybrid construct was confirmed by the presence of chondrocytes housed in lacunae and embedded within the basophilic ground substance, expressing cartilage-specific markers and sGAG production better than in the PLGA group.	Munirah et al. (2008)

Table 16.2 Investigational products or procedures using PLGA in ClinicalTrials.gov

#	Products/procedures	Descriptions (Taken from the relevant web page with minor adjustments)	ClinicalTrials.gov Identifier
1.	The Biosure Regenesorb Interference Screw	An absorbable screw with an open structure design: It consists of biocomposite materials PLGA, β-TCP (tricalcium phosphate) and calcium sulphate. The open structure design enables rapid bone ingrowth. It integrates the screw with adjacent bone tissues to promote healing. In knee surgery, the screw is used to fix ligaments, tendons, soft tissues, or bone-tendon-bone grafts.	NCT04012567
2.	Neuro-Spinal Scaffold™	A porous bioresorbable polymer scaffold: The scaffold is cylindrical and is available in three lengths (Dimension: 3 mm diameter × 6/8/10 mm). It consists of synthetic biomaterials, PLGA-b-poly(L-lysine) (PLGA-PLL), designed for optimal fit in the intraspinal lesion cavity. The scaffold is expected to be resorbed from 4 to 8 weeks post-implantation (using animal studies).	NCT03762655

Next, only 25 interventional studies related to fibrin involved cartilage treatment as the primary purpose, with two trials terminated. On further inspection, three trials are not related to articular cartilage. Accordingly, investigational products or procedures using fibrin are summarised in Table 16.3.

Nonetheless, other identified articles were also considered to enrich the chapter.

16.4 Discussion

16.4.1 Progress In Articular Cartilage Tissue Engineering from Bench to Bedside

As of 2019, the US FDA has approved about 20 PLGA-based formulations through new drug applications, and none of them has generic counterparts on the market yet. The lack of generic PLGA products is partly due to difficulties in reverse engineering (Skidmore et al. 2019). However, the potential of PLGA has been well documented in many online research articles, as presented in the results section. Its disadvantages can be patched up by incorporating many other materials, making it a tunable material for tissue engineering and drug release (Rocha et al. 2022; Gilson et al. 2017). Based on the retrieved literature, the combination of PLGA and fibrin studies is more than the atelocollagen or any combination for articular cartilage tissue engineering. It is assumed that there might be instances where atelocollagen is written as "collagen" only in the literature. It is thus not captured during the search since the keyword used in the literature search is specific to "atelocollagen". In this case, the details were only realised when the authors extended the review to the

Table 16.3 Investigational products or procedures using fibrin in ClinicalTrials.gov

#	Products/procedures	Descriptions (Taken from the relevant web page with minor adjustments)	ClinicalTrials.gov Identifier
1	DeNovo NT and Cartilage Autograft Implantation System (CAIS)	Use morselised autologous hyaline cartilage harvested arthroscopically, affixed onto a synthetic, resorbable implant (fibrin sealant), and implanted in a single surgical procedure designed to address damaged knee cartilage. The system uses the subject's healthy cartilage from a non- or low-weight-bearing region.	NCT01498029 NCT01400607 (Terminated, poor enrolment)
2	Autologous collagen-induced chondrogenesis (ACIC) versus mesenchymal cell-induced chondrogenesis (MCIC)	ACIC and MCIC are single-stage arthroscopic procedures done as a day case procedure. In ACIC, the cartilage defect is debrided and implanted with collagen and fibrin gel mixture under CO_2 insufflation. In MCIC, the cells are harvested intraoperatively, concentrated, mixed with a fibrin gel, and implanted under CO_2 insufflation.	NCT01984450
3	NuTech Affinity™ Membrane	Sharp dissection of the defect will be performed carefully so as not to violate the subchondral bone sparing the calcified cartilage layer. After haemostasis is reached, the defect will be treated with an Affinity™ patch stabilised with fibrin glue.	NCT02837484
4	CHONDRON (Autologous Cultured Chondrocyte)	In one vial (0.4 mL): The main ingredient is ≥1.2 million autologous chondrocytes with an adequate amount of Dulbecco's modified Eagle medium (DMEM). The cells are sufficiently suspended and grafted in the defect with fibrin glue.	NCT01050816 NCT02537067 NCT04236739 NCT02539056 NCT02539069
5	HYALOFAST®	The biodegradable HA-based scaffold (Hyalofast®; Anika Therapeutics, Bedford, MA, USA) will be prepared by cutting it to fit the focal defect. It will be implanted into the defect in a press-fit fashion to the surrounding cartilage. If additional fixation of the scaffold is needed, fibrin glue will be used to secure the scaffold.	NCT04875767

(continued)

Table 16.3 (continued)

#	Products/procedures	Descriptions (Taken from the relevant web page with minor adjustments)	ClinicalTrials.gov Identifier
6	Chondro-Gide®	Debridement of cartilage is followed by placement of small holes in the affected area to fill the debrided area with marrow and cells, and next affixing of Chondro-Gide® with fibrin glue.	NCT04537013 NCT02993510 NCT01458782
7	CartiLife®	An extracellular matrix-associated autologous chondrocyte composed of pellets (1.1–1.8 mm in diameter) in suspension. One pre-filled syringe is implanted per 1 cm^3 of defect volume, and fibrin adhesive is applied to fix pellets in place through minimal arthrotomy.	NCT03517046 NCT03545269 NCT04744402 NCT05051332
8	"Microfracture versus ADSCs for the treatment of articular cartilage defects"	Patients will undergo arthroscopic resection of the infrapatellar fat pad (the source of ADSCs), preparation of the chondral defect and calcified cartilage layer removal. Fibrin glue (Tisseel) will be set at the base of the defect, followed by the acellular collagen dermal matrix, ADSCs, and an additional layer of fibrin glue.	NCT02090140
9	"Autologous cartilage implantation vs. arthroscopic debridement"	Diagnostic arthroscopy of the knee joint: For chondrocyte implantation: A mini open arthrotomy is performed. The lesion is curetted down to the subchondral bone (but not through it; to avoid bleeding). A sterile aluminium foil is used following lesion measurement to cut out a matching piece of collagen sheet to contain the cells in the defect. The collagen sheet/flap is sutured to the lesion. It is sealed with fibrin glue leaving a small opening at the upper part for cell injection purposes. The last opening of the flap is then closed with stitch and fibrin glue.	NCT02636881
10	"All autologous cartilage regeneration in the treatment of the knee cartilage defects"	The harvested, healthy cartilage is shredded directly in situ. It is mixed with the patient's platelet concentrate and may be injected directly into the cartilage defect. The platelet concentrate is used to produce autologous fibrin glue to fix the cartilage chip/platelet concentrate applied to the damaged cartilage.	NCT04785092

(continued)

Table 16.3 (continued)

#	Products/procedures	Descriptions (Taken from the relevant web page with minor adjustments)	ClinicalTrials.gov Identifier
11	"The use of autologous BMSCs in the treatment of articular cartilage defects"	The culture-expanded MSC pellet is implanted into the osteochondral knee defect. It is performed using open surgery or arthroscopically under sterile aseptic conditions and proper anaesthesia. The defect is covered with an autologous periosteal flap from the proximal tibia. Human fibrin glue seals the defect and retains the cell pellet.	NCT00891501
12	"Cartilage tissue engineering"	Extracellular matrix (ECM) is thought to influence the differentiation process of MSCs significantly. The researchers hypothesised that the osteoarthritic cartilage matrix could promote the chondrogenesis of MSCs. It was a favourable method without exogenous growth factors induction.	NCT01301664

materials product sheet, which is beyond this chapter's scope. This aspect of observing the terminology detail is not considered as it requires another in-depth study.

There are many fabrication methods ranging from simple to complex to construct polymeric biomaterial scaffolds in tissue engineering. However, solvent casting and particulate leaching were the most common methods employed to construct PLGA scaffolds in the selected studies (Table 16.1). It may be due to the simplicity of the procedure since it does not need sophisticated devices or tools in the setup. Based on the selected studies, the PLGA is dissolved in an organic solvent, usually dichloromethane, resulting in a homogenous solution. Specific sizes of particles, either salts or sucrose, are then added to the resulting polymeric solution. The mixture is cast into its intended shape using either a glass or plastic plate to produce a membrane scaffold or a 3D silicone mould in specific dimensions to produce a 3D scaffold. The mixture is left to dry and compressed to preserve its shape. A solid composite material comprising particles and the PLGA copolymer is created when the solvent evaporates. This composite material is then placed in distilled water to dissolve the particles, leaving behind a porous PLGA sponge-like scaffold. The scaffold will be freeze-dried and sterilised upon use. Unlike simple PLGA sponges, fibrous PLGA scaffold (polymer fibre) formation used a wet spinning setup (Kim et al. 2019). The wet spinning method is more complex than solvent casting and particulate leaching. The method setup involved a syringe pump and turntable responsible for the fixation of the flow rate of PLGA/solvent solution in a vertical plane and the rotation of the coagulation bath in the horizontal plane, respectively (Abay Akar et al. 2019). Although wet-spinning and electrospinning methods are complicated, they have been recognised as the most efficient techniques for producing polymeric fibrous constructs for various biomedical applications, electronics and the food industry. They have unique features and tuneable architectures that help construct materials resembling desirable biological features (Miranda et al. 2022).

Not much information describes techniques to prepare atelocollagen and fibrin scaffolds in the selected articles. It may be because if they are commercially available, the only step required is to dissolve the atelocollagen and fibrin powders in a miscible solvent, and they are ready to be used. However, in the case of fibrin prepared by researchers, the starting material is the fresh whole blood that would undergo centrifugation to separate red-blood cells and plasma-derived fibrinogen (Mohamad Sukri et al. 2022; Md Nazir et al. 2019, 2020). Fibrin (clot) is acquired through fibrinogen coagulation by dropping thrombin-$CaCl_2$ solution (Batool et al. 2020; Mehdikhani-Nahrkhalaji et al. 2017), while atelocollagen gels after incubation at 37 °C (Gantenbein-Ritter and Sakai 2011). Atelocollagen is soluble below 10 °C but solidifies above this temperature requiring careful handling at room temperature (Shim et al. 2013). It was also reported that atelocollagen is in liquid form at 41 °C (Gantenbein-Ritter and Sakai 2011). Although atelocollagen and fibrin fabrication is not as systematic as PLGA, these two natural biomaterials have shown their potential in promoting cartilage formation. Observations in selected studies indicated that PLGA alone seeded with cells could form cartilage. However, it was evident that PLGA hybrid scaffolds performed better in promoting articular cartilage formation. PLGA's potential or capacity to facilitate cartilage formation was adequate without incorporating other biomaterials. However, as mentioned earlier in this chapter, adding other biomaterials to the PLGA will help reduce inflammatory reactions caused by its acidic by-products. Other than this reason, like in the case of fibrin, its addition to PLGA promoted early cartilage formation (Sha'ban et al. 2008).

Out of many registered trials involving "cartilage" in ClinicalTrials.gov, no specific study addressed articular cartilage repair involving PLGA and atelocollagen. However, in a different random search, there is one PLGA-based microparticle product called Zilretta®. It is developed to treat osteoarthritis pain by intra-articular injection (Rocha et al. 2022; Lim et al. 2022). This FDA-approved Zilretta® is a 75: 25 PLGA microsphere formulation of extended-release injectable corticosteroid suspension. A single injection of extended-release Zilretta® could significantly reduce knee pain for entire 3 months. The use of fibrin seems more common in clinical trials for cartilage treatment than PLGA and atelocollagen. The trend, from bench to bedside, showed that fibrin was used either as a temporary matrix or cell delivery vehicle or as a glue/sealant to tissue flap covering, collagen membrane or patch sutured to the cartilage defect implanted with various cell sources. Fibrin usage, in most cases, is depicted as an auxiliary to stabilise or secure the implant at the treated area, not the primary material. Fibrin is too soft for an implant that requires immediate mechanical strength, like in the weight-bearing area of the joint. It can be pressed-fit into the defect area but unable to hold itself independently

(Munirah et al. 2007). For atelocollagen, as indicated above, there may be instances where atelocollagen is written as "collagen" only; wherefore, atelocollagen-related studies (if any) were not captured during the search.

In "*Technical Enhancements and Update on Chondrocyte Implantation*" (Gomoll et al. 2014), the authors provided a comprehensive overview of technical enhancements in autologous chondrocyte implantation. They include collagen patches, second-generation cell growth on scaffolds, cultured chondrocytes embedded in atelocollagen gel and fourth-generation arthroscopic chondrocyte implantation in a 3D fibrin gel from bench to bedside. Recently, a team of orthopaedic surgeons created an algorithm to treat focal grade 3 or 4 knee cartilage defects using classic and novel cartilage restoration techniques (Hinckel et al. 2021) with and without sealant or fibrin glue. The classic and novel procedures covered by the authors are listed below. The details of each procedure with supporting evidence can be found in the said manuscript. A few overlapping investigational products or procedures can be seen later in this section, stemming from this list. The classic procedures include microfracture or bone marrow-stimulation techniques (MST), matrix-induced autologous chondrocyte implantation (MACI), osteochondral autograft transplant (OAT) and allograft transplant (OCA). In the meantime, the novel procedures include

- AMIC: "autologous matrix-induced chondrogenesis".
- Agili-C: "aragonite biphasic osteochondral scaffolds".
- BMAC implantation: "bone marrow aspirate concentrate".
- CVOCA: "cryopreserved viable osteochondral allograft" (by Cartiform and Prochondrix).
- CARTISEM: "human umbilical cord blood-derived mesenchymal stem cells".
- Biocartilage: "micronised cartilage extracellular matrix".
- PJAC: "particulated juvenile allograft cartilage" (by DeNovo NT).
- PACI: "particulated autologous cartilage implantation".
- CartiMax: "viable cartilage allograft putty".

Although a rapid increase in registered studies using various biologics for cartilage repair and osteoarthritis treatment has been seen recently, most still require late-stage clinical trials to substantiate the effectiveness. Based on ClinicalTrials.gov data, a recent 2022 article reported that there were 365 registered clinical trials using biologics for cartilage repair and osteoarthritis treatment from 2000 to 2022 (Negoro et al. 2018). As the most investigated biologics (38%), mesenchymal stem (or stromal) cells (MSCs), were mainly derived from bone marrow and adipose tissue (70%). A nearly equal number of studies in the early- and late-stage clinical trials involved chondrocytes and platelet-rich plasma (PRP). Some findings on Phase 3 clinical trials investigating the former two items, BMAC, hyaluronic acid, collagen membrane and albumin, have been published. Negoro et al. (2018) indicated that selected studies in ClinicalTrial.gov on cartilage regeneration showed a shift in the utilised cells. Many studies using cartilage as the cell source appeared in Phase 3 trials. While fewer studies using BMSCs and ADSCs in Phase 3 suggest that most clinical developments using BMSCs and ADSCs as cell sources were unsuccessful. Although various approaches have been used to develop TEMPs, only a few are approved and commercialised for clinical uses. Based on the 2021 list of approved TEMPs commercialised in various countries (Sekar et al. 2021), the following cartilage-related TEMPs have been made into clinical. Note that only one TEMP uses autologous cultured chondrocytes on atelocollagen, three use autologous cultured chondrocytes on collagen and one uses chondrocyte spheroids.

1. JACC®: An autologous cultured cartilage made by harvesting the patient's cartilage tissue and mixing it with gel-form *atelocollagen* to mould it into a 3D form and then transplanted at the site of cartilage loss in the patient's knee, commercialised in 2012, based in Japan.
2. MACI®: A cellular sheet (MACI implant) consists of autologous cultured chondrocytes on a resorbable porcine Type I/III *collagen* membrane for treating damaged cartilage tissue, commercialised in 2016, based in the United States (US).
3. Ortho-ACI™: Autologous chondrocytes suspension is applied to a porcine-derived type I/III *collagen* scaffold for implantation into the cartilage defect(s), commercialised in 2017, based in Australia.
4. Novocart: Matrix-supported autologous chondrocyte transplantation using biphasic/bilayered *collagen* type I scaffolds: (1) a porous-spongy side, which allows for cell distribution within the scaffold, and (2) a layer of tough *collagen* membrane cover, commercialised in 2003, based in European Union (EU).
5. Spherox: Human autologous chondrocyte spheroids (spherical aggregates) for treating cartilage defects in adults; the cells are from healthy cartilage prepared from the patient's tissues, commercialised in 2007, Germany and 2017 in Europe.

Other than the above TEMPs, a 2020 article in Tissue Engineering Part B: Reviews (Oberweis et al. 2020) listed two more tissue-based products, namely, Carticel (by Vericel Denmark ApS) and ChondroCelect (by TiGenix N.V.), which did not appear in the more recent 2021 article (Sekar et al. 2021). Both products used cell suspension for implantation without the support of scaffolds. Carticel, authorised for marketing in August 1997, was indicated for repairing "*symptomatic cartilage defects in medial, lateral or trochlea femoral condyle. The Carticel's autologous cultured chondrocytes are harvested from the patient's normal femoral articular cartilage. Cartilage biopsies from a lesser weight-bearing area are isolated, cultured and expanded in vitro. The cells were implanted into articular cartilage defects under an autologous periosteal flap*".

On the other hand, ChondroCelect was withdrawn at the company's request in November 2016. It was initially authorised for marketing in October 2009. The product, ChondroCelect, is a preparation of 10,000 cells/μL implantation suspension. It characterised viable autologous cartilage cells expanded ex vivo expressing specific marker proteins. The product was initially indicated for repairing "*single symptomatic cartilage defects of the femoral condyle of the knee in adults with or without a presence of concomitant asymptomatic cartilage lesions. A randomised* *controlled trial evaluating the ChondroCelect efficacy in patients with lesions between 1.0 and 5.0 cm^2 is performed to demonstrate efficacy*" (Oberweis et al. 2020). A few other products listed in the 2018 article (Negoro et al. 2018) include the following. Only one product uses autologous cultured chondrocytes on fibrin, and one uses allogeneic umbilical blood MSCs on the hyaluronic acid hydrogel. The last one uses a modified transforming growth factor-β (TGF-β)-expressing allogeneic chondrocytes gene therapy approach.

1. Chondron™: An autologous chondrocyte-pre-seeded *fibrin* 3D matrix gel by Sewon Cellontech Korea; received approval in 2001/2008 by the Ministry of Food & Drug Safety (MFDS). It is indicated for articular cartilage defect (knee/ankle).
2. CARTISTEM®: A composite of allogeneic umbilical blood MSCs and *hyaluronic acid* hydrogel by Medipost Korea; received approval in 2012 by the MFDS. It is indicated for osteoarthritis due to degeneration or repetitive trauma.
3. Invossa™ (TissueGene-C): A gene therapy implant using modified TGF-β-expressing allogeneic chondrocytes by Kolon Life Sciences Korea; received approval in July 2017 by the MFDS. It is indicated for moderate knee osteoarthritis with persistent symptoms despite $\geq$3 months of conservative treatment with TGF-β-expressing allogeneic chondrocytes.

Other than the selected articles, detailed information on TEMPs and tissue-based products indicated in this chapter was retrieved from the relevant product information sheet available on the company's website. They can be quickly sought through ClinicalTrials.gov or Google search engine based on their trade name.

16.5 Regulatory and Bioethics Issues

Materials safety is determined based on biocompatibility data, scientific information and immunological responses to products regulated as medical devices (Salazar-Fontana 2022). However, the nature of tissue-engineered products (TEMPs and tissue-based products) designs are more complex than medical devices because they are produced using multiple substances and mixtures, involving not only the cells but signalling factors and biomaterials. They have been classified as cell and gene therapy products (CGTPs) or advanced therapy medicinal products (ATMPs) (Salazar-Fontana 2022; World Health Organization 2021). The diverse nature of CGTPs/ATMPs does not conform to a "one size fits all" concept that applies to product development and regulation (National Pharmaceutical Control Bureau Ministry of Health Malaysia 2015a). Uniqueness in each product merits attention; therefore, many guidance documents are prepared to address these specialised needs and differences between countries worldwide (World Health Organization 2021).

In ASEAN (Association of Southeast Asian Nations), Malaysia is among the first countries to embark on the regulation of CGTPs (Loh et al. 2021). However, CGTPs were regulated earlier in the United States, the European Union and Japan starting in 1993, 2003 and 2014, respectively. Debates on regulating CGTPs have been gathering steam in Malaysia since 2009. Multiple discussion series within the National Pharmaceutical Control Bureau (NPCB), Ministry of Health (MOH) Malaysia and researchers led to the formation of a Technical Working Group (TWG) in 2012. Chaired by the NPCB, the TWG was assigned to formulate strategies for CGTPs regulatory framework, which was later published in 2015 (National Pharmaceutical Control Bureau Ministry of Health Malaysia 2015a) and adopted in 2016. The development of CGTPs guidance documents and guidelines for registration in Malaysia engaged stakeholders from relevant government agencies, higher education and research institutions and industries. It was agreed that the mandatory registration and enforcement of CGTPs would begin on January 1, 2021, though the voluntary registration phase was open from 2016 through 2020. The suggested timeline is to give the industry ample time to meet the regulatory requirements set out in the guidelines. This CGTPs guidance document should be read with the Good Tissue Practice (GTP) Guideline by NPCB Malaysia (National Pharmaceutical Control Bureau Ministry of Health Malaysia 2015b).

Based on the Malaysian CGTPs guidance document, "*labile (fresh) blood and blood components (e.g., fresh frozen plasma) and secreted or extracted human products (e.g., milk, collagen)*" are not included in the framework (National Pharmaceutical Control Bureau Ministry of Health Malaysia 2015a). It also indicated that "*samples of human cells or tissues solely for diagnostic purposes in the same individual*" (autologous procedures) are not included in the framework. Fibrin is produced from fibrinogen, a blood component, and atelocollagen is a unique, highly compatible collagen type with an implant. Furthermore, massive applications of fibrin and collagen in existing surgical procedures (Hinckel et al. 2021) and data gathered from ClinicalTrial.gov indicate that the materials are safe for clinical applications. The same goes for PLGA since this copolymer is already approved by the US FDA and the European Medicines Agency (EMA) (Rocha et al. 2022). However, as a precautionary measure, researchers should be aware that when these materials are combined with cultured or genetically modified cells and the resulting substances undergo steps beyond minor manipulation, the products will fall under CGTPs, requiring adherence to the guidelines. Therefore, substantiating safety and efficacy is essential (Md Nazir 2018).

Some religious or cultural beliefs may have reservations about technological advancements in medical and health care. The most fundamental question is whether employing such advancements beyond norms is ethical. In tissue engineering, it is especially true when there are uncertainties concerning the use of materials taken from animals or non-permissible sources (Radzi et al. 2020; Aa'zamuddin Ahmad Radzi et al. 2020). Some lingering doubts include: Can the materials be applied to humans without risks? Any long-term evidence that it is safe and non-cancerous? If using animal sources—is xenotransplantation downgrading human dignity? Is it safe and efficient to use animal sources since they are relatively short-lived compared to humans? In Islam, for instance, there are flexibilities when a particular pressing condition meets the five essentials of human existence classified as the "higher objectives of Islamic law" or *Maqasid al-Shariah*. Abu al-Darda narrated that the Prophet (ﷺ) said: "*Allah has sent down both the disease and the cure, and He has appointed a cure for every disease, so treat yourselves medically, but use nothing unlawful*" (Dawud n.d.). As seeking treatment is encouraged, the essentialities may sometimes transform certain decisions from non-permissible to permissible. It is so because Islamic rulings aim to protect the five elements from harm, including protection and preservation of faith (*al-Diin*), soul/life (*al-Nafs*), intellect (*al-'Aql*), lineage/progeny (*al-Nasl*) and property (*al-Mal*) (Ghafar Bin Don et al. 2022). These elements can be hierarchical and non-hierarchical, but they are interconnected if appropriately understood.

Conventional bioethics principles include "autonomy" (respect for humans' principles, people understand their own best interests), "beneficence" (do good for people), "non-maleficence" (do no harm) and "justice" (fair distribution of benefits and costs, including health risks) across stakeholders (Beauchamp 2003; Harrison 2008; Beauchamp and Childress 2019). These principles are not meant to hinder scientific progress but to humanise technological advancements. While the authors are contemplating their original publication on "Principles of Biomedical Ethics", its development over four decades, some of its major themes, and persistent misunderstandings (Beauchamp and Childress 2019), *Maqasid al-Shariah* has been proposed as a practical framework for Islamic bioethics (Saifuddeen et al. 2014). It can guide in-depth discussions on many bioethical issues complementing the conventional bioethics principles. Through a deeper understanding of the "higher objectives of Islamic law", researchers can evaluate arising bioethical issues stemming from biotechnology advancements. When applied to the above four conventional bioethical principles, the *Maqasid al-Shariah* can help resolve conflicting moral obligations (Saifuddeen et al. 2014; Ibrahim et al. 2018, 2019) in the West and the Muslim worlds.

Besides *Maqasid al-Shariah*, contemporary Muslim scholars also used the five *Qawāìd al-Fiqhiyyah* (Fiqh methods; Islamic legal maxims): intention, certainty, removal of hardship, elimination of harm and custom (*'urf*), as guiding principles to determine the legal ruling for various contemporary issues in medical and health. The subjects are still open for discussion. In general, it is originally permissible to use technological advancement in medicine with the noble intent of life-saving and therapeutic purposes. However, if the goal is merely for enhancement or cosmetic purposes, it is prohibited in Islam if it is without sound scientific judgments (Radzi et al. 2020; Aa'zamuddin Ahmad Radzi et al. 2020; Syamimi et al. 2022).

16.6 Conclusion

This chapter presented some progress in articular cartilage tissue engineering using fibrin, atelocollagen and PLGA. However, the content and context are still insufficient to explain the entire area. Continuous efforts are necessary to compile and summarise all information from time to time. Although there is still room for improvement in search accuracy, scientific and clinical trial databases such as

Scopus, PubMed, Web of Science (WOS), Google Scholar and ClinicalTrial.gov can help researchers follow up or capture the progress of the field. Deliberations on tissue engineering are never exhaustive or conclusive because of the unique multidisciplinary nature of the field. More meritorious scientific research is needed to translate the discoveries into clinical applications. Using hybrid bioscaffolds based on natural and biocompatible synthetic materials may be one of the valuable options for sustainable materials for biomedical cartilage engineering applications.

Acknowledgement The authors thank the Ministry of Higher Education Malaysia (for TRGS/1/2016/UIAM/02/8/2 Sub-project 02 under TRGS/1/2016/UIAM/02/8 programme), International Islamic University Malaysia (IIUM) and Kulliyyah of Allied Health Sciences (KAHS) for direct and indirect supports for this work.

Conflict of Interest Nil.

References

Aa'zamuddin Ahmad Radzi M, Sha'ban M, Azharuddin M et al (2020) A survey on relevant Malaysian Fatwā about cadaver concerning tissue engineering research. Pertanika J Soc Sci Humanit 28:1451

Abay Akar N, Gürel Peközer G, Torun Köse G (2019) Fibrous bone tissue engineering scaffolds prepared by wet spinning of PLGA. Turk J Biol 43:235–245. https://doi.org/10.3906/biy-1904-63

Abdul Rahman R, Mohamad Sukri N, Md Nazir N et al (2015a) Evaluation of three-dimensional construct engineered from poly(lactic-co-glycolic acid)/fibrin hybrid scaffold using rabbit bone marrow mesenchymal stem cells for osteochondral defect repair. J Teknol 77:77. https://doi.org/10.11113/jt.v77.6743

Abdul Rahman R, Mohamad Sukri N, Md Nazir N et al (2015b) The potential of 3-dimensional construct engineered from poly(lactic-co-glycolic acid)/fibrin hybrid scaffold seeded with bone marrow mesenchymal stem cells for in vitro cartilage tissue engineering. Tissue Cell 47:420. https://doi.org/10.1016/j.tice.2015.06.001

Ali M, Tahir AH, Mohamed Amin MAI, Azhim A, Sha'ban M (2019) Structural and functional properties of neocartilage construct engineered in poly (lactic-co-glycolic acid) (PLGA) based scaffolds. Int J Integr Eng 11:012. https://doi.org/10.30880/ijie.2019.11.03.012

Angel L (2010) Tissue engineering. Embryo Project Encyclopedia

Asawa Y, Sakamoto T, Komura M et al (2012) Early stage foreign body reaction against biodegradable polymer scaffolds affects tissue regeneration during the autologous transplantation of tissue-engineered cartilage in the canine model. Cell Transplant 21:1431–1442. https://doi.org/10.3727/096368912X640574

Barbon S, Contran M, Stocco E et al (2021) Enhanced biomechanical properties of polyvinyl alcohol-based hybrid scaffolds for cartilage tissue engineering. Processes 9:730. https://doi.org/10.3390/pr9050730

Barbuzano J (2017) Organoids: a new window into disease, development and discovery. Harvard Stem Cell Institute News. https://hsci.harvard.edu/organoids#:~:text=Organoids%20are%20tiny%2C%20self%2Dorganized,only%20certain%20types%20of%20cells. Accessed 23 Aug 2022

Batool H, Mohammad M, Maryam B et al (2020) Comparison of fibrin and PLGA, fibrin scaffolds for chondrogenesis of human adipose derived stem cells by icariin. J Kerman Univ Med Sci 27:14–23

Beauchamp TL (2003) Methods and principles in biomedical ethics. J Med Ethics 29:269–274. https://doi.org/10.1136/jme.29.5.269
Beauchamp T, Childress J (2019) Principles of biomedical ethics: marking its fortieth anniversary. Am J Bioeth 19:9–12. https://doi.org/10.1080/15265161.2019.1665402
Bin Mohamed Amin MAI, Tahir AHMA, Azhim A, Sha'ban M (2018) Physical properties and biocompatibility of 3D hybrid PLGA based scaffolds. In: 2018 IEEE-EMBS Conference on Biomedical Engineering and Sciences (IECBES). IEEE, Washington, DC, pp 480–484
Britannica The Editors of Encyclopaedia (2018) Fibrin. Encyclopaedia Britannica. https://www.britannica.com/science/fibrin. Accessed 7 Sep 2022
Cookson C, Kuchler H, Miller J (2022) How science is getting closer to a world without animal testing. The Financial Times. https://www.ft.com/content/7c35e08a-4931-4401-b27e-acabf974bff8. Accessed 23 Aug 2022
Dang LH, Tseng Y, Tseng H, Hung S-H (2021) Partial decellularization for segmental tracheal scaffold tissue engineering: a preliminary study in rabbits. Biomolecules 11:866. https://doi.org/10.3390/biom11060866
SA Dawud (n.d.) Chapter 11: Regarding the disliked remedies. In: Book 29: Medicine (Kitab Al-Tibb). https://sunnah.com/abudawud:3874
Eschweiler J, Horn N, Rath B et al (2021) The biomechanics of cartilage—an overview. Life 11:302. https://doi.org/10.3390/life11040302
Fu L, Yang Z, Gao C et al (2020) Advances and prospects in biomimetic multilayered scaffolds for articular cartilage regeneration. Regen Biomater 7:527–542. https://doi.org/10.1093/rb/rbaa042
Gantenbein-Ritter B, Sakai D (2011) Biomaterials for intervertebral disc regeneration. In: Comprehensive biomaterials. Elsevier, Amsterdam, pp 161–169
Gao Y, Gao J, Li H et al (2019) Autologous costal chondral transplantation and costa-derived chondrocyte implantation: emerging surgical techniques. Ther Adv Musculoskelet Dis 11:1759720X1987713. https://doi.org/10.1177/1759720X19877131
Ghafar Bin Don A, Puteh A, Irdha Mokhtar A (2022) Da'wah Maqasid Al-Syariah in nurturing community well-being. Pengajian Islam J 15:250
Gilson K, Eun YK, Jeong ES et al (2017) Synthetic/PLGA hybrid scaffold for tissue regeneration: update 2015. In: Khang G (ed) handbook of intelligent scaffolds for tissue engineering and regenerative medicine, 2nd edn. Jenny Stanford Publishing, New York, NY
Gomoll AH, Kamei G, Ochi M et al (2014) Technical enhancements and update on chondrocyte implantation. Oper Tech Orthop 24:35–47. https://doi.org/10.1053/j.oto.2014.02.007
Harrison M (2008) Applying bioethical principles to human biomonitoring. Environ Health 7:S8. https://doi.org/10.1186/1476-069X-7-S1-S8
Hinckel BB, Thomas D, Vellios EE et al (2021) Algorithm for treatment of focal cartilage defects of the knee: classic and new procedures. Cartilage 13:473S–495S. https://doi.org/10.1177/1947603521993219
Ibrahim AH, Rahman NNA, Saifuddeen SM (2018) Maqasid al-Shariah as a complementary framework for conventional bioethics: application in Malaysian Assisted Reproductive Technology (ART) Fatwa. Sci Eng Ethics 24:1493–1502. https://doi.org/10.1007/s11948-017-9963-6
Ibrahim AH, Rahman NNA, Saifuddeen SM, Baharuddin M (2019) Maqasid al-Shariah based Islamic bioethics: a comprehensive approach. J Bioeth Inq 16:333–345. https://doi.org/10.1007/s11673-019-09902-8
Kanamoto T, Hikida M, Sato S et al (2021) Integrin α2β1 plays an important role in the interaction between human articular cartilage-derived chondrocytes and atelocollagen gel. Sci Rep 11:1757. https://doi.org/10.1038/s41598-021-81378-2
Kawasaki M, Goyama T, Tachibana Y et al (2022) Farm and companion animal organoid models in translational research: a powerful tool to bridge the gap between mice and humans. Front Med Technol 4:895379. https://doi.org/10.3389/fmedt.2022.895379

Kim HJ, Han MA, Shin JY et al (2019) Intra-articular delivery of synovium-resident mesenchymal stem cells via BMP-7-loaded fibrous PLGA scaffolds for cartilage repair. J Control Release 302: 169–180. https://doi.org/10.1016/j.jconrel.2019.04.002

Kim J, Koo B-K, Knoblich JA (2020) Human organoids: model systems for human biology and medicine. Nat Rev Mol Cell Biol 21:571–584. https://doi.org/10.1038/s41580-020-0259-3

Langer R, Vacanti J (1993) Tissue engineering. Science 260:920. https://doi.org/10.1126/science.8493529

Li B, Li F, Ma L et al (2014) Poly(lactide-co-glycolide)/fibrin gel construct as a 3D model to evaluate gene therapy of cartilage in vivo. Mol Pharm 11:2062–2070. https://doi.org/10.1021/mp5000136

Li X, Ding J, Zhang Z et al (2016) Kartogenin-incorporated thermogel supports stem cells for significant cartilage regeneration. ACS Appl Mater Interfaces 8:5148–5159. https://doi.org/10.1021/acsami.5b12212

Lim YW, Tan WS, Ho KL et al (2022) Challenges and complications of poly(lactic-co-glycolic acid)-based long-acting drug product development. Pharmaceutics 14:614. https://doi.org/10.3390/pharmaceutics14030614

Lind M, Larsen A, Clausen C et al (2008) Cartilage repair with chondrocytes in fibrin hydrogel and MPEG polylactide scaffold: an in vivo study in goats. Knee Surg Sports Traumatol Arthrosc 16: 690–698. https://doi.org/10.1007/s00167-008-0522-1

Liu H, Ding J, Li C et al (2016) Hydrogel is superior to fibrin gel as matrix of stem cells in alleviating antigen-induced arthritis. Polymers 8:182. https://doi.org/10.3390/polym8050182

Loh EYX, Goh PS, Mannan AMM et al (2021) Cell and gene therapy products in Malaysia: a snapshot of the industry's current regulation preparedness. Cytotherapy 23:1108–1113. https://doi.org/10.1016/j.jcyt.2021.06.006

Martínez-Moreno D, Jiménez G, Gálvez-Martín P et al (2019) Cartilage biomechanics: a key factor for osteoarthritis regenerative medicine. Biochim Biophys Acta (BBA) - Mol Basis Dis 1865: 1067–1075. https://doi.org/10.1016/j.bbadis.2019.03.011

Matsubara N, Nakasa T, Ishikawa M et al (2021) Autologous meniscus fragments embedded in atelocollagen gel enhance meniscus repair in a rabbit model. Bone Joint Res 10:269–276. https://doi.org/10.1302/2046-3758.104.BJR-2019-0359.R2

Md Ali Tahir AH, Amin MAIM, Azhim A, Sha'ban M (2019) Evaluation of cartilaginous extracellular matrix production in in vitro "cell-scaffold" construct. In: 2018 IEEE EMBS Conference on Biomedical Engineering and Sciences. IEEE, Washington, DC

Md Nazir N, Sha'ban M (2018) Overview of safety and efficacy of non-viral gene transfer in cartilage tissue engineering from the worldview of Islam. IIUM Med J Malay 17:1029. https://doi.org/10.31436/imjm.v17i1.1029

Md Nazir N, Zulkifly AH, Khalid KA et al (2019) Matrix production in chondrocytes transfected with sex determining region Y-Box 9 and telomerase reverse transcriptase genes: an in vitro evaluation from monolayer culture to three-dimensional culture. Tissue Eng Regen Med 16: 285–299. https://doi.org/10.1007/s13770-019-00191-1

Md Nazir N, Zulkifly AH, Khalid KA et al (2020) The cartilaginous tissue formation using Sry (Sex Determining Region Y)-BOX9 and telomerase reverse transcriptase genes transfected chondrocytes: in vivo approach. Sains Malay 49:1067. https://doi.org/10.17576/jsm-2020-4905-11

Mehdikhani-Nahrkhalaji M, Tavakoli E, Zargar-Kharazi A, Hashemi-Beni B (2017) A novel nano-composite scaffold for cartilage tissue engineering. Sci Iranica 25:1815–1823. https://doi.org/10.24200/sci.2017.4595

Miranda CS, Silva AFG, Pereira-Lima SMMA et al (2022) Tunable spun fiber constructs in biomedicine: influence of processing parameters in the fibers' architecture. Pharmaceutics 14: 164. https://doi.org/10.3390/pharmaceutics14010164

Mohamad Sukri N, Ahmad Radzi MA, Abdul Rahman R et al (2015) Identifying the potential of transcription factor SOX9 gene transfer in chondrocytes differentiation and articular cartilage formation in vitro. J Teknol 77:107. https://doi.org/10.11113/jt.v77.6748

Mohamad Sukri N, Sha'ban M, Ahmad Radzi MA et al (2022) The feasibility of using human primary chondrocytes derived from osteoarthritic patients overexpressed with SOX9 seeded on PLGA-fibrin hybrid scaffolds for cartilage engineering. Malay J Fund Appl Sci 18:304–322. https://doi.org/10.11113/mjfas.v18n3.2344

Mohamad MY, Mohamed Amin MAI, Harun AF et al (2017) Fabrication and characterization of three-dimensional poly(lactic-co-glycolic acid), atelocollagen, and fibrin bioscaffold composite for intervertebral disk tissue engineering application. J Bioact Compat Polym 32:456. https://doi.org/10.1177/0883911516686091

Mohamad MY, Amin MAIM, Sha'ban M, Zulkifly AH (2020) In vitro evaluation of poly (lactic acid-co-glycolic acid)/atelocollagen/fibrin bioscaffold for annulus fibrosus regeneration. Trends Biomater Artif Organs 34:118123

Munirah S, Samsudin OC, Chen HC et al (2007) Articular cartilage restoration in load-bearing osteochondral defects by implantation of autologous chondrocyte-fibrin constructs. J Bone Joint Surg (Br) 89:1099. https://doi.org/10.1302/0301-620X.89B8.18451

Munirah S, Kim S, Ruszymah B, Khang G (2008) The use of fibrin and poly(lactic-co-glycolic acid) hybrid scaffold for articular cartilage tissue engineering: an in vivo analysis. Eur Cell Mater 15: 41–52. https://doi.org/10.22203/eCM.v015a04

National Pharmaceutical Control Bureau Ministry of Health Malaysia (2015a) Guidance document and guidelines for registration of cell and gene therapy products (CGTPs) in Malaysia. Ministry of Health Malaysia, Putrajaya

National Pharmaceutical Control Bureau Ministry of Health Malaysia (2015b) Good tissue practice guideline. Ministry of Health Malaysia, Putrajaya

Negoro T, Takagaki Y, Okura H, Matsuyama A (2018) Trends in clinical trials for articular cartilage repair by cell therapy. NPJ Regen Med 3:17. https://doi.org/10.1038/s41536-018-0055-2

Oberweis CV, Marchal JA, López-Ruiz E, Gálvez-Martín P (2020) A worldwide overview of regulatory frameworks for tissue-based products. Tissue Eng Part B Rev 26:181–196. https://doi.org/10.1089/ten.teb.2019.0315

Operti MC, Bernhardt A, Grimm S et al (2021) PLGA-based nanomedicines manufacturing: technologies overview and challenges in industrial scale-up. Int J Pharm 605:120807. https://doi.org/10.1016/j.ijpharm.2021.120807

Radzi MAA, Zawawi M, Shaban M et al (2020) Legal aspects of articular cartilage tissue engineering experimentation: a review on Malaysian acts, regulations and guidelines. Malay J Med Health Sci 16:272

Ren P, Xu Y, Fan F et al (2021) [In vitro regeneration of tissue engineered cartilage and its clinical application for nasal reconstruction]. Zhongguo Xiu Fu Chong Jian Wai Ke Za Zhi 35:221–226. https://doi.org/10.7507/1002-1892.202009053

Rocha CV, Gonçalves V, da Silva MC et al (2022) PLGA-based composites for various biomedical applications. Int J Mol Sci 23:2034. https://doi.org/10.3390/ijms23042034

Saifuddeen SM, Rahman NNA, Isa NM, Baharuddin A (2014) Maqasid al-Shariah as a complementary framework to conventional bioethics. Sci Eng Ethics 20:317–327. https://doi.org/10.1007/s11948-013-9457-0

Salazar-Fontana LI (2022) A regulatory risk-based approach to ATMP/CGT development: integrating scientific challenges with current regulatory expectations. Front Med 9:855100. https://doi.org/10.3389/fmed.2022.855100

Sekar MP, Budharaju H, Zennifer A et al (2021) Current standards and ethical landscape of engineered tissues—3D bioprinting perspective. J Tissue Eng 12:204173142110276. https://doi.org/10.1177/20417314211027677

Sha'ban M, Ahmad Radzi MA (2020) Scaffolds for cartilage regeneration: to use or not to use? Springer, Singapore

Sha'ban M, Kim SH, Idrus RB, Khang G (2008) Fibrin and poly(lactic-co-glycolic acid) hybrid scaffold promotes early chondrogenesis of articular chondrocytes: an in vitro study. J Orthop Surg Res 3:17. https://doi.org/10.1186/1749-799X-3-17

Sha'ban M, Zainuddin ZI, Abdul Rahman R et al (2014) Exploring the Islamic perspective on tissue engineering principles and practice. Global J Al Thaqafah 4:29–40. https://doi.org/10.7187/GJAT642014.04.02

Sheykhhasan M, Qomi RT, Kalhor N et al (2015) Evaluation of the ability of natural and synthetic scaffolds in providing an appropriate environment for growth and chondrogenic differentiation of adipose-derived mesenchymal stem cells. Indian J Orthop 49:561–568. https://doi.org/10.4103/0019-5413.164043

Shim G, Kim M-G, Park JY, Oh Y-K (2013) Small interfering RNAs (siRNAs) as cancer therapeutics. In: Biomaterials for cancer therapeutics. Elsevier, Amsterdam, pp 237–269

Skidmore S, Hadar J, Garner J et al (2019) Complex sameness: separation of mixed poly(lactide-co-glycolide)s based on the lactide:glycolide ratio. J Control Release 300:174–184. https://doi.org/10.1016/j.jconrel.2019.03.002

Sophia Fox AJ, Bedi A, Rodeo SA (2009) The basic science of articular cartilage: structure, composition, and function. Sports Health Multidiscipl Appr 1:461–468. https://doi.org/10.1177/1941738109350438

Syamimi N, Azharuddin M, Aa'zamuddin M et al (2022) Ethical issues in tissue engineering: a systematic review. Malay J Med Health Sci 18:275

Tsuyuguchi Y, Nakasa T, Ishikawa M et al (2021) The benefit of minced cartilage over isolated chondrocytes in atelocollagen gel on chondrocyte proliferation and migration. Cartilage 12:93–101. https://doi.org/10.1177/1947603518805205

Ude CC, Shamsul BS, Ng MH et al (2018) Long-term evaluation of osteoarthritis sheep knee, treated with TGF-β3 and BMP-6 induced multipotent stem cells. Exp Gerontol 104:43–51. https://doi.org/10.1016/j.exger.2018.01.020

Wang W, Li B, Li Y et al (2010a) In vivo restoration of full-thickness cartilage defects by poly (lactide-co-glycolide) sponges filled with fibrin gel, bone marrow mesenchymal stem cells and DNA complexes. Biomaterials 31:5953–5965. https://doi.org/10.1016/j.biomaterials.2010.04.029

Wang W, Li B, Yang J et al (2010b) The restoration of full-thickness cartilage defects with BMSCs and TGF-beta 1 loaded PLGA/fibrin gel constructs. Biomaterials 31:8964–8973. https://doi.org/10.1016/j.biomaterials.2010.08.018

World Health Organization (2021) WHO considerations on regulatory convergence of cell and gene therapy products (draft for public consultation). WHO, Geneva

Xu W, Zhao R, Wu T et al (2021) Biodegradable calcium carbonate/mesoporous silica/poly(lactic-glycolic acid) microspheres scaffolds with osteogenesis ability for bone regeneration. RSC Adv 11:5055–5064. https://doi.org/10.1039/D0RA09958A

Yin L, Wu Y, Yang Z et al (2018) Characterization and application of size-sorted zonal chondrocytes for articular cartilage regeneration. Biomaterials 165:66–78. https://doi.org/10.1016/j.biomaterials.2018.02.050

Yin Z, Li D, Liu Y et al (2020) Regeneration of elastic cartilage with accurate human-ear shape based on PCL strengthened biodegradable scaffold and expanded microtia chondrocytes. Appl Mater Today 20:100724. https://doi.org/10.1016/j.apmt.2020.100724

Yusof F, Sha'ban M, Azhim A (2019) Engineered meniscus scaffolds using sonication decellularization treatment system. Int J Integr Eng 11:81–89. https://doi.org/10.30880/ijie.2019.11.03.009

Yusof F, Sha'ban M, Azhim A (2021) Decellularization strategies to engineer fibrocartilage bioscaffolds: a review. J Biomater Tissue Eng 11:1435–1451. https://doi.org/10.1166/jbt.2021.2740

17 Sustainable Design of Natural and Synthetic Biomaterials for Wound Healing Applications

Nur Fatiha Ghazalli, Veithesswary Mogan, Norhayati Yusop, Juan Matmin, and Siti Fatimah Noor Mat Johar

Abstract

Various scientists and researchers from prestigious institutions have turned their research to achieve sustainability goals in almost every field, especially materials. The fabrication of new synthetic and natural renewable biomaterials is the fastest-growing field in biomedical applications. In well-known biomedical applications such as tissue engineering, implants, and drug delivery systems, renewable materials now have new opportunities due to the improvement of their bioactivity, compatibility, and mechanical properties. One example is skin tissue engineering which has arisen as an alternative and successful method to restore completely functioning tissue using novel biomaterials or engineered scaffolding. The design of advanced biomaterials with regulated physical, chemical, electrical, and biological properties would also promote functional tissues. Both natural and synthetic biomaterials used in tissue engineering provisionally support and mass transfer to promote system while also regulating regenerated tissue size and shape.

N. F. Ghazalli (✉) · N. Yusop
Basic and Medical Sciences Unit, School of Dental Sciences, Universiti Sains Malaysia, Kota Bharu, Kelantan, Malaysia
e-mail: fatiha85@usm.my

V. Mogan
Postgraduate Unit, School of Dental Sciences, Universiti Sains Malaysia, Kota Bharu, Kelantan, Malaysia

J. Matmin
Department of Chemistry, Faculty of Science, Universiti Teknologi Malaysia, Skudai, Johor, Malaysia

S. F. N. Mat Johar
Reconstructive Sciences Unit, School of Medical Sciences, Universiti Sains Malaysia, Kota Bharu, Kelantan, Malaysia

W. S. Wan Kamarul Zaman, A. A. Nurul (eds.), *Sustainable Material for Biomedical Engineering Application*, https://doi.org/10.1007/978-981-99-2267-3_17

Keywords

Renewable biomaterials · Natural biomaterials · Synthetic biomaterials · Sustainable design · Nanomaterials · Drug delivery · Wound healing

Abbreviations

AGEs	Advanced glycation end products
AgNCs	Silver nanoclusters
AgNPs	Silver nanoparticles
ALA	Alpha lipoic acid
AuAgNSs	Gold-silver nanoshells
AuNPs	Gold nanoparticles
bFGF	Basic fibroblast growth factor
BTB	Bromothymol blue
$CaCl_2$	Calcium chloride
CMC	Carboxymethylcellulose
CNCs	cellulose nanocrystals
CO_2	Carbon dioxide
CS hydrogel	Chitosan hydrogel
DDS	drug-delivery systems
dECM	Decellularized extracellular matrix
DR	Drug resistent
DS	Drug sensitive
DTTC	3,3′-diethylthiatricarbocyanine iodide
ECM	Extracellular matrix
EGCG	Epigallocatechin gallate
EGF	Epidermal growth factor
EPS	Extracellular polymeric substances
FGF-2	Fibroblast growth factor
GAGs	Glycosaminoglycans
GFs	Growth factors
Gly	Glycine
GMS3	Ganglioside-monosialic acid 3 synthase
HA	Hydroxyapatite
KGF	Keratinocyte growth factor
MRSA	Methicillin-resistant Staphylococcus aureus
NIR	Near-infrared
NPs	Nanoparticles
PCL	Polycaprolactone
PCN-224	Porphyrinic metal-organic frameworks
PDA	Polydopamine
PDGF	Platelet derived growth factors
PDT	Photodynamic therapy

PEG	Polyethylene glycol
PEG	Polyethylene glycol
PLA	Poly Lactic acid
PLGA	Poly(lactide-co-glycolide)
PLLA	Poly-l-Lactic acid
PU	Polyurethane
PVA	Polyvinyl alcohol
QCSG/GM/GO	Chitosan/gelatin methacrylate/graphene oxide
RAGE	Receptor for AGEs
ROS	Reactive oxygen species
SERS	Surface-enhanced Raman scattering
TNF	Tumor necrosis factor
VEGF	Vascular endothelial growth factor
ZnO	Zinc oxides

17.1 Introduction

The body's surface or skin is the layer of soft outer tissues, approximately 15% of the total body weight (Raymund et al. 2005). The skin has capabilities that include regulating metabolism and body temperature (Vig et al. 2017). Skin (Fig. 17.1) is the biggest organ in the mammalian species, covering the whole-body surface and interacting with the outside world. It is a shield against harmful biological, mechanical, and chemical attacks as well as serves as a safety net that is critical in maintaining the physiological homeostasis of the human body. Damage to the skin will expose the person to physiological disparity, injuries, or even death. Acute trauma, persistent injury, or surgery can all cause cutaneous damage. Wounds have been a major hazard to human health, resulting in tissue necrosis (Klevens et al. 2007). Mammalian skin cells can destroy pathogenic species while also removing wound debris and attempting to restore anatomical and functional integrity after damage. Skin injury replacement is commonly divided into three stages: inflammation, proliferation, and remodelling. This procedure necessitates different interactions and coordination across numerous cells and mediators.

Various factors can obstruct this process, resulting in a delayed wound. Several local variables can have a significant impact on wound healing. For example, ischemic tissues, wounds with foreign bodies, bacteria, and contamination, all affect wound healing. The common link bacterial species such as *Staphylococcus aureus, Pseudomonas aeruginosa, Streptococcus pyogenes, and methicillin-resistant staphylococcus aureus* can be easily found in wound infections. In addition, the normal healing mechanism can be limited or may not restore completely functioning skin tissue when severe or permanent skin damage is incurred (Enoch and Leaper 2005). Hence, the patients will experience not only physical but also psychological pain, as well as financial pressures. This will not only cause physical and psychological

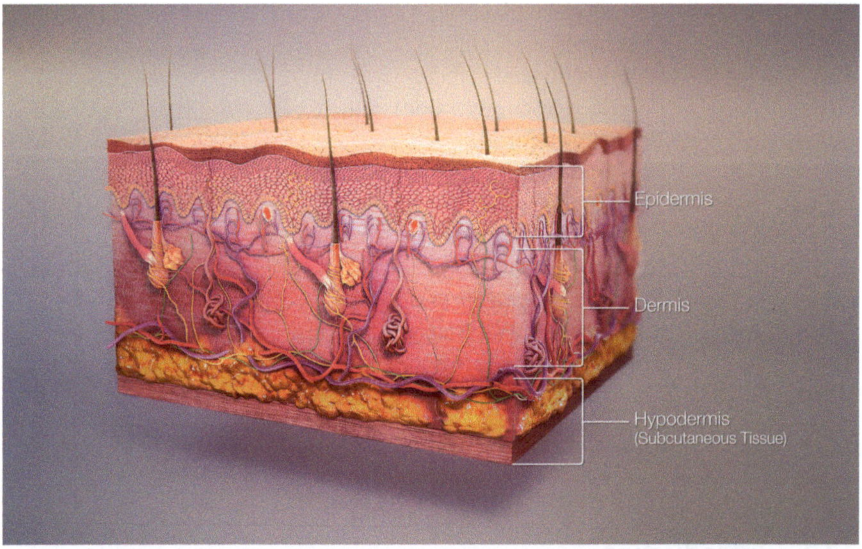

Fig. 17.1 Image of 3D medical illustration showing the major layers of human skin. Reproduced under the Creative Commons BY-SA (Attribution-Share Alike 4.0 International) license from (Scientific Animation 2020)

distress to victims, but also creates economic pressures. Treatments utilizing natural skin substitutes including wound dressings and high-dose antibiotic therapy are a popular clinical procedure for extensive skin infection. However, a tropical micro-environment cannot be afforded by conventional wound dressings such as cotton wool, bandages, and gauzes, which results in wound dehydration and causes trauma upon removal. Moreover, conventional therapies for wound healing like autographs, allografts, and xenografts have certain disadvantages, such as shortage of donors, limited availability, chance of infection, immune rejection, and extreme financial pressure. In addition, the treatment of high dosage antibiotics for medical treatment runs the risk of developing drug resistance and having unfavorable side effects. Bacterial cells may also clump together and form biofilms by encasing in cell membrane of also known as an extracellular matrix (ECM) composed of extracellular polymeric substances (EPS). Bacterial biofilms are regarded as a crucial mechanism of bacterial resistance to antibiotics due to their ability to obstruct medicine penetration into infected wounds (Hall-Stoodley et al. 2004; He et al. 2020a). Therefore, creation of novel and effective antibacterial methods for the treatment of bacterial infections was desperately needed.

In result of that, skin tissue engineering has arisen as an alternative and successful method to restore completely functioning tissue using novel biomaterials or engineered scaffolding (Neethu et al. 2015). Wound healing holds great promise for the fabrication of complex tissues that can mimic native tissue mechanisms. To encourage the growth of functional tissues, well-developed biomaterials with controlled physical, chemical, electrical, and biological properties could be created.

(Fisher et al. 2010). In recent years, special attention has been paid to hydrogels, one of the potential candidates among various biomaterials as they can replicate the physical, chemical, biological, and electrical characteristics of major biological tissues. A highly hydrated polymeric three-dimensional (3D) network found in hydrogels allows them to hold up to 20–40 times more water than their dry weight. Also known as aqua-gels, these chains of hydrophilic polymers are adaptable to moisture environment due to their 99% water retention, which make them up to absorb water and solvent molecules. Hydrogels are sensitive to external stimuli such as pH values due to the ionization of functional groups such as hydroxyl (OH) and carboxyl (COO). Moreover, it also has a strong swelling reaction, water diffusion, osmotic pressure, and capillary action. There are several factors that influence the ability of the gels to swell, types of cross-linking, polymerization processes, and the involvement of functional groups. In terms of the responsiveness improvement, pH value, temperature, electrical field, ion intensity, and light were the main crucial factors of hydrogels (Roy et al. 2010).

On a macroscopic scale, the hydrogels are solid and have a definite form. Moreover, hydrogels have the ability to capture water-soluble molecules. In the biological sciences, hydrogels have therefore been exploited as molecular filters. Hydrogels are also indispensable for electrophoresis. Basically, it can be categorized as natural or synthetic polymer-based and biopolymer-based hydrogels with fascinating properties such as biodegradability and biocompatibility. Examples of biopolymer-based including alginate, chitosan, collagen, chitin, gum Arabic, xanthan gum, and modified biopolymers (Zhang et al. 2011). Recent development also shows a notable and expanding interest in creating nanocomposite hydrogels over hydrogels for various biomedical applications. Nanocomposite hydrogels, also called hybrid hydrogels which can be described as polymer hydrated networks, cross-linked with each other and or with nanoparticles or nanostructures either physically or covalently. Nanocomposite hydrogels have superior chemical, biochemical, electrical, and physical properties (Gaharwar et al. 2014). Due to their increased reaction time and ability to work remotely, nanocomposite hydrogels have lately sparked a lot of attention. Incorporation of synthetic and natural hydrogels through various organic and inorganic parts by forming a network of nanocomposite hydrogels will enhance the mechanical, physical, chemical, biological, swelling, or de-swelling properties (Song et al. 2015).

According to the journal Facile and eco-friendly fabrication of polysaccharides-based nanocomposite hydrogel for photo-thermal treatment of wound infection released back in 2019, a novel nanocomposite hydrogel with NIR responsiveness has been reported (Deng et al. 2020) for efficient photo-thermal sterilization (Deng et al. 2020). The tremendous nanocomposite hydrogel with an optimum condition has shown favorable mechanical properties and superior process as well as displayed excellent photo-thermal conversion capability, which played a crucial role in antibacterial activity for further proving the advantage of using nanocomposite hydrogels over hydrogels.

With their unique physicochemical, optical, and biological features, biomaterial nanoparticles (NPs) have been successfully employed as wound dressings. This chapter thus sums up all the recent scientific advancements of natural and synthetic-based wound patching for detecting and treating wounds. The first part includes the natural and synthetic biomaterials for wound healing; the second part introduces and eleborates the sustainable design of natural and synthetic biomaterials for wound dressings and the final part elucidates nano-based therapeutic drugs for wound healing applications.

17.2 Types of Biomaterials Apply in Wound Healing Applications

Numerous different kinds of wounds and the characteristics of each type depend on a number of factors, including the cause, the patient's stage, the site, and the appearance of infections. Obviously, it shows the challenges in finding a single ideal dressing that can be applied to all types of wounds. In fact, the adaptability of biopolymers and the ability to combine them to take advantage of each material's best features make them useful for treating a variety of wounds. Another crucial factor is the capacity to prevent inflammation or, the function as antibacterial agents to combat invasive pathogens that reside in wounds infection. Last but not the least, improving and accelerating wound healing depends on the material's capacity to maintain moisture at the wound bed and provide permeability to water and gases (Agrawal et al. 2014; Agarwal et al. 2014).

The two main groups of biomaterials for wound healing are synthetic and natural, which can be distinguished by their sources, chemical composition, type of patching, or use. Organic and inorganic polymers make up synthetic biomaterials, whereas natural biomaterials include biological components generated from bacterial, fungal, fungi, or animal origins that may be further categorized as proteins, polysaccharides, or macromolecules that contain both proteins and polysaccharides. Moreover, many properties including immunogenicity as well as flexibility, elasticity, degradation rate, and mechanical strength, can be controlled by engineering the composition and structure of these molecules, as will be emphasized below. Many properties such as immunogenicity, flexibility, mechanical strength, elasticity, and degradation rate can be controlled by modifying the composition and structure of these molecules. Biomaterials exhibit very different behaviors, especially when it comes to evoking a host response. (Sarkar et al. 2017).

17.2.1 Natural Biomaterials

Natural biomaterials come from widely available and accessible microbial, plant, or animal sources. This guarantees a low overall manufacturing cost and a relatively simple separation process. It is different from substances present in the body's natural support structures, such as connective tissues and ECM, in terms of biological activity, biocompatibility, and water affinity. Furthermore, it is highly adaptable and flexible in terms of technological demands due to their chemical structure, which is rich in groups that can be altered using derivatives. Another crucial quality of natural biopolymers is their susceptibility to enzymatic degradation, which results in secondary products that are typically well-accepted by the target organism and do not cause adverse reactions. This is especially true of non-cross-linked varieties.

These molecules lose stability due to the challenging control degradation process (Fishman et al. 2015). Additionally, the separation process is frequently not precise enough to get rid of contamination from other substances, especially those derived from animal sources, which can lead to immunogenicity and safety issues. Due to its inherent bioactive components, which includes cell growth, adhesion, and nontoxicity to healthy skin, natural biomaterials have a wide range of applications in the field of wound healing. Examples of natural biomaterials include alginate, chitin, chitosan, cellulose, collage, dextran, decellularized extracellular matrix (dECM), elastin, fibrinogen, glycosaminoglycans, gelatin, laminin, and silk. Collagen, glycosaminoglycans, elastin, laminin, fibrinogen, and dECM are natural biomaterials that mimic the natural extracellular matrix (ECM) components. Basically, the natural ECM consists of all tissue components except cells and is mostly composed of polysaccharides and proteins, providing an excellent 3D biological environment for live tissue. ECM is crucial in tissue repairing as it provides physical support to cells and controlled cellular activity such as proliferation, growth, migration, homeostasis, differentiation, and morphogenesis and allowed the human body to grow small deficiencies of cells, tissues, and organs (Mukherjee et al. 2010; Theocharis et al. 2016).

Yet, due to a shortage of ECM to fill these flaws, the human body is unable to heal significant abnormalities. Natural biomaterials were therefore manufactured and used in tissue engineering applications. There are three different group materials of naturally derived hydrogels which are protein-based, polysaccharide-based, and those made from decellularized tissue. Proteins and ECM components make up most natural gels, they are naturally biocompatible, bioactive, and may be useful for a wide range of biomedical applications since they support several cellular processes. Their structure and characteristics are similar to those of native soft tissues, as was previously said; however, they do have certain drawbacks, chiefly the difficulty in controlling them due to their significant batch to batch variance.

The ECM structure contains several proteins, including elastin, collagen, and fibrin, which give it the strength and elasticity it needs to operate effectively. Hence, these proteins are particularly attractive materials for tissue regeneration and cell culture systems. These proteins are mostly obtained from animal sources; for instance, collagen is primarily obtained from swine tissue or from the tails of murine animals, whereas fibroin is discovered and obtained from insects (Vieira et al. 2017).

17.2.1.1 Protein Based Biomaterials

17.2.1.1.1 Collagen

The primary ECM protein in the body is collagen, which serves as mechanical support for opposing action forces to fend off recurrent permanent deformation.

Additionally, the type of collagen and the direction of its fibers control the distribution of different cell types within the tissue. Four separate organizational layers characterize the structure of collagen. Around 30% of the amino acids in collagen are glycine (Gly), and typically X is a proline and Y is a hydroxyproline. This tripeptide sequence makes up the main structure. The repeated triplet chains of amino acids provide the secondary structure. The third organizational level is a left triple helix with thousand amino acids in each chain. Collagen fundamental building block is made up of three polypeptide chains joined together in a triple-helix form (tropocollagen). Every third amino acid in the remaining sequence in all the three chains is frequently proline and hydroxyproline in the case of the other two is glycine (Gelse et al. 2003). The 28 varieties of collagen that have been discovered so far result from various combinations of these chains, but respectively, types I, II, and III are the most typical ones which the primary constituents of (I) bone and tendons, (II) cartilage, and (III) skin and vasculature. Basal membranes are mostly made of collagen type IV, a non-fibrillar version of the protein. Typically, the collagen employed in biomedical applications is extracted from rat tails, pig skin, fish or avian sources, or bovine skin and tendons, with behavioral variations according to the species (Chattopadhyay and Raines 2015).

Due to the lack of yield and cost competition between bacterial or heterologous expression of collagen, these sources continue to be preferred despite the danger of allergic responses and disease transmission. Collagen is highly biocompatible, non-immunogenic, non-pyrogenic, and naturally biodegradable due to the action of collagenases. The high degree of covalent cross-linking between chains also makes collagen strong and elastic from a mechanical standpoint. This leads to its main drawback, limited solubility, which calls for the use of acids to produce solutions in usable concentrations. In addition, collagen promotes the development of granulation tissue and acts as a natural substrate for cell migration, making it a very intriguing substance for wound healing. Collagen type I hydrogels are frequently produced by adding sodium chloride into the solution to obtain better optical performance and mechanical properties. The sequential viscoelastic properties and transparency can be improved more by raising pH value near the isoelectric point at a below ionic strength. (Lian et al. 2017). Before gelation, the collagen solution might include living cells and bioactive compounds. Hydrogel's micro and macro structures are crucial for cell growth and, by extension, tissue regeneration. In this situation, 3D technology and gelation process settings can imitate the precise form of the target tissue (Stegemann and Nerem 2003). The positioning of collagen fibrils in various tissues, such nerves, and corneas, is crucial for controlling cell migration and orientation. To achieve this goal, many techniques have been devised, including utilizing keratocytes to build micro-patterned templates (Vrana et al. 2007) and nanoclusters with magnetic force to regulate the direction of collagen fibrils (Antman-Passig and Shefi 2016).

Due to its tremendous versatility, collagen is employed in many tissue targets. Bio performances, which is why several investigations have been conducted were performed, including combining it with various synthetic materials and or biomolecules by means of fabrication strategies. The use of collagen as a biomaterial has been limited owing to the fabrication of amino acids with potential for thrombosis and the activation of the cascade of coagulation as a result of the breakdown product of collagen. Using pure collagen as a cost-effective method in the production of large quantities of biomaterials is also limited by its high price (Nair and Thottappillil 2015).

17.2.1.1.2 Elastin

Elastin is a crucial protein connector in tissues that gives organs their leeway. Tropoelastin, a solubilized precursor that grows into an insoluble polymer, makes up the crosslink network. Besides, the deposited elastic fiber consists of non-crystalline elastin and microfibrils as a scaffold. Moreover, the self-assembly characteristic of tropoelastin and elastin-like peptides under physiological circumstances is very important. Due to its elasticity property, elastin is also a component of interest when it comes to growing scaffolds for skin substitutes and vascular grafts. Elastin or polymer that look-a-like with a high porosity has been developed in drug delivery, 3D cell cultures, and gene delivery systems. It is also naturally occurring as a substance of the ECM and possesses a number of intriguing qualities, including biocompatibility and biodegradability. Since the proteins applied in biomaterials must be refined, elastin must be sorted out from the elastic fibers that produce during its synthesis to be employed in hydrogel manufacture, despite the fact that it is a very crucial protein for the ECM structure. Contaminations frequently occur throughout this purifying procedure, which might subsequently result in immune responses by the body. Different processes can be used to create elastin-based hydrogels. Depending on the formulation: usage of glutaraldehyde agent for crosslinking, electrospinning, self-assembly crosslinking, and suitable functional groups. The desired hydrogel can be created by altering the porosity and mechanical properties of elastin during the synthesis process. For instance, a high porosity hydrogel that is suitable for cell growth is produced when the pore size is increased during the fabrication process using high pressure carbon dioxide (CO_2) (Mithieux et al. 2004). Numerous hydrogels can be made using elastin.

In this process, one technique to increase the elastic qualities of other kinds of natural hydrogels is to add elastin to them. Collagen gels may be enhanced with elastin to create vascular scaffolds, which is of special interest (Boland et al. 2004). Collagen-elastin gels have furthermore demonstrated efficacy in constructing scaffolds for the healing of skin wounds (Lamme et al. 1998). Additionally, research has shown that elastin scaffolds promote and facilitate fibroblast development (Annabi et al. 2009). Although no substantial alterations were made to the gel, elastin may also be combined with gelatin to create hydrogels with a gelatin-elastin foundation (Thomas et al. 2007).

17.2.1.1.3 Fibrin

The last step in the blood coagulation process results in fibrin, which is created when thrombin cleaves fibrinogen. Its primary function is the creation of the clot, which helps to halt the bleeding and forms the foundation for cell migration during wound healing. This molecule plays a significant part in "fibrin glues", often employed in surgical procedures, where pure thrombin and fibrinogen are combined with factor XIII and calcium to simulate the last stage of coagulation in vitro. Fibrin glue has hemostatic properties in addition to speeding up wound healing and lowering bacterial infections (Ahmed et al. 2008). The dormant form of fibrin, fibrinogen, when activated, forms a vast network of fibers that aid in the regeneration of injured tissue. Fibrillar gel made of fibrin clots serves as support in this situation, providing a scaffold that is engaged in the ultimate healing process. Blood can be utilized to produce fibrinogen, which can then be used as an allogenic source for the scaffold to reduce or completely eliminate the problems associated with incompatibility of the immune system (Catto et al. 2014). These gels are viewed as an option to collagen because they can be created easily from patient blood. (Seifu et al. 2013). Additionally, according to some research, cells seeded into a collagen gel produce less collagen and elastin than cells grown into a fibrin gel (Grassl et al. 2002) and fibrin gel outperforms autologous bone grafts when used to transfer cells (osteoblast) or regenerate bone (Kim et al. 2007). Fibrin can also encourage cell growth and angiogenesis while lowering inflammation because it is a natural substrate for platelets, endothelial cells, fibroblasts, and macrophages. (Heher et al. 2018). These are all highly advantageous properties for a wound healing substance. Additionally, fibrin is used for its capacity to bind heparin to entrap and administer growth factors (GFs) in a regulated manner. To specifically bind GFs to fibrin and regulate their release, functionalization is also used (Whelan et al. 2014). Fibrin's weak mechanical properties and rapid biodegradability are its main drawbacks, but they are frequently offset by its reactions with other biopolymers. The use of fibrin from nonautologous source further raises the effect of blood-transmissible diseases.

17.2.1.1.4 Gelatin

Gelatin is produced by the hydrolysis reaction of collagen Type 1, which involved the physical, chemical, or enzymatic mechanism. Type A or Type B gelatin are produced primarily by denaturation by acid or alkali treatment. Due to the broken triple helix and non-covalent interactions between molecules, gelatin and collagen have the same molecular structure from a structural perspective (Echave et al. 2017). In contrast to native collagen, this significantly improves its immunogenic profile and solubility at the expansion of its mechanical stability, which necessitates additional cross-linker to other molecules for use in biomedical applications. A biological perspective, gelatin is attractable to metalloprotease 2 and 9 matrix rather than typical collagen by accelerating the emergence of granulation tissue. Gelatin is one of the greatest solutions for scaffold engineering because of its limited immunogenicity and appreciable degradability. Gelatin is a natural and inexpensive vascular polymer, which can be grouped into two types with different treatments. The first type gelatin A is formed from acids with pH 1 to pH 3 and preferable for scaffold building block, while gelatin B is produced from the alkaline state. Another method of regenerating vascular tissue uses gelatin as a coating agent to improve cell attachment (Kozlov and Burdygina 2016). Gelatin has the ability to thicken, emulsify, foam, and produce gels. Supramolecular structure is important for the mechanical ones (Jao et al. 2016). Based on extraction circumstances and basically for filling molecules by electrostatic contact, anionic or cationic gelatin is synthesized. Since gelatin can withstand a wide pH range and high temperatures, synthetic polymers can be grafted onto the gelatin backbone.

17.2.1.1.5 Silk Fibroin

Naturally occurring silk fibers are produced by a variety of worms and spiders, the most prevalent of which is Bombyx mori. Sericin and fibroin are two of the many proteins that make up the fibers in its cocoon. The linear part of the silk filament is made up of two silk fibroin chains known as brins (Koh et al. 2015), which give it its strong and glossy qualities. Silk proteins have been replicated and formed in heterological systems. However, silkworm synthesis is the most practical and efficient technique because the produced proteins lack the physical characteristics of native proteins. The light weight with 25 kDa and heavy weight with 325 kDa chains in silk fibroin are bonded together by a disulfide linker to form a complex. In the heavy chain, the presence of the sequence of hydrophobic repetitive domains, comprising glycine with two accelerate residues, which primarily were alanine or serine, and forms antiparallel sheet crystalline regions, was responsible for the mechanical properties. Hydrophilic non-repetitive domains also contribute to fiber elasticity. Due to the structure of the material, silk fibroin is strong, adaptable, and incredibly compressible. Its use in surgical sutures is explained by its tensile strength of up to 300-740 MPa (Kim et al. 2017). Fibroin can be degraded by proteases, and if sericin is successfully removed, both the typical protein and its degraded components have a very potent immunogenic profile. Following breakdown, silk fibroin can be shaped into films, hydrogels, and sponges.

Several solvents such as lithium bromide, ionic liquids, formic acid, and trinary solvent system (mixture of $CaCl_2$, ethanol, and water) can be used to dissolve silk fibroin. The preceding essential action is to remove the sericin, an antibacterial and UV-resistant protein which adheres to the silk strands by utilizing degum methods including weak $NaCO_3$ (sodium carbonate) or citric acid, sodium oleate, and urea solution (Han et al. 2013). In some cases, for example to stabilize the film form by fibrin, Silk I and Silk II were mixed together, although Silk I is a free-solvent and Silk II is non-water-soluble because of its antiparallel beta-sheet structure. During the heating process, Silk I is transformed into crystallite Silk II with the increment of mechanical properties such as tensile strength strain and Young's Modulus (Kim et al. 2018). Hence, it can be used in the engineering of vascular tissue due to its superior mechanical qualities, low adverse immunological response, limited thrombogenic, and compatible degradation rates. Furthermore, it works with a variety of production procedures, including electrospinning and gel spinning. A few of the methods used in 3D printing with silk hydrogel include extrusion, lithography, and most currently a digital light technique utilizing grafting methacrylate. (Wang et al. 2017). Moreover, to construct a bioink with better rheology properties, silk is frequently mixed with other polymeric gel like gelatin and alginate (Kirschning et al. 2018).

17.2.1.2 Polysaccharides-Based Biomaterials

The most prevalent and easily accessible biomaterials in nature are polysaccharides, which can make up to 75% of all organic compounds and come from renewable sources. They may have bacterial, fungal, fungi, or plant origins, which can have an impact on the purity and variability of the polymer. In general, manufacture is more expensive than extraction from plants, which involves some risk of contamination from other compounds. Although the availability from bacterial strains is expensive, the purity is still higher. Long linear branched units of saccharide is known as polysaccharides which are bonded to homo (one) or hetero (two or more) monomers of high molecular weight polysaccharides. The most prevalent and easily accessible biomaterials in nature are polysaccharides, which can make up to 75% of all organic compounds and come from renewable sources. Polysaccharides are endowed with numerous physicochemical properties, such as hydrophilicity, biocompatible, biodegradable, water solubility, and biological activity, depending on the types of O-glycoside linkages, branching, molecular weight, and most crucial is the residues that make up the chain. These polysaccharides may be derived from bacteria, fungi, fungi, or plants. Due to the readily accessible hydroxyl, carboxylic, and amine residuals, all of these qualities may be altered chemically (Shelke et al. 2014) by altering the structure of the substance (Hakkarainen et al. 2016).

17.2.1.2.1 Natural Polysaccharides

Cellulose

The primary part of the walls in bacteria, fungi, algae, and plants is cellulose, which is one of the most pervasive substances in nature. This polymer has a highly crystalline structure and is made up of chains of d-glucose units that are -(1,4)-linked and congregate to form fibrils. Although normally insoluble in water, cellulose has a high hydrophilicity. The modulation of its properties is strongly influenced by the source, the size of the fiber, and the proportion of crystalline areas. To give the fibers their stiffness and elasticity, cellulose fibers alternate between crystalline and amorphous regions. The flexible and plastic properties of the fibers are due to their amorphous regions. The capabilities of cellulose are influenced by a second crucial factor called scale. Functionality is greatly increased when a molecule is shrunk to a nanofiber size, like a nanofibril or a shorter nanocrystal. (Picheth et al. 2017). The third factor impacting the characteristics of cellulose is its origin. While cellulose of plant origin may be separated using low-cost methods, it still includes minute amounts of lignin, pectin, and hemicellulose, which necessitates more cost purification operations. These involve strong acid or alkali treatment prior to destruction (de Oliveira Barud et al. 2016).

The most promising methods for producing nanocellulose, which is made up of fermenting glucose, are bacteria from the families *Gluconacetobacter*, *Agrobacterium*, *Pseudomonas*, and *Rhizobium* (Fu et al. 2012; Klemm et al. 2009). Nanocellulose is a pure polymer that can be produced in this way. The finished product has numerous advantages over its parallel plant-based because it is made of nano-fibrils which is linear with highly crystalline and non-woven material (Ullah et al. 2016). It is also enhanced in properties such as tensile strength, decreased degradation, water absorption capacity, and permeability (Lin and Dufresne 2014). Bacterial cellulose is produced in safe conditions with minimal to no solvents, and it is less toxic and more biocompatible than cellulose derived from plants. Overall, bacterial cellulose meets several of the prerequisites for materials used in wound healing. Additionally, as stated by Stumpf et al. (Stumpf et al. 2018), it may be tailored to a great degree by using biotechnological technologies to modify it in situ and ex situ. These modifications might be used to enhance solubility and antibacterial activity (Sulaeva et al. 2015; Kucinska-Lipka et al. 2015), lessen immunogenicity (Da et al. 2017), or both. One of the most widely used cellulose derivatives is carboxymethylcellulose (CMC). The production of wound dressings like hydrocolloids has utilized CMC (Wong and Ramli 2014). Regardless of the level of replacement, CMC improves wound healing and helps control microbial infections (Ng and Jumaat 2013) in burns and ulcers. However, it has a tendency to be too sticky, which may cause discomfort while changing the dressing (Maurer et al. 2013).

Starch, Dextran, and Pullulan

Carbohydrate or starch in the form of granules, the main polysaccharide that used by plants for storage. It was made by 25% linear amylose -(1,4)-linked d-glucose units and 75% branching amylopectin (Kirschning et al. 2018). The branches of amylose are connected by -(1,6)-linkages, respectively. While amylopectin makes up the polymer's crystalline sections, amylose mostly resides in the amorphous form. Starch is also an excellent degradability profile, which is widely accessible, inexpensive, and has a low mechanical resistance due to its superior hydrophilicity. Hence, it is rarely used in biomedical applications, instead, it is frequently combined with another type of polymers or formed chemically after being modified. For example, the use of the nanofibrous starch-based scaffolds which were fabricated by Waghmare et al. (Waghmare et al. 2017) or the hydrogel membrane starch-based founded by Pal et al. (2006).

Bacteria, especially those of the *Lactobacillus* spp., produce a polymeric homosaccharide known as dextran, which is made up of side chains and linearly linked glucose residuals connected by -(1,6)-glycosidic bonds. High biocompatibility and no degradation of the polymer enzymatic disintegration of dextranases yields non-toxic substances like those. As a result of its compatibility with blood and high swelling potential, it can be utilized to prepare wound dressings during emergencies. Proangiogenic in nature, it promotes skin regrowth (Sun et al. 2017). The fungus *Aureobasidium pullulans*, which resembles yeast, produces the exopolysaccharide known as pullulan. It consists of linearly arranged -(1,6)-maltotriose monomers connected by -(1,4)-glycosidic linkages to form a non-polar, dissoluble in water polysaccharide. It is very hydrophilic and soluble, as well as biocompatible and degradable. Pullulan has anti-inflammatory and anticoagulant properties, although it is relatively costly (Tabasum et al. 2018). Its key drawbacks include a lack of antibacterial characteristics, weak mechanical strength, and a limited capacity to assist cell adhesion. These characteristics make it the most commonly employed in conjunction with other polymers.

17.2.1.2.2 Basic Polysaccharides

Chitosan and Chitin

After cellulose, chitin is the polysaccharide that occurs in nature in the greatest abundance. Its main sources are the exoskeletons of arthropods, such as crustaceans and insects, though chitin can also be extracted from fungi. Due to chitin's limited solubility in aqueous solutions, its deacetylated derivatives are often lumped together in the definition of chitosan (Fig. 17.2) and are mostly utilized in pharmaceutical formulations and biological applications.

Chitosan is a chemical compound that is a 2-Amino-2-deoxy-D-glucopyranose linear copolymer containing nitrogen (glucosamine) and N-acetylglucosamine (2-acetamido-2deoxy-D-glucopyranose) units linked by linkages that are -(1,4)-glycosidic. After sucrose, chitin is the second-most prevalent polysaccharide in nature. Lysozyme uses chitosan as a substrate, and it is a GRAS (generally regarded as safe) excipient (Elviri et al. 2017). It has a low profile for acute and chronic toxicity (Baldrick 2017). The body naturally produces glucosamine and

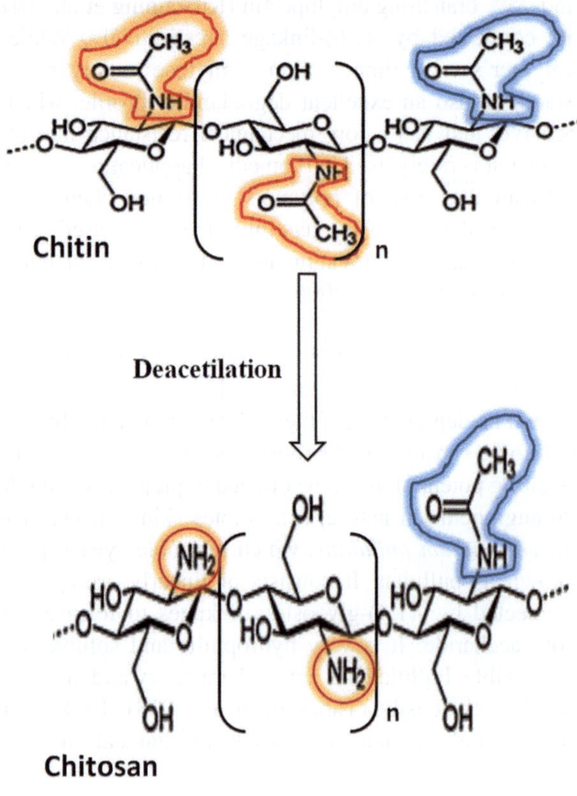

Fig. 17.2 Schematic of deacetylation of chitin to form chitosan. (Reproduced under Creative Commons Attribution 4.0 International license from Vicente (2019))

Nacetylglucosamine, the breakdown products of the glycosaminoglycans that make up ECM and cartilage, as well as HA. Chitosan is regarded as a helpful polymer for tissue regeneration and hence wound healing and the body frequently tolerates it well. Its biocompatibility profile is frequently expressed as a percentage of deacetylation because higher levels of deacetylation enhance cell compatibility and biodegradability while reducing inflammatory reactions. Chitosan's physicochemical and biological properties are determined by the degree of deacetylation, the order of the deacetylated units, and the molecular weight of the polymer. Due to the protonation of amine groups, chitosan, a basic polymer, is more soluble in acidic pH. Its hemostatic function (Patrulea et al. 2015) and antibacterial activity (Sundaram et al. 2006) against fungus and bacteria are both attributed to positive charge density. Chito-oligosaccharides, in particular, are known to have analgesic and anti-inflammatory actions (Kong et al. 2010) via activating macrophages, the complement system, and releasing cytokines. Furthermore, chitosan (Singh et al. 2017), which encourages the development of granulation tissue and neo-angiogenesis in wounds (Intini et al. 2018), is used, which improves the caliber of tissue regeneration.

Chitosan has several benefits that make it a good choice for hydrogel preparations, including antibacterial characteristics, ease of sterilization, cheap cost, bioactivity, and controllable biocompatibility (Ueno et al. 2015). Temperature and pH are only two examples of the variables that may readily impact this sort of hydrogel. Poor mechanical qualities are one of its drawbacks; however, this issue may be solved by crosslinking with chemical substances (Vieira et al. 2017) or gelatin (Huang et al. 2005). Chitosan has enormous potential as a natural polymer for use in wound healing because of its stimulatory effect on leukocytes and antibacterial properties (Hamedi et al. 2018). Chitosan-based hydrogels for wound healing applications can be created using a variety of methods, including photopolymerization, chemical or ionic cross-linking, and polyelectrolyte complex formation when anionic polymers are present. Like cellulose, chitosan units may be modified by adding amino and hydroxyl groups. The chitosan derivative most frequently used in the creation of wound dressings is carboxymethyl chitosan, which has improved solubility and can be formed into films (Hamedi et al. 2018). As previously mentioned, chitosan is combined with other polymers like PEG, PVA, alginate, and gelatin to create wound dressings.

17.2.1.2.3 Sulfated Polysaccharides

Sulfated polysaccharides make up the other saccharidic components of the connective tissues and ECM. Given that they are all extremely acidic and negatively charged, chondroitin and GAGs heparan sulfate, heparin, keratan sulfate, dermatan sulfate, and others frequently referred to as GAGs share structural similarities with each other. These polysaccharides, unlike HA, are typically present in conjunction with structural proteins when proteoglycans are formed. By supplying hydration, supporting structural integrity, and promoting cell migration, all GAGs actively contribute to wound healing (Maver 2018). GAGs are typically connected with other polymers in their natural environment, where they provide structural support. This is also true when GAGs are used to make wound dressings.

A dimer of N-acetylgalactosamine and -glucuronic acid molecules is repeated several times to form the negatively charged sulfated branching GAGs known as chondroitin sulfates. Based on where N-acetylgalactosamine is sulfated, different forms of chondroitin sulfates can be produced, with types A (4-sulfated) and C (6-sulfated) being the most common (Melrose 2016). Chondroitin sulfate, which is a component of cartilage, actively supports joints' resistance to compression. L-iduronate and N-acetylgalactosamine-4-sulfate are dimers that make up dermatan sulfate, which is frequently combined with chondroitin sulfate. While actively promoting FGF-2 activity, which is beneficial for cell growth, it has only mild anticoagulant effects.

Heparan sulfate, which is made up of dimers of l-iduronate-2-sulfate and 2-N- and 6-Osulfated d-glucosamine, is frequently coupled with heparin (Olczyk et al. 2015), the sulfated GAG with the highest density of negative charges. Heparin, which is composed of 6-O-sulfated N-sulfated glucosamine and 2-O-sulfated iduronic acid, has a well-known antithrombogenic activity. Because it has anti-inflammatory qualities, encourages angiogenesis, stimulates the production of collagen, and has an affinity for and protective role against GFs, its fractionated form is beneficial in wound healing.

17.2.1.2.4 Acid Polysaccharides

Alginate

Alginate is a polymer found in some bacterium capsules and the cell walls of brown algae. Two distinct monomers, -D-mannuronate (M) and -L-guluronate (G), arranged into blocks, provide the basis of its structure. Since G blocks create ionic bridges, the quantity and distribution of these blocks affect the mechanical characteristics of alginate. To form gels, alginates may bind divalent cations such as Ca^{2+}, Ba^{2+}, Sr^{2+}, Zn^{2+}, Cu^{2+}, Cd^{2+}, or Co^{2+}. The behavior of the compounds depends on their molecular weights and the proportions of the two subunits. Mannuronic acid oversees the capacity to absorb fluids and provide the chains flexibility, whereas stiffness improves specifically with an increase in molecular weight and proportion of guluronic acid units (Thomas 2000). Depending on the source, the relative proportions of the two units produce alginates with noticeably different chemical and physical characteristics. Alginate is typically isolated from brown seaweeds such as *Macrocystis pyrifera*, *Laminaria*, and *Ascophyllum* spp. (Lee and Mooney 2011), but given its widespread use in industrial and medical settings, some interest has recently been directed toward producing alginate using bacterial strains, which allows for better control of its properties. It was discovered that a few strains of *Pseudomonas* and *Azotobacter* release alginate as exopolysaccharide (Hay et al. 2010) and may be utilized to generate alginate or alter the guluronic content of alginate produced from seaweeds.

Bacterial alginates have higher viscosity, better mechanical stability, a wide range of pore sizes, and osmotic swelling as advantages. Alginate (Fig. 17.3) is a versatile

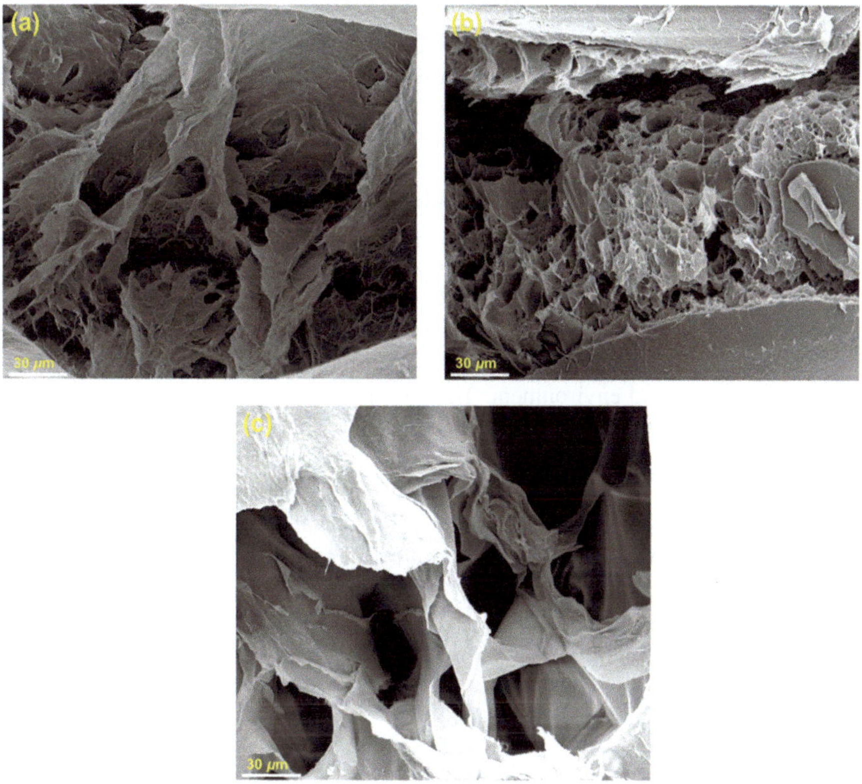

Fig. 17.3 SEM images of pure alginate hydrogel structures with different concentration (**a**) 1.5%, (**b**) 1.8% and (**c**) 2.1% at 3000× magnifications (unpublished image from author)

wound healing material that can be formed into sponges, hydrogels, and film membranes, among other things (Aderibigbe and Buyana 2018). Alginates' capacity to absorb fluids is without doubt what makes them effective for wound treatment. Calcium alginate dressings with wound exudate create a gel by exchanging sodium ions from bodily fluids with calcium ions that help with hemostasis (Boateng and Catanzano 2015). To preserve moisture and promote the development of granulation tissue, the alginate dressing expands and forms a gel on the skin's surface of the wound. Due to its ease of removal and lack of adhesion to the wound, this gel layer also lessens the discomfort associated with changing the dressing (Vowden and Vowden 2014). How the polymer swells is influenced by its content, cross-linker presence, and composition in terms of mannuronic and guluronic units (Liu et al. 2011).

Hyaluronic Acid

Hyaluronic acid is a linear anionic GAG linked together by -(1,3) links and made up of alternating units of 1,4-D-glucuronic acid and 1,3-*N*-acetyl-D-glucosamine.

Traditional extraction techniques that had negative impacts on the molecular weight of the finished product, the potential for contamination, and safety problems were used to separate hyaluronic acid from rooster crest or bovine sources. Because of this, it is currently preferable to synthesis hyaluronic acid utilizing recombinant strains of *Lactococcus lactis*, *Escherichia coli* (*E. coli*), *Agrobacterium* spp., and *Bacillus* spp. (Liu et al. 2011). HA serves as a structural element of connective, epithelial, and basal layer tissues in mammals and acts as a lubricant and anti-shock agent in synovial fluid. It is therefore very well tolerated and non-immunogenic. An extremely viscous gel may be formed by hyaluronic acid, which has a strongly negative charge and can absorb and hold a lot of water. The thickness of this gel grows with the molecular weight of the polymer, which can be anywhere between 5 and 20,000 kDa. This function helps to retain hydration and ensure high exudate absorption in a wound environment. The physiological reactions to hyaluronic acid are also influenced by chain length. Additionally, hyaluronic acid directly promotes wound healing by promoting the migration, adhesion, and proliferation of skin fibroblasts and keratinocytes in several species as well as activating the production of collagen (Longinotti 2014). From a microbiological perspective, hyaluronic acid can lessen or stop bacterial adhesion and functions as a bacteriostatic (Romano et al. 2007; Pirnazar et al. 1999). Due to all of these qualities, hyaluronic acid is frequently used, notably in the form of sodium salt, in the production of wound dressings, either alone or in combination with other molecules.

17.2.1.2.5 Synthetic Polysaccharides

Synthetic biomaterials have the advantages of batch-to-batch uniformity, simple production, and low cost. They are free of contaminants, typically mechanically stable, and degrade under controlled conditions, but there is a possibility that they could be poisonous. Both hydrophilic and hydrophobic polymer systems, which are frequently used in dressings for wound healing, fall into these two categories. Hydrophobic polymers have functional groups with low polarity, making them insoluble in water or bodily fluids. Due to their insolubility in water, these systems are particularly appealing for tissue regeneration in wound healing and many other applications.

Silicones, polyesters (polyethylene terephthalate), polytetrafluorethylene, polyethylene, polypropylene, polycaprolactone (PCL), poly(l-lactic acid), and a number of polyurethanes are examples of materials that are biocompatible and stable (PUs). Polar functional groups like carbonyl, carboxyl, ester, amino, and/or amide increase the polar component of the surface's free energy on synthetic polymer-based scaffolds, making these biomedical devices more wettable and consequently more susceptible to interactions with ECM biomolecules. In most cases, hydrogels are created by physically or chemically cross-linking synthetic hydrophilic polymers, which are hydrophilic in nature. The synthetic polymers that are used most frequently are poly(ethylene glycol), poly(glycolic acid), poly(propylene fumarate), poly(hydroxyethylmethacrylate), poly(vinyl alcohol), poly(aminoamide), poly (aminoacid), and poly(ethylene oxide) (phosphoester). An overview of the synthetic polymers most frequently used in the treatment of wounds is given in the sub-sections follow.

Polyurethanes (PU)

A group of biodegradable biomaterials called PUs are used in applications for wound healing (Lamba et al. 2017). PU is a tunable material because its composition can be adjusted by varying the amounts of its hard di-isocyanate, soft polyether, and preparation extenders. PUs are utilized in wound healing as semipermeable dressings with great flexibility, mechanical strength, oxygen permeability, and transparency to prevent bacterial infections and provide an appropriate moist environment. The organic components of a polyurethane polymer are joined by carbamate bonds. When making wound dressings, polyurethane serves as a semi-permeable barrier that shields the wound from the environment and bacterial invasion. Due to its semipermeability, the wound site has a moist environment, which is ideal. One way to get around the PU polymer's adhesion problem is to cover polyurethane meshwork with collagen or collagen-based peptides, which dramatically improve cell adhesion and tissue biocompatibility. PUs can be used to make films, hydrogels, and hydrocolloid dressings with different water permeabilities and swelling ratios, depending on the synthesis conditions. PU scaffolds can be made using electrospinning, solvent-casting/salt-leaching techniques, and thermophase separation. The preparation method utilized affects the porosity, size, and rate of deterioration. Particularly, it is important to keep an eye out for and limit harmful degradation products. They also have a polyester chemistry, break down hydrolytically, and produce a significant amount of by-products that lower the pH of the immediate environment, which affects how the body reacts to wound healing. PUs can serve as substrates for natural polymers to produce composite wound dressings that enhance the physical characteristics of natural polymers like collagen.

Poly(Ethylene Glycol)

Polyethylene glycol is a versatile, biocompatible, hydrophilic, and non-immunogenic etherbased polymer (PEG). PEG is an oligomer or polymer of ethylene oxide that is water soluble, non-toxic, and immuneogenic. PEG is a polymer that has been authorized for human intravenous, oral, and cutaneous use by the US Food and Drug Administration (FDA). PEGs come in a variety of molecular weights (from 300 Da to 106 Da) and are commercially available as liquids or low-melting-point solids. They are also water soluble. Growth factors, including EGF and PEG macromers, have strong adhesiveness and can be joined. These might be directed at the area of the wound. PEGs have a large volume of hydration and structural flexibility, which minimize cell adsorption. Although the chain length largely has no effect on the polymer's chemical properties, it does have an impact on some of its physical characteristics, such as viscosity. Only low-molecular-weight PEGs are allowed for use in medicine, and the excretory system is the only place where polyether chains can be eliminated biologically. Utilizing PEG improves the hydrophilicity, oxygen permeability, and porosity of the scaffold (Chitrattha and Phaechamud 2013). PEGs have been widely used as matrices for controlling the delivery of medication as well as cell delivery vehicles for promoting tissue regeneration. By reacting the terminal hydroxyl groups with acrylate or methacrylate monomers, they can be chemically changed to produce PEG diacrylate or dimethacrylate, biocompatible polymers suitable for hydrogel preparations. Bioactive scaffolds may also be made by co-polymerizing PEG with the -hydroxy acids of lactic acid or glutamic acid (PLG), which hold particular promise for soft tissue engineering (Rojo et al. 2014).

Polycaprolactone (PCL)

It is possible to create a biocompatible and biodegradable polymer by ring-opening polymerizing ε-caprolactone-PCL. Due to its potential to encourage collagen synthesis and FDA approval for use in humans as a medication delivery system, PCL is a biomaterial for wound healing processes. PCL exhibits strong hydrophobicity, as do other aliphatic polyesters, and gradually deteriorates in physiological settings due to hydrolytic breakage of its ester bonds. By using autocatalyzed bulk hydrolysis and degraded linear aliphatic polyester, polycaprolactone is produced. It is a wonderful example of a biocompatible, biodegradable, and bioresorbable polymer due to its exceptional mechanical qualities. Some of the methods for creating PCL scaffolds include electrospinning, solvent casting, photopolymerization, fused deposition modelling, extrusion deposition, and low-temperature deposition. It is possible to adjust porosity, pore size, and pore interconnectivity using solid free-form methods, which improves cell adhesion and proliferation on PCL. PCL is blended with natural polysaccharides such as chitosan, alginate, or gelatin to improve the hydrophilicity, mechanical resistance, and tissue regeneration processes of scaffolds used in wound healing applications (Bahrami et al. 2016; Ehterami et al. 2018; Zhou et al. 2017). Copolymers of poly(caprolactone fumarate) and poly(ethylene glycol fumarate) have been created to create scaffolds with regulated rheological, physical, and mechanical properties suitable for tissue healing applications (Wang et al. 2006).

Poly Lactic Acid (PLA)

Poly l-lactic acid (PLLA) is an aliphatic polyester that can also be obtained naturally from maize and tapioca roots. It is made from lactic acid and cyclic di-ester lactide monomers. Three isomeric forms result from the monomer's chirality: the racemic (d,l) amorphous form, the semicrystalline d(−), and the d(+) form. Because of the chirality of lactic acid, polymerization of l,l-lactide results in polymerization of poly-l-lactide, a slowly crystallizing, semicrystalline polymer with melting and glass transition temperatures that vary depending on the polymer processing parameters in the ranges of 40–50 °C, 55–80 °C, and 170–180 °C, respectively (DeStefano et al. 2020). PLA is insoluble in water but soluble in chlorinated solvents, tetrahydrofuran, and dioxane, and it is biodegradable and in vivo absorbable. PLA also has high mechanical strength and is suitable for use as a wound dressing (Danhier et al. 2012). The stereochemical makeup of PLA has a considerable impact on the melting point of polymers, but no changes were seen in the temperature at which glass transitions occur. Several factors influence the rate of PLA degradation, including scaffold surface porosity, sterilizing method, molecular weight (lower molecular weight distribution equals faster degradation), and degree of crystallinity (lower degradation rate for higher degree of crystallinity). Despite its poor hydrophilic properties, PLA is frequently combined with natural polymers (such as chitosan, alginate, and others) or co-polymerized with other monomers to produce co-polymers with tunable properties for a variety of biomedical applications, such as poly(lactic acid-co-glycolic acid) (PLGA), poly(lactic acid-co-caprolactone), poly(lactic acid-co-ethylene glycol), and poly(lactic acid (lactic acid).

Poly(Lactic-co-Glycolic Acid) (PLGA)

A copolymer known as poly(lactide-co-glycolide) (PLGA) is created when polyglycolic acid (PGA) and polylactic acid (PLA) are combined. Numerous crucial characteristics of this copolymer, including biocompatibility, mechanical strength, and ease of shaping into suitable forms and sizes, are present. PLGA is a biodegradable, biocompatible co-polymer composed of two distinct monomers—cyclic dimers (1,4-dioxane-2,5-diones) of glycolic acid and lactic acid in varying ratios. It has a glass transition temperature of around 37 °C. It can be made in a variety of forms, including hydrogels, nanoparticles [NPs], microspheres, and nanofibers, and is frequently used to make a variety of wound-healing dressings. It also has tunable mechanical properties and is suitable for encapsulating drugs with high efficiency for local controlled delivery (Jain 2000; Machado et al. 2013; Aoki et al. 2017). PLGA has been approved by the FDA and the European Medicines Agency as an excipient for parenteral administration, making it commercially available and GMP-compliant.

PLGA is a versatile, cost-effective, and scalable material that can accelerate wound healing when used alone or in combination with other growth factors (GFs) such as VEGF and bFGF, antibiotics (vancomycin, gentamicin), analgesics (lidocaine), and biologically active and antiinflammatory substances (Chen et al. 2015; Liu et al. 2018; Reis et al. 2017). The benefits of PLGA include its ability to protect drugs from degradation (pH and/or enzyme degradation) and to increase their stability, solubility, and bioavailability, which improves therapeutic wound healing in a locally controlled manner with less medication administration. Other PLGA drawbacks include polymer aggregate formation, byproduct production in acidic conditions, and poor drug loading (roughly 1 mg active chemical per 100 mg of polymer). Several articles have described the use of PLGA polymer as an active component in applications for wound healing. The importance of PLGA in supplying lactate as an active agent in angiogenesis processes was reviewed by Porporato et al. (Porporato et al. 2012) with a noticeably enhanced reduction in wound healing time (almost 60% decrease in injured area compared with the control group after 10 days from damage). When electrospun membranes made of PLGA/collagen were tested for antimicrobial activity in a rat wound model, they showed a biphasic release pattern with exceptional activity against Staphylococcus aureus (*S. aureus*) and *E. coli*. A useful technique for delivering hydrophobic (poorly soluble) medications is to create PLGA nanoparticles (NPs). On diabetic mice (db/db), VEGF and bFGF-loaded PLGA NPs (also known as scaffold/GF-loaded NPs) were tested for use in wound healing. The results revealed complete re-epithelialization, as well as improved granulation tissue formation/maturity and collagen deposition when compared to other groups.

17.3 Nanoparticle-Based Smart Wound Dressing for Detection and Therapy

Wound dressings based on nanoparticles are extensively employed as successful solutions for providing therapeutic substances as well as a moist environment for bacteria treatment and wound healing. They cannot, however, deliver therapeutic chemicals in a highly regulated manner, which may result in adverse effects and unneeded treatments. On the other hand, they lack the ability to detect bacterial infection during the healing process. As a result, creating innovative smart wound dressings capable of monitoring and treating bacterial infections is critical. Smart wound dressings have evolved as a novel wound care technique. Smart wound dressings, as the name indicates, are a class of wound dressings that can interact with wounds, perceive, and react to changes in wound state or surroundings by utilizing built-in sensors and smart materials such as stimuli-responsive materials and self-healing materials. The wound healing process has several stages, each with its own set of traits and physiological processes (Rodrigues et al. 2009). Smart wound dressings are anticipated to meet the demands of each stage and to respond effectively and appropriately to environmental changes. To date, various categories of smart wound dressings have been developed, including biomechanical wound dressing (Li et al. 2020), stimuli-responsive wound dressing, self-healing wound dressing for emotional wounds, self-removable wound dressing (Liang et al. 2021), and monitoring wound dressing. These wound coverings can interact with the wound environment, reflect changes in the wound healing process, manage wound infection, prevent chronic wound formation, and encourage autonomous wound healing.

17.3.1 Environment-Responsive Wound Dressing for Bacterial Therapy

Bacteria invade the skin, numerous physical and biochemical indicators in the wound change, including pH, temperature, enzymes, and toxins generated by bacteria (Halder et al. 2016). For example, a bacterial infection will generate an inflammatory reaction, with the temperature at the wound site being 1–2 °C higher than normal skin (Porporato et al. 2012). Thus, various NPs-based wound dressings that are responsive to bacterium environment, such as temperature-, pH, and enzyme responsive wound dressings, have been created for the on-demand release of therapeutic drugs during bacterial therapy in wounds. They can transition between solid and liquid phases in response to changes in temperature, thermo-responsive hydrogels are attractive candidates for precisely delivering and releasing therapeutic chemicals to wound sites (Cedervall et al. 2007). For example, thermo-responsive hydrogels containing gold nanoparticles demonstrated better regulated and sustained NPs release at the wound site via the sol-gel transition of the hydrogel.

Thermo-sensitive hydrogels based on light-triggerable NPs can be employed as drug reservoirs to provide on-demand antibiotic release as well as photothermal bacterial inactivation. By integrating ciprofloxacin (Cip)-loaded PDA (polydopamine) NPs into glycol chitosan (CS) hydrogel, researchers created an injectable hydrogel (Gel-Cip). Moreover, the polymer NPs underwent a unique surface charge conversion in response to changes in environment pH. When placed in an acidic environment, the pH-responsive NPs changed their surface charge from neutral to positive, and thus facilitated their potent electrostatic interactions with *Methicillin-resistant Staphylococcus aureus* (MRSA) since the cell wall is negatively charged, eventually resulting in situ aggregation of bacteria.

In addition, pH and enzyme-responsive wound dressings can also be used as drug reservoirs for bacterial therapy. In addition, silver nanoparticles (AgNPs) clusters (silver nanoclusters AgNCs) have been shown to be pH-responsive, with the ability to collapse and reassemble into AgNPs when exposed to an acidic environment in bacterial infections (Zuo et al. 2020). This action contributed considerably to AgNCs' improved bactericidal efficacy against MRSA and *E. coli*, which was also observed in vivo. Yu et al. also reported a PDA-modified chitin hydrogel containing a bacteria-triggered hyaluronan/AgNPs/gentamicin nanocarrier (HA/AgNPs/Gen) that demonstrated controlled release of antibiotic and AgNPs via hyaluronidase or pH stimulation, resulting in synergistic bactericidal activity in vitro and in vivo (Xie et al. 2020).

17.3.2 Wound Dressing for Point-of-Care Detection and Therapy for Bacterial Infection

In addition to infection therapy, smart wound dressings have the ability to identify bacterial infections at the point of care in relation to wound requirement. Currently, identification is frequently performed visually based on clinical signs such as swelling, redness, and purulent discharge (Qiao et al. 2020). However, the procedure should be carried out by changing dressings, which may slow down infection treatment and potentially raise the danger of subsequent harm. In situ detection of wound infection in real time can significantly enhance patient care pathways and provide promising clinical outcomes. As a result, the development of novel wound dressings for in situ diagnosis and on-demand treatment of wound infections is crucial. Many types of smart dressings with sensitive antibacterial capabilities depending on endogenous and external stimuli, such as light and toxin released by bacteria, (Sun et al. 2020) have been effectively manufactured to monitor and treat bacterial infection in wounds. Toxins and lipases, for example, are secreted in significant numbers by pathogenic bacteria (*S. aureus* and *P. aeruginosa*) but not by non-pathogenic *E. coli* species (Qin et al. 2022).

Furthermore, bacterial-responsive wound dressing combined with surface-enhanced Raman scattering (SERS) imaging can enable a real-time evaluation of therapeutic impact in skin wound. On the other hand, the AuAgNSs-DTTC enabled exceptionally sensitive detection (down to 300 CFU/mL for MRSA) and a lengthy tracking duration for at least 8 days of leftover germs, allowing for non-invasive monitoring and estimation of the treatment impact. Concurrently, a multifunctional SERS band-aid enabling real-time quorum sensing and bacterial film inactivation was disclosed. The sticky note was formed like a sandwich, with Au nano stars encased between hexagonal boron nitride layers, allowing SERS imaging to track the released signaling chemical from biofilm in real time. The sticky note demonstrated a quick reaction to the secretion as early as 1-h biofilm formation using *P. aeruginosa* and its pyocyanin secretion as a model (He et al. 2020b).

The antibiotic-loaded sticky note demonstrated the potential to successfully inactivate bacterial films at the same time. Paper-based wound dressings have been widely utilized for bacterial detection and therapy due to its low cost, biosafety, and ease of customization. Sun et al. created a pH- and enzyme-responsive paper-based band-aid based on porphyrinic metalorganic frameworks (PCN-224) for monitoring and treating drug-resistant bacteria. Drug-sensitive (DS) and drug-resistant (DR) bacteria, with CS-coated ampicillin-loaded PCN-224 (known as CP) as a treatment agent. When the band-aid was placed on the surface of a drug-sensitive bacteria, the color of the bromothymol blue (BTB) changed from green to yellow in an acidic environment, and the yellow intensity increased with higher bacteria concentration; meanwhile, CS-coated ampicillin-loaded metal–organic framework (MOF) lured to the surface of the bacteria released antibiotic in an acid-responsive manner, causing the DS bacteria to die. Furthermore, nitrocefin would be activated by the -lactamase released by drug-resistant (DR) bacteria when the band-aid came into touch with them. Because MOF has excellent photodynamic therapy (PDT) properties, reactive oxygen species (ROS) would be created during near-infrared (NIR) irradiation to kill bacteria. Moreover, in an *E.coli* infection in mouse skin wounds, a paper-based band-aid was applied to the site for 4 h. *E. coli* infection, and the treated infected wound was nearly healed on the third day, implying that the released antibiotic was sufficient for killing DS *E. coli*, whereas PDT abuse would cause side effects that would inhibit the healing process. In the groups of *E. coli*-infected wounds, all of the examined paper became red, and the combination of antibiotics and PDT hastened bacteria ablation and enhanced wound healing. As a result, the dressings contain a “sensing and treat” capability that may be used to determine whether DR bacteria present based on color and subsequently treat them (Oh et al. 2018).

17.4 Nanoparticle Delivery of Therapeutic Drugs for Wound Healing

It is well known that administering therapeutic active components such as growth factors, nitric oxide, nucleic acid, antioxidants, and antibiotics to injured tissue can increase cell proliferation, migration, angiogenesis, and collagen production, as well as suppress microorganisms, impacting healing wounds that are persistent (Hu et al. 2019). Incidence of chronic wounds nanofibers have gotten a lot of interest because of their structural resemblance to the natural environment (Wang et al. 2013). Nanofibers promote wound healing by having a high surface area to volume ratio, tunable mechanical properties, increased porosity, and the ability to encapsulate nanoparticles and bioactive compounds for controlled release, which allows cells to interact with the matrix during functionalization and remodelling (Zhou et al. 2011). Hydrogels are hydrophilic 3D polymer networks with established properties.

More importantly, nanoparticles and biomolecules can be incorporated in hydrogels, allowing for more advanced topical drug delivery with unique benefits such as improved tissue localization, minimized burst release, and controlled sequential drug release by preserving nanoparticle structural integrity. Non-polymeric nanoparticles, such as silver nanoparticles (AgNPs) and gold nanoparticles (AuNPs), are commonly utilized as therapeutic agents, owing to their anti-infective and anti-inflammatory properties. There is an unmet need for a novel antibiofilm method and effective antimicrobial chemicals, and silver nanotechnology-based treatments have attracted the interest of health care practitioners for improving health care (Goh et al. 2016). AgNPs are utilized in clinical practice for a variety of therapies, including antibiotic-resistant burns, chronic ulcers, and diabetic wounds, as well as hospital acquired bacterial infection. Moreover, AuNPs are widely employed in tissue regeneration, targeted medication delivery, and wound healing. Furthermore, because of its biocompatibility, zinc oxide (ZnO) nanoparticles have been investigated as drug delivery vehicles and treatment techniques for human biological applications (Rakhshaei and Namazi 2017). ZnO nanoparticles have been proven to offer therapeutic properties against bacterial infection and inflammation, as well as the potential for wound healing applications. Ceramic nanoparticles with inorganic components have essential therapeutic properties and can transport medicines to injured locations.

In addition to being safe, lipid-based nanoparticles are widely employed to deliver both hydrophilic and hydrophobic medicines. Liposomes maintain long-term medication release by decreasing the toxicity caused by massive drug release via traditional administration. Chitosan is a natural polymer to utilize in the case of polymeric nanoparticles owing to its biocompatibility and antibacterial activity. Aloe vera, vitamin E, and curcumin all have potential benefits for skin wound healing (Ezhilarasu et al. 2020a). The Food and Drug Administration (FDA) has approved the synthetic polymers poly (lactic-co-glycolic acid), poly (−caprolactone), poly (lactic acid), and poly (ethylene glycol). The best biodegradable polymer among them is thought to be PLGA due to its propensity to emit lactate, a breakdown by-product. It has been demonstrated that PLGA nanoparticles promote cell proliferation and shorten the time it takes for wounds to heal. The following is a list of the therapies that nanoparticles can offer.

17.4.1 Growth Factor

Physiologically, every healing process is regulated by growth factors. Growth factors stimulate inflammatory response, angiogenesis, granulation of tissue, and modelling. In the wound healing process, growth factors play an important role by stimulating inflammatory response. Growth factor in an accessible form may be inadequate to accelerate wound healing in chronic wound healing. Growth factor encapsulation in nanoparticles has been widely employed to solve protein delivery limitations by enhancing half-life, encapsulation of more than one biomolecule, and protection against protease degradation in the wound bed via nanoparticle protective properties. In comparison to PLGA nanoparticles and recombinant human epidermal growth factor (EGF) alone, nanoparticle-loaded vascular endothelial growth factor (VEGF) has been shown to accelerate wound closure in both diabetic and non-diabetic mice with chronic wounds (Veith et al. 2019).

Encapsulating growth factor in nanoparticles may protect against enzymatic hydrolysis due to regulated administration of growth factor from the encapsulated nanoparticles. Lai et al. created a collagen (Col)-hyaluronic acid (HA) electrospun with gelatin nanoparticles that may release numerous angiogenic growth factors at the excision wound site, including VEGF, PDGF, bFGF, and EGF (Li et al. 2019). Furthermore, other than that, Li et al. also created KGF-AuNPs to test the stability and binding affinity of KGF for diabetic wound healing. They discovered that by conjugating KGF with AuNPs, KGF retained its bioactive effect at the wound site while being more stable and resistant to proteolytic degradation, promoting keratinocyte proliferation and migration, and generating a greater binding effect to its physiological receptor than unmodified KGF (Li et al. 2019). Clinical investigations on topically applied exogenous growth factors such as VEGF or bFGF in the healing of chronic diabetic wounds were conducted, and local administration of growth factors was shown to be well tolerated. The free form of growth factor administration has encountered issues such as quick leaking from the wound bed, a short biological half-life, and rapid enzymatic degradation, making adequate concentration difficult to attain.

17.4.2 Antibiotics

Infection is the most prevalent symptom of delayed chronic wound healing. Surface infections in chronic wound patients cause the formation of biofilms superficially inside the wound, resulting in protracted inflammation. To provide antimicrobial medications to infected wounds and promote normal wound healing, an effective therapy is necessary. Nanoparticles can be used to particularly target and eradicate infections in this regard. AgNPs are embedded in cellulose nanocrystals (CNCs) obtained from Syzygium cumini leaves, which help keep the wound moist. These CNCs induce re-epithelization when combined with EGF, leading in faster wound healing in diabetic mice with chronic wound (Singla et al. 2017).

The encapsulation of nanoparticles with antimicrobial drugs has emerged as a unique and capable approach to treating diabetic wound infection with minimum negative side effects. Antibiotic resistance is a key issue in antibiotic treatment. *Methicillin-resistant Staphylococcus aureus* infections have been linked to a 64% increase in mortality when compared to non-resistant strains. As a paradigm for producing nano-antibiotics, Kalita et al. created lysozyme-capped gold nanoclusters functionalized with ampicillin, a commonly used antibiotic. The rise of bacterial resistance to conventional antibiotics is a broad issue that must be addressed in clinical trials (Kalita et al. 2018). Dai et al. created an AgNPs-coated bacterial nanocomposite (EPL-g-butyl@AgNPs) that shown no resistance to the most widely used antibiotic, levofloxacin, in a culture of Gram-negative and Gram-positive bacteria (Dai et al. 2016).

Chronic wound infections are usually polymicrobial, meaning that many microorganisms coexist in the same niche. Polymicrobial wound infections typically need higher dosages of antibiotics and fungicides. Continued antimicrobial therapy is associated with the possibility of systemic adverse effects as well as the danger of drug-resistant bacteria. Chitosan bandages incorporating nanoparticles-loaded antibiotics have shown substantial antibacterial activity against co-culture of *S. aureus*, *E. coli*, and *Candida albicans* (Lin et al. 2017). Infrared irradiation of QCSG/GM/GO hydrogels for more than 10 min resulted in death rates of about 100% for all three bacteria, confirming the effectiveness of near-infrared-assisted photothermal antibacterial capabilities. Injectable conductive nanocomposite hydrogel dressings based on GO and cationic polymer for wound healing show great promise (Akturk et al. 2016). In comparison to commercially available Tegaderm, injectable QCSG/GM/GO hydrogels with antibiotics hastened infectious skin defect wounds. The inhibition of S. aureus and MRSA persisted 9 days, confirming the drug's long-term release. Cell compatibility results indicated that increasing the incubation period for the hydrogel groups resulted in increased L929 cell viability. Bacterial infection and persistent inflammation are major factors in the failure of diabetic wound healing interventions. The research investigations listed above assessed the antibacterial properties of nanoparticles loaded with antibiotics. The researchers discovered that nanoparticulate systems against major drug-resistant bacteria might lead to innovative therapeutic uses in the near future.

17.4.3 Nucleic Acid

Chronic wound healing can be accelerated by using nucleic acid-encapsulated particles containing gene therapy and nanotechnology to deliver particular proteins to damaged areas (Mast and Schultz 1996). Non-viral vectors, such as nanoparticles, are a superior alternative for delivering nucleic acid since viral vectors can elicit immunological responses and should always be used with caution.

Nanoparticle-based technology has enabled tailored siRNA delivery and degradation prevention (Dizaj et al. 2014). Although clinical experiments delivering siRNA to heal illnesses have been promising, early clinical trials have failed owing to insufficient effectiveness or substantial off-target effects. Prior to widespread clinical application, RNAi technology must be refined further. Degradation of siRNAs by enzymes in the wound environment and siRNAs not readily taken up by cells due to electrostatic constraints, as the negatively charged cell wall will not easily allow penetration of negatively charged siRNAs into the cells, are barriers to successful siRNA delivery for efficient therapy. To address this, a wide range of nanoparticle-based delivery methods have been pre-clinically evaluated. Delivery vehicles for SiRNAs, such as those listed below, have demonstrated varied degrees of effectiveness, including topical and intravenous formulations.

Firstly, ganglioside (GM3) is a major insulin resistance intermediate that has been demonstrated to be substantially expressed in human and mouse skin. Knocking down GM3S expression in diabetic mice using AuNPs linked with GMS3rsiRNA-based spherical nucleic acids (SNAs) improved wound healing while causing minimal harm (Randeria et al. 2015). Secondly, TNF-a is an inflammatory cytokine that is needed in small levels to speed up wound healing. TNF-a inhibits normal wound healing in diabetic wounds by increasing cell apoptosis, ROS, and matrix degradation (Liu et al. 2006). To accelerate wound healing in diabetic mice, Kasiewicz et al. created lipid nanoparticles encapsulated with specific TNF- siRNA. The same application of liposuction loaded with TNF/SiR reduced TNF expression in diabetic mice wounds by 40–50%. Moreover, another SiRNAs-based delivery vehicle is the nuclear factor erythroid 2–related factor 2 (Nrf2) binds to Keap1 in the absence of oxidative stress (in the cytoplasm). Nrf2 reaches the nucleus after dissociating from Keap 1's repressor site and binding to the antioxidant response element (ARE). By enhancing the removal of harmful electrophiles and ROS, these enzymes are vital for cellular defense. Diabetes produces an imbalance of ROS and overproduction of KEap1, which leads to the degradation of NRF2, which then controls diabetic oxidative stress (Kasiewicz and Whitehead 2016).

Lastly, hyperglycemia also activates redox-sensitive transcription factors, primarily NFkB, leading to an increase in the production of pro-inflammatory cytokines such as IL-6 and IL-8. Because of its capacity to scavenge free radicals, cerium oxide nanoparticles (CNP) can be used as a therapeutic treatment for oxidative stress. In diabetic wounds, miR-146a inhibits interlukin-1 receptor associated kinase 1 (IRAK1) and tumor necrosis factor receptor associated kinase 6 (TRAF6) (Ambrozova et al. 2017).

17.4.4 Antioxidant

During the inflammatory phase of wound healing, physiologically active mediators draw neutrophils, leucocytes, and monocytes to the wound sites, where they assault bacteria and foreign debris. This will result in the formation of reactive oxygen radicals (ROS), which are deadly and can cause cell damage. The accumulation of

ROS causes considerable damage of stem cells, growth factors, and nucleic acids in the injured tissue, reducing their regeneration capacity significantly. Nanoparticle-based wound healing therapy has demonstrated encouraging results in increasing antioxidant activity in animals. Advanced glycation end products (AGEs) production has been identified as a key pathophysiological mechanism in diabetic ulcers which is a chronic wound. When compared to the diabetic wound group, wounds treated with FA-loaded polymeric nanoparticles dispersion showed quicker epithelization, leading to successful wound closure on day 14 (Huijberts et al. 2008).

In a murine model, the antioxidants EGCG, ALA, and AuNPs dramatically reduced RAGE protein production in cultured fibroblasts (Hs68) and diabetic mouse skin. This implies that combining these antioxidants significantly accelerated diabetic wound healing via inflammatory and angiogenesis regulation of antioxidants (Ezhilarasu et al. 2020b). The inhibition of RAGE by intraperitoneal soluble RAGE dramatically reduced TNF-a and IL-6 levels while improving cutaneous wound healing in mice. On diabetic mice, topical gas-injection of an EGCG and AuNP liquid mixture (AuE) using the GNT GoldMedTM liquid DDS resulted in a considerably greater rate of wound closure. In another work, Ponnanikajamideen et al. demonstrated 50% free radical inhibition utilizing the leaf extract powder of a plant, Chamaecostus Cuspidatus, and produced green manufactured AuNPs. The nanofibrous wound dressings exhibited soft tissue-like mechanical properties, free radical scavenging ability, antibacterial activity, and biocompatibility. Because diabetic foot ulcers remain a challenging therapeutic problem, the results significantly support the positive benefits of antioxidants and nanoparticles on diabetes patients with cutaneous wounds. The findings lay the groundwork for the future therapeutic use of AuEA and PLGA nanoparticles in chronic wound treatment (Ponnanikajamideen et al. 2019).

17.5 Conclusion

We have highlighted recent advances in sustainable design using natural and synthetic biomaterials toward for wound healing treatment for this chapter. Due to the presence of a vast type of biomaterials, it is impossible to have the diverse quality of material fabrication, shortage to compare different studies of materials for specific applications as well as determining the type of polymer. Biomaterials from natural sources are more intrinsically biocompatible than synthetic biomaterials which are less superior, and even though it cannot be altered mechanically, structurally, or chemically. In fact, mixing various biomaterials via methods involves such as blended, grafted, and cross-linking synthesis has been the focus, regardless, the most encouraging finding of recent studies on polymers for biomedical applications. Moreover, the methods involved such as electrospinning allowed the creation of many types of nanoscale polymers. Those nanomaterials with different structures such as tubes, fibers, and spheres showed highly constructive combinations with the limitless potential of enhancing the properties such as their biocompatibility, degradation, and physicochemical properties, respectively. On the other hand, the development of nanotechnology for drug delivery has gained the attention of researchers. There are concerns with employing oral or intravenous administration for traditional drug delivery systems for the delivery of proteins and nucleic acids. The old system keeps on struggling with the side effects of the drugs, effectiveness, and patient consent. Hence, the exploitation of nanotechnology has exhibited in the number of novel drug delivery such as not only a broader range of molecules, proteins, peptides, and nucleic acids, but also the more certain targets of controlled release. Particularly nanoparticles made of polymer from biodegradable and biocompatible properties demonstrated potential in cancer therapy and as supported drug delivery systems. Nevertheless, they are capable of not only co-deliver multiple drugs at the same time but also managing to reduce drug resistance and the possible combinatorial effects of various medications. The wound treatment or healing process can be done by using the delivery of a bioactive agent to complete the procedure of wound healing. As of now, clinical applications for wound healing are considerably more compliant as enhancement of this technology would heal the wound better and faster. Moreover, increase in new knowledge would help create novel nanosized drug delivery systems which will definitely exhibit considerable benefits over conventional drug-delivery systems.

Acknowledgments This research was funded by Universiti Sains Malaysia (USM) Short Term Grant (034/PPSG/6315509). We would like to extend our gratitude to the staff of Craniofacial Science laboratory, School of Dental Sciences, USM.

Conflict of Interest The authors declare no conflict of interest.

References

Aderibigbe BA, Buyana B (2018) Alginate in wound dressings. Pharmaceutics 10(2):42. https://doi.org/10.3390/pharmaceutics10020042

Agarwal A, McAnulty JF, Schurr MJ (2014) Polymeric materials for chronic wound and burn dressings. In: Farrar D (ed) Advanced wound repair therapies. Woodhead Publishing, Cambridge, pp 186–208. https://doi.org/10.1002/jbm.b.34861

Agrawal P, Soni S, Mittal G, Bhatnagar A (2014) Role of polymeric biomaterials as wound healing agents. Int J Low Extrem Wounds 13(3):180–190. https://doi.org/10.1177/1534734614544523

Ahmed TA, Dare EV, Hincke M (2008) Fibrin : a versatile scaffold for tissue engineering applications. Tissue Eng Part B Rev 14(2):199–215. https://doi.org/10.1089/ten.teb.2007.0435

Akturk O, Kismet K, Yasti AC, Kuru S, Duymus (2016) Collagen/gold nanoparticle nanocomposites: a potential skin wound healing biomaterial. J Biomater Appl 31:283–301. https://doi.org/10.1177/0885328216644536

Ambrozova N, Ulrichova J, Galandakova A (2017) Models for the study of skin wound healing. The role of Nrf2 and NF-κB. Biomed Pap Med Fac Univ Palacky Olomouc Czech Repub 161(1):1–13

Annabi N, Mithieux SM, Boughton EA, Ruys AJ, Weiss AS, Dehghani F (2009) Synthesis of highly porous crosslinked elastin hydrogels and their interaction with fibroblasts in vitro. Biomaterials 30:4550–4557. https://doi.org/10.1016/j.biomaterials.2009.05.014

Antman-Passig M, Shefi O (2016) Remote magnetic orientation of 3D collagen hydrogels for directed neuronal regeneration. Nano Lett 16:2567–2573. https://pubs.acs.org/doi/abs/10.1021/acs.nanolett.6b00131

Aoki S, Fujii M, Fujie T (2017) The efficacy of basic fibroblast growth factor-loaded poly(lactic-co-glycolic acid) nanosheet for mouse wound healing. Wound Repair Regen 25(6):1008–1016. https://doi.org/10.1111/wrr.12604

Bahrami H, Keshel SH, Chari AJ, Biazar E (2016) Human unrestricted somatic stem cells loaded in nanofibrous PCL scaffold and their healing effect on skindefects. Artif Cells Nanomed Biotechnol 44(6):1556–1560. https://doi.org/10.3109/21691401.2015.1062390

Baldrick P (2017) The safety of chitosan as a pharmaceutical excipient. Regul Toxicol Pharmacol 56(3):290–299. https://doi.org/10.1016/j.yrtph.2009.09.015

Boateng J, Catanzano O (2015) Advanced therapeutic dressings for effective wound healing: a review. J Pharm Sci 104(11):3653–3680. https://doi.org/10.3390/pharmaceutics10020042

Boland ED, Matthews JA, Pawlowski KJ, Simpson DG, Wnek GE, Bowlin GL (2004) Electrospinning collagen and elastin: preliminary vascular tissue engineering. Front Biosci 9: 1422–1432. https://doi.org/10.2741/1313

Catto V, Farè S, Freddi G, Tanzi MC (2014) Vascular tissue engineering: recent advances in small diameter blood vessel regeneration. ISRN Vasc Med 2014:1–27. https://doi.org/10.1155/2014/923030

Cedervall T, Lynch I, Lindman S, Berggård T, Thulin E, Nilsson H, Dawson KA, Linse S (2007) Understanding the nanoparticle – protein corona using methods to quantify exchange rates and affinities of proteins for nanoparticles. Proc Natl Acad Sci Unit StatesAm 104(7):2050–2055. https://doi.org/10.1073/pnas.0608582104

Chattopadhyay S, Raines RT (2015) Review collagen-based biomaterials for wound healing. Biopolymers 101(8):821–833. https://doi.org/10.1002/bip.22486

Chen MM, Cao H, Liu YY (2015) Sequential delivery of chlorhexidine acetate and bFGF from PLGA-glycol chitosan core-shell microspheres. Colloids Surf B Biointerfaces 151:189–195. https://doi.org/10.1016/j.colsurfb.2016.05.045

Chitrattha S, Phaechamud T (2013) Modifying poly(L-lactic acid) matrix film properties with high loaded poly(ethylene glycol). Key Eng Mat 545:57–62. https://doi.org/10.4028/www.scientific.net/KEM.545.57

Da LC, Huang YZ, Xie HQ (2017) Progress in development of bioderived materials for dermal wound healing. Regen Biomater 4(5):325–334

Dai X, Guo Q, Zhao Y, Zhang P, Zhang T, Zhang X, Li C (2016) Functional silver nanoparticle as a benign antimicrobial agent that eradicates antibiotic-resistant bacteria and promotes wound healing. ACS Appl Mater Interfaces 8(39):25798–25807. https://doi.org/10.1021/acsami.6b09267. Epub 2016 Sep 26. PMID: 27622986

Danhier F, Ansorena E, Silva JM (2012) PLGA-based nanoparticles: an overview of biomedical applications. J Control Release 161(2):505–522. https://doi.org/10.1016/j.jconrel.2012.01.043

de Oliveira Barud HG, da Silva RR, da Silva Barud H (2016) A multipurpose natural and renewable polymer in medical applications: bacterial cellulose. Carbohydr Polym 153:406–420. https://doi.org/10.1016/j.carbpol.2016.07.059

Deng H, Zhongpeng Y, Chen S, Fei L, Sha Q, Zhou N, Chen Z, Xu C (2020) Facile and eco-friendly fabrication of polysaccharides-based nanocomposite hydrogel for photo thermal treatment of wound infection. Carbohydr Polym 230:115565. https://doi.org/10.1016/j.carbpol.2019.115565

DeStefano V, Khan S, Tabada A (2020) Applications of PLA in modern medicine. Engineered Regeneration 1:76–87. https://doi.org/10.1016/j.engreg.2020.08.002

Dizaj SM, Jafari S, Khosroushahi AY (2014) A sight on the current nanoparticle-based gene delivery vectors. Nanoscale Res Lett 9(1):252. https://doi.org/10.1186/1556-276X-9-252

Echave MC, Burgo LS, Pedraz JL, Orive (2017) Gelatin as biomaterial for tissue engineering. Curr PharmDe 23(24):3567–3584. https://doi.org/10.2174/0929867324666170511123101

Ehterami A, Salehi M, Farzamfar S (2018) In vitro and in vivo study of PCL/COLL wound dressing loaded with insulin-chitosan nanoparticles on cutaneous wound healing in rats model. Int J Biol Macromol 117:601–609. https://doi.org/10.1016/j.ijbiomac.2018.05.184

Elviri L, Bianchera A, Bergonzi C, Bettini R (2017) Controlled local drug delivery strategies from chitosan hydrogels for wound healing. Expert Opin DrugDeliv 14(7):897–908. https://doi.org/10.1080/17425247.2017.1247803

Enoch S, Leaper DJ (2005) Basic science of wound healing. Surgery 23:37–42. https://doi.org/10.1383/surg.23.2.37.60352

Ezhilarasu H, Ramalingam R, Dhand C, Lakshminarayanan R, Sadiq A, Gandhimathi (2020a) Biocompatible aloe vera and tetracycline hydrochloride loaded hybrid nanofibrous scaffolds for skin tissue engineering. Int J Mol Sci 20:5174. https://doi.org/10.3390/ijms20205174

Ezhilarasu H, Vishalli D, Dheen ST, Bay B-H, Srinivasan DK (2020b) Nanoparticle-based therapeutic approach for diabetic wound healing. Nano 10(6):1234. https://doi.org/10.3390/nano10061234

Fisher OZ, Khademhosseini A, Langer R, Peppas NA (2010) Bioinspired materials for controlling stem cell fate. Acc Chem Res 43(3):419–428. https://doi.org/10.1021/ar900226q

Fishman JM, Wiles K, Wood KJ (2015) The acquired immune system response to biomaterials, including both naturally occurring and synthetic biomaterials. In: Badylak SF (ed) Host response to biomaterials. Academic Press, pp 151–187. https://doi.org/10.1016/C2013-0-15382-2

Fu L, Zhang J, Yang G (2012) Present status and applications of bacterial cellulose-based materials for skin tissue repair. Carbohydr Polym 92(2):1432–1442. https://doi.org/10.1016/j.carbpol.2012.10.071

Gaharwar AK, Peppas NA, Khademhosseini A (2014) Nanocomposite hydrogels for biomedical applications. Biotechnol Bioeng 111:441–453. https://doi.org/10.1002/bit.25160

Gelse K, Poschl E, Aigner T (2003) Collagens: structure, function, and biosynthesis. Adv Drug Deliv Rev 55(12):1531–1546. https://doi.org/10.1016/j.addr.2003.08.002

Goh ET, Kirby G, Jayakumar R, Liang XJ, Tan A (2016) Accelerated wound healing using nanoparticles. In: Hamblin MR, Avci P, Prow TW (eds) Nanoscience in dermatology. Elsevier, Amsterdam

Grassl ED, Oegema TR, Tranquillo RT (2002) Fibrin as an alternative biopolymer to typeI collagen for the fabrication of a media equivalent. J Biomed MaterRes 60:607–612. https://doi.org/10.1002/jbm.10107

Hakkarainen T, Koivuniemi R, Kosonen M (2016) Nanofibrillar cellulose wound dressing in skin graft donor site treatment. J Control Release 244(Pt B):292–301. https://doi.org/10.1016/j.jconrel.2016.07.053

Halder S, Yadav KK, Sarkar R, Mukherjee S, Saha P, Haldar S (2016) Alteration of Zeta potential and membrane permeability in bacteria: a study with cationic agents. Springerplus 4:672. https://doi.org/10.1186/s40064-015-1476-7

Hall-Stoodley L, Costerton JW, Stoodley P (2004) Bacterial biofilms: from the natural environment to infectious diseases. Rev Microbiol 2:95–108. https://doi.org/10.1038/nrmicro821

Hamedi H, Moradi S, Hudson SM, Tonelli AE (2018) Chitosan based hydrogels and their applications for drug delivery in wound dressings: a review. CarbohydrPolym 199:445–460. https://doi.org/10.1016/j.carbpol.2018.06.114

Han F, Liu S, Liu X, Pei Y, Bai S, Zhao H (2013) Woven silk fabric-reinforced silk nanofibrous scaffolds for regenerating loadbearing soft tissues. Acta Biomater 10:921–930. https://doi.org/10.1016/j.actbio.2013.09.026

Hay ID, Rehman ZU, Ghafoor A, Rehm BHA (2010) Bacterial biosynthesis of alginates. J Chem Technol Biotechnol 85(6):752–759

He J, Qiao Y, Zhang H, Zhao J, Li W, Xie T, Zhong D, Wei Q, Hua S, Yu Y et al (2020a) Gold-silver nanoshells promote wound healing from drug-resistant bacteria infection and enable monitoring via surface-enhanced raman scattering imaging. Biomaterials 234:119763. https://doi.org/10.1016/j.biomaterials.2020.119763

He J, Qiao Y, Zhang H, Zhao J, Li W, Xie T, Zhong D, Wei Q, Hua S, Yu Y, Yao K, Santos H, Zhou M (2020b) Gold-silver nanoshells promote wound healing from drug-resistant bacteria infection and enable monitoring via surface-enhanced Raman scattering imaging. Biomaterials 234:119763. https://doi.org/10.1016/j.biomaterials.2020.119763

Heher P, Muhleder S, Mittermayr R (2018) Fibrin-based delivery strategies for acute and chronic wound healing. Adv Drug Deliv Rev 129:134–147. https://doi.org/10.1016/j.addr.2017.12.007

Hu C, Zhang F, Kong Q, Lu Y, Zhang B, Wu C, Luo R, Wang Y (2019) Synergistic chemical and photodynamic antimicrobial therapy for enhanced wound healing mediated by multifunctional light-responsive nanoparticles. Biomacromolecules 20:45814592. https://doi.org/10.1021/acs.biomac.9b01401

Huang Y, Onyeri S, Siewe M, Moshfeghian A, Madihally SV (2005) In vitro characterization of chitosan–gelatin scaffolds for tissue engineering. Biomaterials 26:7616. https://doi.org/10.1016/j.biomaterials.2005.05.036

Huijberts MS, Schaper NC, Schalkwijk CG (2008) Advanced glycation end products and diabetic foot disease. Diabetes Metab Res Rev 24(S1):S19–S24

Intini C, Elviri L, Cabral J (2018) 3D-printed chitosan-based scaffolds: an in vitro study of human skin cell growth and an in-vivo wound healing evaluation in experimental diabetes in rats. Carbohydr Polym 199:593–602. https://doi.org/10.1016/j.carbpol.2018.07.057

Jain RA (2000) The manufacturing techniques of various drug loaded biodegradable poly(lactide-coglycolide) (PLGA) devices. Biomaterials 21(23):2475–2490. https://doi.org/10.1016/s0142-9612(00)00115-0

Jao D, Mou X, Hu X (2016) Tissue regeneration: a silk road. J Funct Biomater 7(3):22–40. https://doi.org/10.3390/jfb7030022

Kalita S, Kandimalla R, Bhowal AC, Kotoky J, Kundu S (2018) Functionalization of β-lactam antibiotic on lysozyme capped gold nanoclusters retrogress MRSA and its persisters following awakening. Sci Rep 8(1):5778. https://doi.org/10.1038/s41598-018-22736-5

Kasiewicz LN, Whitehead KA (2016) Silencing TNFα with lipidoid nanoparticles downregulates both TNFα and MCP-1 in an in vitro co-culture model of diabetic foot ulcers. Acta Biomater 32:120–128. https://doi.org/10.1016/j.actbio.2015.12.023

Kim S-J, Jang J-D, Lee S-K (2007) Treatment of long tubular bone defect of rabbit using autologous cultured osteoblasts mixed with fibrin. Cytotechnology 54:115–120. https://doi.org/10.1007/s10616-007-9084-1

Kim HJ, Kim MK, Lee KH, Nho SK, Han MS, Um IC (2017) Effect of degumming methods on structural characteristics and properties of regenerated silk. Int J Biol Macromol 104:294–302. https://doi.org/10.1016/j.ijbiomac.2017.06.019

Kim SH, Yeon YK, Lee JM, Chao JR, Lee YJ, Seo YB (2018) Precisely printable and biocompatible silk fibroin bioink for digital light processing 3D printing. Nat Commun 9:1620. https://doi.org/10.1038/s41467-018-03759-y

Kirschning A, Dibbert N, Drager G (2018) Chemical functionalization of polysaccharides: towards biocompatible hydrogels for biomedical applications. Chemistry 24(6):1231–1240. https://doi.org/10.1002/chem.201701906

Klemm D, Schumann D, Kramer F (2009) Nanocellulose materials: different cellulose, different functionality. Macromol Symp 280:60–71. https://doi.org/10.1002/masy.200950608

Klevens RM, Morrison MA, Nadle J et al (2007) Invasive methicillin-resistant Staphylococcus aureus infections in the United States. JAMA 298(15):1763–1771. https://doi.org/10.1001/jama.298.15.1763

Koh LD, Cheng Y, Teng CP (2015) Structures, mechanical properties, and applications of silk fibroin materials. Prog Polym Sci 46:86–110. https://doi.org/10.1016/j.progpolymsci.2015.02.001

Kong M, Chen XG, Xing K, Park HJ (2010) Antimicrobial properties of chitosan and mode of action: a state of the art review. Int J Food Microbiol 144(1):51–63. https://doi.org/10.1016/j.ijfoodmicro.2010.09.012

Kozlov PV, Burdygina GI (2016) The structure and properties of solid gelatin and the principles of their modification. Polymer 24:651–666. https://doi.org/10.1016/00323861(83)90001-0

Kucinska-Lipka J, Gubanska I, Janik H (2015) Bacterial cellulose in the field of wound healing and regenerative medicine of skin: recent trends and future prospectives. Polym Bull 72(9): 2399–2419. https://doi.org/10.1007/s00289-015-1407-3

Lamba NMK, Woodhouse KA, Cooper SL (2017) Polyurethanes in biomedical applications. CRC Press, Boca Raton, FL. https://doi.org/10.5772/24219

Lamme EN, van Leeuwen RTJ, Jonker A, van Marle J, Middelkoop E (1998) Living skin substitutes: survival and function of fibroblasts seeded in a dermal substitute in experimental wounds. J Investigative Dermatol 111:989–995. https://doi.org/10.1046/j.15231747.1998.00459.x

Lee KY, Mooney DJ (2011) Alginate: properties and biomedical applications. Prog Polym Sci 37(1):106–126. https://doi.org/10.1016/j.progpolymsci.2011.06.003

Li S, Tang Q, Xu H, Huang Q, Wen Z, Liu Y, Peng C (2019) Improved stability of KGF by conjugation with gold nanoparticles for diabetic wound therapy. Nanomedicine 14:2909–2923. https://doi.org/10.2217/nnm-2018-0487

Li M, Liang YP, He JH, Zhang HL, Guo BL (2020) Perspective on theoretical methods and modeling relating to electro-catalysis processes. Chem Commun 56:9937–9949. https://doi.org/10.1007/s40820-021-00681-9

Lian J, Mansel BW, Ingham B, Prabakar S, Williams MAK (2017) Controlling chain flexibility in collagen networks to produce hydrogels with distinct properties. Soft Mater 15:145–152. https://doi.org/10.1080/1539445X.2016.1268626

Liang Y, Li Z, Huang Y, Yu R, Guo B (2021) Dual-dynamic-bond cross-linked antibacterial adhesive hydrogel sealants with on-demand removability for post- wound-closure and infected wound healing. ACS Nano 15:7078–7093. https://doi.org/10.1021/acsnano.1c00204

Lin N, Dufresne A (2014) Nanocellulose in biomedicine: current status and future prospect. Eur Polym J 59:302–325. https://doi.org/10.1016/j.eurpolymj.2014.07.025

Lin YH, Lin JH, Hong YS (2017) Development of chitosan/poly-γ-glutamic acid/pluronic/curcumin nanoparticles in chitosan dressings for wound regeneration. J Biomed Mater Res B Appl Biomater 105:81–90. https://doi.org/10.1002/jbm.b.33394

Liu R, Bal HS, Desta T, Behl Y, Graves DT (2006) Tumor necrosis factor-alpha mediates diabetes-enhanced apoptosis of matrix-producing cells and impairs diabetic healing. Am J Pathol 168(3): 757–764. https://doi.org/10.2353/ajpath.2006.050907

Liu L, Liu Y, Li J (2011) Microbial production of hyaluronic acid: current state, challenges, and perspectives. Microb Cell Factories 10:99. https://doi.org/10.1186/1475-2859-1099

Liu X, Nielsen LH, Klodzinska SN (2018) Ciprofloxacin-loaded sodium alginate/poly (lactic-co-glycolic acid) electrospun fibrous mats for wound healing. Eur J Pharm Biopharm 123:42–49. https://doi.org/10.1016/j.ejpb.2017.11.004

Longinotti C (2014) The use of hyaluronic acid based dressings to treat burns: a review. Burns Trauma 2(4):162–168. https://doi.org/10.1007/978-3-319-44785-8

Machado HA, Abercrombie JJ, You T (2013) Release of a wound-healing agent from PLGA microspheres in a thermosensitive gel. Biomed Res Int 2013:387863. https://doi.org/10.1155/2013/387863

Mast BA, Schultz GS (1996) Interactions of cytokines, growth factors, and proteases in acute and chronic wounds. Wound Repair Regen 4:411–420. https://doi.org/10.1046/j.1524475x.1996.40404.x

Maurer RJ, Sax AF, Ribitsch V (2013) Molecular simulation of surface reorganization and wetting in crystalline cellulose I and II. Cellulose 20(1):25–42. https://doi.org/10.1007/s10570-012-9835-9

Maver T (2018) Polysaccharide based wound care materials. In: Navard P (ed) Bioactive polysaccharide materials for modern wound healing. Springer, p 9. https://doi.org/10.1007/978-3-319-89608-3_2

Melrose J (2016) Glycosaminoglycans in wound healing. Bone Tissue Regen Insights 7:BTRI. S38670. https://doi.org/10.4137/BTRI.S38670

Mithieux SM, Rasko JEJ, Weiss AS (2004) Synthetic elastin hydrogels derived from massive elastic assemblies of self-organized human protein monomers. Biomaterials 25:4921. https://doi.org/10.1016/j.biomaterials.2004.01.055

Mukherjee S, Reddy Venugopal J, Ravichandran R, Ramakrishna S, Raghunath M (2010) Evaluation of the biocompatibility of PLACL/collagen nanostructured matrices with cardiomyocytes as a model for the regeneration of infarcted myocardium. Adv Funct Mater 21:2291–2300. https://doi.org/10.1002/adfm.201002434

Nair P, Thottappillil N (2015) Scaffolds in vascular regeneration: current status. Vasc Health Risk Manag 2015:79–91. https://doi.org/10.2147/vhrm.s50536

Neethu N, Muthunarayanan M, In-Kyu P, Tin WW, Sabu T, Yves G (2015) Natural polymer/inorganic material based hybrid scaffolds for skin wound healing. Polym Rev 55(3):453–490. https://doi.org/10.1080/15583724.2015.1019135

Ng SF, Jumaat N (2013) Carboxymethyl cellulose wafers containing antimicrobials: a modern drug delivery system for wound infections. Eur J Pharm Sci 2014(51):173–179. https://doi.org/10.1016/j.ejps.2013.09.015

Oh SY, Hong SY, Jeong YR, Yun J, Park H, Jin SW, Lee G, Oh JH, Lee H, Lee SS (2018) Skin-attachable, stretchable electrochemical sweat sensor for glucose and pH detection. ACS Appl Mater Inter 10:13729–13740. https://doi.org/10.1021/acsami.8b03342

Olczyk P, Mencner L, Komosinska-Vassev K (2015) Diverse roles of heparan sulfate and heparin in woundrepair. Biomed Res Int 2015:549417. https://doi.org/10.1155/2015/549417

Pal K, Banthia AK, Majumdar DK (2006) Preparation of transparent starch-based hydrogel membrane with potential application as wound dressing. Trends Biomater Artif Organs 20(1): 59–67

Patrulea V, Ostafe V, Borchard G, Jordan O (2015) Chitosan as a starting material for wound healing applications. Eur J Pharm Biopharm 97:417–426. https://doi.org/10.1016/j.ejpb.2015.08.004

Picheth GF, Pirich CL, Sierakowski MR (2017) Bacterial cellulose in biomedical applications: a review. Int J Biol Macromol 104(Pt A):97–106. https://doi.org/10.1016/j.ijbiomac.2017.05.171

Pirnazar P, Wolinsky L, Nachnani S (1999) Bacteriostatic effects of hyaluronic acid. J Periodontol 70(4):370–374. https://doi.org/10.1902/jop.1999.70.4.370

Ponnanikajamideen M, Rajeshkumar S, Vanaja M, Annadurai G (2019) In vivo type 2 diabetes and wound-healing effects of antioxidant gold nanoparticles synthesized using the insulin plant Chamaecostus cuspidatus in albino rats. Can J Diabetes 43(2):82–89. https://doi.org/10.1016/j.jcjd.2018.05.006

Porporato PE, Payen VL, De Saedeleer CJ (2012) Lactate stimulates angiogenesis and accelerates the healing of superficial and ischemic wounds in mice. Angiogenesis 15(4):581–592. https://doi.org/10.1007/s10456-012-9282-0

Qiao B, Pang Q, Yuan P, Luo Y, Ma L (2020) Smart wound dressing for infection monitoring and nirtriggered antibacterial treatment. Biomater Sci 8:164. https://doi.org/10.1039/C9BM02060H

Qin S, Xiao W, Zhou C, Pu Q, Deng X, Lan L, Liang H, Song X, Wu M (2022) Pseudomonas aeruginosa: pathogenesis, virulence factors, antibiotic resistance, interaction with host, technology advances and emerging therapeutics. Signal Transduct Target Ther 7(1):199. https://doi.org/10.1038/s41392-022-01056-1

Rakhshaei R, Namazi H (2017) A potential bioactive wound dressing based on carboxymethyl cellulose/ZnO impregnated MCM-41 nanocomposite hydrogel. Mat Sci Eng C-Mater 73:456–464. https://doi.org/10.1016/j.msec.2016.12.097

Randeria PS, Seeger MA, Wang XQ, Wilson H, Shipp D, Mirkin CA, Paller AS (2015) siRNA-based spherical nucleic acids reverse impaired wound healing in diabetic mice by ganglioside GM3 synthase knockdown. Proc Natl Acad Sci U S A 112(18):5573–5578. https://doi.org/10.1073/pnas.1505951112

Raymund EH, Jürgen K, Ulrich K, Bach AD (2005) Tissue engineering of cultured skin substitutes. J Cell Mol Med 9:592–608. https://doi.org/10.1111/j.1582-4934.2005.tb00491.x

Reis MB, Pereira PAT, Caetano GF (2017) Lipoxin A4 encapsulated in PLGA microparticles accelerates wound healing of skin ulcers. PLoS One 12(7):e0182381. https://doi.org/10.1371/journal.pone.0182381

Rodrigues M, Kosaric N, Bonham CA, Gurtner GC (2009) Wound healing: a cellular perspective. Physiol Rev 99:665–706. https://doi.org/10.1152/physrev.00067.2017

Rojo L, Vázquez B, Román JS (2014) Biomaterials for scaffolds: synthetic polymers. In: Migliaresi C, Motta A (eds) Scaffolds for tissue engineering: biological design, materials, and fabrication. Pan Stanford Publishing, Singapore, pp 263–263. https://doi.org/10.4324/9780429422652-1

Romano CL, De Vecchi E, Bortolin M (2007) Hyaluronic acid, and its composites as a local antimicrobial/antiadhesive barrier. J Bone Joint Infect 2(1):63–72. https://doi.org/10.7150/jbji.17705

Roy D, Cambre JN, Sumerlin BS (2010) Future perspectives and recent advances in stimuli-responsive materials. Prog Polym Sci 35:278–301. https://doi.org/10.1016/j.progpolymsci.2009.10.008

Sarkar K, Xue Y, Sant S (2017) Host response to synthetic versus natural biomaterials. In: Corradetti B (ed) The immune response to implanted materials and devices. International Publishing, Cham, pp 81–105. https://doi.org/10.3389/fbioe.2021.664592

Scientific Animation (2020) Tustin, Canada. https://www.scientificanimations.com/wikiimages/. Accessed 15 Jan 2020

Seifu DG, Purnama A, Mequanint K, Mantovani D (2013) Small diameter vascular tissue engineering. Nature reviews. Cardiology 10:410–421. https://doi.org/10.1038/nrcardio.2013.77

Shelke NB, James R, Laurencin CT, Kumbar SG (2014) Polysaccharide biomaterials for drug delivery and regenerative engineering. Polym Adv Technol 25(5):448–460. https://doi.org/10.1002/pat.3266

Singh R, Shitiz K, Singh A (2017) Chitin, and chitosan: biopolymers for wound management. Int Wound J 14(6):1276–1289. https://doi.org/10.1111/iwj.12797

Singla R, Soni S, Patial V, Kulurkar PM, Kumari A, Mahesh S, Padwad YS, Yadav SK (2017) Cytocompatible anti-microbial dressings of Syzygium cumini cellulose nanocrystals decorated with silver nanoparticles accelerate acute and diabetic wound healing. Sci Rep 7:10457. https://doi.org/10.1038/s41598-017-08897-9

Song F, Li X, Wang Q (2015) Nanocomposite hydrogels and their applications in drug delivery and tissue engineering. J Biomed Nanotechnol 11:40–52. https://doi.org/10.1166/jbn.2015.1962

Stegemann JP, Nerem RM (2003) Altered response of vascular smooth muscle cells to exogenous biochemical stimulation in two- and three-dimensional culture. Exp Cell Res 283:146–155. https://doi.org/10.1016/s0014-4827(02)00041-1

Stumpf TR, Yang X, Zhang J, Cao X (2018) In situ and ex situ modifications of bacterial cellulose for applications in tissue engineering. Mater Sci Eng C Mater Biol Appl 82:372–383. https://doi.org/10.1016/j.msec.2016.11.121

Sulaeva I, Henniges U, Rosenau T, Potthast A (2015) Bacterial cellulose as a material for wound treatment: properties and modifications. A review. Biotechnol Adv 33(8):1547–1571. https://doi.org/10.1016/j.biotechadv.2015.07.009

Sun G, Zhang X, Shen YI (2017) Dextran hydrogel scaffolds enhance angiogenic responses and promote complete skin regeneration during burn wound healing. Proc Natl Acad Sci U S A 108(52):20976–20981. https://doi.org/10.1073/pnas.1115973108

Sun Y, Zhao C, Niu J, Ren J, Qu X (2020) Colorimetric band-aids for pointof-care sensing and treating bacterial infection. ACS Central Sci 6:207–212. https://doi.org/10.1021/acscentsci.9b01104

Sundaram N, Mony U, Jayakumar R (2006) Chitin and chitosan as hemostatic agents. In: Mark HF (ed) Encyclopedia of polymer science and technology. Wiley, pp 1–12. https://doi.org/10.1002/0471440264.PST647

Tabasum S, Noreen A, Maqsood MF (2018) A review on versatile applications of blends and composites of pullulan with natural and synthetic polymers. Int J Biol Macromol 120 (Pt A):603–632. https://doi.org/10.1016/j.ijbiomac.2018.07.154

Theocharis AD, Skandalis SS, Gialeli C, Karamanos NK (2016) Extracellular matrix structure. Adv Drug Deliv Rev 97:4–27. https://doi.org/10.1016/j.addr.2015.11.001

Thomas S (2000) Alginate dressings in surgery and wound management – part 1. J Wound Care 9(2):56–60. https://doi.org/10.12968/jowc.2000.9.2.26338

Thomas V, Zhang X, Catledge SA, Vohra YK (2007) Functionally graded electrospun scaffolds with tunable mechanical properties for vascular tissue regeneration. Biomed Mater 2:224–232. https://doi.org/10.1088/1748-6041/2/4/004

Ueno H, Mori T, Fujinaga (2015) Topical formulations, and wound healing applications of chitosan. Adv Drug Deliv Rev 52(2):105–115. https://doi.org/10.1016/s0169409x(01)00189-2

Ullah H, Wahid F, Santos HA, Khan T (2016) Advances in biomedical and pharmaceutical applications of functional bacterial cellulose-based nanocomposites. Carbohydr Polym 150: 330–352. https://doi.org/10.1016/j.carbpol.2016.05.029

Veith AP, Henderson K, Spencer A, Sligar AD, Baker AB (2019) Therapeutic strategies for enhancing angiogenesis in wound healing. Adv Drug Deliv Rev 146:97–125. https://doi.org/10.1016/j.addr.2018.09.010

Vicente N (2019) Chitin deacetylation. https://commons.wikimedia.org/wiki/File:Chitin_and_Chitosan.jpg. Accessed 29 Jan 2019

Vieira S, da Silva Morais A, Silva-Correia J, Oliveira JM, Reis RL (2017) Natural-based hydrogels: from processing to applications. In: Encyclopedia of polymer science and technology. Wiley, Hoboken, NJ, pp 1–27

Vig K, Chaudhari A, Tripathi S, Dixit S, Sahu R, Pillai S, Dennis VA, Singh SR (2017) Advances in skin regeneration using tissue engineering. Int J Mol Sci 18:789. https://doi.org/10.3390/ijms18040789

Vowden K, Vowden P (2014) Wound dressings: principles and practice. Surg Oxford Inter Edn 32(9):462–467

Vrana NE, Elsheikh A, Builles N, Damour O, Hasirci V (2007) Effect of human corneal keratocytes and retinal pigment epithelial cells on the mechanical propertiesof micropatterned collagen films. Biomaterials 28:4303–4310. https://doi.org/10.1016/j.biomaterials.2007.06.013

Waghmare VS, Wadke PR, Dyawanapelly S, Deshpande A, Jain R, Dandekar P (2017) Starch based nanofibrous scaffolds for wound healing applications. Bioact Mater 3(3):255–266. https://doi.org/10.1016/j.bioactmat.2017.11.006

Wang S, Lu L, Gruetzmacher JA (2006) Synthesis and characterizations of biodegradable and crosslinkable poly(epsilon-caprolactone fumarate), poly(ethylene glycol fumarate), and their amphiphilic copolymer. Biomaterials 27(6):832–841. https://doi.org/10.1016/j.biomaterials.2005.07.013

Wang X, Meier RJ, Wolfbeis OS (2013) Fluorescent pH-sensitive nanoparticles in an agarose matrix for imaging of bacterial growth and metabolism. Angew Chem Int Edit 52:406–409. https://doi.org/10.1002/anie.201205715

Wang Q, Han G, Yan S, Zhang Q (2017) 3D printing of silk fibroin for biomedical applications. Materials 12:504–504. https://doi.org/10.3390/ma12030504

Whelan D, Caplice NM, Clover AJP (2014) Fibrin as a delivery system in wound healing tissue engineering applications. J Control Release 196:1–8. https://doi.org/10.1016/j.jconrel.2014.09.023

Wong TW, Ramli NA (2014) Carboxymethylcellulose film for bacterial wound infection control and healing. Carbohydr Polym 112:367–375. https://doi.org/10.1016/j.carbpol.2014.06.002

Xie X, Sun T, Xue J, Miao Z, Yan X, Fang W, Li Q, Tang R, Lu Y, Tang L et al (2020) Ag nanoparticles cluster with pH-triggered reassembly in targeting antimicrobial applications. Adv Funct Mater 30:2000511. https://doi.org/10.1002/adfm.202000511

Zhang X, Pint CL, Lee MH (2011) Optically- and thermally responsive programmable materials based on carbon nanotube-hydrogel polymer composites. Nano Lett 11:3239–3244. https://doi.org/10.1021/nl201503e
Zhou J, Tun TN, Hong SH, Mercer-Chalmers JD, Laabei M, Young AE, Jenkins AT (2011) Development of a prototype wound dressing technology which can detect and report colonization by pathogenic bacteria. Biosens Bioelectron 30:67–72. https://doi.org/10.1016/j.bios.2011.08.028
Zhou X, Wang H, Zhang J (2017) Functional poly(epsilon-caprolactone)/chitosan dressings with nitric oxide-releasing property improve wound healing. Acta Biomater 54:128–137. https://doi.org/10.1016/j.actbio.2017.03.011
Zuo YM, Yan X, Xue J, Guo LY, Fang WW, Sun TC, Li M, Zha Z, Yu Q, Wang Y (2020) Enzyme-responsive Ag nanoparticle assemblies in targeting antibacterial against methicillin-resistant Staphylococcus aureus. ACS Appl Mater Inter 12:4333–4342. https://doi.org/10.1021/acsami.9b22001

Polysaccharide-Based Injectable Nanocomposite Hydrogels for Wound Healing Application

18

Vieralynda Vitus, Nasrul Anuar Abd Razak, Tan Qing Hao, Ehsan Zeimaran, Nur Aina Syafiqah Suhaimi, Wan Safwani Wan Kamarul Zaman, and Wan Syamimi Wan Kamarul Zaman

Abstract

Research on injectable hydrogel is rising due to its injectability that offers minimal surgical invasion during application. Current research focuses its application on wound healing due to the hydrophilicity nature of hydrogels that could provide a moist environment and structural components like the extracellular matrix niche of tissue. Therefore, this chapter discusses the ability of injectable hydrogel in wound healing and its characteristics. The regulatory aspect and the commercialisation potential of the injectable hydrogel are also discussed in this chapter.

Keywords

Injectable hydrogels · Polysaccharide · Wound Healing · Regulatory · Commercialisation Potential

V. Vitus · N. A. A. Razak (✉) · T. Q. Hao · E. Zeimaran · N. A. S. Suhaimi
Department of Biomedical Engineering, Faculty of Engineering, Universiti Malaya, Kuala Lumpur, Malaysia
e-mail: nasrul.anuar@um.edu.my

W. S. Wan Kamarul Zaman
Department of Biomedical Engineering, Faculty of Engineering, Universiti Malaya, Kuala Lumpur, Malaysia

Centre for Innovation in Medical Engineering (CIME), Department of Biomedical Engineering, Faculty of Engineering, Universiti Malaya, Kuala Lumpur, Malaysia

W. S. W. K. Zaman
Institute of Biological Sciences, Faculty of Science, Universiti Malaya, Kuala Lumpur, Malaysia

W. S. Wan Kamarul Zaman, A. A. Nurul (eds.), *Sustainable Material for Biomedical Engineering Application*, https://doi.org/10.1007/978-981-99-2267-3_18

Abbreviations

Alg-Gel-NCC	Alginate-gelatin-nanocrystalline cellulose
Am-β-Chn-Sq/ADA	Acrylamide-modified β-chitin/alginate dialdehyde
$CaCl_2$	Calcium chloride
CDRH	Center for Medical Device and Radiological Health
Ch	Chitosan
CLP	Classification, Labelling, and Packaging of Substances and Mixtures
CNF-PPy	Cellulose nanofibers-polypyrrole
CNF-PPy/PB	Cellulose nanofibers-polypyrrole/polyvinyl alcohol-borax
EC	European Commisions
ECM	Extracellular matrix
FDA	Food and Drug Administration
G-DFO-SA/BG	Grafted deferoxamine-sodium alginate/bioglass
GelDA/DA-PPy	Dopamine gelatin/dopamine functionalized polypyrrole
Gel-DA/GG@Ag	Dopamine-modified gelatin/guar gum@Ag nanoparticles
H_2O_2	Hydrogen peroxide
HA	Hyaluronate
HA-CHO	Oxidized sodium hyaluronic acid
HHA	Hydrazided hyaluronic acid
HRP	Horseradish peroxidase
ISO	International Standard Operation
MDR	Medical Devices Regulation
N,O-CS/ODA	N,O-carboxymethyl chitosan/oxidized dextran
OCMC/G3KP	Oxidized carboxymethyl cellulose/generation 3 lysine and poly(ethyleneglycol)
OHC-SA-CS	Aldehyde-sodium alginate-chitosan
OHC-SA-CS/FB	Aldehyde-sodium alginate-chitosan/nanofibers
PB	Polyvinyl alcohol-borax
PCEC/Alg	Poly(ε-caprolactone)-b-poly(ethyleneglycol)-b-poly (ε-caprolactone)/alginate
PL-CD@ODA	ε-poly(L-lysine) carbon dot and oxidized dextran
PNIPAAm/CNC	Poly(N-isopropylacylamide)/cellulose nanocrystals
REACH	Registration, Evaluation, Authorization, and Restrictions of Chemicals
SA/BG	Sodium alginate/bioglass
SAP	Superabsorbent polymer
UV	Ultraviolet

18.1 Introduction

Hydrogel is made of either a natural or synthetic material. However, the sustainability, renewability, and economic values of nature-based hydrogel have shown to be more advantageous compared to synthetics-based hydrogels. These natural materials are commonly known as a biopolymer. Polysaccharides are the most common macromolecules in the biosphere among biopolymers. These polysaccharides are made up of a wide range of organic polymers that are produced naturally in algae, plants, and wood, as well as by microorganisms (Bertolino et al. 2020). Plant polysaccharides (such as pectin, inulin, and gums), marine polysaccharides (such as hyaluronic acid, chitosan, and alginate), and microbial exopolysaccharides (including levan, glucans, pullulan, dextran, curdlan, and gellan) are the three main categories of natural polysaccharides (Lopez-Sanchez et al. 2020; Wan et al. 2020; Koyyada and Orsu 2021; Sun et al. 2021; Tang et al. 2019).

Polysaccharides are naturally occurring, renewable, nontoxic, and intrinsically biodegradable carbohydrates whose molecules are made up of many connected sugar moieties. They are distinguished by hydroxyl, amino, and carboxyl groups that might offer particular functionality and biological recognition traits or work to adjust the parent molecules' biological or mechanical qualities (Rimondo et al. 2019). Through chemical modification of polysaccharides, superior mechanical characteristics, physiological signalling, and programmable tissue response have been attained. Additionally, a proper formulation method enables the scaffold to be placed spatially at a desired location. These preparations can now be specifically designed to react to changes in temperature, pH, or other physiological stimuli thanks to the development of the latest technology (Tiwari et al. 2018). Many polysaccharide-based biomaterials are still employed in preclinical research, and interest is growing because of their hydrophilicity, which makes them suitable sources for the synthesis of hydrogels.

Additionally, the advancement of technology has opened the ability to fabricate a multifunctional injectable hydrogel with enhanced effectiveness. Injectable hydrogel can be prepared from natural polymers through various mechanisms such as chemical and physical crosslinking methods into different types. Furthermore, injectable hydrogel has been designed with a rapid sol-gel phase transition or in situ chemical polymerisation activity, which allows them to solidify after application, thus making it an implantable material into the human body (Chao et al. 2020; Gao et al. 2020). In addition, injectable hydrogel has also been recognised as an excellent drug carrier due to its unique ability to chemically cross link with different types of drugs or other beneficial materials, providing a stable mode of drug delivery for wound healing applications (Rizzo and Kehr 2021). Thus, injectable hydrogel has received wide attention as a potential strategy in wound healing applications as it mimics the biological properties of human tissue and offers minimally invasive application during surgery (Fig. 18.1) (Tavakoli and Klar 2020; Dimatteo et al. 2018).

In this chapter, we seek to highlight the design and development of polysaccharide-based injectable nanocomposite hydrogel for wound treatment applications. We begin by giving a brief overview of the wound healing process

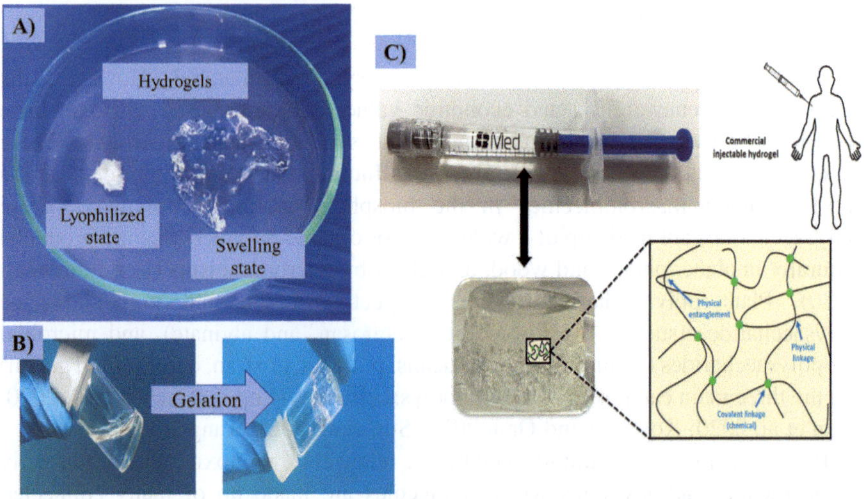

Fig. 18.1 (**a**) Swelling ability of hydrogel from the lyophilised state. (**b**) Example of a gelation time measurement by inverted vial method. (**c**) Physical interactions and chemical linkages of the chemical structure of an injectable hydrogel (Alonso et al. 2021). Adapted under the Creative Common (CC-BY 4.0) Licence from Polymer, MDPI

and how hydrogel could enhance the healing progression. Next, the properties of injectable hydrogel are described. Lastly, we describe the regulatory aspects and commercial potential of injectable hydrogel products as medical devices, particularly for wound healing applications.

18.2 Injectable Hydrogel in Wound Healing

The normal biological process of wound healing of the body is in charge of regenerating new cells to replace the injured ones. Blood clotting (haemostasis), inflammation, tissue development (proliferation), and remodelling are four separate and well-defined stages (cell differentiation). All the four stages must take place in the right sequence and window of time for the wound healing process to be fully completed. During the haemostasis phase, platelets clump together and stop the flow of blood. After that, neutrophils phagocytose foreign objects and dead cells to start the inflammatory phase. Macrophages assume control of phagocytosis when they arrive and start secreting cytokines that draw in and activate additional cells. The proliferative phase starts as fibroblasts and endothelial cells are attracted to the wound site. Next, endothelial cells create microvascular structures as fibroblasts multiply and produce extracellular matrix (ECM) elements such as collagen. The remodelling phase follows, during which the dermal ECM contracts to seal the wound, and then the collagen fibres are crosslinked. The healing time for wounds ranges from 4 to 6 weeks. Chronic wounds are those that take longer than this to heal (Wallace et al. 2022).

Recent years have shown considerable efforts dedicated to the development of injectable hydrogel that facilitates wound healing. This is due to the favourable properties of injectable hydrogel, including the ability to cover or infiltrate the asymmetry wound area and to be biodegradable with minimal invasion. These abilities have prompted its usage in the treatment of various wound types. Additionally, the hydrogel-based wound dressing has gained much interest due to its natural structure that resembles the native ECM of tissue that can maintain the moist wound environment, regulate fluid balance at the wound area, perform autolytic debridement, and enhance wound healing (Du et al. 2019). Additionally, to improve the healing process of chronic non-healing wounds, various macromolecules and cells can be crosslinked to provide superior therapeutic properties. Therefore, various injectable hydrogels based on polysaccharides have been developed for a variety of wound types. For example, alginate-boronic acid (Hong et al. 2018), cellulose nanofibers-polypyrrole/polyvinyl alcohol-borax (CNF-PPy/PB) (Ding et al. 2018), alginate/gelatin/nanocrystalline cellulose (Wang et al. 2016), poly(ε-caprolactone)–b-poly(ethyleneglycol)–b-poly(ε-caprolactone)/Alginate (PCEC/Alg) (Liao et al. 2017), alginate/nanocellulose (Siqueira et al. 2019), and strontium-ion-doped sodium alginate aldehyde/polyetherimide (Lu et al. 2020) have shown different extents of wound healing rate.

Additionally, a study reported that the dopamine-modified gelatine with silver nanoparticles (Ag NPs) incorporated in guar gum-based hydrogels (Gel-DA/GG@Ag hydrogel) could flawlessly cover the wound surface and show excellent wound regeneration on a 10 mm full-thickness skin wound on a mouse model (Zhang et al. 2021a). Also, the grafting of deferoxamine sodium alginate/bioglass hydrogel induces the development of blood vessels in a rat full-thickness excision model, which enhances the wound healing process (Zhang et al. 2021b).

Another study demonstrated the wound-healing abilities of alginate/polyetherimide (AP) and SAP. The study investigated the properties of wound healing both in vivo and in vitro. With the addition of strontium ions, the pace at which the hydrogel heals the wounds dramatically increases, especially in the SAP group (Lu et al. 2020). Due to its distinctive qualities such as extremely high absorbability, capacity to create a moist environment for healing, adequate oxygen permeability, ease of replacement, and controlled pharmaceutical drug release, the hydrogel has attracted more attention than other wound-dressing materials (Shojaeiarani et al. 2019).

Additionally, a study used a hydrogel patch made of dopamine-functionalised gelatine, polypyrrole (GelDA/DA-PPy), and injectable hydrogel (oxidised sodium hyaluronic acid/hydrazided hyaluronic acid, or HA-CHO/HHA). When treated with the injectable hydrogel and the patch, L929 cells and H9C2 cardiomyocytes could sustain viabilities above 90%, showing little cell cytotoxicity. Plus, intramyocardial injection of the hydrogel followed by the application of the hydrogel patch to the surface of the heart resulted in a significant enhancement of cardiac function (Wu et al. 2019).

Meanwhile, Qu et al. reported a curcumin-loaded hydrogel with quaternised chitosan and benzaldehyde-terminated Pluronic F-127, which was developed specifically for joint skin wound healing. The hydrogel exhibited stable rheological properties, good adhesion properties, comparative modulus to human soft tissue, good pH-dependent degradation ability, tunable gelation time, sensitive to acidic skin environment drug release, antibacterial against *Escherichia coli* and *Streptococcus aureus*, haemostasis, biocompatibility with L929 cells, and free radical scavenging capabilities. In addition, in a full-thickness skin defect model, balanced inflammatory infiltration, prominent granulation tissue thickness, denser collagen, and fibroblast deposition were invigorated by the hydrogel (Qu et al. 2018).

Additionally, calcium peroxide was used as an oxygen-releasing substance that can be used in combination with polymers to speed up the healing of wounds by creating an oxygen-rich environment (Akhavan-Kharazian and Izadi-Vasafi 2019). A CNF-PPy/PB hydrogel was also said to be capable of self-healing in both air and water environments. However, recovery time was substantially slower in the aquatic environment. This led to a delayed self-healing process underwater than in the air because of low borate concentrations and less hydrogel contact (Ding et al. 2018; Chen et al. 2017).

18.3 Synthesis of Polysaccharide-Based Injectable Nanocomposites Hydrogel for Wound Application

The synthesis method of an injectable hydrogel influences its physicochemical properties. Therefore, the synthesis method is a crucial manufacturing strategy that could lead to the success of the injectable hydrogel production specific to its intended use. In the past decades, various synthesis methods have been explored and applied to different types of biomaterials. Additionally, an ideal injectable hydrogel should have the following characteristics: (1) the hydrogel should have sufficient mechanical properties as these properties influence its injectability and uphold the mechanical dynamic conditions of the target site (De France et al. 2019); (2) the hydrogel could adhere and completely seal the wound area (Chen et al. 2018); (3) the hydrogel can be completely degraded to allow the growth of natural tissue over a period of time (Xu et al. 2021); (4) the hydrogel should be blood-compatible for haemorrhage control that could induce clotting to stop the wound from further bleeding (Lokhande et al. 2018; Zhai et al. 2019); and (5) the hydrogel must be biocompatible with minimal cytotoxicity to facilitate positive cells behaviour such as wound healing (Pourshahrestani et al. 2020).

The concept of injectable hydrogel is generally based on biomaterials that can be injected as a liquid on the skin or into the human body where the hydrogel will eventually solidify in place. Additionally, the structural features of injectable hydrogel such as porosity allow exchange of nutrients and waste, flow of oxygen, cell penetration, and development of functional tissue (Wang et al. 2020). In this regard, the ideal characteristics of the injectable hydrogel as mentioned above can be achieved by the crosslinking strategies during the synthesis process. Commonly, the crosslinking mechanisms used to develop an injectable hydrogel are chemical crosslinking, physical crosslinking, and external stimuli. Examples of chemical crosslinking-based reactions are Michael-type additions (Pupkaite et al. 2019; Fu et al. 2021), Schiff base reaction (Liu et al. 2020; Amirian et al. 2021; Qiu et al. 2021a), Diels Alder click (Hu et al. 2019; Wei et al. 2022a; Bi et al. 2019), enzyme-mediation (Zhong et al. 2019; Pham et al. 2020), or photo-initiated processes (Bian et al. 2019; Popescu et al. 2021). On the other hand, the physical crosslinking can occur through non-covalent interactions such as π-interactions, hydrophobic interactions, van der Waals forces, including London dispersion forces and dipole-dipole interactions, electrostatic interactions between two oppositely charged ions such as hydrogen bonds, ionic bonds, etc. (Lee 2018). In addition, the injectable hydrogel can be triggered by external stimuli (e.g., pH, light, temperature, electric field, magnetic field, and ultrasound), and this hydrogel exhibits good mechanical stability and physicochemical properties with significant durability over time because of the strong covalent bond formation (Alonso et al. 2021).

Additionally, the antibacterial properties, thixotropic or shear-thinning, self-healing or self-assembling, antioxidant, and responsiveness towards external stimuli such as pH, temperature, and enzyme have become the desired functions in the development of injectable hydrogel, which has been extensively investigated to enhance its effectiveness in wound healing and other applications (Qiu et al. 2021b; Jalalvandi and Shavandi 2019; Nair and Choudhury 2020; Wei et al. 2022b; Ma et al. 2021; Thambi et al. 2016).

18.4 Characteristics of Injectable Hydrogel for Wound Application

An injectable hydrogel is modifiable with the currently available synthetic techniques to produce one with the desired physical, chemical, and biological qualities. For that reason, to obtain an injectable hydrogel with the desired qualities that fulfil the necessary specifications for its intended future application, the characterisation of injectable hydrogel is important. Extensive research on the injectable hydrogel for over a decade has produced a variety of available characterisation approaches. Among all the approaches, the common characterisation techniques that can be used are physicochemical, morphological/structural, mechanical, and biological characteristics.

18.4.1 Physicochemical Characteristics

18.4.1.1 Rheology

Hydrogel properties including viscosity and storage and loss moduli, which are crucial determinants of hydrogel injectability, can be quantitatively measured using rheology. The relative ability of hydrogel formulations to adapt to variations in shear stress during injection is directly measured by viscosity, which is a direct indicator of a material's capacity to resist deformation in response to stress. Storage and loss moduli reveal details about a hydrogel's elastic and viscous response characteristics in response to oscillatory shear, respectively. They specifically assess a hydrogel's capacity to respond to stress by either absorbing energy (storage modulus) or relaxing under stress to release energy (loss modulus) (Chen et al. 2017). Additionally, the hydrogel's shear-thinning properties majorly influence the potential of minimally invasive delivery (Uman et al. 2020). Thus, rheology data are important to better understand the behaviour of hydrogels during the injection process.

18.4.1.2 Gelation Time

A solubilised polymer's metamorphosis into a poorly distributed network with contractive macromolecular chains during gelation is known as the process of solubilisation. Physical gelation and chemical gelation are the two types of gelation. The physical gelation process is frequently reversible because of the transient connections that break down when external stimuli (e.g., temperature variations, pH, ionic interactions, stress, and solvent polarity) are present on the hydrogel. On the other hand, the chemical gelation produces strong, stable covalent bonds that are irreversible unless additional catalysts are applied (Miao et al. 2019). In situ bioadhesives require enough gelation time. The hydrogel will not be able to execute haemostasis at the proper time if the gelation process takes too long. However, clogs and not uniform coverage of the damaged tissue might happen if the gelation process is too fast, instead (Zhu et al. 2020).

A few studies have reported improved gelation properties by incorporating suitable biomolecules into the injectable hydrogel. For example are strontium ions, Sr^{2+} (Zubik et al. 2017), oxidised carboxymethyl cellulose/generation 3 lysine and PEG (OCMC/G3KP) (Zhu et al. 2020), calcium chloride ($CaCl_2$) (Al-Sabah et al. 2019; Shin et al. 2018), glycopolypeptide hydrogels crosslinking with hydrogen peroxide (H_2O_2) and horseradish peroxidase (HRP) (Ren et al. 2015), and the addition of sodium hydroxide solution into boronic acid-modified alginate to perform gelation of hydrogels (Hong et al. 2018; Pettignano et al. 2017). Most of these studies demonstrated that the biomolecule concentrations can influence the gelation of hydrogels. However, this only happened if there was a sufficient substrate (hydrogels composite) to initiate a chemical reaction.

18.4.1.3 Swelling Ability

In addition to analysing the hydrogel's structures, it can also be characterised to assess its qualities. Hydrogel characterisation includes calculating the crosslinking density and quality, strength, fibre orientation, composition, distance between crosslinked fibres, and amount of bound and free water in the system. Different criteria can be used to describe the hydrogel. One of the strategies for characterisation is the swelling ratio. The percentage increase in the weight of the hydrogel from water absorption is known as the swelling ratio. The proportion of the polymer that is not crosslinked following a crosslinking procedure is known as the sol fraction. High swelling ratios are crucial for retaining a moist wound environment and soaking up tissue exudates (Zhao et al. 2017; Bai et al. 2018). This is a fantastic quality that hydrogel wound dressings must have. Lower swelling ratios, though, might mean that fewer sections are not gel-like. This thus creates a stiffer and more highly branched network for the hydrogel. In other words, it can increase the hydrogel's structural strength and durability (Zhu et al. 2020).

As indicated in earlier sections, new fabrication techniques or the inclusion of foreign chemicals by crosslinking might change the characteristics of a hydrogel. This was demonstrated in a study by Morais et al. (Morais et al. 2013), who examined the crosslinking of chitosan and hyaluronate with hydrogel based on alginate. The results showed that, compared to alginate-chitosan and alginate-hyaluronate hydrogel, the alginate hydrogel had a larger swelling ratio. Another study examined the swelling ratio of various poly(N-isopropylacrylamide)/cellulose nanocrystals (PNIPAAm/CNC) hydrogel concentrations at different temperatures. The outcomes demonstrated that CNC enhanced the hybrid hydrogel's swelling ratio (Zubik et al. 2017).

Another study on the swelling ratio of alginate-gelatine-nanocrystalline cellulose (Alg-Gel-NCC) found that the swelling ratio of NCC-free hydrogel was significantly higher than that of hydrogel-containing NCC (Wang et al. 2016). However, the point of view of both studies seemed to contradict each other where Wang et al. in their work compared the properties of the hydrogel with the absence and presence of CNC (Wang et al. 2016) while Zubik et al. in their work focused on analysing the properties of hydrogel by varying the concentrations of CNC (Zubik et al. 2017). Therefore, by increasing the concentrations of CNC in hydrogel, the swelling ratio can be enhanced. Nevertheless, the hydrogel without CNC has a much higher swelling ratio. Additionally, many procedures for the swelling test follow a similar pattern, which requires the hybrid hydrogel to be submerged in phosphate-buffered saline (PBS, pH 7.4) solution for more than 24 hours prior to testing. The surface water is then removed from the hybrid hydrogel using a filter paper after it is taken out of the buffer solution. Lastly, the swelling ratio is calculated using the initial and final weights of hybrid hydrogels (Zhu et al. 2020; Al-Sabah et al. 2019; Ren et al. 2015; Morais et al. 2013; Ghaderinejad et al. 2021).

18.4.1.4 Injectability

One of the key objectives in developing an injectable hydrogel is to produce one with properties that would make it biodegradable and biocompatible, could be used for cell induction, and could be injected at or below room temperature. Additionally, it is anticipated that the injectable hydrogel would be able to provide the aforementioned properties while also enabling it to cover irregularly shaped wound sites (Abasalizadeh et al. 2020; Xie et al. 2022). Generally, the injectability assessment is conducted by loading the synthesised hydrogels into a syringe, and then the hydrogel is extruded through the needle by hand pressure. After extrusion, some of the hydrogel exhibits self-healing properties (Fig. 18.2) (Xuan et al. 2021; Yang et al. 2021).

Fig. 18.2 Self-healing and injectability assessment of hydrogels. (**A**) Images of ε-poly(L-lysine) carbon dot and oxidised dextran (PL-CD@ODA) hydrogel: (a) being crushed and healed in the mould; (b) incision and restoration of PL-CD@ODA hydrogel (under UV light); (c) surface of PL-CD@ODA hydrogel (i) scratched by a tweezer tip, (ii) and (iii) showed the gel surface after 20 min and 30 min, respectively; and (d) The injectability of PL-CD@ODA gel. Adapted with permission from ref. (Yang et al. 2021). Copyright 2021 Springer Nature. (**B**) Self-healing procedure of (a) Commercial chitosan hydrogel and (b) aldehyde-SA-chitosan (OHC-SA-CS) and (c) aldehyde-SA-chitosan/nanofibers (OHC-SA-CS/FB) (1 mg/mL) hydrogels; (d) Continuously injectable and (e) self-healing demonstration of OHC-SA-CS/FB hydrogels. (Adapted with permission from Xuan et al. (2021). Copyright 2021 Springer Nature)

18.4.1.5 Degradability

The controlled degradability of biomaterial is frequently an important factor to consider in wound dressings. Physical or chemical techniques can be used to regulate alginate gel disintegration in a manner similar to how they can be used to modify the dynamics of alginate gel. This is because the rapid degradation rate of hydrogel suggests that the hydrogel has a low crosslinking density. The degradation can thus be controlled by altering the molecular weight and content of polymers (Zhang et al. 2021b; Augst et al. 2006). The divalent cation crosslinkers deteriorate in the presence of monovalent cations (such as sodium and potassium) as a result of ion exchange with monovalent ions (Smith and Senior 2021). Therefore, the degradation rate of hydrogel should be carefully planned to coincide with the hydrogel's function during the designation phase (Zhu et al. 2020).

Zhang et al. carried out in vitro degradation experiment on a hydrogel made of grafted deferoxamine-sodium alginate/bioglass (G-DFO-SA/BG). The mass loss and breakdown rate of the hydrogel were quicker. As a result, the process of healing wounds was enhanced (Zhang et al. 2021b). Another study concluded that changing the composition and concentration of the hydrogel could control how quickly it degraded (Zhu et al. 2020). Another study demonstrated how changing the content of the material affected the rate of degradation. Alginate, alginate-chitosan, and alginate-hyaluronic acid hydrogels were investigated for degradability in the experiment. The findings revealed that the hydrogel with additional components had a faster rate of degradation than the pure alginate hydrogel (Morais et al. 2013).

18.4.2 Microstructure Characteristics

Ideally, the structural characteristics of an injectable hydrogel intended for wound dressing should exhibit a porous structure to allow penetration and migration of cells, growth of tissue, and exchange of nutrients and biological wastes, as well as oxygen diffusion (Li et al. 2020). Additionally, the pore sizes and porosity of the injectable hydrogel can be influenced by different crosslinking densities and or ratios of materials used (Fig. 18.3). For example, a study showed that the variation of aldehyde dialdehyde concentration in acrylamide-modified β-chitin/alginate dialdehyde (Am-β-Chn-Sq/ADA) hydrogel provided a varied quantity of aldehyde functional groups available for crosslinking (Balitaan et al. 2020). In addition, the hydrogel network characteristics may affect the absorption and degradation behaviours of the injectable hydrogel (Balitaan et al. 2020).

18.4.3 Mechanical Properties

The mechanical properties of injectable hydrogel are a crucial factor in determining whether it will function as such. The value of hydrogel's mechanical qualities lies in its ability to resist deformation under highly dynamic conditions inside the body during healing, as well as its ability to avoid disassembling or breaking during transit (Zhu et al. 2020; López Vélez and Palencia 2021). Besides, it can affect cellular mechanotransduction, which influences stem cell proliferation, migration, and differentiation (Martino et al. 2018).

According to a study, the mechanical strength of alginate-gelatine hydrogel was dramatically increased when nanocrystalline cellulose (NCC) was added. The hydrogel containing NCC had a disproportionately larger proportion of osteoblasts. Additionally, the live/dead assay showed that some dead cells were present in the NCC-containing hydrogel. This demonstrated osteoblast differentiation and proliferation. NCC can therefore be taken into account as a potential material for an injectable hydrogel (Wang et al. 2016). The mechanical characteristics of alginate hydrogel made from poly(−caprolactone)-b-poly(ethyleneglycol)-b-poly(−caprolactone) (PCEC) microspheres were also found to be exponentially improved in another investigation that performed in vivo formability and absorption test (Liao et al. 2017).

In another experiment, the NCCs were employed to produce an injectable poly (N-isopropylacrylamide) (PNIPAAm) hybrid hydrogel as a reinforcing material. The study found outstanding mechanical capabilities and a volume phase transition

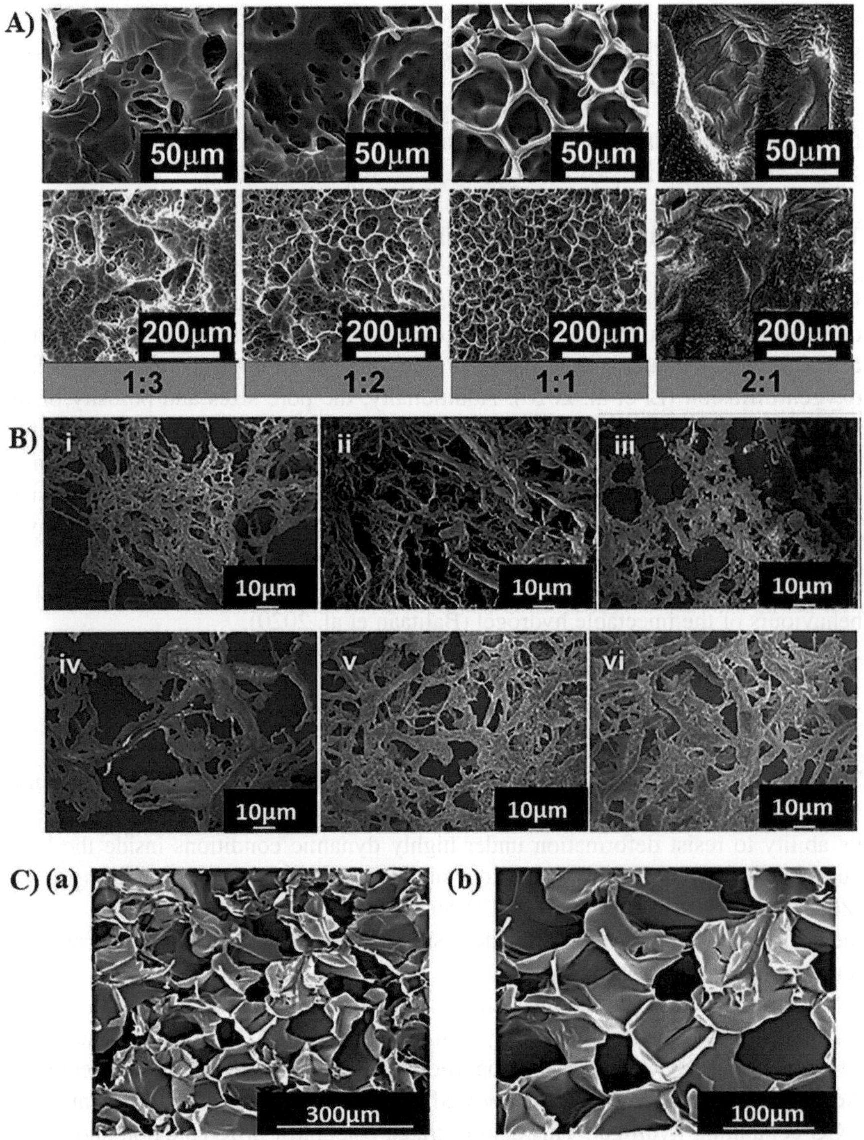

Fig. 18.3 Morphological characteristics of injectable hydrogels. (**A**) SEM images of ε-poly (L-lysine) carbon dot and oxidised dextran (PL-CD@ODA) hydrogels. Adapted with permission from ref. (Yang et al. 2021). Copyright 2021 Springer Nature. (**B**) SEM micrographs of freeze-dried acrylamide-modified β-chitin/alginate dialdehyde (Am-β-Chn-Sq/ADA) hydrogels with different ratios. Adapted with permission from ref. (Balitaan et al. 2020). Copyright 2020 Elsevier. (**C**) SEM images of N, O-carboxymethyl chitosan/oxidised dextran (N, O-CS/ODA) hydrogels at 400× (a) and 800× (b). (Adapted with permission from Li et al. (2021). Copyright 2021 Elsevier)

temperature value of 36–39 °C), which was similar with the average body temperature of an adult person. Additionally, the hydrogel was also found to have the finest stability upon injection and the largest equilibrium swelling ratio (Zubik et al. 2017).

18.4.4 Biological Characteristics

Any product or device that is intended for medical purposes, particularly one that comes into direct or indirect contact with bodily tissue or organ, should be biocompatible. Thus, biocompatibility assessment is crucial to evaluate any potentially unacceptable immunological response due to the material composition of the device with the body (FDA FDA 2021a). Hence, the biocompatibility evaluation of injectable hydrogel that is specific for wound application is required. Concerning that, most studies have shown excellent biocompatibility characteristics of polysaccharide-based injectable hydrogel either in vitro or in vivo, including haemocompatibility. For example, a study developed a dopamine-modified gelatine with silver nanoparticles (Ag NPs) incorporated in guar gum-based hydrogel (Gel-DA/GG@Ag) hydrogel, which exhibited good haemocompatibility to mouse erythrocytes and cytocompatibility with mouse embryonic fibroblast cells (Zhang et al. 2021a). Meanwhile, another study group utilised zebrafish as a skin wound model to assess the toxicity and wound regeneration of acrylamide-modified β-chitin/alginate dialdehyde (Am-β-Chn-Sq/ADA) hydrogel. The study reported that the hydrogel promoted significant wound contraction (Balitaan et al. 2020).

18.5 Clinical Investigation Status of Injectable Hydrogel

According to the National Library of Medicine by the National Institute of Health, United States, 59 clinical investigations on the injectable hydrogel based on the search word "hydrogel+injection" were identified. One of the studies was an investigation entitled *Efficacy and Safety of Hymovis ONE® Intraarticular Injection in Active Patients With Knee Overuse Syndrome* (knee osteoarthritis) with an identifier number NCT04661111, conducted in Rome, Italy, in 2016 and completed in 2018 (NIH 2020). Based on their published findings, the Hymovis was safe and effective in improving knee function and reducing pain over time with low adverse event risks such as joint swelling and arthralgia (Bernetti et al. 2021). Similar results were obtained on the use of Hymovis on hip osteoarthritis (Migliore et al. 2020). Thus, the Hymovis has been approved by the Food and Drug Administration (FDA) and is now commercially available.

18.6 Regulatory Aspects of Injectable Hydrogel

In the United States, injectable hydrogel products that are "intended for use in the cure, mitigation, treatment, or prevention of disease" are considered medical devices as stated in the definitions of medical device per Section 201(h) of the Food, Drug, and Cosmetic Act (FDA FDA 2022). The injectable hydrogel products should comply with the FDA's Centre for Medical Devices and Radiological Health (CDRH). In addition, the injectable hydrogel products should also comply with the regulation's requirements such as the Code of Federal Regulations and Title 21 for CDRH to be approved and cleared for sale in the United States (FDA FDA 2021b).

Meanwhile, in Europe, the injectable hydrogel products should comply with The Classification, Labelling, and Packaging of Substances and Mixtures (CLP) Regulation (European Regulation (EC) No. 1272/2008) and the Registration, Evaluation, Authorisation and Restrictions of Chemicals (REACH) Regulation, which both contain specific requirements for products from all industries that contain or are composed of chemical substances and mixtures (HSE HaSE 2022; HSE HaSE UK REACH 2022). Apart from CLP and REACH, the injectable hydrogel products claimed as a medical device are also subjected to the Medical Devices Regulations (MDR) EU/2017/745. The MDR EU/2017/745 regulates the requirement for assessing and ensuring the safety and quality of medical devices prior to marketing and providing accurate information about the medical devices to users (i.e., practitioners and consumers) (The National Archieves 2017). In the case where the injectable hydrogel products containing cells, tissue, or viable components are considered advanced therapy medicinal products, they are required to comply with Regulations (EC) No 1394/2007 (EUR-Lex Regulation (EC) 2022).

Additionally, to ensure the safety of the finished product, it is advised that pharmaceutical-grade raw materials be used when creating the injectable hydrogel products as medical devices. A pharmaceutical-grade compound is any medication, biological compound, or reagent—active or inactive—for which a recognised pharmacopoeia has defined a chemical purity standard. Manufacturers utilise these standards to help ensure that the items are stable, safe, and effective, and that they are of the right chemical purity and quality in the right solution or compound (Alonso et al. 2021). Currently, there is a wide selection of FDA's approved commercial polysaccharides such as alginate, collagen, and chitosan that are of pharmaceutical grade and can be readily used. Thus, this eases the manufacturers on raw materials regulatory issues.

Finally, the safety evaluation and biocompatibility of injectable hydrogel products can be evaluated according to ISO 10993, which provides guidelines to guarantee the safety and risk of injectable hydrogel products that are important as part of the information required before registration as a medical device.

18.7 Commercialisation Potential of Polysaccharide-Based Injectable Nanocomposite Hydrogel

Technological advancement has opened wide opportunities to the industry players in the development of new healthcare technology, particularly for wound treatment applications. The current effort focuses on the development of an injectable hydrogel as it offers minimally invasive delivery of therapeutic payloads, has the ability to fill any irregular shapes and depth of the wound, and induces the regeneration of the damaged site of the body. Furthermore, the polysaccharide-based hydrogel provides biocompatible materials having similar characteristics to the native ECM (Dimatteo et al. 2018). Additionally, the introduction of injectable hydrogel by insertion into a living body or implanted on skin with minimal invasion may further uplift the demand to investigate this technology as a mode of treatment of choice.

According to the Market Analysis by Grand View Research, the market size value of hydrogel dressing is USD849.23 million with an estimated compound annual growth rate of 4.28% from 2022 to 2030. The rising of acute and chronic wound cases is the major key to the hydrogel dressing demands worldwide (Grand View Research 2022). Additionally, it has been predicted that the population of people aged 60 years and older will double to 2.1 billion by 2050 compared to 1 billion in 2020 (World Health Organization 2022). The increasing number of elderly people will promote higher usage of hydrogel dressing due to this geriatric population having a slow healing ability. In a nutshell, the demands for injectable hydrogel will rise in the coming future as one of the strategies in minimally invasive wound healing applications.

18.8 Conclusion and Future Directions

Rapid progress is being made over the years to produce a hydrogel with various target applications such as wound dressing and wound healing, which undoubtedly has been the centre of attention of many researchers these recent years. The injectable hydrogel has shown significant advancements, and many studies have been focused on the continuous development of injectable hydrogel. Other than that, studies have also focused on finding the most optimum parameters and concentration to produce the best possible hydrogel with excellent mechanical properties, biocompatibility, injectability, and self-healing properties. Future works should be focused towards in vivo studies to ensure that this hybrid hydrogel is safe to be used for further developments.

Furthermore, the quality of product development including the safety and efficacy towards the intended application will affect the regulatory approval and commercial success of the polysaccharide-based injectable nanocomposite hydrogel. Therefore, other than the major commercial key factors, an early settlement on development strategies is essential as well. Additionally, the polysaccharide-based injectable hydrogel is made of naturally occurring polymers such as agar, alginate, collagen, chitosan, hyaluronic acid, and gelatine that are biodegradable, biocompatible, and mostly have integrin binding sites that allow for adhesion and influence cellular behaviour. These attributes show the advantages of polysaccharide materials over synthetic materials. However, significant batch-to-batch variation and potential immune response have become the limitations of these materials. Therefore, further investigations to overcome these limitations are essential.

Acknowledgement The authors' acknowledgement goes to the Universiti Malaya (UM), Kuala Lumpur.

Conflicts of Interest The authors declare that they have no conflict of interest.

References

Abasalizadeh F, Moghaddam SV, Alizadeh E et al (2020) Alginate-based hydrogels as drug delivery vehicles in cancer treatment and their applications in wound dressing and 3D bioprinting. J Biol Eng 14(1):1–22

Akhavan-Kharazian N, Izadi-Vasafi H (2019) Preparation and characterization of chitosan/gelatin/nanocrystalline cellulose/calcium peroxide films for potential wound dressing applications. Int J Biol Macromol 133:881–891

Alonso JM, Andrade Del Olmo J, Perez Gonzalez R et al (2021) Injectable hydrogels: from laboratory to industrialization. Polymers (Basel) 13(4):650

Al-Sabah A, Burnell SE, Simoes IN et al (2019) Structural and mechanical characterization of crosslinked and sterilised nanocellulose-based hydrogels for cartilage tissue engineering. Carbohydr Polym 212:242–251

Amirian J, Zeng Y, Shekh MI et al (2021) In-situ crosslinked hydrogel based on amidated pectin/oxidized chitosan as potential wound dressing for skin repairing. Carbohydr Polym 251:117005

Augst AD, Kong HJ, Mooney DJ (2006) Alginate hydrogels as biomaterials. Macromol Biosci 6(8): 623–633

Bai Z, Dan W, Yu G et al (2018) Tough and tissue-adhesive polyacrylamide/collagen hydrogel with dopamine-grafted oxidized sodium alginate as crosslinker for cutaneous wound healing. RSC Adv 8(73):42123–42132

Balitaan JNI, Hsiao C-D, Yeh J-M et al (2020) Innovation inspired by nature: biocompatible self-healing injectable hydrogels based on modified-β-chitin for wound healing. Int J Biol Macromol 162:723–736

Bernetti A, Agostini F, Alviti F et al (2021) New viscoelastic hydrogel hymovis MO.RE. single intra-articular injection for the treatment of knee osteoarthritis in sportsmen: safety and efficacy study results. Front Pharmacol 12:673988

Bertolino V, Cavallaro G, Milioto S et al (2020) Polysaccharides/Halloysite nanotubes for smart bionanocomposite materials. Carbohydr Polym 245:116502

Bi B, Ma M, Lv S et al (2019) In-situ forming thermosensitive hydroxypropyl chitin-based hydrogel crosslinked by Diels-Alder reaction for three dimensional cell culture. Carbohydr Polym 212: 368–377

Bian S, Zheng Z, Liu Y et al (2019) A shear-thinning adhesive hydrogel reinforced by photo-initiated crosslinking as a fit-to-shape tissue sealant. J Mater Chem B 7(42):6488–6499

Chao Y, Chen Q, Liu Z (2020) Smart injectable hydrogels for cancer immunotherapy. Adv Funct Mater 30(2):1902785

Chen MH, Wang LL, Chung JJ et al (2017) Methods to assess shear-thinning hydrogels for application as injectable biomaterials. ACS Biomater Sci Eng 3(12):3146–3160

Chen T, Chen Y, Rehman HU et al (2018) Ultratough, self-healing, and tissue-adhesive hydrogel for wound dressing. ACS Appl Mater Interfaces 10(39):33523–33531

De France KJ, Cranston ED, Hoare T (2019) Mechanically reinforced injectable hydrogels. ACS Appl Poly Mater 2(3):1016–1030
Dimatteo R, Darling NJ, Segura T (2018) In situ forming injectable hydrogels for drug delivery and wound repair. Adv Drug Deliv Rev 127:167–184
Ding Q, Xu X, Yue Y et al (2018) Nanocellulose-mediated electroconductive self-healing hydrogels with high strength, plasticity, viscoelasticity, stretchability, and biocompatibility toward multifunctional applications. ACS Appl Mater Interfaces 10(33):27987–28002
Du X, Liu Y, Wang X et al (2019) Injectable hydrogel composed of hydrophobically modified chitosan/oxidized-dextran for wound healing. Mater Sci Eng C 104:109930
EUR-Lex Regulation (EC) No 1394/2007 of the European Parliament and of the Council of 13 November 2007 on advanced therapy medicinal products and amending Directive 2001/83/EC and Regulation (EC) No 726/2004 (Text with EEA relevance). (Cited 25 November 2022.); Available from: https://eur-lex.europa.eu/legal-content/en/ALL/?uri=CELEX%3A32007R1394
FDA FDA (2021a) Basics of Biocompatibility: Information Needed for Assessment by the FDA. (Cited 28 November 2022); Available from: https://www.fda.gov/medical-devices/biocompatibility-assessment-resource-center/basics-biocompatibility-information-needed-assessment-fda
FDA FDA (2021b) Products and medical procedures. (cited 25 November 2022); Available from: https://www.fda.gov/medical-devices/products-and-medical-procedures
FDA FDA (2022) How to determine if your product is a medical device. (cited 24 November 2022); Available from: https://www.fda.gov/medical-devices/classify-your-medical-device/how-determine-if-your-product-medical-device
Fu H, Yu C, Li X et al (2021) Facile engineering of ECM-mimetic injectable dual crosslinking hydrogels with excellent mechanical resilience, tissue adhesion, and biocompatibility. J Mater Chem B 9(48):10003–10014
Gao Y, Li Z, Huang J et al (2020) In situ formation of injectable hydrogels for chronic wound healing. J Mater Chem B 8(38):8768–8780
Ghaderinejad P, Najmoddin N, Bagher Z et al (2021) An injectable anisotropic alginate hydrogel containing oriented fibers for nerve tissue engineering. Chem Eng J 420:130465
Grand View Research (2022) Hydrogel dressing market size, share & trends analysis report by product (amorphous hydrogel, impregnated hydrogel), by application (acute wounds, chronic wounds), by end use (hospitals, specialty clinics), by region, and segment forecasts, 2022–2030. (Cited 25 November 2022); Available from: https://www.grandviewresearch.com/industry-analysis/hydrogel-dressing-market-report
Hong SH, Kim S, Park JP et al (2018) Dynamic bonds between boronic acid and alginate: hydrogels with stretchable, self-healing, stimuli-responsive, remoldable, and adhesive properties. Biomacromolecules 19(6):2053–2061
HSE HaSE (2022) The GB CLP Regulation. (Cited 25 November 2022); Available from: https://www.hse.gov.uk/chemical-classification/legal/clp-regulation.htm
HSE HaSE UK REACH explained. (Cited 25 November 2022); Available from: https://www.hse.gov.uk/reach/about.htm
Hu X, Gao Z, Tan H et al (2019) An injectable hyaluronic acid-based composite hydrogel by DA click chemistry with pH sensitive nanoparticle for biomedical application. Front Chem 7:477
Jalalvandi E, Shavandi A (2019) Shear thinning/self-healing hydrogel based on natural polymers with secondary photocrosslinking for biomedical applications. J Mech Behav Biomed Mater 90: 191–201
Koyyada A, Orsu P (2021) Natural gum polysaccharides as efficient tissue engineering and drug delivery biopolymers. J Drug Deliv Sci Technol 63:102431
Lee JH (2018) Injectable hydrogels delivering therapeutic agents for disease treatment and tissue engineering. Biomater Res 22:27
Li J, Yu F, Chen G et al (2020) Moist-retaining, self-recoverable, bioadhesive, and transparent in situ forming hydrogels to accelerate wound healing. ACS Appl Mater Interfaces 12(2): 2023–2038

Li H, Wei X, Yi X et al (2021) Antibacterial, hemostasis, adhesive, self-healing polysaccharides-based composite hydrogel wound dressing for the prevention and treatment of postoperative adhesion. Mater Sci Eng C 123:111978

Liao J, Wang B, Huang Y et al (2017) Injectable alginate hydrogel cross-linked by calcium gluconate-loaded porous microspheres for cartilage tissue engineering. ACS Omega 2(2): 443–454

Liu J, Li J, Yu F et al (2020) In situ forming hydrogel of natural polysaccharides through Schiff base reaction for soft tissue adhesive and hemostasis. Int J Biol Macromol 147:653–666

Lokhande G, Carrow JK, Thakur T et al (2018) Nanoengineered injectable hydrogels for wound healing application. Acta Biomater 70:35–47

López Vélez J, Palencia M (2021) An approximation to mechanics and mass transport in polymer-based hydrogel systems. J Sci Technol Appl 10:66–91

Lopez-Sanchez P, Martinez-Sanz M, Bonilla M et al (2020) Nanostructure and poroviscoelasticity in cell wall materials from onion, carrot and apple: roles of pectin. Food Hydrocoll 98:105253

Lu W, Bao D, Ta F et al (2020) Multifunctional alginate hydrogel protects and heals skin defects in complex clinical situations. ACS Omega 5(28):17152–17159

Ma Z, Yang X, Ma J et al (2021) Development of the mussel-inspired pH-responsive hydrogel based on Bletilla striata polysaccharide with enhanced adhesiveness and antioxidant properties. Colloids Surf B Biointerfaces 208:112066

Martino F, Perestrelo AR, Vinarský V et al (2018) Cellular mechanotransduction: from tension to function. Front Physiol 9:824

Miao L, Zhang M, Tu Y, et al (2019) Stimuli-responsive cellulose-based hydrogels. Cellulose-based super-absorbent hydrogels, polymers and polymeric composites: a reference series, pp 269–308

Migliore A, Frediani B, Gigliucci G et al (2020) Efficacy of a single intra-articular HYMOVIS ONE injection for managing symptomatic hip osteoarthritis: a 12-month follow-up retrospective analysis of the ANTIAGE register data. Orthop Res Rev 12:19–26

Morais D, Rodrigues M, Silva T et al (2013) Development and characterization of novel alginate-based hydrogels as vehicles for bone substitutes. Carbohydr Polym 95(1):134–142

Nair R, Choudhury AR (2020) Synthesis and rheological characterization of a novel shear thinning Levan gellan hydrogel. Int J Biol Macromol 159:922–930

NIH (2020) Efficacy and safety of Hymovis ONE® (32mg/4ml) intrarticular injection in active patients with knee overuse syndrome

Pettignano A, Grijalvo S, Haering M et al (2017) Boronic acid-modified alginate enables direct formation of injectable, self-healing and multistimuli-responsive hydrogels. Chem Commun 53(23):3350–3353

Pham T-N, Jiang Y-S, Su C-F et al (2020) In situ formation of silver nanoparticles-contained gelatin-PEG-dopamine hydrogels via enzymatic cross-linking reaction for improved antibacterial activities. Int J Biol Macromol 146:1050–1059

Popescu I, Turtoi M, Suflet DM et al (2021) Alginate/poloxamer hydrogel obtained by thiol-acrylate photopolymerization for the alleviation of the inflammatory response of human keratinocytes. Int J Biol Macromol 180:418–431

Pourshahrestani S, Zeimaran E, Kadri NA et al (2020) Polymeric hydrogel systems as emerging biomaterial platforms to enable hemostasis and wound healing. Adv Healthc Mater 9(20): 2000905

Pupkaite J, Rosenquist J, Hilborn J et al (2019) Injectable shape-holding collagen hydrogel for cell encapsulation and delivery cross-linked using thiol-michael addition click reaction. Biomacromolecules 20(9):3475–3484

Qiu W, Han H, Li M et al (2021a) Nanofibers reinforced injectable hydrogel with self-healing, antibacterial, and hemostatic properties for chronic wound healing. J Colloid Interface Sci 596: 312–323

Qiu W, Wang Q, Li M et al (2021b) Peptidoglycan-inspired peptide-modified injectable hydrogels with enhanced elimination capability of bacterial biofilm for chronic wound healing. Compos Part B Eng 227:109402

Qu J, Zhao X, Liang Y et al (2018) Antibacterial adhesive injectable hydrogels with rapid self-healing, extensibility and compressibility as wound dressing for joints skin wound healing. Biomaterials 183:185–199

Ren K, He C, Xiao C et al (2015) Injectable glycopolypeptide hydrogels as biomimetic scaffolds for cartilage tissue engineering. Biomaterials 51:238–249

Rimondo S, Perale G, Rossi F (2019) 6 - Polysaccharide-based scaffold for tissue-regeneration. In: Maiti S, Jana S (eds) Functional polysaccharides for biomedical applications. Woodhead Publishing, pp 189–212

Rizzo F, Kehr NS (2021) Recent advances in injectable hydrogels for controlled and local drug delivery. Adv Healthc Mater 10(1):2001341

Shin JU, Gwon J, Lee S-Y et al (2018) Silver-incorporated nanocellulose fibers for antibacterial hydrogels. ACS Omega 3(11):16150–16157

Shojaeiarani J, Bajwa DS, Hartman K (2019) Esterified cellulose nanocrystals as reinforcement in poly (lactic acid) nanocomposites. Cellulose 26(4):2349–2362

Siqueira P, Siqueira É, De Lima AE et al (2019) Three-dimensional stable alginate-nanocellulose gels for biomedical applications: towards tunable mechanical properties and cell growing. Nano 9(1):78

Smith AM, Senior JJ (2021) Alginate hydrogels with Tuneable properties. In: Lavrentieva A, Pepelanova I, Seliktar D (eds) Tunable hydrogels. Springer, Cham, pp 37–61

Sun Y, Ma X, Hu H (2021) Marine polysaccharides as a versatile biomass for the construction of nano drug delivery systems. Mar Drugs 19(6):345

Tang C, Ding R, Sun J et al (2019) The impacts of natural polysaccharides on intestinal microbiota and immune responses–a review. Food Funct 10(5):2290–2312

Tavakoli S, Klar AS (2020) Advanced hydrogels as wound dressings. Biomol Ther 10(8):1169

Thambi T, Phan VG, Lee DS (2016) Stimuli-sensitive injectable hydrogels based on polysaccharides and their biomedical applications. Macromol Rapid Commun 37(23): 1881–1896

The National Archieves (2017) Regulation (EU) 2017/745 of the European Parliament and of the Council

Tiwari S, Patil R, Bahadur P (2018) Polysaccharide based scaffolds for soft tissue engineering applications. Polymers (Basel) 11(1):1

Uman S, Dhand A, Burdick JA (2020) Recent advances in shear-thinning and self-healing hydrogels for biomedical applications. J Appl Polym Sci 137(25):48668

Wallace HA, Basehore BM, Zito PM (2022) Wound healing phases

Wan X, Guo H, Liang Y et al (2020) The physiological functions and pharmaceutical applications of inulin: a review. Carbohydr Polym 246:116589

Wang K, Nune K, Misra R (2016) The functional response of alginate-gelatin-nanocrystalline cellulose injectable hydrogels toward delivery of cells and bioactive molecules. Acta Biomater 36:143–151

Wang L, Dong S, Liu Y et al (2020) Fabrication of injectable, porous hyaluronic acid hydrogel based on an in-situ bubble-forming hydrogel entrapment process. Polymers 12(5):1138

Wei H, Li S, Liu Z et al (2022a) Preparation and characterization of starch-cellulose interpenetrating network hydrogels based on sequential Diels-Alder click reaction and photopolymerization. Int J Biol Macromol 194:962–973

Wei Q, Chen K, Zhang X et al (2022b) Facile preparation of polysaccharides-based adhesive hydrogel with antibacterial and antioxidant properties for promoting wound healing. Colloids Surf B Biointerfaces 209:112208

World Health Organization (2022) Ageing and health. (Cited 25 November 2022); Available from: https://www.who.int/news-room/fact-sheets/detail/ageing-and-health

Wu T, Cui C, Huang Y et al (2019) Coadministration of an adhesive conductive hydrogel patch and an injectable hydrogel to treat myocardial infarction. ACS Appl Mater Interfaces 12(2): 2039–2048

Xie Y, Gao P, He F et al (2022) Application of alginate-based hydrogels in hemostasis. Gels 8(2): 109

Xu C, Chang Y, Wu P et al (2021) Two-dimensional-germanium phosphide-reinforced conductive and biodegradable hydrogel scaffolds enhance spinal cord injury repair. Adv Funct Mater 31(41):2104440

Xuan H, Wu S, Fei S et al (2021) Injectable nanofiber-polysaccharide self-healing hydrogels for wound healing. Mater Sci Eng C 128:112264

Yang X, Li P, Tang W et al (2021) A facile injectable carbon dot/oxidative polysaccharide hydrogel with potent self-healing and high antibacterial activity. Carbohydr Polym 251:117040

Zhai Z, Xu K, Mei L et al (2019) Co-assembled supramolecular hydrogels of cell adhesive peptide and alginate for rapid hemostasis and efficacious wound healing. Soft Matter 15(42):8603–8610

Zhang H, Sun X, Wang J et al (2021a) Multifunctional injectable hydrogel dressings for effectively accelerating wound healing: enhancing biomineralization strategy. Adv Funct Mater 31(23): 2100093

Zhang X, Li Y, Ma Z et al (2021b) Modulating degradation of sodium alginate/bioglass hydrogel for improving tissue infiltration and promoting wound healing. Bioact Mater 6(11):3692–3704

Zhao X, Wu H, Guo B et al (2017) Antibacterial anti-oxidant electroactive injectable hydrogel as self-healing wound dressing with hemostasis and adhesiveness for cutaneous wound healing. Biomaterials 122:34–47

Zhong Y, Wang J, Yuan Z et al (2019) A mussel-inspired carboxymethyl cellulose hydrogel with enhanced adhesiveness through enzymatic crosslinking. Colloids Surf B Biointerfaces 179:462–469

Zhu H, Mei X, He Y et al (2020) Fast and high strength soft tissue bioadhesives based on a peptide dendrimer with antimicrobial properties and hemostatic ability. ACS Appl Mater Interfaces 12(4):4241–4253

Zubik K, Singhsa P, Wang Y et al (2017) Thermo-responsive poly (N-isopropylacrylamide)-cellulose nanocrystals hybrid hydrogels for wound dressing. Polymers 9(4):119

Roles of Sustainable Biomaterials in Biomedical Engineering for Ischemic Stroke Therapy

19

Farah Amna Othman, Asmaa' Mohd Satar, and Suat Cheng Tan

Abstracts

Ischemic stroke is the leading cause of adult disability which imposes high socioeconomic burden worldwide. Natural and synthetic biomaterials have been elucidated as viable strategies for the treatment of ischemic stroke since decades. Nowadays, due to the advanced development in biomedical engineering, biomaterial applications have been perfected for sustainable usage in different phases of in vitro, in vivo and clinical studies. This chapter reviews a broad range of sustainable biomaterials and major therapeutic roles of sustainable biomaterials for ischemic stroke treatment. Furthermore, the neurological recovery effects triggered by these engineered biomaterials after being implanted into in vivo ischemic stroke models are highlighted as well. In brief, this chapter presents the roles and potentials of sustainable biomaterials in biomedical engineering for ischemic stroke therapy.

Keywords

Ischemic stroke · Biomaterial-based treatments · Biomedical engineering · Neurological recovery

F. A. Othman · A. M. Satar · S. C. Tan (✉)
School of Health Sciences, Health Campus, Universiti Sains Malaysia, Kubang Kerian, Kelantan, Malaysia
e-mail: tansc@usm.my

W. S. Wan Kamarul Zaman, A. A. Nurul (eds.), *Sustainable Material for Biomedical Engineering Application*, https://doi.org/10.1007/978-981-99-2267-3_19

Abbreviations

ASP	Spermine-modified amylose
ATP	Adenosine triphosphate
BBB	Blood brain barrier
BCSFB	Blood-cerebrospinal-fluid barrier
bFGF	Basic fibroblast growth factor
CAM	Cell adhesion molecule
CDI	1,1′-carbonyldiimidazole
CNS	Central nervous system
CS-maleimide	Chitosan-maleimide
ECM	Extracellular matrix
ECs	Endothelial cells
FDA	Food and Drug Administration
Fuco-SPs	Fucoidan-functionalized dextran submicroparticles
GFP	Green fluorescent protein
HA	Hyaluronic acid
HUVECs	Human umbilical vein endothelial cells
I/R	Ischemia/reperfusion
IV	Intravenous
LBD	Laminin-binding domain
MC	Methylcellulose
MCA	Middle cerebral artery
MCAO	Middle cerebral artery occlusion
MRI	Magnetic resonance imaging
MSCs	Mesenchymal stem cells
Ngn2	Neurogenin-2
NSCs	Neural stem cells
OD	Optical density
OGD	Oxygen glucose deprivation
PA	Palmitic acid
PCNs	Polyelectrolyte complex nanoparticles
PEG	Polyethylene glycol
pH	Potential hydrogen
PLA	Poly lactic acid
PLGA	Poly (lactic-co-glycolic acid)
PTD	Protein transduction domain
ROS	Reactive oxygen species
rtPA	Recombinant tissue plasminogen activator
RUI	Research University Individual
SDF-1α	Stromal derived factor 1α
SPION	Superparamagnetic iron oxide nanoparticles
SVZ	Subventricular zone
TAT	Transactivator of transcription

19.1 Introduction

19.1.1 Stroke

Stroke is a cerebrovascular disease that can damage the brain. Stroke occurs when the blood supply to a part of the brain is interrupted, causing the affected part of the brain to be deprived of oxygen and nutrients, and subsequently leading to irreversible brain cell death. Stroke is the single-most common cause of adult disability and remains a serious health problem throughout the world. In Malaysia, stroke is one of the five leading factors of death and one of the major factors of hospitalization apart from ischemic heart disease, pneumonia, septicemia, and chronic lower respiratory disease (Tan and Venketasubramanian 2022). One notable recent finding published in 2022 has highlighted a worrying rise in the trend of ischemic stroke incidence among the younger Malaysian population. The data has reported an increasing trend from 1.9% to 2.6% of ischemic stroke incidence among 35–39-year-old patients; 11.9% to 13.0% among 55–59-year-old patients and 13.2% to 14.1% among 60–64 years old group (Hwong et al. 2021). In addition, the rate is expected to increase continuously for the next 10 years in Malaysia (Hwong et al. 2021). Worldwide, a similar increasing trend of hospitalization rate for ischemic stroke among patients under 65 years also has been documented in administrative databases and population-based stroke registries (Olié et al. 2022). The socio-economic burden induced by the increasing trend of stroke incidence, compounded with an increase observed among the younger population have desperately called for immediate actions to be taken to strengthen stroke disease prevention, as well as better management of stroke patients worldwide.

19.1.2 Types of Stroke

19.1.2.1 Ischemic Stroke

There are two main types of stroke: ischemic stroke and hemorrhagic stroke. About 87% of stroke incidence are ischemic (Feske 2021). An ischemic stroke occurs when a clot disrupts the blood flow to a particular region of the brain. There are three subtypes of ischemic stroke, namely thrombotic stroke, embolic stroke, and lacunar stroke (Fig. 19.1). In a thrombotic stroke, the artery is blocked by a thrombus (blood clot) that has formed due to atherosclerosis. The thrombus is made up of hardened buildup of cholesterol and other substances called plaque (Fig. 19.1, middle panel). The other type of ischemic stroke is embolic stroke. In this case, the blood clot called an embolus would form in another part of the body (e.g., heart) and travel to the brain through the bloodstream where it gets stuck and blocks the artery. Embolic strokes are often resulted from heart diseases such as atrial fibrillation or heart surgery and it can occur rapidly without any warning signs (Fig. 19.1, left panel). In contrast, lacunar strokes occur due to the occlusion of the minimally sized, deeply penetrating cerebral vessel branches from the circle of Willis. These include the branches from the middle cerebral artery (MCA), or the basilar, anterior cerebral or posterior

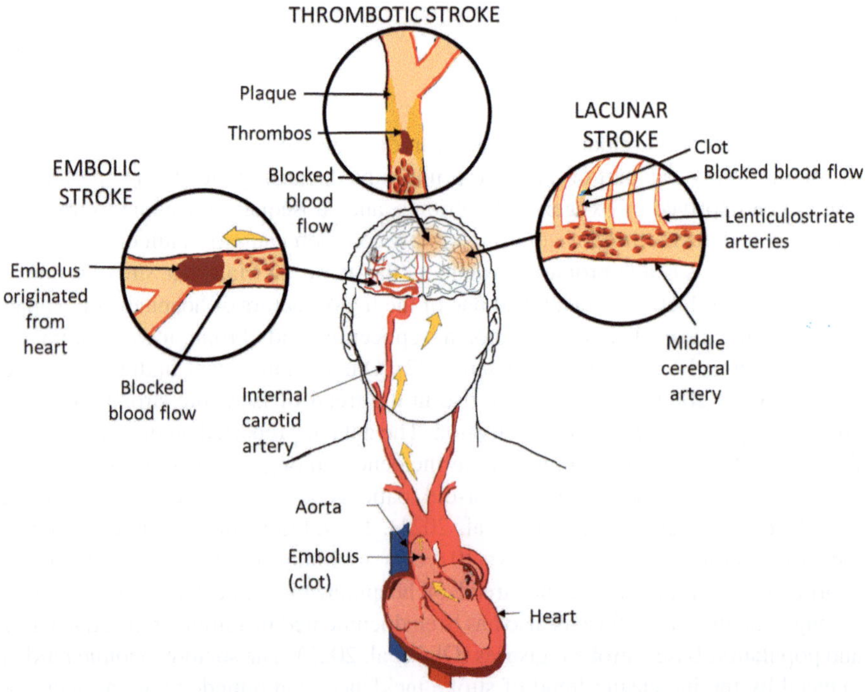

Fig. 19.1 Types of ischemic stroke. Left panel: Embolic stroke occurs when the artery within the brain is blocked by the embolus travelled from the heart. Middle panel: Thrombotic stroke occurs when the artery within the brain is blocked by the thrombus formed due to plaque. Right panel: Lacunar stroke occurs when small and deep penetrating branches of the cerebral vessels (e.g., lenticulostriate arteries branch from MCA) are occluded

cerebral arteries (Gore et al. 2022). One of the examples of lacunar stroke is lenticulostriate infarcts, resulting from ischemia within the territory that is supplied by the deep perforating branches of the MCA (Fig. 19.1, right panel). Lacunar infarcts are usually asymptomatic due to their small size. However, the accumulation of multiple lacunar infarctions at a certain critical location in the brain can lead to significant physical and cognitive disabilities (Decavel et al. 2012).

19.1.2.2 Hemorrhagic Stroke

Hemorrhagic stroke is the second major type of stroke which has accounted for approximately 13% of all stroke incidents. Hemorrhagic stroke occurs when the blood vessel in the brain ruptures and causes bleeding inside or around brain tissue (Fig. 19.2). Then, the leaching blood will diffuse into the brain tissue and increase the pressure inside the brain. This will push the adjacent brain tissue against the skull, resulting in further damage. Furthermore, the occurrence of vasospasm (secondary constriction) of arteries adjacent to the bleeding site may take place due to irritation by-products of blood leakage, which can interrupt blood flow and cause a secondary ischemic stroke (Montaño et al. 2021; An et al. 2017). Hemorrhagic

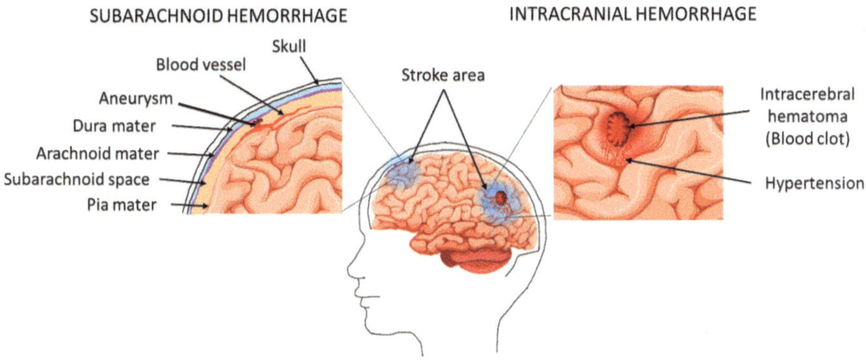

Fig. 19.2 Types of hemorrhagic stroke. Left panel: Subarachnoid hemorrhage occurs when a blood vessel would rupture upon the surface of the brain, causing blood to spill into the subarachnoid space. Right panel: Intracerebral hemorrhage occurs when the artery within the brain ruptures and bleeding occurs inside the brain

stroke is subdivided into two main types based on the location of bleeding. Subarachnoid hemorrhage occurs when blood leaks into the subarachnoid space, which is an area between the arachnoid mater and pia mater of the brain (Fig. 19.2, left panel). It is usually caused by aneurysm, an outpouching of a weak spot in a vessel wall that has ruptured, resulting in profuse bleeding in the brain. On the other hand, intracerebral hemorrhage occurs when blood leaks inside the brain tissue (Fig. 19.2, right panel). It is usually caused by uncontrolled high blood pressure that weakens the wall of the small vessel inside the brain, causing the vessel to rupture easily.

19.2 Biomaterial-Based Therapies for Ischemic Stroke

In recent years, advances in biomaterial-based regenerative medicine and tissue engineering have emerged as a promising option for stroke alternative treatment. Up to date, there are many biomaterials being engineered from various resources including natural compounds, synthetic compounds, or a mix of both with the aim to enhance the therapeutic potential for stroke treatment. In this type of strategy, biomaterials incorporated with chosen therapeutic agent(s) are used to structurally support the regeneration of tissue or assist the targeted delivery of therapeutic agents into the brain. Furthermore, some of the materials are able to support healing or reduce the secondary damage post stroke, thereby contributing to tissue regeneration and endogenous repair processes. This chapter reviews several types of biomaterials and their specific features for the applications in ischemic stroke therapy based on the selected literature data from PubMed, ScienceDirect, and Scopus databases using keywords consisting of "biomaterial" and "ischemic stroke". There are nine major applications of biomaterials for ischemic stroke treatment identified and summarized in this chapter (Fig. 19.3). The description of these selected biomaterials and the

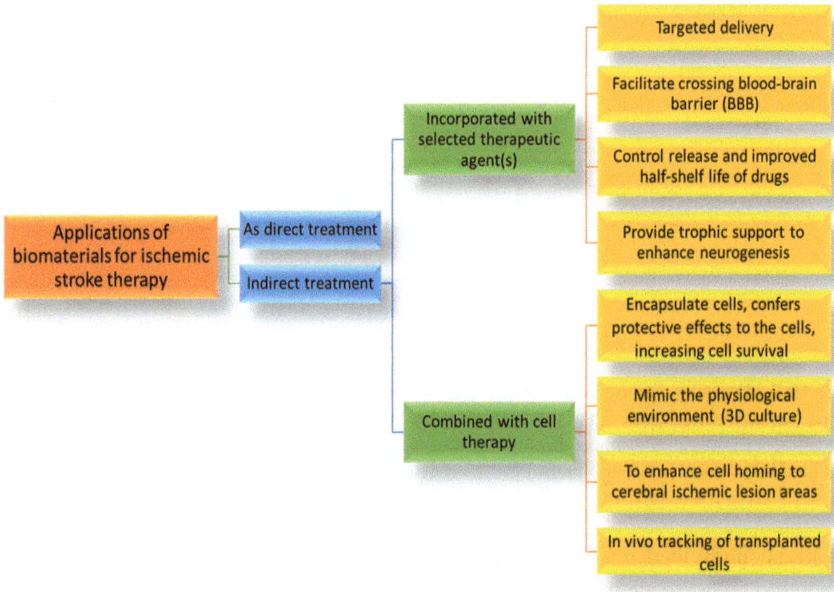

Fig. 19.3 Major roles of biomaterial in enhancing therapeutic efficacy for ischemic stroke treatment

in vivo therapeutic benefits gained from the ischemic stroke studies were discussed in detail in the subsequent section.

19.3 Major Therapeutic Roles of Sustainable Biomaterials for Ischemic Stroke Treatment

19.3.1 As a Direct Treatment for Ischemic Stroke

Ischemic stroke results from cerebral artery occlusion that can interrupt blood flow and limit oxygen and glucose supply to the brain for the neuronal cell survival. Till date, alteplase is the only recombinant tissue plasminogen activator (rtPA) approved by the United States Food and Drug Administration (US FDA) as the intravenous (IV) thrombolytic medication for acute ischemic stroke. As compared to placebo, rtPA administered within 4.5 hours after the onset of acute ischemic stroke symptoms has significantly improved the clinical outcomes of the patients (Reed et al. 2022). Nonetheless, the administration of rtPA beyond the golden therapeutic window of 4.5 h would increase the risk of intracerebral hemorrhage. Furthermore, rtPA also has been reported to trigger complications on patients including reperfusion injury with edema and angioedema, reocclusion, and secondary embolization due to ineffective thrombolysis or redistribution of the lysed clot (Balami et al. 2013;

Hughes et al. 2022). Therefore, there is an urgent necessity to discover other alternative resources for effective ischemic stroke treatment.

Recently, naturally-derived biomaterials such as melatonin, that is secreted from the human pineal gland, has been explored for the treatment of ischemic stroke due to its potential as free radical scavenger and antioxidant agent (Yip et al. 2021). Melatonin is a naturally occurring biomacromolecule that possesses anti-apoptotic, anti-excitotoxicity, and anti-inflammatory properties, making it a suitable candidate as an alternative for the ischemic treatment approach (Sadanandan et al. 2020). More importantly, melatonin also has been found to act as a mitochondrial-targeted antioxidant to protect the mitochondria against oxidative-stress damage that could occur during an ischemic stroke attack (Reiter et al. 2017; Tan et al. 2016). Mitochondria is the powerhouse of a cell. Its role is critical in maintaining energy homeostasis, especially in the brain as brain tissue is the most energy-consuming organ in the human body. During an ischemic stroke attack, the adenosine triphosphate (ATP) synthesis by the mitochondria is disrupted due to the shortage of oxygen supply, causing disturbance to the energy balance and inevitably leading to neuronal death (Liu et al. 2018). An accumulation of evidence has demonstrated that mitochondrial dysfunction is one of the typical hallmarks of ischemia/reperfusion (I/R) injury (He et al. 2020). Therefore, Yip et al. had highlighted a strategy for stroke treatment through a method which utilized melatonin, a naturally derived biomaterial that can enhance the neuroprotective bioactivity of mitochondria against cerebral ischemic (Yip et al. 2021). It was reported that the treatment with melatonin had significantly increased mitochondrial transfer and antioxidants activity, as well as inhibited apoptosis activity. Moreover, the administration of melatonin-pretreated mitochondria into the ischemic-induced rodent stroke models (400 μg in 10 μL saline/rat) was found to significantly reduce the brain infarction area, with improved neurological function in rodent acute ischemic stroke model.

19.3.2 As a Delivery Agent for Targeted Delivery of Pharmacological Drugs to a Specific Area in Injured Brain

Apart from supporting direct healing or reducing the secondary damage post stroke as described in an earlier section, biomaterials can also be used to incorporate with other potential therapeutic agents to enhance their efficacy for ischemic stroke therapy. One of the examples is to package thrombolytic agents such as rtPA and/or anticoagulant drugs into functionalized biomaterial that can specifically travel toward the thrombus for efficiently targeted thrombolytic therapy. Targeted delivery of thrombolytic agent is very critical in ischemic stroke therapy as the occurrence of side effects such as systemic fibrinolysis could take place if the delivery deviates from the targeting site. This will lead to the occurrence of dangerous side effects such as internal bleeding, intracranial hemorrhage, and heart arrhythmia in patients (Ma et al. 2019).

In 2018, Juenet et al. had fabricated a novel thrombus-targeted nanocarrier that was made of biocompatible and biodegradable fucoidan-functionalized dextran submicroparticles (Fuco-SPs) (Juenet et al. 2018). Fucoidan is a naturally-occurring algae-derived sulfated polysaccharide that exhibits strong and specific tropism for P-selectin, a cell adhesion molecule (CAM), which lines the inner surfaces of blood vessels and activated platelets (Wang et al. 2019). P-selectin is usually overexpressed at the sites of thrombus as it plays a significant role in platelet aggregation and platelet-leukocyte interactions, both of which are important mechanisms in the development of arterial thrombosis (Ma et al. 2019). Therefore, the sub microparticles that are functionalized with fucoidan can accurately travel to the thrombus sites as they possess high and specific affinity to the P-selectin. Furthermore, there are several articles that have claimed on the ability of fucoidan in exerting some antithrombotic action, making it an excellent starting biomaterial for the development of an ideal carrier for precise targeted delivery of thrombolytic drugs to ischemic stroke brain (Choi et al. 2018).

In 2021, Choi et al. had developed a method to immobilize rtPA onto the fabricated Fuco-SPs by using the adsorption strategy (Zenych et al. 2021). They reported a high drug encapsulation ($81.04 \pm 1.86\%$) efficiency due to strong physical electrostatic interaction between the negatively charged polysaccharide Fuco-SPs and the positively charged amphoteric rtPA protein molecule in water, with a pH of less than the rtPA isoelectric point ($IP = 7.7$). A preclinical stroke model was used to test the rtPA-loaded Fuco-SPs' thrombolytic activity in which they found that each animal which had received rtPA-loaded Fuco-SPs was able to outperform those that had received the same concentration of free rtPA. In addition, the animals that received treatment with rtPA-loaded Fuco-SPs were seen to exhibit faster clot dissolution and MCA reperfusion, which led to injuries to the brain being minimized and, thereafter, smaller post-stroke infarct areas.

19.3.3 To Facilitate Crossing Through Blood-Brain Barrier (BBB)

Over the years between 1995 and 2015, there were a total of 430 potential drugs evaluated in treating ischemic stroke. Of these, just 19 (4%) were able to achieve market reach in various countries (Chen and Wang 2016). Of these 19 drug products on the market, most (11) were agents to dissolve thrombus such as rtPAs, antithrombotic agents, and inhibitors of platelet aggregation. Although several other neuroprotective agents had been proven effective in a variety of animal ischemic stroke models, they had failed to show consistent efficacy signal or in compliance to safety requirements when tested in patients in phase III clinical trials (Min et al. 2013). A major challenge when attempting to develop successful clinical pharmacological drug products to treat forms of brain disease is the tightly regulated transport from the blood to the brain and back. There are three structural features involved, which are blood-cerebrospinal-fluid barrier (BCSFB), blood-brain barrier (BBB), and arachnoid membrane (Rhim et al. 2013). These main barriers actively function in regulating molecules that are supplied to the brain to preserve the neural environment within a narrow homeostatic range. It shields the brain from toxic substances in the blood and filters detrimental compounds from the brain back to the blood vessel (Persidsky et al. 2006). Such tight regulation poses a critical challenge to deliver drugs into the brain, contributing to more failures of many neuroprotective agents in clinical trials (Zhang et al. 2022).

Researchers have discovered a few methods to address the challenges associated with drug delivery crossing the BBB. One of them is by utilizing BBB-permeable biomaterials as delivery agents for potential drugs. In a study conducted by Deng et al., the potential of neurogenin-2 (Ngn2) was examined. This gene plays critical roles in regulating neural precursor cells survival, proliferation, and differentiation within the central nervous system (CNS) (Deng et al. 2013). In the study, laminin-binding domain (LBD) and transactivator of transcription (TAT) were combined and fused with Ngn2 to construct TAT-LBD-Ngn2 to enhance the capability of this gene across BBB. The transduction domain of the TAT protein was an 11-amino acid peptide with a positive charge in which studies had demonstrated, able to pull different molecules across the membranes of cells in vitro and in vivo (Moore and Payne 2004) while the LBD was a domain with high laminin affinity (Han et al. 2011). These two short amino acids acted as sustainable biomaterials to enhance the delivery of Ngn2 to the ischemic brain region, which was overexpressed inside the ischemic region. Moreover, since the protein transduction domain (PTD) of TAT was BBB permeable, the amount of the Ngn2 gene delivered through the synthesized TAT-LBD-Ngn2 construct had markedly elevated. This has greatly improved the Ngn2-based gene therapy as described in the study. The functions of these biomaterials were tested in vitro using hippocampal neurons that were exposed to oxygen glucose deprivation (OGD) and in vivo using middle cerebral artery occlusion (MCAO) rat model. The results from the in vitro study demonstrated increased survival of the hippocampal neurons and outgrowth of neuronal neurites, while the in vivo studies displayed promising outcomes, as seen in the attenuation of neuronal degradation, reduced infarcted volume and improvement of neurological functions among the treated rats.

19.3.4 To Control Release and Improved Half-Shelf Life of Ischemic Stroke Drugs

As mentioned earlier, thrombolytic is the main strategy in ischemic stroke treatment strategies. Treatment using rtPA in thrombolytics is a gold standard treatment of ischemic stroke as approved by FDA (Jinatongthai et al. 2017). Apart from a brief therapeutic window of less than 4.5 h, the application of rtPA is also limited due to its short half-life (4–8 min) in the circulation due to rapid drug elimination by autolysis and other inhibitors, enzymes, and antibodies in the blood (Jilani and Siddiqui 2023). As a consequence, high dose intravenous infusion or continuous dosing treatment is required to maintain the effective drug concentration (Chen et al. 2020). Since rtPA has minimal bioavailability and the administration of a high dosage is required, linking a biocompatible nanocarrier to this enzyme could enable the rtPA release to be controlled and improve its circulation half-life, which might address the issues. In this case, Chen et al. had fabricated poly(lactic-co-glycolic acid) (PLGA) material, a strategy which efficiently encapsulated rtPA (Chen et al. 2020). Poly lactic acid (PLA) and poly glycolic acid (PGA) are the main components of PLGA copolymer. Over the last 20 years, PLGA has been considered as a highly appealing polymeric option due to its favorable degradation characteristics and possibilities for sustained drug delivery (Makadia and Siegel 2011). Studies have recently indicated that the release of rtPA drugs can be sustained in the required amounts and there is no repetition of administration processes needed by employing PLGA degradation. Notably, the PLGA-rtPA developed in the study did not only enhance the fibrinolytic activity, as shown in the agar plate clot lysis assays of OD_{415}, but it also prolonged the circulation time of rtPA. As outlined by the researchers in an embolic stroke model using a rat, rtPA delivery was found to be preferable when using PLGA compared to free rtPA administration. This is due to the clot lysis reperfusion time having experienced a 40% reduction with the application of the PLGA-rtPA in comparison to the free rtPA when the dose of drugs was identical. In a nutshell, the in vivo study has envisioned a targeted strategy for rtPA delivery that is feasible to minimize the required dose of rtPA; hence, decreases the hemorrhagic side effects in thrombolytic therapy for treatment.

19.3.5 As Supporting Matrix at Ischemic Cavity to Provide Physical and Trophic Support for Post-Stroke Neurogenesis

Research has demonstrated that a brain is capable of stimulating neurogenesis to activate endogenous self-repair procedure so that the damaged areas can be repopulated when they encounter a pathological situation such as post ischemic stroke (Cuartero et al. 2021). However, post-stroke neurogenic response remains insufficient for complete recovery. Due to this, large tissue cavities remain in stroke survivors (Ghuman et al. 2016). Till date, the reported total replacement of lost tissue following stroke injury remains unavailable. The limitations of post-stroke neurogenesis are multifactorial. One of the main factors could be attributed to the loss of stromal support and trophic factors provided by extracellular matrix (ECM) at the injury site (Jin et al. 2010).

ECM has been known to regulate many aspects of neural development in the human brain. This includes roles in regulating neural progenitor cell proliferation and differentiation as well as dendritic elongation and connectivity which are crucial for nervous system function (Long and Huttner 2022). It also has been reported to create a functional microenvironment for neurogenesis by regulating the diffusion of signaling factors to direct the endogenous neural progenitor cell migration route to a specific discrete area. Collectively, the brain ECM consists of multimeric proteins that participate in cellular migration and differentiation as well as functioning as a support system for brain tissue development (Edwards and Bix 2019). Therefore, any defects in ECM post ischemic stroke injury can result in obstacles for brain tissue regeneration. Hence, biomaterials that can mimic the native brain tissue ECM can be introduced to ischemic cavity areas where it can serve as a core template to facilitate the regeneration after ischemic stroke attack (Wang et al. 2012).

This strategy has been applied by Jian and colleagues (Jian et al. 2018) where they had developed nanohybrid hydrogels using hyaluronic acid (HA) hydrogel matrices that were incorporated with stromal derived factor 1α (SDF-1α) and basic fibroblast growth factor (bFGF)-loaded polyelectrolyte complex nanoparticles (PCNs). HA was utilized in this study as it was a primary component of brain ECM. On the other hand, PCNs are the nanoscale complexes formed due to electrostatic interaction between oppositely charged particles (e.g., polymer-polymer, polymer-drug, and polymer-drug-polymer) (Lankalapalli and Kolapalli 2009). In this study, the PCNs were composed of chitosan that was conjugated with maleimide (CS-maleimide) as it had been reported as an ideal precursor for bioconjugation (Matsumoto et al. 2016). SDF-1α and bFGF were conjugated on the PCNs for endogenous neural stem cells (NSCs) recruitment and proliferation. In this study, delivering SDF-1α and bFGF via PCNs and nanohybrid hydrogel of HA had resulted in successful elevation of the amount of neural precursor cells migration from their location in the brain's subventrizular zone (SVZ) to the peri-infarction tissue and the stroke's core region. The study also had discovered that the HA-PCN nanohybrid hydrogel could act as a supporting matrix to promote immature neuron invasion, enhance neurogenesis and angiogenesis process in the infarcted region, reduce the infarct cavity volume, and promote neurological behavior recovery in rat model (Jian et al. 2018).

19.3.6 To Encapsulate Cells and Confers Protection to the Cells

Besides activating the endogenous stem cell recruitment and proliferation, biomaterial also can be utilized to enhance the exogenous stem cell-based treatment for ischemic stroke. Recent research and development of ischemic stroke treatment have focused on stem cell transplantation as one of the alternative treatments for this disease. Stem cells therapy is a promising strategy to exert therapeutic effects of ischemic stroke treatment. However, the application of this therapy in clinical application is hampered by several limitations that include poor cell survival and low engraftment rate that is caused by hostile microenvironment after the ischemic insult (Hong et al. 2013). To overcome this issue, the encapsulation of stem cells in biomaterials to confer protection to the stem cells during transplantation into ischemic stroke models are applied (Tang et al. 2018). Encapsulation of cells with a selected porous polymeric matrix would allow extra protection for the cells from the host's immune system and provide localization of transplanted cells to the target region. Moreover, the transplanted cells can maintain their survival by allowing the bi-directional exchange of nutrients and waste products between the encapsulated cells and the biomaterial surrounding them (Leslie et al. 2017). Over the years, several cell encapsulation technologies have been developed such as micromolding, electrostatic droplet extrusion, microfluidics, and bioprinting (Kim et al. 2019).

In a study done by Ghuman and colleagues, polyethylene glycol (PEG) microsphere was used to encapsulate NSCs and endothelial cells (ECs) prior to transplantation into the affected lesion cavities of ischemic stroke-induced rats (Ghuman et al.

2021). Microsphere encapsulation of cells not only allows rapid growth of both cell types, it also confers effective protection to the cells against host enzymatic degradation, allowing the cells to be delivered to the target site (Singh et al. 2010). Furthermore, the cells encapsulated in the microspheres had exhibited high viability and survived under ischemic conditions. The cells also were more widely distributed in the cavity of damaged tissue in the ischemic stroke-induced rat brains. Co-transplantation of NSCs with ECs also enhanced the angiogenesis and survival of cells due to the interaction between these two cells that upregulated the trophic factor release (Ghuman et al. 2021). In conclusion, the constructed biomaterial had enhanced the stem cells-based efficiency by providing a viable cellular microenvironment that can activate or potentiate intrinsic host repair after engraftment (Zhao et al. 2021). This biomaterial innovation has high potential to further advance the clinical setting for the treatment of stroke using stem cell therapy.

19.3.7 To Mimic the Physiological Environment (3D Culture) for Optimum Cell Growth

Mimicry of the physiological environment is critical for transplanted cells to maintain its phenotype, adhesion, metabolism, and response signal, allowing the potential of their capacity for viability and differentiation into targeted lineage in damaged areas to be maximized (Baker and Chen 2012). Cumulative evidence have shown that scaffolds that are made up of synthetic and natural polymers in the form of three-dimensional (3D) spheroids are able to fill the cavity that is formed post ischemic stroke and promote a better interaction between the implanted cell and the host brain tissue in the penumbra region after the ischemic insult (Duncan et al. 2015). The basis of 3D spheres concept is to create structures that imitate the parental tissue architecture with greater accuracy than models in 2D culture.

The majority of cell biology studies have utilized 2D in vitro cell cultures which have numerous drawbacks such as disturbance of interactions between the cellular and extracellular environments, changes in cell morphology, polarity, and method of cell division (Kapałczyńska et al. 2018). Therefore, in 2021, Hsu et al. had developed a temperature-responsive methylcellulose (MC) hydrogel system that was derived from umbilical cord blood and human umbilical vein endothelial cells (HUVECs) 3D spheroids to construct mesenchymal stem cells (MSCs) (Hsu et al. 2021). According to this paper, the microsphere encapsulation had shielded the transplanted cells from shear stress which resulted from the movement through a narrow-bored needle while the cells were being transplanted. Moreover, this approach had employed biologically active components that mimicked the in vivo conditions of the cells, to regulate them similarly to the events that would occur in their natural cellular microenvironment. In comparison to conventional suspensions of mixed cell suspensions, the rates at which cells survived and retained were considerably enhanced by pre-assembling such cells into 3D spheroids. Moreover, the study also demonstrated that the mouse ischemic stroke model had experienced improved recovery in terms of structural and motor functions after it was treated with

3D MSC/HUVECs. This might be due to the neuroprotective, pro-neurogenic, pro-angiogenic, and anti-scarring potential qualities, which was evident in the formation of clear ECM structure as well as their expression and secretion of related paracrine factors.

19.3.8 To Enhance Cell Homing to Cerebral Ischemic Lesion Areas

Cell-based therapy has emerged as a new modality with considerable potential in treating ischemic stroke. Numerous basic science and clinical studies have extensively reported the potential of stem cell-based therapies for ischemic stroke models/ patients at different phases of severity by using different stem cell types, doses, and delivery routes (Nagpal et al. 2017). The reported outcomes in these studies however are inconsistent (Ouyang et al. 2019; Boncoraglio et al. 2019). This is because the therapeutic efficacy of stem cell-based therapy is mainly mediated by the homing capacity of transplanted cells to the ischemic brain lesion areas where they can differentiate into viable neurons or astrocytes to replace the damaged cells. There is ample evidence which highlights the fact that there is only a small percentage of systemically administered stem cells that have actually reached the target lesion tissue thus, limiting the stem cell potential for regenerative and reparative procedure (De Becker and Van Riet 2016). The limitation for exogenous stem cell transplantation to reach the target site is due to several key events induced by ischemic stroke such as inflammation, loss of homeostasis, free radical-mediated toxicity, impairment of BBB, and many more (Woodruff et al. 2011; Suh et al. 2008; Wang et al. 2007). Therefore, optimization of cell migration and homing capacity to the target site could definitely enhance the stem cell-based therapeutic potential for ischemic stroke treatment.

Biomaterial-based brain targeting method has emerged as a promising approach in targeting cell delivery to brain tissue. Study by Huang and team had utilized peptide-based biomaterial to enhance MSCs homing to ischemic core in MCAO rats (in vivo study) as well as in OGD model (in vitro study) (Huang et al. 2017). In both the studies, palmitic-acid (PA) peptide was coated on the cell membrane of MSCs prior to the introduction of the MSCs to the ischemic stroke models. A peptide is a biopolymer of great versatility that comprises 2–100 amino acid residues. It presents an extensive array of biological functions and can be utilized as a form of therapy for many diseases, partially because of its capacity for cell membrane penetration (Boaro et al. 2020; Lotze et al. 2016). Meanwhile, PA is a saturated fatty acid that has been found to stimulate the expression of inflammatory cytokines in macrophages (Wu et al. 2014). The use of targeted PA-peptide to modify cell surfaces is a transient cell-coating method which can be effective in labeling cells and increasing the localization of cells in myocardial ischemia cells, as demonstrated in past studies. Furthermore, this particular strategy of transient modification would prevent the cells from being affected in the long-term period. The study also reported that the PA–peptide-coated MSCs had shown significant migration capacity and increased engraftment on targeted injured brain tissue. With enhanced homing ability of MSCs, more cells could be accumulated in the diseased tissue and reduced their distribution in non-target organs, thus lowering the side effects on non-targeted sites. Huang et al. also had reported the absence of cytotoxicity and other side effects of PA–peptides on MSC proliferation and differentiation potential, which further validated its safety profile.

19.3.9 To Label Cells for In Vivo MRI Tracking

Stem cell-based therapy could be a highly effective method of treating ischemic stroke due to its innate capability to renew cells, home, and differentiate multi-lineages. The long-term survival of cells and the dynamics of migration following stem cell engraftment in models of animals could be understood more clearly by monitoring the animals used in these experiments in the long-term by using imaging that is not invasive. For the last few years, various imaging techniques have been applied to monitor exogenous cell population in vivo which include magnetic resonance imaging (MRI). MRI is considered suitable for long-term non-invasive imaging due to its high spatial resolution in 3D, superior delineation of host tissue morphology, and absent exposure to radiation (Gera et al. 2010). However, the MRI sensitivity is considered relatively low than other imaging systems. Therefore, the cells must be labeled magnetically through the usage of superparamagnetic iron oxide nanoparticles (SPION) for cell labeling. Nonetheless, these nanoparticles have low intracellular labeling efficiency (Gera et al. 2010). With that, other materials must be conjugated to SPIONs to enhance the MRI detection sensitivity.

These requirements could be fulfilled by the innovative biocompatible nanocomplex, superparamagnetic iron oxide nanoparticles-complexed cationic amylode-spermine-modified amylose (ASP-SPIONs), as created by Lin et al. (2017). Cationic ASP-SPIONs were synthesized by utilizing the hydrophobic amylose that was modified with spermine through N,N′-carbonyldiimidazole (CDI) activation method. This was followed by fabrication using one-pot synthesis in a solution at high temperature. Then, the transgenic green fluorescent protein (GFP)-MSCs were labeled using this nanocomplex prior to its injection into the ipsilateral striatum. Upon testing in vitro, it was found that the stem cells uptake had modified SPIONs more efficiently compared to unmodified SPIONs. Moreover, the uptake of modified SPIONs did not detrimentally affect the viability of the stem cells, proliferation, apoptosis, the levels of reactive oxygen species (ROS) within the cells, the potential for mitochondrial transmembrane, or their differentiation. These outcomes have further highlighted the potential to label stem cells more efficiently, rapidly, and safely so that the cellular MRI can be tracked in regenerative medical procedures.

19.4 Conclusion

This chapter highlights the current state of biomaterial utilization in the direct or indirect treatment for ischemic stroke. The key point of biomaterial-based therapy is to select a suitable biomaterial with inherent biophysical and biochemical properties that can match the functional requirements for brain lesion recovery without causing additional harmful effects to the body. Prior reports have confirmed that these biomaterials are particularly excellent as a targeted delivery agent, facilitate drug crossing of BBB, control the release of drugs, and provide trophic support in ischemic stroke brain for the enhancement of brain tissue regeneration process. Moreover, when they are combined with stem cell-based treatment, these biomaterials would exert high potential to create 3D microenvironment that can mimic the cells original physiological conditions, enhance the cell homing to ischemic site, and trace the fate of the cells after transplantation for better understanding of the cell-based reparative mechanism in vivo. Future research should focus on the clinical translation and large-scale manufacturing technologies for these biomaterials to be marketed and benefited by ischemic stroke patients in every aspect of life.

Acknowledgement This work was supported by Universiti Sains Malaysia (USM) Research University Individual (RUI) Grant, No: 1001/PPSK/8012273. All authors thank the Universiti Sains Malaysia libraries, particularly Perpustakaan Hamdan Tahir and Perpustakaan Hamzah Sendut, for providing the resources to write this book chapter. The authors also appreciate all the scientists whose previous work contributed to this review article.

Conflict of Interest All authors declare no conflict of interest.

References

An SJ, Kim TJ, Yoon BW (2017) Epidemiology, risk factors, and clinical features of intracerebral hemorrhage: an update. J Stroke 19:3–10. PMID: 28178408. https://doi.org/10.5853/JOS.2016.00864

Baker BM, Chen CS (2012) Deconstructing the third dimension: how 3D culture microenvironments alter cellular cues. J Cell Sci 125:3015–3024

Balami JS, Sutherland BA, Buchan AM (2013) Complications associated with recombinant tissue plasminogen activator therapy for acute ischaemic stroke. CNS Neurol Disord Drug Targets 12: 155–169. PMID: 23394532. https://doi.org/10.2174/1871527311312020003

Boaro A, Ageitos L, Torres M, Bartoloni FH, de la Fuente-Nunez C (2020) Light-emitting probes for labeling peptides. Cell Rep Phys Sci 1(12):100257. https://doi.org/10.1016/j.xcrp.2020.100257

Boncoraglio GB, Ranieri M, Bersano A, Parati EA, Del Giovane C (2019) Stem cell transplantation for ischemic stroke. Cochrane Database Syst Rev 5:CD007231

Chen X, Wang K (2016) The fate of medications evaluated for ischemic stroke pharmacotherapy over the period 1995–2015. Acta Pharm Sin B 6:522. PMID: 27818918. https://doi.org/10.1016/J.APSB.2016.06.013

Chen H, Ma Y, Hsu T, Chen J (2020) Preparation of peptide and recombinant tissue plasminogen activator conjugated poly (lactic-co-glycolic acid) (PLGA) magnetic nanoparticles for dual targeted thrombolytic therapy. Int J Mol Sci 21(8):2690

Choi Y, Min SK, Usoltseva R, Silchenko A, Zvyagintseva T, Ermakova S, Kim JK (2018) Thrombolytic fucoidans inhibit the tPA-PAI1 complex, indicating activation of plasma tissue-type plasminogen activator is a mechanism of fucoidan-mediated thrombolysis in a mouse thrombosis model. Thromb Res 161:22–25. https://doi.org/10.1016/j.thromres.2017.11.015

Cuartero MI, García-Culebras A, Torres-López C, Medina V, Fraga E, Vázquez-Reyes S, Jareño-Flores T, García-Segura JM, Lizasoain I, Moro MÁ (2021) Post-stroke neurogenesis: friend or foe? Front Cell Develop Biol 9:657846. PMID: 33834025. https://doi.org/10.3389/FCELL.2021.657846

De Becker A, Van Riet I (2016) Homing and migration of mesenchymal stromal cells: how to improve the efficacy of cell therapy? World J Stem Cells 8:73–87. https://doi.org/10.4252/wjsc.v8.i3.73

Decavel P, Vuillier F, Moulin T (2012) Lenticulostriate infarction. Front Neurol Neurosci 30:115–119. PMID: 22377876. https://doi.org/10.1159/000333606

Deng B, Gou X, Chen H, Li L, Zhong H, Xu H, Jiang F, Zhao Z, Wang Q, Xu L (2013) Targeted delivery of Neurogenin-2 protein in the treatment for cerebral ischemia-reperfusion injury. Biomaterials 34:8786–8797. https://doi.org/10.1016/j.biomaterials.2013.07.076

Duncan K, Gonzales-Portillo GS, Acosta SA, Kaneko Y, Borlongan CVTN (2015) Stem cell-paved biobridges facilitate stem transplant and host brain cell interactions for stroke therapy. Brain Res 1623:160–165

Edwards DN, Bix GJ (2019) Roles of blood-brain barrier integrins and extracellular matrix in stroke. Am J Physiol Cell Physiol 316:C252–C263. PMID: 30462535. https://doi.org/10.1152/AJPCELL.00151.2018/ASSET/IMAGES/LARGE/ZH00011983690003.JPEG

Feske SK (2021) Ischemic stroke. Am J Med 134:1457–1464. . PMID: 34454905. https://doi.org/10.1016/J.AMJMED.2021.07.027

Gera A, Steinberg GK, Guzman R (2010) In vivo neural stem cell imaging: current modalities and future directions. Regen Med 5:73–86

Ghuman H, Massensini AR, Donnelly J, Kim SM, Medberry CJ, Badylak SF, Modo M (2016) ECM hydrogel for the treatment of stroke: characterization of the host cell infiltrate. Biomaterials 91:166. PMID: 27031811. https://doi.org/10.1016/J.BIOMATERIALS.2016.03.014

Ghuman H, Matta R, Tompkins A, Nitzsche F, Badylak SF, Gonzalez AL, Modo M (2021) ECM hydrogel improves the delivery of PEG microsphere-encapsulated neural stem cells and endothelial cells into tissue cavities caused by stroke. Brain Res Bull 168:120–137. https://doi.org/10.1016/j.brainresbull.2020.12.004

Gore M, Bansal K, Asuncion RMD (2022) Lacunar stroke. StatPearls (e-pub ahead of print)

Han Q, Li B, Feng H, Xiao Z, Chen B, Zhao Y, Huang J, Dai J (2011) The promotion of cerebral ischemia recovery in rats by laminin-binding BDNF. Biomaterials 32:5077–5085. PMID: 21529937. https://doi.org/10.1016/J.BIOMATERIALS.2011.03.072

He Z, Ning N, Zhou Q, Khoshnam SE, Farzaneh M (2020) Mitochondria as a therapeutic target for ischemic stroke. Free Radic Biol Med 146:45–58. . PMID: 31704373. https://doi.org/10.1016/J.FREERADBIOMED.2019.11.005

Hong KU, Li QH, Guo Y et al (2013) A highly sensitive and accurate method to quantify absolute numbers of c-kit+ cardiac stem cells following transplantation in mice. Basic Res Cardiol 108:346

Hsu T, Lu Y, Lin Y, Huang Y, Hsieh L, Wu B, Lin Y, Chen L, Wang H, Chuang J, Fang Y, Huang C (2021) Transplantation of 3D MSC/HUVEC spheroids with neuroprotective and proangiogenic potentials ameliorates ischemic stroke brain injury. Biomaterials 272:120765. https://doi.org/10.1016/j.biomaterials.2021.120765

Huang AB, Jiang X, Zhang T (2017) Peptide modified mesenchymal stem cells as targeting delivery system transfected with MiR-133b for the treatment of cerebral ischemia. Int J Pharm 531(1): 90–100. https://doi.org/10.1016/j.ijpharm.2017.08.073

Hughes RE, Tadi P, Bollu PC (2022) TPA therapy. StatPearls, pp 1–6. https://www.ncbi.nlm.nih.gov/books/NBK482376/. Accessed 11 September 2022

Hwong WY, Ang SH, Bots ML, Sivasampu S, Selvarajah S, Law WC, Latif LA, Vaartjes I (2021) Trends of stroke incidence and 28-day all-cause mortality after a stroke in Malaysia: a linkage of national data sources. Global Heart 16:39. PMID: 34211825. https://doi.org/10.5334/GH.791

Jian W, Wang H, Kuan C, Chen M, Wu H, Sun J, Wang T (2018) Glycosaminoglycan-based hybrid hydrogel encapsulated with polyelectrolyte complex nanoparticles for endogenous stem cell regulation in central nervous system regeneration. Biomaterials 174:17–30. https://doi.org/10.1016/j.biomaterials.2018.05.009

Jilani TN, Siddiqui AH (2023) Tissue plasminogen activator. StatPearls, Treasure Island, FL

Jin R, Yang G, Li G (2010) Inflammatory mechanims in ischemic stroke: role of inflammatory cells. J Leukoc Biol 87:779–789

Jinatongthai P, Kongwatcharapong J, Foo CY, Phrommintikul A, Nathisuwan S, Thakkinstian A, Reid CM, Chaiyakunapruk N (2017) Comparative efficacy and safety of reperfusion therapy with fibrinolytic agents in patients with ST-segment elevation myocardial infarction: a systematic review and network meta-analysis. Lancet 390:747–759. PMID: 28831992 DOI: 10.1016/S0140-6736(17)31441-1

Juenet M, Aid-Launais R, Li B, Berger A, Aerts J, Ollivier V, Nicoletti A, Letourneur D, Chauvierre C (2018) Thrombolytic therapy based on fucoidan-functionalized polymer nanoparticles targeting P-selectin. Biomaterials 156:204–216. https://doi.org/10.1016/j.biomaterials.2017.11.047

Kapałczyńska M, Kolenda T, Przybyła W, Zajączkowska M, Teresiak A, Filas V, Ibbs M, Bliźniak R, Łuczewski Ł, Lamperska K (2018) 2D and 3D cell cultures – a comparison of different types of cancer cell cultures. Arch Med Sci 14:910. PMID: 30002710. https://doi.org/10.5114/AOMS.2016.63743

Kim H, Bae C, Kook Y, Koh WG, Lee K, Park MH (2019) Mesenchymal stem cell 3D encapsulation technologies for biomimetic microenvironment in tissue regeneration. Stem Cell Res Ther 10:51

Lankalapalli S, Kolapalli VRM (2009) Polyelectrolyte complexes: a review of their applicability in drug delivery technology. Indian J Pharm Sci 71:481. PMID: 20502564. https://doi.org/10.4103/0250-474X.58165

Leslie SK, Kinney RC, Schwartz Z, Boyan BD (2017) Microencapsulation of stem cells for therapy. Methods Mol Biol 1479:251–259

Lin B, Zhang J, Lu L, Mao J, Cao M (2017) Superparamagnetic iron oxide nanoparticles-complexed cationic amylose for in vivo magnetic resonance imaging tracking of transplanted stem cells in stroke. Nanomaterials (Basel) 7(5):107. https://doi.org/10.3390/nano7050107

Liu F, Lu J, Manaenko A, Tang J, Hu Q (2018) Mitochondria in ischemic stroke: new insight and implications. Aging Dis 9:924. PMID: 30271667. https://doi.org/10.14336/AD.2017.1126

Long KR, Huttner WB (2022) The role of the extracellular matrix in neural progenitor cell proliferation and cortical folding during human neocortex development. Front Cell Neurosci 15:563. https://doi.org/10.3389/FNCEL.2021.804649/BIBTEX

Lotze J, Reinhardt U, Seitz O, Beck-Sickinger AG (2016) Peptide-tags for site-specific protein labelling in vitro and in vivo. Mol BioSyst 12:1731–1745. PMID: 26960991. https://doi.org/10.1039/C6MB00023A

Ma YH, Liu CH, Liang Y, Chen JP, Wu T (2019) Targeted delivery of plasminogen activators for thrombolytic therapy: an integrative evaluation. Molecules 24(18):3407. PMID: 31546842. https://doi.org/10.3390/MOLECULES24183407

Makadia HK, Siegel SJ (2011) Poly lactic-co-glycolic acid (PLGA) as biodegradable controlled drug delivery carrier. Polymers 3:1377. PMID: 22577513. https://doi.org/10.3390/POLYM3031377

Matsumoto M, Udomsinprasert W, Laengee P, Honsawek S, Patarakul K, Chirachanchai S (2016) A water-based chitosan-maleimide precursor for bioconjugation: an example of a rapid pathway for an in situ injectable adhesive gel. Macromol Rapid Commun 37:1618–1622. PMID: 27461078. https://doi.org/10.1002/MARC.201600257

Min J, Farooq MU, Gorelick PB (2013) Neuroprotective agents in ischemic stroke: past failures and future opportunities. Clin Invest 3:1167–1177. https://doi.org/10.4155/cli.13.91

Montaño A, Hanley DF, Hemphill JC (2021) Hemorrhagic stroke. Handb Clin Neurol 176:229–248. PMID: 33272397. https://doi.org/10.1016/B978-0-444-64034-5.00019-5

Moore VDG, Payne RM (2004) Transactivator of transcription fusion protein transduction causes membrane inversion. J Biol Chem 279:32541–32544. PMID: 15169776. https://doi.org/10.1074/JBC.M405930200

Nagpal A, Choy FC, Howell S, Hillier S, Chan F, Hamilton-Bruce MA, Koblar SA (2017) Safety and effectiveness of stem cell therapies in early-phase clinical trials in stroke: a systematic review and meta-analysis. Stem Cell Res Ther 8:191

Olié V, Grave C, Tuppin P, Duloquin G, Béjot Y, Gabet A (2022) Patients hospitalized for ischemic stroke and intracerebral hemorrhage in France: time trends (2008–2019), in-Hospital Outcomes, Age and Sex Differences. J Clin Med 11:1669. https://doi.org/10.3390/jcm11061669

Ouyang Q, Li F, Xie Y, Han J, Zhang Z, Feng Z, Su D, Zou X, Cai Y, Zou Y, Tang Y, Jiang X (2019) Meta-analysis of the safety and efficacy of stem cell therapies for ischemic stroke in preclinical and clinical studies. Stem Cells Dev 28:497–514

Persidsky Y, Ramirez SH, Haorah J, Kanmogne GD (2006) Blood-brain barrier: structural components and function under physiologic and pathologic conditions. J Neuroimmune Pharmacol 1:223–236. PMID: 18040800. https://doi.org/10.1007/s11481-006-9025-3

Reed M, Kerndt CC, Nicolas D (2022) Alteplase. In: xPharm: the comprehensive pharmacology reference, pp 1–5. PMID: 29763152. https://doi.org/10.1016/B978-008055232-3.61202-4

Reiter RJ, Rosales-Corral S, Tan DX, Jou MJ, Galano A, Xu B (2017) Melatonin as a mitochondria-targeted antioxidant: one of evolution's best ideas. Cell Mol Life Sci 74:3863–3881. https://doi.org/10.1007/s00018-017-2609-7

Rhim T, Lee DY, Lee M (2013) Drug delivery systems for the treatment of ischemic stroke. Pharm Res 30:2429–2444. PMID: 23307348. https://doi.org/10.1007/s11095-012-0959-2

Sadanandan N, Cozene B, Cho J, Park YJ, Saft M, Gonzales-portillo B, Borlongan CV (2020) Melatonin—a potent therapeutic for stroke and stroke-related dementia. Antioxidants 9:1–15

Singh MN, Hemant KSY, Ram M, Shivakumar HG (2010) Microencapsulation: a promising technique for controlled drug delivery. Res Pharm Sci 5:65–77

Suh SW, Shin BS, Ma H, Van Hoecke M, Brennan AM, Yenari MA, Swanson RA (2008) Glucose and NADPH oxidase drive neuronal superoxide formation in stroke. Ann Neurol 64:654–663

Tan KS, Venketasubramanian N (2022) Stroke burden in Malaysia. Cerebrovasc Dis Extra 12:58. PMID: 35325896. https://doi.org/10.1159/000524271

Tan DX, Manchester LC, Qin L, Reiter RJ (2016) Melatonin: a mitochondrial targeting molecule involving mitochondrial protection and dynamics. Int J Mol Sci 17:2124. https://doi.org/10.3390/ijms17122124

Tang J-N, Cores J, Huang K, Cui XL, Luo L, Zhang JY, Li TS, Qian L, Cheng K (2018) Concise review: is cardiac cell therapy dead? Embarrassing trial outcomes and new directions for the future. Stem Cells Transl Med 7:354–359. https://doi.org/10.1002/sctm.17-0196

Wang J, Fields J, Zhao C, Langer J, Thimmulappa RK, Kensler TW, Yamamoto M, Biswal S, Doré S (2007) Role of Nrf2 in protection against intracerebral hemorrhage injury in mice. Free Radic Biol Med 43:408–414

Wang X, He J, Wang Y, Cui F-Z (2012) Hyaluronic acid-based scaffod for central neural tissue engineering. Interf Focus 2:278–291

Wang Y, Xing M, Cao Q, Ji A, Liang H, Song S (2019) Biological activities of Fucoidan and the factors mediating its therapeutic effects: a review of recent studies. Mar Drugs 17(3):183. PMID: 30897733. https://doi.org/10.3390/MD17030183

Woodruff TM, Thundyil J, Tang S-C, Sobey CG, Taylor SM, Arumugam TV (2011) Pathophysiology, treatment, and animal and cellular models of human ischemic stroke. Mol Neurodegener 6:11

Wu D, Liu J, Pang X, Wang S, Zhao J, Zhang X, Feng L (2014) Palmitic acid exerts pro-inflammatory effects on vascular smooth muscle cells by inducing the expression of C-reactive protein, inducible nitric oxide synthase and tumor necrosis factor-α. Int J Mol Med 34:1706–1712. PMID: 25242580. https://doi.org/10.3892/IJMM.2014.1942/HTML

Yip H, Kumar N, Lin K, Sung P (2021) Melatonin rescues cerebral ischemic events through upregulated tunneling nanotube-mediated mitochondrial transfer and downregulated mitochondrial oxidative stress in rat brain. Biomed Pharmacother 139:111593. https://doi.org/10.1016/j.biopha.2021.111593

Zenych A, Jacqmarcq C, Aid R, Fournier L, Forero Ramirez LM, Chaubet F, Bonnard T, Vivien D, Letourneur D, Chauvierre C (2021) Fucoidan-functionalized polysaccharide submicroparticles loaded with alteplase for efficient targeted thrombolytic therapy. Biomaterials 277:121102. https://doi.org/10.1016/j.biomaterials.2021.121102

Zhang S, Zhou Y, Li R, Chen Z, Fan X (2022) Advanced drug delivery system against ischemic stroke. J Control Release 344:173–201

Zhao X, Li Q, Guo Z, Li Z (2021) Constructing a cell microenvironment with biomaterial scaffolds for stem cell therapy. Stem Cell Res Ther 12:583. https://doi.org/10.1186/s13287-021-02650-w

Sustainable Materials for Biomedical Engineering Application in Dentistry

20

Sook-Luan Ng, Jia Xian Law, and Farinawati Yazid

Abstract

Recent biomaterials have shown promising potential to be used as tissue replacement or vehicles for cell transplantation and tissue regeneration in dental research and clinical applications of periodontics, endodontics and oral and maxillofacial surgery. Sustainable biomaterials are obtained from natural, renewable and recyclable resources, which promote green dental application. The biologic performance and biocompatibility of dental biomaterials are crucial to improve the treatment outcome of dental tissue regeneration. The regulatory and bioethics issues, which are specific to the proposed sustainable materials (hydroxyapatite, collagen, chitosan and cellulose), and their sustainability, quality, limitations, translation and commercialisation potential must be considered from the clinic to the market.

Keywords

Dental biomaterials · Sustainable biomaterials · Clinical translation · Commercialisation · Hydroxyapatite · Collagen · Chitosan · Cellulose

S.-L. Ng (✉)
Department of Craniofacial Diagnostics and Biosciences, Faculty of Dentistry, Universiti Kebangsaan Malaysia, Kuala Lumpur, Malaysia
e-mail: ngsookluan@ukm.edu.my

J. X. Law
Centre for Tissue Engineering and Regenerative Medicine, Faculty of Medicine, Universiti Kebangsaan Malaysia, Kuala Lumpur, Malaysia

F. Yazid
Department of Family Oral Health, Faculty of Dentistry, Universiti Kebangsaan Malaysia, Kuala Lumpur, Malaysia

W. S. Wan Kamarul Zaman, A. A. Nurul (eds.), *Sustainable Material for Biomedical Engineering Application*, https://doi.org/10.1007/978-981-99-2267-3_20

Abbreviations

β-TCP	β-tricalcium phosphates
ATMPs	Advanced therapy medicinal products
BC	Bacterial cellulose
BCM	Bacterial cellulose membrane
CAL	Clinical attachment level
CGTPs	Cell and gene therapy products
CMC	Carboxymethyl chitosan
CT	Computed tomography
CTO	Cells, tissues, and organs
DPSC	Dental pulp stem cells
EI	Electron beam irradiation
EMA	European Medicines Agency
EnHAP	Eggshell-derived nano-hydroxyapatite
EU	European Union
FDA	Food and Drug Administration
GLP	Good laboratory practices
HA	Hyaluronic acid
HAP	Hydroxyapatite
ICH	International Council on Harmonisation of Technical Requirements for Registration of Pharmaceuticals for Human
MDA	Medical Device Authority
MFDS	Ministry of Food and Drug Safety
NMPA	National Medical Product Administration
PMDA	Pharmaceutical and Medical Device Agency
PPD	Probing pocket depth
PRF	Platelet-rich fibrin
rhCOL1	Recombinant human type 1 collagen
ROI	Return of investment
TGA	Therapeutic Goods Administration
US	United States

20.1 Introduction

Sustainable biomaterials can be defined as safe, non-hazardous, reusable, recyclable and environmental-friendly materials that have been manufactured from natural and renewable resources (Agnieray et al. 2021). Ideal dental biomaterials are materials with good mechanical properties, biocompatibility, osteoconductive, osteoinductive, antimicrobial activity and anti-inflammation, which can be potentially applied for dentin, periodontal ligament, dental pulp and enamel regeneration. Biomaterials can be divided into two categories: synthetic substances and natural derivatives. The synthetic biomaterials are designed by manipulating their physical and chemical properties based on the specific applications. However, they often induce inflammatory responses due to their heterogeneity, surface topography, physical and chemical properties (Patel et al. 2020). On the other hand, natural biomaterials are more biocompatible, biodegradable and low-cost than synthetic biomaterials (Aguilar-Ayala et al. 2022). Nonetheless, their limitations include low mechanical strength and poor scaffold formation.

Most biofabrication businesses are still in an early transitional stage of innovation, where the development of highly variable products and services is between the stages of laboratory and market. The challenges of biofabrication are good manufacturing facilities, manufacturing optimisation, standardised quality control, consistency of culture cocktails and formulation. The customisable biodegradable implants are required to showcase their ability to regenerate the tissues rather than replace them as compared to the pre-existing standard metal implants (Costa 2019).

The process of product translation requires regulatory, prototype, patenting and registration, business models, scale-up production and competitive assessments (Taylor et al. 2021). The therapeutic efficacy and safety of a translational product are the two major qualifications, which contribute to the success of product commercialisation. Most of the translation failures are related to clinical failures, where the therapeutic efficacy of a product is inauspicious in the clinical trials but shows potentials in the in vivo models. Therefore, the fundamental characteristics (e.g., characterisation methods, physiochemical properties), clinical potential and regulatory aspects of a product should be well understood in order to achieve successful product translation and commercialisation (Agrahari and Hiremath 2017; Agrahari and Agrahari 2018).

20.2 Regulatory and Bioethics Issues of the Sustainable Materials

There are many new sustainable biomaterials/scaffolds that show a great promise as tissue replacements or as vehicles for cell transplantation to promote dental tissue regeneration, protect the tooth structure and prevent periodontitis. Some of these biomaterials/scaffolds also possess antibacterial activities against oral pathogens. In addition, these biomaterials/scaffolds also serve as drug delivery vehicles to deliver drugs such as antibiotics to treat dental tissue infections (Chi et al. 2019). Following the success of in vitro and in vivo testing of these biomaterials, there have been several clinical trials conducted, specifically to evaluate their safety and efficacy in humans. As obtaining ethical approval is critical prior to the commencement of clinical trials, the main aim is to determine the safety and efficacy of these biomaterials/scaffolds in humans. Once the clinical trial is complete, the manufacturer will be able to register the product with the regulatory agency to obtain a license for commercialisation. The biomaterials/scaffolds are regulated as medical devices by the Food and Drug Administration (FDA) in United States (US), European

Medicines Agency (EMA) in European Union (EU), Health Canada in Canada, Medical Device Authority (MDA) in Malaysia, National Medical Product Administration (NMPA) in China, The Pharmaceutical and Medical Device Agency (PMDA) in Japan, The Ministry of Food and Drug Safety (MFDS) in South Korea and Therapeutic Goods Administration (TGA) in Australia. The regulation of medical devices is highly important to ensure the existence of premarket control over their safety as well as their efficacy and quality. Furthermore, it helps to assure the public and health professionals to make an informed decision prior to usage and ensure the establishment of a post-market reporting system to identify and monitor any problems that may arise post commercialisation.

Generally, recent biomaterials/scaffolds have minimal ethical and regulatory issues. In obtaining approval for clinical trials, the funders must show that the trials are conducted in accordance with the international standards such as ISO 14155, ISO 13485, ISO 14971 and International Council on Harmonisation of Technical Requirements for Registration of Pharmaceuticals for Human (ICH) guidelines (Deep 2022a). Simultaneously, the preclinical safety and toxicology studies should be completed under good laboratory practices (GLP). Additionally, the production must be completed in an accredited cGMP facility, fulfilling all the required guidelines by using a standardised protocol that can produce reproducible results to fabricate products, which can be used clinically. It is highly crucial to source raw materials from a supplier to minimise variation, despite the risk of price hike. Nevertheless, the synthetic biomaterials usually have less batch-to-batch variation as the production process is more controlled and possesses a lower risk of pathogen transmission (Volpi et al. 2022). Meanwhile, natural biomaterials have higher batch-to-batch variation (inconsistency in compositions and properties) and higher risk of infection transmission. Recombinant expression systems have been utilised to prepare natural biomaterials without interspecies variation and risk of pathogen contamination (Davison-Kotler et al. 2019). Thus, the sources of these natural biomaterials must be carefully evaluated and monitored and the raw materials should go through rigorous quality control and quality assurance processes.

The registration of biomaterials/scaffolds as medical devices is not an easy process to go through as the regulatory agency will scrutinise every aspect related to them. This includes the conception and development, manufacturing, packaging and labelling, information on the placement in the market, including advertising and sale as well as post-market information on the usage and disposal. The stringent screening is necessary as it will assure the entrance of safe and beneficial medical devices into the market. The principle of the regulation is that the regulatory control level is directly proportional to the degree of risk (Deep 2022b). This means that a medical device with a lower degree of risk will require a lower level of regulatory control and vice versa. In the EU, the medical devices are categorised as Class I (low risk), Class IIa (medium risk), Class IIb (medium/high risk) and Class III (high risk). More details on the regulation of medical devices can be obtained from the websites of the regulatory agencies in the respective countries.

The registration of engineered dental tissues consisting of biomaterials/scaffolds incorporated with stem cells is more complicated. As stem cells can undergo genetic and phenotypic changes, they might develop into tumours and be rejected by the immune system (only for allogeneic and xenogeneic stem cells) (Patel et al. 2020). Furthermore, tissue-engineered products are relatively new in the market, with several countries having just enforced regulations for these products. For example, Malaysia started the mandatory registration and enforcement of cell and gene therapy products (CGTPs) on January 1, 2021 (Loh et al. 2021). The regulations for engineered dental tissue differ in each country. They are regulated as advanced therapy medicinal products (ATMPs) in the EU, as human cell, tissue and tissue-based products (HCT/Ps) within the human medicinal products in the US, as cells, tissues and organs (CTO) for transplantation under the Food and Drugs Act in Canada, as biological in Australia, as regenerative medicine products in Japan, as cell therapy products within the biological products in Korea and as CGTPs in Malaysia (Oberweis et al. 2020). Similarly, the level of regulation depends on the degree of risk associated with the products. Oberweis et al. have provided a worldwide overview of the regulatory frameworks for tissue-engineered products (Oberweis et al. 2020).

Overall, the development cost of a novel biomaterial/scaffold is extremely high due to the requirement of conducting comprehensive in vitro, in vivo and clinical studies (phases 1 to 3) as required by most regulatory agencies. On top of that, there are other significant costs required in bringing these products to the market. Thus, it is very important to predict the return on investment (ROI) of these products to ensure their long-term survival in the market, a necessary aspect required by the regulatory agency during the product registration (Hench and Jones 2015).

20.3 Clinical Trials of the Sustainable Materials in Dental Application

Biomaterials are part of the tissue engineering triad along with stem cells and growth factors. Regenerative dentistry will be able to provide a better treatment option for patients with oral diseases such as caries, periodontal disease, and bone defects. Currently, active research to understand the interaction between different biomaterials in dentistry application, from pre-clinical studies to clinical trials, is still ongoing. Clinical trials of these biomaterials will evaluate the clinical outcome and radiographic findings of the regenerated tissues. Research areas of the biomaterials in dentistry are bone regeneration, dentin pulp regeneration and periodontal regeneration.

20.3.1 Bone Regeneration

Congenital deformity, trauma and malignancy can cause bone defects in dentistry. Current treatment involves the placement of bone grafts either autologous, allogeneic, xenograft or bone substitutes. However, there are pros and cons concerning these graft options, which can affect the outcome of the treatment. Due to this, bone tissue regeneration is the alternative treatment option to repair bone defects. Ideally, the biomaterial used in bone regeneration should have the following properties: osteoconductive, osteoinductive, biocompatible, bioresorbable and structural and mechanical properties that are similar to those of the native bone. Ceramics, metals, polymers, and composites are some of the types of biomaterials developed for bone tissue engineering.

Collagen sponge biomaterial loaded with dental pulp stem cells (DPSC) has been used in socket preservation to extract non-erupted or partially erupted third molars (d'Aquino et al. 2009). A clinical trial was conducted with 17 patients, and its clinical outcomes and radiographic findings were assessed 15 months later. Based on this study, patients' alveolar bone showed optimal vertical repair and complete periodontal tissue restoration after the second molar. Another study also utilised a collagen sponge scaffold with pulp micrograft in a sinus lift elevation procedure (Brunelli et al. 2013). The findings revealed that the bone density of the newly formed bone was double the amount of the native bone which was assessed with computed tomography (CT) at 4-months postsurgery.

In another study on socket preservation of third molar postextraction, a pulp micrograph with collagen matrix was transplanted in the required area of 32 patients (Barbier et al. 2018). However, the CT revealed no significant difference in the bone density and interdental septum height between the collagen matrix with and without cells 6 months postextraction.

20.3.2 Dentine Pulp Regeneration

Caries with pulp involvement require endodontic treatment either with a pulpotomy or a pulpectomy. Conventional endodontic treatment involves the placement of material in the root canal system. Dentin pulp regeneration requires biomaterials that can provide an extracellular matrix for cell attachment in the regeneration process.

Dental pulp stem cells supplemented with granulocyte colony-stimulating factors in atelocollagen were placed in patients with a canal of irreversible pulpitis with no periapical lesion (Nakashima et al. 2017). This pilot clinical study involved five healthy patients who were monitored for up to 24 weeks. Safety evaluations including blood and urine chemistry investigations were performed for all the patients. The outcome of this study showed positive results from the electric pulp test after 4 weeks of treatment. The regenerated tissue in the root-treated tooth had shown similar signal intensity from magnetic resonance imaging with the normal dental pulp in the untreated tooth after 24 weeks.

Platelet-rich fibrin (PRF) has also been actively used in regenerative endodontics for the treatment of necrotic and inflamed pulp. PRF is prepared from human-derived blood via the centrifugation process. Various cytokines are available, such as vascular endothelial growth factor and fibroblast growth factor that can contribute to angiogenesis and osteoblast proliferation. A clinical study involving 15 patients diagnosed with apical periodontitis was treated with a PRF matrix placed in the root canal. A 3- to 12-months follow-up showed a positive response of the sensibility test (Nageh et al. 2018). Another study used advanced platelet-rich fibrin and PRF as a scaffold for regenerative endodontics on 30 traumatised immature non-vital maxillary incisors (Jayadevan et al. 2021). Advanced PRF is prepared using the low-speed centrifugation concept and has been shown to have more growth factors. This study showed positive periapical healing and apical response after 12 months of follow-up.

20.3.3 Periodontal Regeneration

Periodontal disease can cause inflammation and degradation of periodontal tissue, leading to tooth loss. The progression of periodontal disease requires removing the source of infection and reducing inflammation. However, the loss of periodontal tissue would require periodontal regenerative therapies via enamel matrix derivatives, bone grafts, guided tissue regeneration, growth factor delivery and a combination of biomaterials, cells and growth factors.

A pulp micrograft with a collagen sponge scaffold was used in a patient with periodontal disease with a probing pocket depth (PPD) of 9 mm (Aimetti et al. 2014). A year after the treatment, the probing pocket depth was reduced to 3 mm. The radiographic finding had shown the filling of the intraosseous component of the defect by a bonelike tissue. Another study for intrabony defects in 29 chronic periodontitis patients also utilised pulp micrografts that was seeded on the collagen sponge (Ferrarotti et al. 2018). The patients were observed until 12 months post-treatment with clinical and radiographic examination. The results reported a reduction of the PPD from 4.9 mm to 3.4 mm and an increase in the clinical attachment level from 2.9 mm to 4.5 mm.

A study by Hernández-Monjaraz on a periodontally compromised tooth placed a lyophilised collagen-polyvinylpyrrolidone sponge that was seeded with DPSC at the affected site (Hernández-Monjaraz et al. 2018). The patient showed no signs of immunological rejection at 3 and 6 months after surgical intervention. Furthermore, cone beam volumetric tomography had revealed periodontal bone defect of 18.3–4.58 mm after 6 months. Apart from that, an increase in the bone mineral density on the walls of the defected area was observed from the densitometry assays, suggesting bone regeneration. Another study of intrabony defects used collagen biomaterial that was loaded with DPSCs in 11 chronic periodontitis patients (Aimetti et al. 2018). After a year of follow-up, the measurement of PDD had a 5 mm reduction and the clinical attachment level (CAL) had increased to 4.7 mm.

A periodontal ligament (PDL)-derived cell sheet and deproteinised bovine bone mineral biomaterials were used in 20 patients with periodontal intrabony bone defects (Chen et al. 2016). Clinical and radiographic examination had shown no significant difference in decreased bone defect depth, increased probe depth and increased CAL between beta-tricalcium phosphate with and without PDL-derived cell sheet. Another study had utilised beta-tricalcium phosphate biomaterials with PDL-derived cell sheets in periodontal intrabony bone defects in 10 patients (Iwata

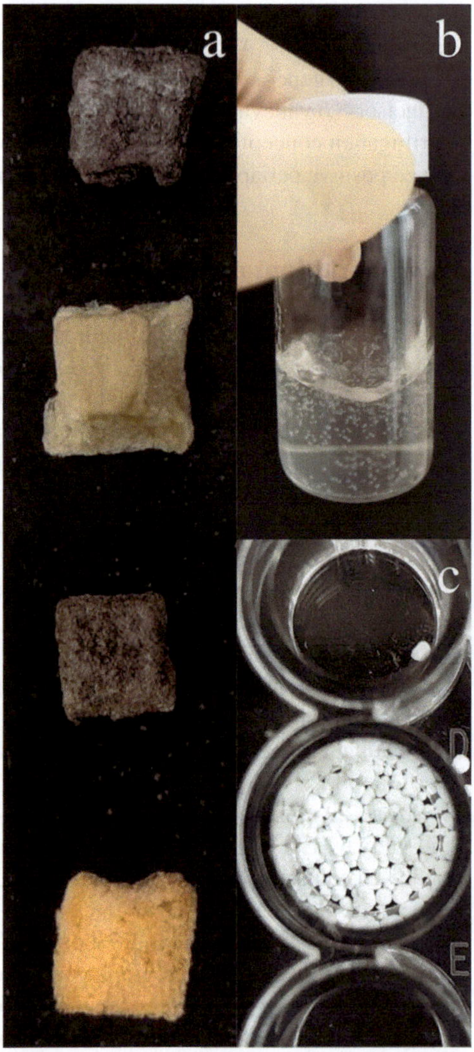

Fig. 20.1 Potential biomaterials from laboratory to market. (**a**) Carbon chitosan sponge, (**b**) Hydrogel, (**c**) Hydroxyapatite marine source

et al. 2018). This study had shown a reduction in the probing depth, an increase in the level of clinical attachment and radiographic bone height.

Most of the clinical trials reported the use of collagen sponges as biomaterials in the transplanted area. Collagen-based biomaterials were able to support the stem cells' attachment, proliferation and differentiation. They were also biocompatible and biodegradable. Thus, it is important to select biomaterials that are stem cell friendly and able to provide a suitable extracellular matrix for tissue regeneration. Figure 20.1 shows the current potential biomaterials (our research groups), such as carbon chitosan sponge, hydrogel and hydroxyapatite (HAP) marine sources that have been investigated for their biocompatibility and therapeutic efficacy in product translation from laboratory to market.

20.4 The Sustainability, Quality, Limitations, Translational and Commercialisation Potential of the Sustainable Materials

There are three factors related to the sustainability criteria of biomaterials: environment, economy and society. The source of sustainable materials must be ideally abundant, renewable, biodegradable, reusable, recyclable and low cost. The manufacturing processes of these biomaterials should also reduce carbon footprint and pollution. In addition, these biomaterials should demonstrate high therapeutic values to society (Agnieray et al. 2021). Of late, scientists and clinicians have actively investigated the sustainable biomaterials with high potential for dental regeneration through in vitro, in vivo and clinical studies. Figure 20.2 illustrates the biomaterials involved in the dental tissue regeneration.

20.4.1 Bioceramics

Bioceramic materials, which mainly consist of calcium phosphate, have been widely used for implantation due to their similarities to inorganic components of bone. Bioceramics are favourable bone substitutes due to their low risk of health issues. There are risks of immunologic rejection and disease transmission in the application of allografts and xenografts. The chemically synthesised HAP ceramics had shown poor durability and stability, which led to the emerging of the natural

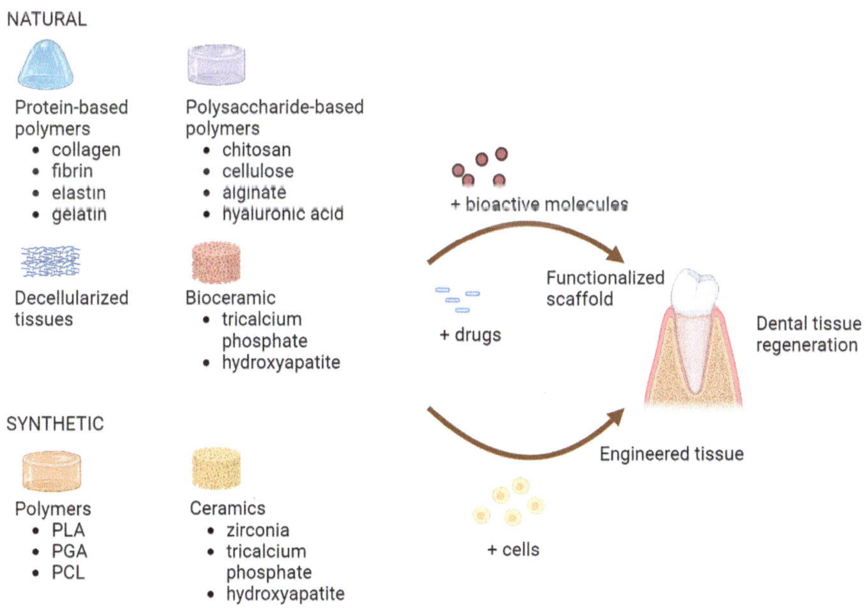

Fig. 20.2 Biomaterials involved in dental tissue regeneration

biowaste-derived HAP. The HAP is changed into a secondary phase, namely β-tricalcium phosphates (β-TCP) at the high sintering temperature (Wu et al. 2017). Previous studies mostly reported the application of synthetic β-TCP in dental regeneration (Cheah et al. 2021).

20.4.1.1 Natural Hydroxyapatite

Natural HAP is a promising bioactive ceramic material in bone regeneration for several reasons, which include its similarity to the HAP in human bone, excellent biocompatibility, low immunogenicity, economical, abundant availability and environmental-friendly. HAP can be derived from various bio-wastes such as eggshells, seashells, fish and bovine bones (Terzioğlu et al. 2018; Bee and Hamid 2020). Chicken eggshells are available abundantly and can be classified as low-cost bio-wastes, especially in the food industry. The biocompatibility of eggshell-derived nano- HAP (EnHAP) was proven with in vitro degradation when soaking the solutions with the same human body pH and non-cytotoxicity with a yield of 70% and above cell viability (Rohmadi et al. 2021). EnHAP powder also revealed a better trend of cell viability than other xenograft powders, which were Bio-Oss®, totally deproteinised cancellous bovine bone and Gen-Os®, partially deproteinised mixture of cortical and cancellous porcine bone (Dumitrescu et al. 2021). EnHAP had been used as bone substitutes to enhance bone regeneration after removing the mandibular impacted third molar in the randomised controlled clinical study. With properties of nanoscale size and porosity similar to the native bone HAP and commercially available synthetic HAP, EnHAP promoted bone regeneration by increasing bone density, reducing bone resorption and probing depth at the grafted site (Kattimani et al. 2019a). EnHAP promoted mineralisation and restored the lost microhardness of bleached human enamel in vitro (Kunam et al. 2019). EnHAP demonstrated its potential to treat dentin hypersensitivity with its ability to produce an efficient occlusion of dentin tubule (Onwubu et al. 2019). However, EnHAP consists of other crystalline carbonate structures after synthesis.

Osteoconductive natural HAP is often combined with other polymers to enhance its mechanical strength (brittleness and fragility), stability, reliability and mineralisation processes (Radulescu et al. 2022). Eggshell-derived HAP incorporated with silicon and poly(lactic-co-glycolic) acid had demonstrated higher bioactivity and compatibility with osteoblastic cells from the perspective of increased cell viability and adhesion due to the improvement of hydrophilicity of materials (Gutiérrez-Prieto et al. 2019). EnHAP was grafted with autologous PRF membrane to preserve the socket in order to prevent bone resorption post-tooth extraction. Thick gingival re-epithelisation and moderate to abundant bone formation with interconnecting trabeculae were observed in the patients' grafted site. The hydrophilicity of EnHAP had allowed for the manipulation of HAP mixtures with saline and fibrin (Kattimani et al. 2019b). Baskar and team members had investigated the potential of dentine regeneration by using EnHAP incorporated with carboxymethyl chitosan (CMC) scaffold. The human DPSCs that were cultured on the optimised 1:5 EnHAP-CMC scaffold showed significant higher cell proliferation, expression of dentin sialophosphoprotein and vascular endothelial growth factor on day 21 (Baskar et al. 2022).

Particulate coral-derived HAP with the characteristics of a slower degradation rate and good osteointegration has been used together with anti-inflammatory titanium mesh for alveolar rehabilitation without any exposure to exterior growth factors. Coral-derived HAP had constructed the bone within 6 months, followed by a successful dental implant without infection or other complications within a 3-year follow up (Zhou et al. 2018).

Crab shell can be considered one of the natural sources of HAP, as crab shell-derived HAP powder was found biocompatible with human gingival fibroblast cultures at the lowest concentration of 25 ppm (Kamadjaja et al. 2019). The combination of crab shell-derived HAP and chitosan gel as a bone graft has shown an anti-inflammatory effect by reducing the production of interleukin-1 as well as an osteogenic effect by increasing the production of bone morphogenic protein 2 in rats (Gani et al. 2022).

Natural sources of HAP are easily available biowastes from human activities; therefore, they can be easily sourced at a low cost. The manufacturing and commercialisation potential of natural HAP is extremely high, especially eggshell-derived HAP. The biggest challenge of natural HAP processing is to ensure the calcination and furnishings of HAP have a similar calcium/phosphate ratio to human bone apatite. The emitted carbon dioxide and other greenhouse gases from the calcination process can be minimised and recovered by using chemical hoods or other procedures (Agbeboh et al. 2020). Eggshell-derived HAP is more sustainable than other sources such as seashells or corals; therefore, it is recommended to be used as a bioceramic. Eggshells are daily disposed of food wastes, whereas some of the seashell and coral species have greatly reduced in number and are in danger of extinction.

20.4.2 Natural Polymers

The development of sustainable polymers from natural polymers, biomass or other renewable resources through chemical modification has been growing rapidly due to the fear of depletion of the fossil resources for synthetic polymer production (Papageorgiou 2018). Natural polymers can be produced from plants, animals and microorganisms. Natural polymers have been reported to have good biocompatibility, biodegradability, low immunogenic adverse effects and environment-friendly production. However, there are several limitations of natural polymers, which include poor mechanical strength, a high degradation rate, low reproducibility, and lack of variability among sources (Paradowska-Stolarz et al. 2021). The use of natural polymers in dental research and application can be categorised into protein-based biomaterials (collagen, fibrin, elastin and gelatin) and polysaccharide-based biomaterials (chitosan, cellulose, alginate and hyaluronic acid) (Patel et al. 2020).

20.4.2.1 Protein-Based Material: Collagen

Collagen is an important extracellular matrix in bone and aplenty available in oral tissues. Collagen-based biomaterials have been popularly used in the form of haemostatic sponge, membrane, scaffold and bone grafting matrix for dental regeneration (Binlateh et al. 2022). The commercially available collagen-based materials are mostly derived from bovine and porcine sources and their effectiveness as a barrier to promote guided bone regeneration in rat calvaria has been demonstrated. Porcine-collagenous membrane had exhibited a greater bone neoformation area on day 60 (Ramires et al. 2021). The application of collagenated bovine xenograft (Bio-Oss®) with a thicker particle size had a higher success rate in sinus augmentation simultaneously with dental implant placement as compared to the collagenated porcine xenografts (Apatos-Cortical® and Gen-Os®) with a smaller particle size (Küçükkurt and Moharamnejad 2021). Although autogenous grafts were the gold standard of dental treatment, its limited availability and risk of clinical complications associated with harvesting procedures have led to the development of xenograft application. Autogenous graft was more effective than porcine collagen matrix (Mucograft®) to improve the buccal peri-implant soft tissue thickness in order to prevent the bone resorption. However, the patients had higher satisfaction rate with the xenograft (Mucograft®) procedures due to shorter surgical times, less postoperative pain and less consumption of anti-inflammatory tablets (Cairo et al. 2017).

Fish collagen is an attractive alternative collagen source to bovine and porcine-derived collagen due to a lack of concern for infectious disease transmission and religious restriction. Fish collagen is abundantly available in fish skin, scales and bones, which can be easily obtained as biowastes from the fish-processing industry. A cross-linked electrospun tilapia skin-derived collagen nanofibrous membrane was produced for bone guided regeneration purposes. The study demonstrated that the fish-derived collagen membrane had better osteogenic biocompatibility and lower long-term inflammatory responses than the commercialised porcine-derived collagen membrane (Bio-Gide®). However, the current fabricated membrane must be improved due to its lower mechanical strength (shrunken and less porosity) than the Bio-Gide® membrane via the modification of crosslinking or membrane thickness (Hassanbhai et al. 2017). Suzuki and team members had formed a fully differentiated and stratified epithelium on the microstructured fish scale collagen-derived scaffold but the dimensions of micropattern were not maintained. Therefore, the mechanical properties of the scaffold must be enhanced through the modification of cross-linking, so that the scaffold can be manufactured as a durable tissue-engineered oral mucosa for clinical use in the future (Suzuki et al. 2020). Another research study had developed a collagen-based bioactive composite from *rohu* fish scale-derived type 1 collagen and synthetic HAP. The COL40%/HAP60% composite exhibited non-cytotoxic and angiogenic properties in vitro (Ilyas et al. 2018).

Recent studies have shown that tobacco plant-extracted recombinant human type 1 collagen (rhCOL1) is potentially an alternative to animal-based collagen through a recombinant expression system. The application of rhCOL1 prevents issues of pathogenic transmission, immune response, religious constraint and environmental burdens. The intact triple helix structures of rhCOL1 resemble those of native collagen. The prototype photocurable dermal filler containing rhCOL1 and cross-linked hyaluronic acid (HA) has remarkably promoted cell proliferation and tissue regeneration as compared to the standard HA-based filler (Seror et al. 2021). The plant-based rhCOL1 is a new biomaterial and currently, there is no study reporting on its application in the dental field (Binlateh et al. 2022). Overall, fish-derived collagen and plant-based collagen have the potential to be manufactured and commercialised for dental applications with more efforts required to elucidate their mechanisms in tissue regeneration and clinical-evidence based studies.

20.4.2.2 Polysaccharide-Based Materials: Chitosan and Cellulose

Chitosan is a naturally abundant polysaccharide-based derivative from chitin in the crustacean exoskeleton. Recent studies have revealed the potential of chitosan for preventive and regenerative dental applications due to its sustainability, low immunogenicity, antimicrobial, anti-inflammatory, biocompatible, biodegradable, bioadhesive and osteoconductive properties (Gayen et al. 2021). Chitosan is mucoadhesive due to its cationic and swelling properties, which allow it to bind to the moist anionic surface of oral mucosal epithelial cells. A randomised, double-blinded study found that the chitosan-containing oral mucoadhesive film can be used as a drug carrier, promote wound healing of oral ulcers and reduce pain among recurrent aphthous stomatitis patients (Shao and Zhou 2020). Chitosan-injectable hydrogels also promoted periodontal tissue regeneration with their properties of enhancing cell adhesion, proliferation, differentiation and vascularisation. However, the main challenges of chitosan-based hydrogel production are standardising the origins of the raw material, molecular weight distribution and manufacturing procedures (Tang et al. 2020).

Natural cellulosic material is a sustainable polysaccharide with adjustable physical properties, flexibility, biocompatibility and biodegradability. It can be produced from economically and environmentally-friendly biomass due to the newly developed cellulose solvent with high solubility and low dissolution temperature as well as the process of cellulose modification via esterification and etherification. Cellulose-based hydrogels can be formed by cross-linking the cellulose nanofibers with other polymers (Yan et al. 2020). The chitosan/CMC/HAP hydrogels were reported to have the capacity to hold thyroxine (an angiogenic agent) for periodontal regeneration due to their features such as rapid absorption, high swelling ratio, shape-transformability and stretchability (Malik et al. 2020). Nanocellulose hydrogel derived from rice straw agricultural wastes was developed to form an injectable, bioactive, and biocompatible bone graft materials to fill irregular-shaped bone defects. This rice straw-derived nanocellulose hydrogel had strong gel properties, capability to induce ossification and less cytotoxicity when compared to its mineralisation with the synthetic biphasic calcium phosphate (HAP and β-TCP) (Safwat et al. 2018).

Bacterial cellulose membrane (BCM) is a natural polymer-based membrane that is synthesised by bacteria such as *Kumagatabacter hansenii* and *Gluconacetobacter hansenii*. Bacterial cellulose (BC) has a high degree of crystallinity, purity, biocompatibility, water-holding capability and suspension stability (Zhong 2020). The commercial BCM, Gengiflex® is a non-resorbable cellulose membrane for guided bone regeneration (Mensah et al. 2021). The additional surgical procedure is required to remove the non-resorbable BCM as human body does not have the cellulase to degrade the cellulose (Cho et al. 2017). Inoue and team members had developed a bioabsorbable and bactericidal BCM through the oxidation method for potential periodontal treatment. Bactericide (chlorhexidine) was included in the BCM to kill the remaining bacteria after periodontal treatment (Inoue et al. 2020). Although the mechanical properties of BCM (wet tensile stress, wet tensile strain and Young's modulus) were reduced due to the chain cleavage of BC nanofibers by high-energy electron beam irradiation (EI), EI-BCM became more resorbable and promoted cell interaction and bone regeneration on rat calvarial defects (An et al. 2017).

Both plant-derived cellulose and bacterial cellulose are sustainable, translational and commercialisable biomaterials for dental treatment. The industrial production and purification of BC fermentation are simple and cause less pollution but they are costly and consume high energy. The utilisation of plant-based cellulose obtained from agricultural and industrial wastes are more economical and environmental-friendly (Zhong 2020).

20.5 Conclusion

In future studies, the combination of bioceramics with biopolymers would be a good direction for the development of biomaterials for regenerative dentistry. The formulated biomaterials made of both natural ceramics and polymers will be more biomimetic with better mechanical strength and biocompatibility. The biomaterials can also be incorporated with induced stem cells and specified growth factors that could promote cell proliferation, osseointegration, vascularisation, dentinogenesis, osteogenesis, etc. for dental tissue regeneration and restoration. Preclinical verification and validation should be performed to characterise a newly developed biomaterial for its physicochemical properties, biocompatibility and functionality through in vitro and in vivo studies. Subsequently, the preclinical scientific information can be used in clinical trials to prove the safety and efficacy of biomaterials in patients.

Resorption pattern of the materials, graft material origin, material availability or self-sustainability, cost-effectiveness and evidence for tissue regeneration should be taken into consideration when selecting the biomaterial in future clinical applications in dentistry. Sustainable biomaterials derived from natural sources have emerged as highly potential candidates for dental tissue engineering applications as they possess excellent biocompatibility with modifiable physicochemical characteristics. The success rate of transitioning the sustainable biomaterial-based products to the market requires collaboration and integration among scientists, clinicians, industries, and regulatory authorities. It is expected that more sustainable biomaterial-based products for regenerative dentistry will be tested clinically and commercialised due to the positive preclinical and clinical outcomes as well as the high demand for these products.

Acknowledgements We thank the editors, Dr. Wan Safwani Wan Kamarul Zaman and Assoc. Prof. Dr. Nurul Asma Abdullah for giving us the opportunity to contribute to this chapter. We also thank Dr. Farinawati Yazid's research team for providing the photos of biomaterials from their research studies.

Conflict of Interest The authors declare no conflict of interest.

References

Agbeboh NI, Oladele IO, Daramola OO, Adediran AA, Olasukanmi OO, Tanimola MO (2020) Environmentally sustainable processes for the synthesis of hydroxyapatite. Heliyon 6(4): e03765. https://doi.org/10.1016/j.heliyon.2020.e03765

Agnieray H, Glasson JL, Chen Q, Kaur M, Domigan LJ (2021) Recent developments in sustainably sourced protein-based biomaterials. Biochem Soc Trans 49(2):953–964. https://doi.org/10.1042/BST20200896

Agrahari V, Agrahari V (2018) Facilitating the translation of nanomedicines to a clinical product: challenges and opportunities. Drug Discov Today 23(5):974–991. https://doi.org/10.1016/j.drudis.2018.01.047

Agrahari V, Hiremath PJN (2017) Challenges associated and approaches for successful translation of nanomedicines into commercial products. Nanomed J 12(8):819–823

Aguilar-Ayala FJ, Aguilar-Pérez FJ, Nic-Can GI, Rojas-Herrera R, Chuc-Gamboa G, Aguilar-Pérez D, Rodas-Junco BAJAS (2022) A molecular view on biomaterials and dental stem cells interactions: literature review. Appl Sci 12(12):5815

Aimetti M, Ferrarotti F, Cricenti L, Mariani GM, Romano F (2014) Autologous dental pulp stem cells in periodontal regeneration: a case report. Int J Periodontics Restorative Dent 34(Suppl 3): s27–s33. https://doi.org/10.11607/prd.1635

Aimetti M, Ferrarotti F, Gamba MN, Giraudi M, Romano F, Dentistry R (2018) Regenerative treatment of periodontal intrabony defects using autologous dental pulp stem cells: a 1-year follow-up case series. Int J Periodontics Restorative Dent 38(1):51

An SJ, Lee SH, Huh JB, Jeong SI, Park JS, Gwon HJ, Kang ES, Jeong CM, Lim YM (2017) Preparation and characterization of Resorbable bacterial cellulose membranes treated by electron beam irradiation for guided bone regeneration. Int J Mol Sci 18(11):2236. https://doi.org/10.3390/ijms18112236

Barbier L, Ramos E, Mendiola J, Rodriguez O, Santamaria G, Santamaria J, Arteagoitia I (2018) Autologous dental pulp mesenchymal stem cells for inferior third molar post-extraction socket healing: a split-mouth randomised clinical trial. Med Oral Patol Oral Cir Bucal 23(4):e469–e477. https://doi.org/10.4317/medoral.22466

Baskar K, Saravana Karthikeyan B, Gurucharan I, Mahalaxmi S, Rajkumar G, Dhivya V, Kishen A (2022) Eggshell derived nano-hydroxyapatite incorporated carboxymethyl chitosan scaffold for dentine regeneration: a laboratory investigation. Int Endod J 55(1):89–102. https://doi.org/10.1111/iej.13644

Bee S-L, Hamid ZAJCI (2020) Hydroxyapatite derived from food industry bio-wastes: syntheses, properties and its potential multifunctional applications. Ceram Int 46(11):17149–17175

Binlateh T, Thammanichanon P, Rittipakorn P, Thinsathid N, Jitprasertwong P (2022) Collagen-based biomaterials in periodontal regeneration: current applications and future perspectives of plant-based collagen. Biomimetics (Basel, Switzerland) 7(2):34. https://doi.org/10.3390/biomimetics7020034

Brunelli G, Motroni A, Graziano A, D'Aquino R, Zollino I, Carinci F (2013) Sinus lift tissue engineering using autologous pulp micro-grafts: a case report of bone density evaluation. J Indian Soc Periodontol 17(5):644

Cairo F, Barbato L, Tonelli P, Batalocco G, Pagavino G, Nieri M (2017) Xenogeneic collagen matrix versus connective tissue graft for buccal soft tissue augmentation at implant site. A randomized, controlled clinical trial. J Clin Periodontol 44(7):769–776

Cheah CW, Al-Namnam NM, Lau MN, Lim GS, Raman R, Fairbairn P, Ngeow WCJM (2021) Synthetic material for bone, periodontal, and dental tissue regeneration: where are we now, and where are we heading next? Materials 14(20):6123

Chen F-M, Gao L-N, Tian B-M, Zhang X-Y, Zhang Y-J, Dong G-Y, Lu H, Chu Q, Xu J, Yu Y (2016) Treatment of periodontal intrabony defects using autologous periodontal ligament stem cells: a randomized clinical trial. Stem Cell Res Ther 7(1):1–11

Chi M, Qi M, Lan A, Wang P, Weir MD, Melo MA, Sun X, Dong B, Li C, Wu J, Wang L, Xu HHK (2019) Novel bioactive and therapeutic dental polymeric materials to inhibit periodontal pathogens and biofilms. Int J Mol Sci 20(2):278. https://doi.org/10.3390/ijms20020278

Cho Y-D, Seol Y-J, Lee Y-M, Rhyu I-C, Ryoo H-M, Ku Y (2017) An overview of biomaterials in periodontology and implant dentistry. Adv Mater Sci Eng 2017:1948241

Costa PF (2019) Translating biofabrication to the market. Trends Biotechnol 37(10):1032–1036. https://doi.org/10.1016/j.tibtech.2019.04.013

d'Aquino R, De Rosa A, Lanza V, Tirino V, Laino L, Graziano A, Desiderio V, Laino G, Papaccio GJECM (2009) Human mandible bone defect repair by the grafting of dental pulp stem/progenitor cells and collagen sponge biocomplexes. Eur Cells Mater 18(7):75–83

Davison-Kotler E, Marshall WS, Garcia-Gareta E (2019) Sources of collagen for biomaterials in skin wound healing. Bioengineering (Basel) 6(3):56. https://doi.org/10.3390/bioengineering6030056

Deep A (2022a) Chapter 2 - Ethics of clinical trials of medical devices. In: Deep A-M (ed) Medical device regulations, 1st edn. Academic Press, pp 13–21

Deep A (2022b) Chapter 1 - Introduction of medical devices. In: Deep A-M (ed) Medical device regulations, 1st edn. Academic Press, pp 1–11

Dumitrescu CR, Neacsu IA, Surdu VA, Nicoara AI, Iordache F, Trusca R, Ciocan LT, Ficai A, Andronescu E (2021) Nano-hydroxyapatite vs. xenografts: synthesis, characterization, and in vitro behavior. nanomaterials (Basel, Switzerland). Nanomaterials (Basel) 11(9):2289. https://doi.org/10.3390/nano11092289

Ferrarotti F, Romano F, Gamba MN, Quirico A, Giraudi M, Audagna M, Aimetti M (2018) Human intrabony defect regeneration with micrografts containing dental pulp stem cells: a randomized controlled clinical trial. J Clin Periodontol 45(7):841–850. https://doi.org/10.1111/jcpe.12931

Gani A, Yulianty R, Supiaty S, Rusdy M, Dwipa Asri G, Eka Satya D, Rahayu Feblina A, Achmad H (2022) Effectiveness of combination of chitosan gel and hydroxyapatite from crabs shells (Portunus pelagicus) waste as bonegraft on periodontal network regeneration through IL-1 and BMP-2 analysis. Int J Biomater 2022:1817236

Gayen K, Pabale S, Shirolkar S, Sarkar S, Roychowdhury S (2021) Chitosan biomaterials: natural resources for dentistry. Int J Oral Health Sci 11(2):75

Gutiérrez-Prieto SJ, Fonseca LF, Sequeda-Castañeda LG, Díaz KJ, Castañeda LY, Leyva-Rojas JA, Salcedo-Reyes JC, Acosta AP (2019) Elaboration and biocompatibility of an eggshell-derived hydroxyapatite material modified with Si/PLGA for bone regeneration in dentistry. Int J Dent 2019:5949232

Hassanbhai AM, Lau CS, Wen F, Jayaraman P, Goh BT, Yu N, Teoh SH (2017) In vivo immune responses of cross-linked electrospun tilapia collagen membrane. Tissue Eng Part A 23(19–20):1110–1119. https://doi.org/10.1089/ten.tea.2016.0504

Hench LL, Jones JR (2015) Bioactive glasses: frontiers and challenges. Front Bioeng Biotechnol 3:194. https://doi.org/10.3389/fbioe.2015.00194

Hernández-Monjaraz B, Santiago-Osorio E, Ledesma-Martínez E, Alcauter-Zavala A, Mendoza-Núñez VM (2018) Retrieval of a periodontally compromised tooth by allogeneic grafting of mesenchymal stem cells from dental pulp: a case report. Int J Med Res 46(7):2983–2993

Ilyas K, Qureshi SW, Afzal S, Gul R, Yar M, Kaleem M, Khan AS (2018) Microwave-assisted synthesis and evaluation of type 1 collagen–apatite composites for dental tissue regeneration. J Biomater Appl 33(1):103–115

Inoue BS, Streit S, Dos Santos Schneider AL, Meier MM (2020) Bioactive bacterial cellulose membrane with prolonged release of chlorhexidine for dental medical application. Int J Biol Macromol 148:1098–1108. https://doi.org/10.1016/j.ijbiomac.2020.01.036

Iwata T, Yamato M, Washio K, Yoshida T, Tsumanuma Y, Yamada A, Onizuka S, Izumi Y, Ando T, Okano T (2018) Periodontal regeneration with autologous periodontal ligament-derived cell sheets–a safety and efficacy study in ten patients. Regen Ther 9:38–44

Jayadevan V, Gehlot P-M, Manjunath V, Madhunapantula SV, Lakshmikanth J-S, Dentistry E (2021) A comparative evaluation of advanced platelet-rich fibrin (a-prf) and platelet-rich fibrin (prf) as a scaffold in regenerative endodontic treatment of traumatized immature non-vital permanent anterior teeth: a prospective clinical study. J Clin Exp Dent 13(5):e463

Kamadjaja MJK, Abraham JF, Laksono HJMA (2019) Biocompatibility of portunus pelagicus hydroxyapatite graft on human gingival fibroblast cell culture. Med Arch 73(6):378

Kattimani V, Lingamaneni KP, Yalamanchili S, Mupparapu M (2019a) Use of eggshell-derived nano-hydroxyapatite as novel bone graft substitute–a randomized controlled clinical study. J Biomater Appl 34(4):597–614

Kattimani VS, Lingamaneni KP, Kreedapathi GE, Kattappagari KK (2019b) Socket preservation using eggshell-derived nanohydroxyapatite with platelet-rich fibrin as a barrier membrane: a new technique. J Korean Assoc Oral Maxillofac Surg 45(6):332–342

Küçükkurt S, Moharamnejad N (2021) Comparison of the effects of three different xenogeneic bone grafts used in sinus augmentation simultaneous with dental implant placement on the survival of the implants and the dimensional changes of the region. Minerva Dent Oral Sci 70(6):248–256

Kunam D, Sampath V, Manimaran S, Sekar M (2019) Effect of indigenously developed nano-hydroxyapatite crystals from chicken egg Shell on the surface hardness of bleached human enamel: an in vitro study. Contemp Clin Dent 10(3):489–493. https://doi.org/10.4103/ccd.ccd_810_18

Loh EYX, Goh PS, Mannan AMM, Mohd Sani N, Ab Ghani A (2021) Cell and gene therapy products in Malaysia: a snapshot of the industry's current regulation preparedness. Cytotherapy 23(12):1108–1113. https://doi.org/10.1016/j.jcyt.2021.06.006

Malik MH, Shahzadi L, Batool R, Safi SZ, Khan AS, Khan AF, Chaudhry AA, Rehman IU, Yar M (2020) Thyroxine-loaded chitosan/carboxymethyl cellulose/hydroxyapatite hydrogels enhance angiogenesis in in-ovo experiments. Int J Biol Macromol 145:1162–1170. https://doi.org/10.1016/j.ijbiomac.2019.10.043

Mensah A, Chen Y, Christopher N, Wei Q (2021) Membrane technological pathways and inherent structure of bacterial cellulose composites for drug delivery. Bioengineering (Basel) 9(1):3. https://doi.org/10.3390/bioengineering9010003

Nageh M, Ahmed GM, El-Baz AA (2018) Assessment of regaining pulp sensibility in mature necrotic teeth using a modified revascularization technique with platelet-rich fibrin: a clinical study. J Endod 44(10):1526–1533

Nakashima M, Iohara K, Murakami M, Nakamura H, Sato Y, Ariji Y, Matsushita K (2017) Pulp regeneration by transplantation of dental pulp stem cells in pulpitis: a pilot clinical study. Stem Cell Res Ther 8(1):61. https://doi.org/10.1186/s13287-017-0506-5

Oberweis CV, Marchal JA, Lopez-Ruiz E, Galvez-Martin P (2020) A worldwide overview of regulatory frameworks for tissue-based products. Tissue Eng Part B Rev 26(2):181–196. https://doi.org/10.1089/ten.TEB.2019.0315

Onwubu SC, Mhlungu S, Mdluli PS (2019) In vitro evaluation of nanohydroxyapatite synthesized from eggshell waste in occluding dentin tubules. J Appl Biomater Funct Mater 17(2):2280800019851764. https://doi.org/10.1177/2280800019851764

Papageorgiou GZJP (2018) Thinking green: sustainable polymers from renewable resources. Polymers 10(9):952

Paradowska-Stolarz A, Wieckiewicz M, Owczarek A, Wezgowiec J (2021) Natural polymers for the maintenance of Oral health: review of recent advances and perspectives. Int J Mol Sci 22(19):10337. https://doi.org/10.3390/ijms221910337

Patel E, Choonara Y, Pillay VJSADJ (2020) Dental biomaterials: challenges in the translation from lab to patient. S Afr Dent J 75(1):16–28

Radulescu D-E, Neacsu IA, Grumezescu A-M, Andronescu EJP (2022) Novel trends into the development of natural hydroxyapatite-based polymeric composites for bone tissue engineering. Polymers 14(5):899

Ramires GA, Helena JT, Oliveira J, De C, Faverani LP, Bassi APF (2021) Evaluation of guided bone regeneration in critical defects using bovine and porcine collagen membranes: histomorphometric and immunohistochemical analyses. Int J Biomater 2021:8828194

Rohmadi R, Harwijayanti W, Ubaidillah U, Triyono J, Diharjo K, Utomo P (2021) In vitro degradation and cytotoxicity of eggshell-based hydroxyapatite: a systematic review and meta-analysis. Polymers (Basel) 13(19):3223. https://doi.org/10.3390/polym13193223

Safwat E, Hassan ML, Saniour S, Zaki DY, Eldeftar M, Saba D, Zazou M (2018) Injectable TEMPO-oxidized nanofibrillated cellulose/biphasic calcium phosphate hydrogel for bone regeneration. J Biomater Appl 32(10):1371–1381. https://doi.org/10.1177/0885328218763866

Seror J, Stern M, Zarka R, Orr NJP, Surgery R (2021) The potential use of novel plant-derived recombinant human collagen in aesthetic medicine. Plast Reconstr Surg 148(6S):32S–38S

Shao Y, Zhou H (2020) Clinical evaluation of an oral mucoadhesive film containing chitosan for the treatment of recurrent aphthous stomatitis: a randomized, double-blind study. J Dermatolog Treat 31(7):739–743. https://doi.org/10.1080/09546634.2019.1610548

Suzuki A, Kato H, Kawakami T, Kodama Y, Shiozawa M, Kuwae H, Miwa K, Hoshikawa E, Haga K, Shiomi A, Uenoyama A, Saitoh I, Hayasaki H, Mizuno J, Izumi K (2020) Development of microstructured fish scale collagen scaffolds to manufacture a tissue-engineered oral mucosa equivalent. J Biomater Sci Polym Ed 31(5):578–600. https://doi.org/10.1080/09205063.2019.1706147

Tang G, Tan Z, Zeng W, Wang X, Shi C, Liu Y, He H, Chen R, Ye X (2020) Recent advances of chitosan-based injectable hydrogels for bone and dental tissue regeneration. Front Bioeng Biotechnol 8:587658. https://doi.org/10.3389/fbioe.2020.587658

Taylor D, Yoshida M, Fuller K, Giannobile W, Sfeir C, Wagner W, Kohn D (2021) Translating dental, oral, and craniofacial regenerative medicine innovations to the clinic through interdisciplinary commercial translation architecture. J Dent Res 100(10):1039–1046

Terzioğlu P, Öğüt H, Kalemtaş A (2018) Natural calcium phosphates from fish bones and their potential biomedical applications. Mater Sci Eng 91:899–911

Volpi M, Paradiso A, Costantini M, Swieszkowski W (2022) Hydrogel-based fiber biofabrication techniques for skeletal muscle tissue engineering. ACS Biomater Sci Eng 8(2):379–405. https://doi.org/10.1021/acsbiomaterials.1c01145

Wu S-C, Hsu H-C, Hsu S-K, Tseng C-P, Ho W-FJAPT (2017) Preparation and characterization of hydroxyapatite synthesized from oyster shell powders. Adv Powder Technol 28(4):1154–1158

Yan G, Chen B, Zeng X, Sun Y, Tang X, Lin L (2020) Recent advances on sustainable cellulosic materials for pharmaceutical carrier applications. Carbohydr Polym 244:116492. https://doi.org/10.1016/j.carbpol.2020.116492

Zhong C (2020) Industrial-scale production and applications of bacterial cellulose. Front Bioeng Biotechnol 8:605374. https://doi.org/10.3389/fbioe.2020.605374

Zhou M, Li SY, Terheyden H, Cao SS, Che YJ, Geng YM (2018) Particulate coral hydroxyapatite sheltered by titanium mesh for localized alveolar rehabilitation after Onlay graft failure: a case report. J Oral Implantol 44(2):147–152. https://doi.org/10.1563/aaid-joi-D-17-00109

Glass Ionomer Cements as Sustainable Material for Restorative Dentistry

21

Norhayati Luddin

Abstracts

One of the most biocompatible dental restorative materials is glass ionomer cement (GIC), which has seen extensive use in clinical settings. However, due to the inadequacies of conventional GIC, such as their poor physical and mechanical properties, nano-hydroxyapatite-silica (nano-HA-silica) fillers were developed. It has received some investigation in terms of its characteristic whereby these fillers are added to conventional GIC to boost the material's strength. This chapter discusses the GIC material and provides insight into the synthesis and characterisation of nano-HA-silica and its addition to GIC in order to improve and sustain its application in clinical settings. Furthermore, this chapter discusses how the addition of these nano-HA-silica fillers to GIC affects its behaviour in terms of mechanical properties, biocompatibility and dentinogenic differentiation potential.

Keywords

Nanohydroxyapatite · Nanosilica · Glass ionomer cement · Mechanical · Biocompatibility · Dentinogenic differentiation

N. Luddin (✉)
Prosthodontic Unit, School of Dental Sciences, Health Campus, Universiti Sains Malaysia, Kubang Kerian, Kota Bharu, Kelantan, Malaysia
e-mail: norhayatikck@usm.my

W. S. Wan Kamarul Zaman, A. A. Nurul (eds.), *Sustainable Material for Biomedical Engineering Application*, https://doi.org/10.1007/978-981-99-2267-3_21

Abbreviations

CaOH	Calcium hydroxide
cGIC	Conventional glass ionomer cement
etOH	Ethanol
FTIR	Fourier transform infrared spectroscopy
GIC	Glass ionomer cement
Nano-HA-SiO_2-GIC	Nanohydroxyapatite-silica-GIC
SEM	Scanning electron microscopy
TEM	Transmission electron microscopy
TEOS	Tetraethylorthosilicate
XRD	X-ray diffraction

21.1 Introduction

Glass ionomer cement (GIC) is a dental restorative material that is frequently employed in clinical settings due to its therapeutic application. This tooth-coloured material (Craig 2002) was developed by Wilson et al. (Wilson and Kent 1972) in 1972 and is known as "water-based cement" (Crisp et al. 1975) as it was developed through an acid-based chemical reaction between aluminium silicate glass powder and polyacrylic acid (Wilson and Nicholson 1993).

The setting reaction of the GIC would occur by mixing glass powder and polyalkenoic acid. During the mixing process, the polyalkenoic acid would be attacked, thus damaging the glass surface. Following that, Ca^{2+} and Al^{3+} ions are leached into the aqueous medium. These ions are combined with acid to form complexes thus forming a solid gel. The silica and phosphate ions released from the glass during the hardening process would undergo condensation to create a gel matrix (Nicholson 1998a; Griffin and Hill 2000).

Phosphorus and silicon would form an inorganic network within the matrix of glass ionomers based on their presence in the matrix of GIC. There are two different types of networks identified in the structure of set cement: an organic network that is made up of polyacid chains (copolymers) and connected by metallic ions (polysalt bridges) (Nicholson 1998a,b; Bowen and Marjenhoff 1992) and an inorganic network that is made up of silica and phosphate ions.

An optimal restorative material must be able to adhere well to tooth structure (dentine and enamel), have anti-bacterial effects on oral cariogenic microorganisms, be resilient and be capable of withstanding the traumas of occlusion (Mickenautsch et al. 2011). There are several distinctive qualities of GIC, which make the materials prevalent in dentistry, such as good biocompatibility, long-term fluoride release that also works as an anticariogenic agent, modulus of elasticity akin to dentin and direct bonding to the tooth. Clinically, they would serve as lining and base materials, fissure sealants and atraumatic restorative therapy (ART) materials in paediatric dentistry (Sidhu and Nicholson 2016). In ART procedures, high viscosity GIC (HVGIC) such as Fuji IX has been utilised as the preferred material (Zanata et al. 2011). As HVGIC can be utilised in larger cavities as well as early phases of caries formation, it is considered as perfectly suited to control dental caries. Furthermore, the restoration of teeth becomes simpler as the dentine-pulp complex is able to react in opposition to the caries process (Ericson et al. 2003).

Despite these positive attributes, there are several limitations of GIC, which include a higher propensity for dehydration, poor physical qualities (high solubility and slow setting rate) and poor mechanical properties (low-wear resistance) (Pelka et al. 1996). As such, they are rarely used as restorative materials in stress-bearing areas. The most commercially available GIC are considered clinical grade GIC because they meet the minimum requirements of the ISO standard (ISO 9917–1 2003). In certain cases, however, the utilisation under high occlusal loads, such as on the posterior teeth, is not advised as they are still mechanically fragile. As a result, numerous researchers have attempted to alter the chemical composition of ionomer glass or polyalkenoic acid to enhance the mechanical properties of GIC (Davidson and Mjor 1999). A number of materials including alumina, zirconia, silicon carbide, hydroxyapatite (HAp), glass fibre and bioactive glass (BAG) have been employed to enhance the mechanical and physical properties of GIC (Wilson and McLean 1988; Lucas et al. 2003; Gu et al. 2005; Yli-Urpo et al. 2005a, b; Hammouda 2009). These efforts, however, have not successfully increased their mechanical strength.

In the form of minerals, hydroxyapatite (HA) naturally occurs from calcium apatite. It exhibits great biological behaviour, has a hardness that is comparable to a normal tooth, and has some degree of radiopacity (Arcís et al. 2002; Arita et al. 2003; Wang and Shaw 2009). Additionally, nanosized HA crystals have been proven to boost the remineralisation of enamel (Huang et al. 2009; Najeeb et al. 2016). Due to its excellent properties, HA has been utilised in many areas of dentistry, including implant dentistry (Javed et al. 2014), caries prevention (Ong and Chan 2000), bone void fillers (Pryor et al. 2009), restoration of periodontal defects (Kaushick et al. 2011), alveolar ridge augmentation (Singh et al. 2014), endodontic treatment (Haghgoo et al. 2015), repair of mechanical furcation perforations (Kakani et al. 2015), desensitising agent and as a remineralising agent in toothpastes (Kantharia et al. 2014). Several studies have been carried out to determine whether the addition of HA to cGIC would affect the results of the GIC. It has been demonstrated that HA could enhance the mechanical strength, adhesion to tooth and improve the release of fluoride in cGIC (Arita et al. 2003; Gu et al. 2005; Yamamoto 1984). On top of that, the incorporation of HA (Gu et al. 2005) and silica (Felemban and Ebrahim 2016) in the micron size range as an added filler in GIC has been reported to produce superior mechanical properties of the nano-hydroxyapatite-silica (nano-HA-silica) GIC compared to cGIC. As such, the use of nano-fillers into GIC is growing in popularity due to their higher mechanical performance than cGIC (Najeeb et al. 2016; Ching et al. 2018; Dionysopoulos et al. 2022). Dental hard tissues and nano-HA have a similar chemical composition and are thus often recognised as common fillers in restorative materials. Furthermore, the inclusion of nano-HA has proven to enhance the set GIC's crystallinity (Moshaverinia et al.

2008). It also creates a favourable environment for enamel remineralisation (Moshaverinia et al. 2008; Huang et al. 2011). Aside from that, a report has shown that the utilisation of nanosilica as a GIC additive has drawn more attention (Mabrouk et al. 2012). According to the study, the addition of nanosilica to GIC has increased its bioactivity and would help in preventing the development of marginal gaps inside the teeth (Mabrouk et al. 2012), Ezz et al. (2022), on the other hand, added nano-fluoroapatite to cGIC and reported higher bioactivity as compared to cGIC (Ali Ezz et al. 2022). A combination of nano-HA-silica was utilised as filler for GIC for the first time, as demonstrated by Rahman et al. (Ab Rahman et al. 2014). In the study, the one-pot sol-gel method was used in the synthesis of nano-HA-silica and added to the commercially available GIC Fuji IX GP (GC International, Japan). Following that, there were several other investigations carried out on the newly developed material to further evaluate their properties.

Hence, the objective of this chapter is to provide an insight on the synthesis and characterisation of nano-HA-silica and how this material would improve the properties of GIC in terms of its mechanical, biocompatibility and dentinogenic differentiation for wider application in clinical dentistry.

21.2 Synthesis of Nano-Hydroxyapatite Silica

Following Rahman et al. (Ab Rahman et al. 2014) methodology, the one pot sol-gel process (Fig. 21.1) was used to create nano-hydroxyapatite-silica. 100 ml of distilled water was used to dissolve the calcium hydroxide powder for 30 minutes prior to the addition of phosphoric acid drop-wise to the calcium hydroxide suspension. By adding liquid ammonia drop by drop, the pH of the suspension was regulated between 11 and 12. Following that, the magnetic stirrer was used to agitate the calcium hydroxide suspension for 48 h until it turned into a white viscous sol. In ethanol (etOH), tetraethyl orthosilicate (TEOS) was dissolved and applied drop-wise to the calcium hydroxide mixture, followed by a 12-h magnetic mixing period. After 48 hours, the sol was generated and underwent centrifugation (Eppendorf Centrifuge 5804, Germany) in which the viscous white sol was spun at 7000 rpm for 7 min while being washed with pure ethanol.

After centrifugation, the white sol was vacuum freeze-dried (ScanVac CoolSafe, Denmark) for 24 h before it was placed in a freezer at −80 °C for an hour. After freeze drying, the powder was crushed for 10 min using a mortar and pestle. Then, the grounded powder was heated in a furnace oven for 1 hour at 600 °C and calcined (WiseTherm, Germany). Afterwards, the calcined powder was once again manually crushed in a mortar and pestle for 10 min. All of the powder samples were kept at room temperature in a glass desiccator filled with silica gel (Moheet et al. 2018).

Fourier transform infrared spectroscopy (FTIR) was one of the characterisation tests performed on the synthesised nano-HA-Si powder. In addition, X-ray diffraction (XRD) was used to analyse the nano-HA-silica powder while scanning electron microscopy (SEM) and transmission electron microscopy (TEM) were used to examine the powder's microstructure.

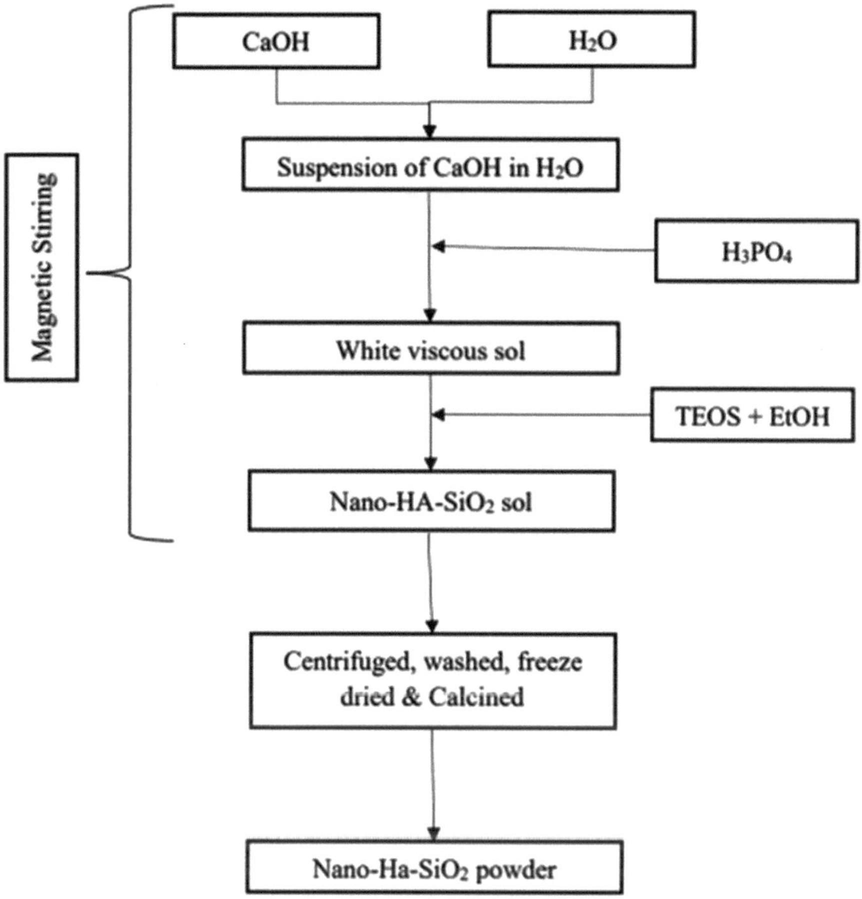

Fig. 21.1 One-pot synthesis process with magnetic stirring

21.3 Incorporation of Nano-Ha Silica into Glass Ionomer Cement

Nano-HA-Si powder was added to cGIC (Fuji IX, GC Company, Japan) and the GIC was prepared by mixing it with powder and liquid in a 1:1 ratio as advised by the manufacturer. The same ratio of powder to liquid was used to create nanohydroxyapatite-silica-GIC (nano-HA-Si-GIC). To achieve homogeneity, the powder mixture (cGIC and nano-HA-Si powder) was manually mixed for 10 min by using a mortar and pestle.

The nano-HA-silica GIC material was then subjected to several tests to investigate its mechanical, biocompatibility and dentinogenic differentiation properties. In particular, vickers hardness, compressive strength, flexural strength and shear bond strength have been evaluated for mechanical strength, while cytotoxicity testing and cell attachment properties as well as odontogenic and dentinogenic gene expression have been investigated for biocompatibility and gene expression studies, respectively.

21.4 Glass Ionomer Cement: Outcome and Update

21.4.1 Material Characterisation

In accordance with earlier research that proposed the successful synthesis of nano-HA-silica composites, nano-HA-silica material is successfully produced through the one-pot sol-gel method (Ab Rahman et al. 2014; Shiekh et al. 2014; Lin et al. 2007). In particular, nano-HA-silica spectra were found by Moheet et al. (Moheet et al. 2018) at stretching frequencies of 1023, 873, 809, 667 and 605 cm^1, respectively. Additionally, the XRD patterns had a prominent peak at 32 degrees that corresponded to (211) hkl planes in the crystalline structure of hydroxyl-apatite and the quantity of counts demonstrated that the samples had a higher degree of crystallinity (Moheet et al. 2018).

It was discovered that the hardness of the material increased to 73% when nano-HA-silica was added to cGIC (Ab Rahman et al. 2014). These positive findings are related to enhanced packing density due to nanosilica particles filling the spaces between the nano-HA in the GIC matrix (Ab Rahman et al. 2014). Similar findings were also reported by Moheet et al. (Moheet et al. 2018) whereby the characterisation of nano-HA-silica powder from SEM and TEM showed that the shape of the nano-powder was comprised of elongated HA particles and encircled by silica particles, which were spherical in shape (Fig. 21.2).

21.4.2 Mechanical Properties

Strength is one crucial factor that is often associated with the success of any restorative material in the oral cavity. In comparison to cGIC, nano-HA-silica-GIC has reportedly demonstrated greater Vickers hardness, compressive strength, flexural strength and shear bond strength (Moheet et al. 2018; Murugan et al. 2022). In this study, GIC powder was incorporated with nano-HA-silica at weight percentages of 0% (control), 5%, 10%, 15% and 20%, respectively. When nano-HA-silica was added to GIC, the Vickers hardness increased by about 36% as opposed to that of cGIC. In comparison to cGIC, it was discovered that the addition of 10% nano-HA-silica to the GIC increased the GIC's compressive, flexural and shear bond strength by 19.7%, 53.4% and 17.34%, respectively. It was claimed that the improved mechanical qualities could be justified by the tighter packing of the cement matrix that was made up of HA, silica and glass. As nano-HA-silica filler particles reached a specific percentage threshold, the material strength was found to be reduced at other percentages (Moheet et al. 2018).

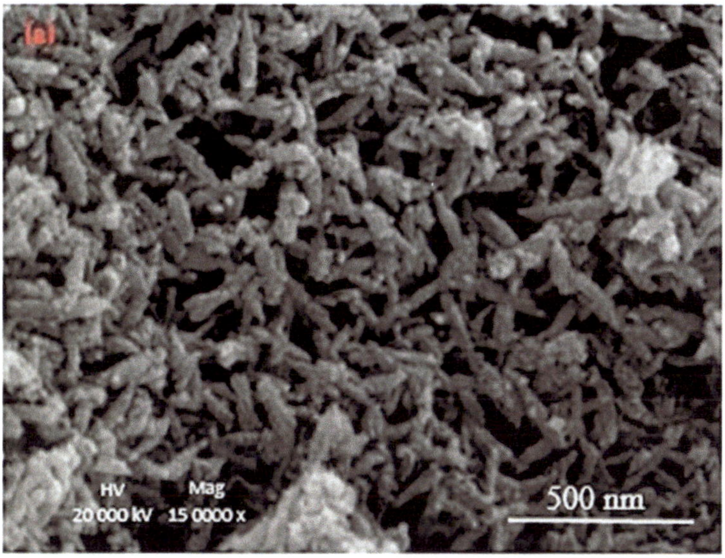

Fig. 21.2 SEM images of nano-HA-SiO_2 powder synthesised magnetic stirring at high magnification (150,000×)

21.4.3 Biocompatibility Study

Another ideal property of a dental material is that it should not injure oral tissues and should not include any poisonous or leachable substances that might get into the oral cavity, which can cause harm to patients, clinicians and other health personnel. As such, it is mandatory to conduct biocompatibility testing to ensure that the dental materials used are safe to use. Apart from cytotoxicity studies, one of the criteria that can be utilised to evaluate the biological properties of a material is to observe how the cells adhere to it. The behaviour of cells that come into contact with biomaterials is strongly related to their biocompatibility and it has been demonstrated that cell adhesion or cell attachment to material surfaces contributes to cell proliferation, migration and differentiation (Shie and Ding 2012). SEM has been proposed by Ahmed et al. (Ahmed et al. 2014) as a viable method of cell attachment and viability assessment. It is suggested that through the study of cell morphology and interactions between materials and cells, visualisation can be improved to provide information on assessing biocompatibility (Ahmed et al. 2014). Human dental pulp stem cells (DPSC) were utilised to assess the cytotoxicity of the nano-HA-silica GIC through the MTT test (Noorani et al. 2017; Hii et al. 2019). Due to its sensitivity, the MTT test is commonly used as a screening assay and is considered as the gold standard for cytotoxicity assays (Mosmann 1983). DPSCs were selected because they not only served as the target cells to mimic the clinical condition but also represented the crucial populations in the pulp tissue of the tooth that are typically in close proximity with the dental restorations. In addition, DPSCs played a key role in the healing process of damaged pulp. It was proven that nano-HA-silica-GIC had favourable and equivalent cell viability to cGIC and the cells grown on each material had healthy spindle morphology (Noorani et al. 2017). According to Ching et al. (Hii et al. 2019), as the concentration of the material extract decreased the cell viability of both nano-HA-silica GIC and cGIC increased. At lower concentrations which were 3.125, 6.25 and 12.5 mg/mL, the cell viability for cGIC was reported to be >89%, indicating that they were non-cytotoxic and slightly cytotoxic. Meanwhile, as the cell viability was greater than 90%, nano-HA-silica GIC proved to be non-cytotoxic to the cells at lower concentrations of extracts. The analysis of SEM micrographs revealed that DPSCs displayed a fibroblastic phenotype on both nano-HA-silica GIC and cGIC, which was a typical morphology of mesenchymal stem cell (MSC). It was reported that lamellipodia, filopodia and membrane ruffles were seen in the SEM micrographs of both nano-HA-silica GIC and cGIC (Hii et al. 2019). According to Small et al. (Small et al. 2002), these structures were crucial for cell motility, membrane domain organisation, phagocytosis and the establishment of substrate adherence, which essentially suggested favourable biocompatibility.

21.4.4 Dentinogenic and Odontogenic Differentiation Study

Researchers have long investigated the impact of nano-HA-silica-GIC on the development of DPSCs into odontogenic lineage (Ching et al. 2020). Real-time reverse transcription polymerase chain reaction (rRT-PCR) was used to assess the odontogenic differentiation of DPSCs. The following genes were examined in this study: osteocalcin (OCN), osteopontin (OPN), alkaline phosphatase (ALP), collagen type I (COL1A1), dentin sialophosphoprotein (DSPP), dentin matrix protein 1 (DMP1) and runt-related transcription factor 2 (RUNX2) (Ching et al. 2020). These genes were compared to the housekeeping gene glyceraldehyde-3-phosphate dehydrogenase (GAPDH). The investigated genes were believed to be associated and linked to certain roles. The development of dentin is regulated by DSPP, a small integrin-binding ligand N-linked glycoprotein (SIBLING) that is expressed more in the dentin than in the bone and it is typically regarded as a unique hallmark of odontoblast (Suzuki et al. 2009). On the other hand, DMP1 regulates the formation of the collagen matrix and the mineralisation of the dentin. It is expressed at the beginning of the odontoblast development process (Paduano et al. 2016; Eslaminejad et al. 2013). Gene OCN, a protein containing gamma-carboxyglutamic acid, is usually produced in the late stages of odontoblast and osteoblast differentiation (Cutarelli et al. 2016). OCN, which contains gamma-carboxyglutamic acid is usually expressed in the late stages of odontoblast and osteoblast differentiation (Cutarelli et al. 2016) while OPN has strong expression during the last phase of bone mineralisation (Khanna-Jain et al. 2012). The early stages of differentiation are marked by ALP, which is important for mineral deposition (Park et al. 2009; Hong et al. 2010). ALP is a significant marker during the early stage of differentiation and plays a significant role in mineral deposition (Park et al. 2009; Hong et al. 2010). COL1A1, a major collagen in dentin provides the structural foundation for cellular migration, proliferation and mineralisation. Osteoblastic and odontoblastic cells are expressed at all phases of development and throughout life (Yang et al. 2007; Sandberg et al. 1988). Furthermore, this particular gene is responsible for providing a template that regulates the deposition of calcium phosphate (Arana-Chavez and Massa 2004). On top of that, the early phases of tooth and bone formation are regulated by the RUNX2 gene. Apart from directly controlling the growth of teeth and bones, RUNX2 also controls these processes indirectly via signalling pathways associated with it, such as Osterix (Osx) (Chen et al. 2009). Based on the findings by Ching et al. (Ching et al. 2020), DSPP and DMP1 expression were higher on days 7 and 10, while OCN expression was higher on day 10. On the other hand, OPN and ALP expression was higher on day 14, while RUNX2 expression was higher on day 1. Meanwhile, from day 7 to day 14, COL1A1 showed a time-dependent rise in its gene expression. Despite the aforementioned time-dependent differences, the expressions between the nano-HA-silica-GIC and cGIC groups were equivalent at a dose of 6.25 mg/mL, indicating that nano-HA-silica-GIC promoted the odontogenic differentiation of DPSC. Figure 21.3 illustrates the testing that nano-HA-silica and/or nano-HA-silica GIC had undergone, which is discussed in this chapter.

21.5 Future Perspective

Apart from the investigations and findings stated earlier, this nano-HA-silica GIC material has undergone some other in-vitro studies related to its physical and chemical properties such as microleakage, colour stability, sorption testing, surface roughness, fluoride release, solubility testing and ionic exchange, respectively. All these investigations revealed favourable findings for nano-HA-silica GIC as compared to conventional GIC. An ongoing animal study is currently being conducted to investigate the effect of this nano-HA-silica GIC material on the Wistar rat together with a genotoxicity study. Preliminary unpublished data demonstrated that nano-HA-silica GIC displayed good tissue biocompatibility and better tertiary dentin deposition as compared to conventional GIC. These favourable findings would encourage and enhance the commercialisation potential of this material as it is stronger compared to conventional GIC and has other added-value properties that could be used as an alternative sustainable material in clinical dentistry. Nevertheless, up to date, the limitation of this material is that its clinical study has yet to be conducted. It is expected that the clinical studies would commence soon following the completion of the animal studies.

21.6 Conclusions

In conclusion, conventional GIC has been widely used in clinical dentistry as it exhibits great properties in terms of fluoride release, chemical adhesion to the tooth surface and a low coefficient of thermal expansion. Nevertheless, it has some

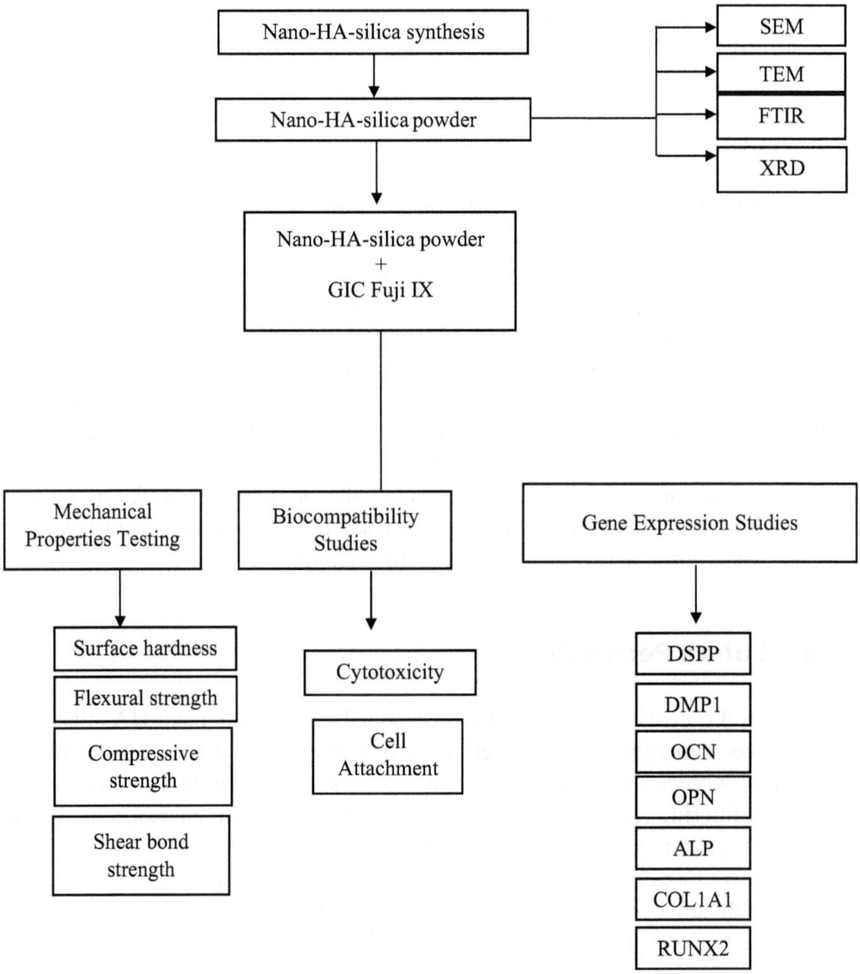

Fig. 21.3 Testings for nano-HA-silica and/or nano-HA-silica GIC

shortcomings that limit its use in clinical dentistry, especially in the high-stress areas. The shortcomings of conventional GIC have been greatly improved by the addition of specific additives, namely, nano-HA-silica. The incorporation of nano-HA-Si into the GIC matrix provided an outstanding biocompatible restorative material with enhanced mechanical properties, good cell survival and favourable cell attachment of dental pulp stem cells (DPSC). Furthermore, the findings from current study on gene expression provided factual support that nanoHA-silica-GIC contributed to the odontogenic differentiation of DPSCs therefore, it has the potential to be commercialised and used in clinical dentistry as a sustainable restorative material. However, in order to provide a final recommendation, a thorough assessment on the addition of nano-HA-silica into GIC is required in the future whereby the evaluation should entail both in vivo and clinical investigations.

Acknowledgement Several particular projects that involve the addition of nano-HA-silica into conventional GIC were supported by Research University Grant USM: 1001/PPSG/8012355, Universiti Sains Malaysia (USM), Malaysia.

Conflict of Interest The author declares no conflict of interest.

References

Ab Rahman I, Sam'an MM, Luddin N, Shiekh RA (2014) One-pot synthesis of hydroxyapatite–silica nanopowder composite for hardness enhancement of glass ionomer cement (GIC). Bull Mater Sci 37(2):213–219

Ahmed HM, Luddin N, Kannan TP, Mokhtar KI, Ahmad A (2014) Cell attachment properties of Portland cement–based endodontic materials: biological and methodological considerations. J Endod 40:1517–1523

Ali Ezz A, Ali MS, Hassan MY, Shaban AM (2022) Bioactivity, water sorption and solubility of sol-gel formulated glass-ionomer cement modified with nano-fluorapatite. J Int Med Dent Sci Res 5(1):11–20

Arana-Chavez VE, Massa LF (2004) Odontoblasts: the cells forming and maintaining dentine. Int J Biochem Cell Biol 36:1367–1373

Arcís RW, López-Macipe A, Toledano M, Osorio E, Rodríguez-Clemente R, Murtra J, Fanovich MA, Pascual CD (2002) Mechanical properties of visible light-cured resins reinforced with hydroxyapatite for dental restoration. Dent Mater 18(1):49–57

Arita K, Lucas ME, Nishiho M (2003) The effect of adding hydroxyapatite on the flexural strength of glass ionomer cement. Dent Mater J 22(2):126–136

Bowen RL, Marjenhoff WA (1992) Dental composites/glass ionomers: the materials. Adv Dent Res 6(1):44–49

Chen S, Gluhak-Heinrich J, Wang Y, Wu Y, Chuang H, Chen L, Yuan GH, Dong J, Gay I, MacDougall M (2009) Runx2, Osx, and Dspp in tooth development. J Dent Res 88:904–909

Ching HS, Luddin N, Kannan TP, Ab Rahman I, Abdul Ghani NRN (2018) Modification of glass ionomer cements on their physical-mechanical and antimicrobial properties. J Esthet Restor Dent 30:1–15

Ching HS, Kannan TP, Luddin N, Ab Rahman I, Ghani NRNA (2020) Early odontogenic differentiation of dental pulp stem cells treated with nanohydroxyapatite–silica–glass. J Polytech:1–14

Craig RG (2002) Restorative dental materials, 11th edn. Mosby, St. Louis, MO

Crisp S, Ferner AJ, Lewis BG, Wilson AD (1975) Properties of improved glass-ionomer cement formulations. J Dent 3:125–130

Cutarelli A, Marini M, Tancredi V, D'arcangelo G, Murdocca M, Frank C, Tarantino M (2016) Adenosine triphosphate stimulates differentiation and mineralization in human osteoblast-like Saos-2 cells. Develop Growth Differ 58:400–408

Davidson CL, Mjor IA (1999) Advances in glass ionomer cements. Quintessence Publishing Co, Berlin

Dionysopoulos D, Gerasimidiou O, Papadopoulos C (2022) Modifications of glass ionomer cements using nanotechnology: recent advances. Recent Prog Mater 4(2):1–17

Ericson D, Kidd E, McComb D, Mjör I, Noack MJ (2003) Minimally invasive dentistry—concepts and techniques in cariology. Oral Health Prev Dent 1(1):59–72

Eslaminejad MB, Bordbar S, Nazarian H (2013) Odontogenic differentiation of dental pulp-derived stem cells on tricalcium phosphate scaffolds. J Dent Sci 8:306–313

Felemban NH, Ebrahim MI (2016) Effects of adding silica particles on certain properties of resin-modified glass-ionomer cement. Eur J Dent 10:225–229

Griffin SG, Hill RG (2000) Influence of glass composition on the properties of glass polyalkenoate cements. Part II: influence of phosphate content. Biomaterials 21:399–340

Gu YW, Yap AUJ, Cheang P, Khor KA (2005) Effects of incorporation of HA/ZrO2 into glass ionomer cement (GIC). Biomaterials 26:713–720

Haghgoo R, Asgary S, Abbas FM, Hedeshi RM (2015) Nano-hydroxyapatite and calcium-enriched mixture for pulp capping of sound primary teeth: a randomized clinical trial. Iranian Endo J 10(2):107

Hammouda IM (2009) Reinforcement of conventional glass ionomer restorative material with short glass fibers. J Mech Behav Biomed Mater 2:73–81

Hii SC, Luddin N, Kannan TP, Ab Rahman I, Nik Abdul Ghani N (2019) The biological evaluation of conventional and nano-hydroxyapatite-silica glass ionomer cement on dental pulp stem cells: a comparative study. Contemp Clin Dent 10:324–332

Hong D, Chen HX, Yu HQ, Liang Y, Wang C, Lian QQ, Deng HT, Ge RS (2010) Morphological and proteomic analysis of early stage of osteoblast differentiation in osteoblastic progenitor cells. Exp Cell Res 316:2291–2300

Huang S, Gao S, Yu H (2009) Effect of nano-hydroxyapatite concentration on remineralization of initial enamel lesion in vitro. Biomed Mater 4(3):034104

Huang S, Gao S, Cheng L, Yu H (2011) Remineralization potential of nano-hydroxyapatite on initial enamel lesions: an in vitro study. Caries Res 45:460–468

ISO 9917–1 (2003) Dental water based cements. International Organization for Standardization, Geneva

Javed F, Vohra F, Zafar S, Almas K (2014) Significance of osteogenic surface coatings on implants to enhance osseointegration under osteoporotic-like conditions. Implant Dent 23(6):679–686

Kakani AK, Veeramachaneni C, Majeti C, Tummala M, Khiyani L (2015) A review on perforation repair materials. J Clin Diagn Res 9(9):ZE09

Kantharia N, Naik S, Apte S, Kheur M, Kheur S, Kale B (2014) Nano-hydroxyapatite and its contemporary applications. Bone 34(15.2):1–71

Kaushick BT, Jayakumar N, Padmalatha O, Varghese S (2011) Treatment of human periodontal infrabony defects with hydroxyapatite+ β tricalcium phosphate bone graft alone and in combination with platelet rich plasma: a randomized clinical trial. Ind J Dent Res 22(4):505

Khanna-Jain R, Vanhatupa S, Vuorinen A, Sandor G, Suuronen R, Mannerstrom B, Miettinen S (2012) Growth and differentiation of human dental pulp stem cells maintained in fetal bovine serum, human serum and serum-free/xeno-free culture media. J Stem Cell Res Ther 2:2

Lin KL, Chang J, Cheng RM, Ruan ML (2007) Hydrothermal microemulsion synthesis of stoichiometric single crystal hydroxyapatite nanorods with mono-dispersion and narrow-size distribution. Mater Lett 61:1683–1687

Lucas ME, Arita K, Nishino M (2003) Toughness, bonding and fluoride-release properties of hydroxyapatite-added glass ionomer cement. Biomaterials 24:3787–3794

Mabrouk SM, Beherei H, El-Gohary MI (2012) Effect of incorporation of nano bioactive silica into commercial glass ionomer cement (GIC). J Genet Eng Biotechnol 10:113–119

Mickenautsch S, Mount G, Yengopal V (2011) Therapeutic effect of glass-ionomers: an overview of evidence. Aust Dent J 56(1):10–15

Moheet IA, Luddin N, Ab Rahman I, Ponnuraj KT, Ghani NRNA (2018) Evaluation of mechanical properties and bond strength of nano-hydroxyapatite-silica added glass ionomer cement. Ceram Int 44(8):9899–9906

Moshaverinia A, Ansari S, Moshaverinia M, Roohpour N, Darr JA, Rehman I (2008) Effects of incorporation of hydroxyapatite and fluoroapatite nanobioceramics into conventional glass ionomer cements (GIC). Acta Biomater 4:432–440

Mosmann T (1983) Rapid colorimetric assay for cellular growth and survival: application to proliferation and cytotoxicity assays. J Immunol Methods 65:55–63

Murugan R, Farinawati Y, Nasruddin NS, Mohamad Anuar NN (2022) Effects of nanohydroxyapatite incorporation into glass ionomer cement (GIC). Fortschr Mineral 12(9): 1–10
Najeeb S, Khurshid Z, Zafar MS, Khan AS, Zohaib S, Martí JMN, Sauro S, Matinlinna JP, Rehman IU (2016) Modifications in glass ionomer cements: nano-sized fillers and bioactive nanoceramics. Int J Mol Sci 17(7):1134
Nicholson JW (1998a) Chemistry of glass-ionomer cements: a review. Biomaterials 6:485–494
Nicholson J (1998b) Glass-ionomers in medicine and dentistry. Proc Inst Mech Eng H 212(2): 121–126
Noorani TY, Luddin N, Rahman IA, Masudi SM (2017) In vitro cytotoxicity evaluation of novel nano-hydroxyapatite-silica incorporated glass ionomer cement. J Clin Diagn Res 11:ZC105–ZC109
Ong JL, Chan DC (2000) Hydroxyapatite and their use as coatings in dental implants: a review. Crit Rev Biomed Eng 28(5–6):667–707
Paduano F, Marrelli M, White LJ, Shakesheff KM, Tatullo M (2016) Odontogenic differentiation of human dental pulp stem cells on hydrogel scaffolds derived from decellularized bone extracellular matrix and collagen type I. PLoS One 11:e0148225
Park BW, Hah YS, Choi MJ, Ryu YM, Lee SG, Kim DR, Kim JR, Byun JH (2009) In vitro osteogenic differentiation of cultured human dental papilla-derived cells. J Oral Maxillofac Surg 67:507–514
Pelka M, Ebert J, Schneider H (1996) Comparison of two- and three-body wear of glass-ionomers and composites. Eur J Oral Sci 104:132–137
Pryor LS, Gage E, Langevin CJ, Herrera F, Breithaupt AD, Gordon CR, Afifi AM, Zins JE, Meltzer H, Gosman A (2009) Review of bone substitutes. Craniomax Traum Recon 2(3):151
Sandberg M, Autio-Harmainen H, Vuorio E (1988) Localization of the expression of types I, III, and IV collagen, TGF-β1 and c-fos genes in developing human calvarial bones. Dev Biol 130: 324–334
Shie MYHCC, Ding SJ (2012) Effects of altering the Si/Ca molar ratio of a calcium silicate cement on in vitro cell attachment. Int Endod J 45:337–345
Shiekh RA, Rahman IAB, Masudi SM, Luddin N (2014) Modification of glass ionomer cement by incorporating hydroxyapatite-silica nanopowder composite: sol-gel synthesis and characterization. Ceram Int 40:3165–3170
Sidhu SK, Nicholson JW (2016) A review of glass-ionomer cements for clinical dentistry. J Funct Biomater 7(3):16
Singh A, Daing A, Anand V, Dixit J (2014) Two dimensional alveolar ridge augmentation using particulate hydroxyapatite and collagen membrane: a case report. J Oral Biol Cranio Res 4(2): 151–154
Small JV, Stradal T, Vignal E, Rottner K (2002) The lamellipodium: where motility begins. Trends Cell Biol 12:112–120
Suzuki S, Sreenath T, Haruyama N, Honeycutt C, Terse A, Cho A, Kohler T, Müller R, Goldberg M, Kulkarni AB (2009) Dentin sialoprotein and dentin phosphoprotein have distinct roles in dentin mineralization. Matrix Biol 28:221–229
Wang J, Shaw LL (2009) Nanocrystalline hydroxyapatite with simultaneous enhancements in hardness and toughness. Biomaterials 30(34):6565–6572
Wilson AD, Kent BE (1972) A new translucent cement for dentistry. The glass ionomer cement. Br Dent J 132:133–135
Wilson AD, McLean JW (1988) Glass–ionomer cement. Quintessence, Berlin
Wilson AD, Nicholson JW (1993) Polyalkenoate cements, their biomedical and industrial applications. Cambridge University Press, pp 90–196
Yamamoto Y (1984) Study on hydroxyapatite-polyacrylic acid composite cement (hydroxyapatite-glass ionomer cement). J Jpn So Dent Mater Dev 3(6):787

Yang X, van den Dolder J, Walboomers XF, Zhang W, Bian Z, Fan M, Jansen JA (2007) The odontogenic potential of STRO-1 sorted rat dental pulp stem cells in vitro. J Tissue Eng Regen Med 1:66–73

Yli-Urpo H, Narhi M, Narhi T (2005a) Compound changes and tooth mineralization effects of glass ionomer cements containing bioactive glass (S53P4), an in vivo study. Biomaterials 26:5934–5941

Yli-Urpo H, Lassila LV, Narhi T, Vallittu PK (2005b) Compressive strength and surface characterization of glass ionomer cements modified by particles of bioactive glass. Dent Mater 21:201–209

Zanata RL, Magalhaes AC, Lauris JR (2011) Microhardness and chemical analysis of high viscous glass-ionomer cement after 10 years of clinical service as ART restorations. J Dent 39:834–840

Mapping the Ethical and Regulatory Issues of 3D Bioprinting Using Biomaterials in a Low- and Middle-Income Nation: Malaysian Perspectives

22

Mohammad Firdaus Bin Abdul Aziz

Abstract

The emergence of 3D bioprinting has the promise of transforming clinical medicine to alleviate a wide range of medical conditions and diseases. As with other emerging technologies, such as cell-based therapies, this advancement also poses ethical and regulatory issues that need to be examined to ensure the technology develops in an ethical fashion and protects the interests and safety of society. This study focuses on the potential applications of 3D bioprinting to create organs and tissues using biomaterials in Malaysia. Whilst this advancement promises to be clinically useful, especially in rectifying organ shortage issues and addressing the increasing number of elderly populations with degenerative diseases in the country, little is known if the existing regulatory framework is adequate to address ethical and regulatory issues that have been widely debated by scholars. This study aims to map out the associated ethical and regulatory issues from the perspective of a low- and middle-income nation such as Malaysia.

Keywords

Bioprinting · Biomaterials · Ethics · Regulation · LMICs · Malaysia

Abbreviations

2D	Two-dimensional
3D	Three-dimensional

M. F. B. A. Aziz (✉)
Centre for Law and Ethics Law in Science and Technology (CELEST), Faculty of Law, Universiti Malaya, Kuala Lumpur, Malaysia
e-mail: firdausaziz@um.edu.my

W. S. Wan Kamarul Zaman, A. A. Nurul (eds.), *Sustainable Material for Biomedical Engineering Application*, https://doi.org/10.1007/978-981-99-2267-3_22

AM	Additive manufacturing
CAD	Computer-aided design
CPR	Cardiopulmonary resuscitation
GMP	Good manufacturing practice
iPSCs	Induced pluripotent stem cells
LMICs	Low- and middle-income countries
NPRA	National Pharmaceutical Regulatory Agency
UDBHR	Universal Declaration of Human Rights

22.1 Introduction

The advancement of biomedical research has introduced many breakthrough discoveries and cutting-edge technologies that could potentially have a great impact on society, particularly in addressing and improving health. Particularly in regenerative medicine, where one of the advanced technologies is bioprinting. Differing from three-dimensional (3D) printing, bioprinting uses living cells as donor cells in developing artificial organs (Gilbert et al. 2018). This technology involves "the use of computer-aided transfer processes for patterning and assembling living and non-living materials with a prescribed two-dimensional (2D) or 3D organisation in order to produce bio-engineered structures serving in regenerative medicine and in pharmacokinetic and basic cell biology studies" (Guillemot et al. 2010, p. 1). It is worth noting that the concept of 3D printing was discovered in the 1980s by Charles Hull, who created a model with a layer-by-layer formation of resin, which is also known as additive manufacturing (AM) (ACFoAM Technologies and ACFoAMTSFo Terminology 2012). However, the concept has been further advanced by the adoption of computer-aided design (CAD) to construct replicas. The promising benefits of this technology are mainly due to the ability to carry out different print functions and structures, which have attracted many researchers and investors to apply this technology in different fields such as agricultural, food processing, biomedical and manufacturing industries (Pugliese et al. 2018; Garcia et al. 2018; Mason et al. 2019; Crisostomo and Dizon 2021). The advancement of this technology has also resulted in it being regarded as the new industrial revolution (Berman 2012).

As with many other countries, Malaysia also aspires to develop this technology, especially in the field of regenerative medicine. Even though Malaysia can be regarded as still far behind compared to developed nations in keeping with the current advances of regenerative medicine, there are numerous researchers in the country exploring the possibility of adopting regenerative medicine therapy in replacing conventional medicine that is currently unable to cure degenerative diseases. Regenerative medicine involves tissue engineering, cell therapy and biomaterials, also known as scaffolds that consist of natural and synthetic polymers. The utilisation of this technology can be regarded as an apt approach to address the increasing public health costs in addressing degenerative diseases in the country due to the increasing number of elderly population. In addition, given that researchers have shown that 3D bioprinting can be used to print solid organs such as heart, liver and lungs, this technology will also be useful in addressing the issue of organ shortages for organ transplantation, which is a global crisis and not unique to the Malaysian context.

It is also worth highlighting that Malaysia is committed to becoming one of the competitors in medical tourism. It is known as a hub of medicine in the region. With its sought-after local cuisines, well-known holiday destinations around the country, more than 35 well-established and reputable private hospitals equipped with a wide variety of state-of-the-art facilities, and well-trained and highly qualified healthcare professionals, it is in the interest of Malaysia to invest in and develop emerging technologies such as 3D bioprinting to ensure the country remains competitive in medicine. Notwithstanding its benefits, it has also been argued that it is a great challenge to utilise 3D printing technology to create biomedical products or devices using biomaterials which are biocompatible and of good quality to ensure that they are safe for use in the human body. Also, this technology is surrounded by an array of ethical and regulatory issues, which have been widely discussed by scholars. This chapter aims to explore the importance of this technology for Malaysia's medical industry in addressing regenerative diseases and organ shortages in the country and map out some of the ethical and regulatory issues from the Malaysian perspective, which may be shared by other low- and middle-income nations.

22.2 3D Bioprinting in Malaysia

3D printing and bioprinting have made their mark in Malaysia for a number of years, especially in the field of medicine. The development of this technology in Malaysia involves both foreign and local stakeholders. There have been a number of reports progress in developing bioprinting in Malaysia over the years, which involve players from different sectors, including universities, research institutions and industries. For instance, Stratasys, which is an international entity, introduced 3D printing technology to the Centre for Biomedical and Technology Integration, Universiti Malaya (Kumar 2013). The aim was to create custom biomodels that could be used by researchers for their research work and surgical training. Another example is Materialise, a foreign entity that has collaborated with Tuanku Mukhris Hospital of the National University of Malaysia in performing reconstruction of a patient's skull using its 3D-printed surgical guide technology (Lee 2015). A custom-made 3D-printed titanium identical to the patient's skull was implanted. Such a procedure was the first of its kind to be performed in the country. Materialise also collaborated with Pantai Medical Hospital, an established private hospital in Kuala Lumpur, the capital city of Malaysia (Park 2013). The collaboration has led to the utilisation of a virtual surgical planning and 3D-printed surgical guides to reposition a patient's fractured arm.

Another instance worth highlighting involves researchers from Universiti Malaya, a premier university in the country, who have been working on the development of bioprinting skin by looking at the suitability of the use of fibrin-gelatin in the technique (Hakam et al. 2016). Another researcher at Universiti Teknologi Malaysia explores the development of a blood vessel that can be beneficial for coronary heart surgery, which can be regarded as a life-saving technique and beneficial to the local population given that heart disease is one of the leading causes of death (Supriyanto 2018). In addition, 3D bioprinting can also be useful for physician training purposes. Researchers at Universiti Putra Malaysia created an anatomical model that has the same texture and structure as a real human organ, known as Medical Grade 3D Print, which can be used to train doctors in carrying out surgery (Zakaria 2018).

22.2.1 Moral Obligation to Develop 3d Bioprinting in Malaysia

Based on the reported statistics for 2020, there are 586,558 people with disabilities in the country. The breakdown of the disabilities includes 52,111 visually impaired cases; 40,319 hearing cases; 213,716 physical cases; 201,355 learning disabilities cases; 2996 speech cases; 48,644 mental cases; and the remaining 27,417 cases recorded as others (Malaysia DoS 2021). Based on these statistics, it is apparent that physical disability cases are the highest. It is important to note that these numbers are not representative of the actual cases in the country. This is because the registration of disabilities in Malaysia is not a mandatory requirement. The recorded cases are based on voluntary registrations. Therefore, it can be argued that the actual number of cases would be higher than the reported cases.

Given the promising benefits of 3D bioprinting and the statistics above, it can be argued that there exists a moral obligation among the researchers and the state to develop the technology, as it will be able to alleviate the suffering of people with disabilities, particularly those with damaged or injured organs or body parts. Among the people with disabilities that can benefit from this technology are those who suffer from chronic and end-stage kidney diseases as well as heart diseases. By applying the principle of utilitarianism, the development of this technology would be able to bring benefits to many of those in the population who suffer from physical disabilities. The statistics show that disabilities are not isolated cases in Malaysia and there are many people who suffer from them. Therefore, it is ethically justified and a moral obligation for the state to utilise public funds and attract private investors to invest and facilitate researchers in exploring and advancing this technology for the benefit of the population.

22.2.2 Prospect of 3D Bioprinting in Organ Transplant

As with many other countries across the globe, Malaysia has not been able to address the issue of organ shortages for many years. There are still many patients on the waiting list with no clarity as to when they will receive organ transplantation. Malaysia is reported as among being the lowest countries for kidney donation globally (Islahudin et al. 2021). Currently, there are more than 21,000 patients on the waiting list and given the low rate of donation in Malaysia; many of these patients will not be able to undergo the transplantation procedure. It was reported that the estimated cost incurred to treat end-stage kidney disease patients through dialysis in Malaysia by 2040 is approximately RM 4,381,708,760 (USD 1,062,490,000) annually, and the estimated cost for transplant procedure is RM 3,286,281,570 (USD 796,867,500). This figure shows that a transplant procedure will be more cost effective than relying on dialysis, which has some disadvantages such as infection and may incur other medical costs.

Even though kidney transplantation is regarded as the gold standard of treatment for end-stage kidney disease, it is of no use if the organ shortage issue still exists. Therefore, bioprinting kidneys would be a progress that is crucially needed (Hunsberger et al. 2016). It would be able to help individuals who are currently suffering from long-term kidney failure and required to undergo lifetime dialysis treatment with a very slim to almost no chance of receiving an organ transplant. This would be a pivotal motivating factor for the government to invest in and develop bioprinting technology so that patients with malfunctioning organs do not have to rely on organ transplant and have access to bioprinted versions of organs that can be created using their own cells as alternatives. Such an option will also avoid the issue of immune rejection, which is a material risk in organ transplantation. In view of the highlighted dire needs, one would argue that 3D bioprinting would be an option that should be explored and developed to address the issue of organ shortages.

22.2.3 Addressing Degenerative Diseases with 3D Bioprinting

The number of patients with degenerative diseases is on the rise given the increasing number of elderly populations in Malaysia (Academy of Sciences Malaysia 2013). Regenerative diseases that are common among the elderly patients include heart diseases, diabetes, Alzheimer's, Parkinson disease and renal disease. This situation has resulted in the rise of public healthcare costs in the country. Realising this issue, the Malaysian government acknowledged the importance of developing regenerative medicine therapy to address degenerative diseases with the hope that in return such an effort would reduce the expenditure in public healthcare (Academy of Sciences Malaysia 2013).

According to the Department of Statistics Malaysia, ischaemic heart diseases remain the principal cause of death in Malaysia, with 17% of the 109,155 medically certified deaths recorded in 2020 (Malaysia DoS 2021). The number increased from 15% of cases reported a year before, in 2019. Patients with heart diseases are also regarded as people with disabilities and can be registered with the Social Welfare Department, Malaysia. Other types of heart diseases commonly reported in Malaysia include heart valve disease, heart failure, and heart arrhythmia (Sentral 2022). It can be argued that with the application of bioprinting technology, this number could have been reduced, if heart patients' condition can be treated with bioprinted hearts. Also, patients would not have to wait for a long time for a heart donor to become available. Similar to the argument made for kidney patients, bioprinted hearts can be created using cells donated by heart patients and this will therefore be able to minimise the risk of heart rejection as compared to heart transplant.

22.3 Ethical Issues

Notwithstanding the promising benefits, this technology is surrounded by an array of ethical issues. This includes ethical questions such as whether or not patients' autonomy is respected; the morality issue of using certain sources, such as stem cells for organ biofabrication; the safety issue of using certain sources such as induced pluripotent stem cells (iPSCs); the benefits and risks of the technology; whether or not the application of this technology has crossed the boundary between therapy and enhancement; and distributive justice.

22.3.1 Consent

The principle of consent is the cornerstone of research ethics involving human subjects. Any research activity in advancing or developing a technology needs to observe this principle. It is particularly important after what has transpired in the past within the realm of medical research whereby the autonomy of patients was not respected (Reverby 2012). They were treated as a means to an end in research activities with the focus only given to the research goal. This unethical issue has led to the introduction of a number of international conventions and declarations that highlight the importance of safeguarding and protecting the autonomy and well-being of patients and research subjects (Code 1949; Vollmann and Winau 1996; Friesen et al. 2017).

Even though consent is a recurring issue in any medical research, it is worth highlighting that such an issue in the realm of 3D bioprinting is unique due to the "digitalisation" of a human person and therefore, it extends the application of consent from a physical world to a digital space. This is because the biofabrication technique used to print organs is based on digital models. Therefore, one would argue that in the future, the advancement of this technique will lead to actual organs being replaced with digital models. Thus, such a notion has resulted in a concern over the possibility of disappearing boundaries between the real and physical worlds and the digital space. There are a number of potential ethical conundrums involving consent that could materialise. For instance, there will be an issue of ownership of digital models. Who has the ownership of the digital models and the rights over the use of the models? If the models are owned by its developer under the pretext of patent rights, can they be used without the consent of the patients from which the model was created? Is this morally acceptable? Can the models be patented and commercialised without informing or getting consent from patients? If the patenting of the models is acceptable, another question that could arise is, can the patients claim any benefits from the commercialisation activities and patent? In addition, the digital models will contain personalised human genetic data. This raises another important issue namely the privacy and confidentiality of the patients involved. Are researchers aware of this issue? What are the necessary measures needed to address this issue?

These questions ought to be considered by any state because they touch upon the concept of trust towards researchers and physicians, as well as human dignity. States have the duty to ensure that the principle of consent is adhered to. It has been clearly stipulated in many conventions and declarations that informed consent is a vital requirement for any medical or research interventions. If these principles are not respected, patients would lose the trust that they have towards the state and other relevant key stakeholders such as researchers and physicians. Without the trust, researchers would lose a very important group of stakeholders in developing the technology.

Another important ethical issue is the practice of informed consent. Given that the utilisation of this technology is 'personalised', it would be difficult to ascertain the benefits and risks of this technology prior to its application (the issue of benefits and risks is further discussed in the latter part of this chapter). Therefore, the approach to the practice of informed consent would be complex and difficult (Vermeulen et al. 2017). The true meaning of 'informed consent' would not be observed, if the material risks and potential adverse events were unknown. Currently, there is no legislation specifically established to regulate medical research in Malaysia. In ensuring research is being developed ethically, the existing regulatory measure relies on the research ethics committee. Since this technology is still in its early stage of its development in the country, it is uncertain whether or not the ethics committee members are well versed to effectively review research protocols to ensure consent is being adequately adhered to. It can be argued that this issue would be a shared concern in any LMICs whereby the technology is still underdeveloped.

22.3.2 Sources of Bioinks

Stem cells are important sources of bioinks in the biofabrication of human tissues and organs to determine the characteristics of the desired bioprinted tissues and organs. As has been widely debated in the field of stem cell research and therapy, the most versatile stem cells are embryonic stem cells. Notwithstanding, the use of these cells has generated the most contentious legal and ethical arguments in the area of stem cell technology. This is due to the fact that embryos will be destroyed in the course of the research activities in order to retrieve embryonic stem cells. Arguments against such an activity involve legal and moral questions on the status of human embryos—whether or not human embryos possess legal and moral status equivalent to an actual human person. Suffice to say that the question on the morality of the use of human embryos and its inevitable destruction has been the subject of never-ending debate, which has not reached any consensus view to this date. In addition, there are other associated arguments, which include the issue of abortion, embryo donation and consent, the use of embryonic stem cells for commercial purposes and patents.

In an effort to address the ethically complex issues surrounding the use of embryonic stem cells, it has been proposed that induced pluripotent stem cells (iPSCs) can be an alternative option. However, this technique is still not an established or fault-proof because the use of cell reprogramming such as iPSCs has the risk of tumorigenicity. Researchers are still unable to control its differentiation, and this may lead to the formation of cancerous tissue (Vermeulen et al. 2017). It is uncertain whether such a technique is an option in Malaysia because many of the local stem cell researchers mainly use adult stem cells (Academy of Sciences Malaysia 2013). Other options that have been proposed that can be used as bioink include animal stem cells or xenogeneic cells. Even though the use of these cells can save researchers from the ethical conundrum associated with embryonic stem cells, it does raise another set of ethical and safety issues. For instance, one would argue against it because of religious perspectives and it is also uncertain whether it is safe to insert animal cells into the human body due to the immune system.

Currently, the legal and moral status of human embryos in Malaysia is not clearly stated in any legislation. The primary regulatory document that regulates the use of human embryos and their derived stem cells is the national guideline on stem cell research and therapy 2019. The guideline clearly stipulates that only the use of surplus human embryos for research purposes is permitted. Any creation of human embryos and admixed embryos for research purposes is prohibited. It can be argued that there is no clarity on the status of human embryos in the country but it does suggest that human embryos are provided with some level of regulatory protection and deserve moral regard that they can't be treated as just another research material in a laboratory. As for the use of animal cells, the national guideline clearly stipulates that xenotransplantation is prohibited. This shows that 3D bioprinted organs created using animal cells are not allowed to be transplanted into a human body. Nonetheless, the monitoring and enforcement of the guiding rules on research work involving stem cells is also of ethical concern, for instance, a private clinic in Malaysia has been offering stem cell injection using animal cells and this has caused great concerns over the enforcement and monitoring system in the country, which has been reported in an empirical study (Aziz et al. 2018).

22.3.3 Benefits Versus Risks

In any research activity and clinical translation or intervention, it is a moral and legal duty of researchers or physicians to inform their patients or research participants of the benefits and risks involved. It is also their responsibility to ensure that the benefits of an intervention weigh over the potential risks before it can be carried out.(World Medical Association 1964) (para 16).

In a conventional pathway for introducing a new clinical intervention or application, it will usually be tested through clinical trials. However, this usual pathway is not applicable or appropriate to developing 3D bioprinting technology and determining its safety and efficacy. This is due to the nature of its development and application. The biofabrication of 3D printed organs is individually tailored for a particular individual depending on his or her condition and the biomaterials are personalised. Therefore, the design of any study is different in each case and tailored to a specific person. This means each 3D bioprinted organ is unique and therefore, the result of a trial or study cannot be generalised. Based on its personalised nature, it is not practical or appropriate to test it using the standard clinical trial. The only option is to test its safety and efficacy in a personalised trial. For this reason, it has been argued that the development of experimental studies on 3D bioprinted organs is promising in theory, however, it is a difficult task in a real life because the safety measures designed for the custom-made organs to mitigate associated risks cannot be applied or tested on other individuals and need to be evaluated on a case-to-case basis (Gilbert et al. 2018; Otto et al. 2016).

This in turn, will affect the practice of informed consent as highlighted earlier and raises further questions with regard to patients' safety, which include how to determine the ratio of benefits and risks in a personalised trial that is traditionally determined in the standard clinical trial; what would be the approach to the usual inclusion and exclusion criteria required by the ethics approval; is the withdrawal provision applicable given the 3D bioprinted organs will be implanted into a patient's body and it would be impractical and harmful to the patients to reverse the procedure, and last but not the least, are the research ethics committees in the country equipped with appropriate knowledge and training to review a study of this nature?

22.3.4 Therapeutic Versus Enhancement

Any attempt to develop emerging technologies for therapeutic application is often being associated with the issue of "human enhancement" and an ethical question as to whether or not the aspiration to develop these technologies are morally valid. In the area of 3D bioprinting, which enables humans to create their own organs have led to a number of questions that require ethical and possibly religious discourse: Do we really need to print new organs and what is the purpose? Are we going against nature? Does this revolutionary technology not disrupt nature? Are we not playing God?

On the one hand, one would argue that the answer to these questions is rather obvious. This technology is rather useful and beneficial to help people who are suffering from degenerative diseases and other medical conditions, which are incurable by conventional medicine. For instance, a heart or kidney disease patient would need to undergo a number of invasive procedures in his or her lifetime because the diseased organ would continue deteriorating. The patient would 1 day reach a stage where there is no further treatment that can be offered to save his or her organ and life. The only option is to let nature take its course, but in some cases, patients in this situation may suffer physically and emotionally throughout the period.

In view of this scenario, one would then argue that it should be morally acceptable or even obligatory to make use of the knowledge and technology to help those who are in need by alleviating their suffering. One would not have to experience pain and suffer, if a diseased organ could be replaced with a new one. This is even more so paramount given the fact that the organ transplantation, which is the usual pathway to address diseased organs is unreliable because of the never-ending challenges of organ shortage for transplantation purposes (Saidi and Kenari 2014). As mentioned earlier, there are many patients who are dying while being put on a waiting list for an organ and this is a global issue.

On the other hand, there is also a concern of 'slippery slope' in allowing the creation of man-made organs for 'therapeutic use' as this could also lead to 'human enhancement'. How do we distinguish the use of 3D bioprinted organs between the two aspects? Technically, it is possible for bioprinting technology to be designed as 'technological immortality' program and therefore fall within the concept of 'human enhancement'. It can be used not only to rejuvenate one's internal organs by printing new organs to replace the old organs, but it can also be used to stop someone from ageing on the outside by producing new skin to maintain a youthful look. This could 1 day replace the existing procedures commonly used, such as botox and laser treatment. It is therefore important to closely monitor the application of this technology. The scope of its use could be wider than just for therapeutic purposes and this could lead to complexity in distinguishing between therapy and human improvement. Such a possibility is a valid concern, especially for countries such as Malaysia that do not have established regulation for emerging technologies.

22.3.5 Distributive Justice

Another associated ethical concern surrounding an emerging and cutting-edge therapeutic procedure is accessibility: whether or not members of the public at large will be able to afford the treatment or if it is only accessible by the few privileged segments of society. Article 27 of the Universal Declaration of Human Rights (UDBHR) asserts that "everyone has the right to share in scientific advancement and its benefits" (Nations 1948). This suggests that new technologies must be made available to everyone at a reasonable price. Therefore, it is the duty of the government to intervene, if a scientific product or new treatment is being made available in the market at an exorbitant price.

While 3D bioprinting is being developed in Malaysia with promising benefits to cure illnesses and help people with damaged and diseased organs or body parts, it is not clear how much the procedure involving biofabricated organs may cost. Given the many different forms of materials used in the biofabrication of an organ and the complexity of the process that may require individual assessment and interventions, this in turn will be reflected in the price of the end product. It is understandable that the cost will be high and may vary depending on the type of materials used. Therefore, the end product would only be available to people of means (Vermeulen et al. 2017; Vijayavenkataraman et al. 2016). As a developing country, Malaysia needs to ensure that the cost of bioprinted organs and tissues is lower than the existing procedures of organ transplant and reconstructive surgery so that the process is viable and the end products will be affordable. For instance, in the year 2015, the reported average lifetime discounted costs for kidney transplantation for adults in Malaysia from pre transplant work up to death were USD 119702 (MYR 409921) for adults (living kidney transplant) and USD 147152 (MYR 503922) for adults (diseased kidney transplant) (Bavanandan et al. 2015).

The Special Rapporteur of the UN General Assembly states:

> developing countries should prioritise the development, importation and dissemination of simple and inexpensive technologies that can improve life or marginalised populations. Industrialised states should comply with their international legal obligations by means of direct aid and the development of international collaborative models of research and development. (UNGA 2012, p. 20)

It can be argued that it is morally imperative for the government and researchers to ensure the types of biomaterials and bioprinters used are affordable and supersede the current price of conventional treatment, especially if public taxpayers' money is spent to develop the technology. A consideration must also be given to the issue of ownership due to patenting if the development of the product involves a commercial entity, as this may as well determine if there is any restricted use. It is vital for low- and middle-income nations such as Malaysia to ensure the principle of distributive justice is adhered to when developing emerging therapies so that the benefits are available to the public at large and not restricted to certain groups in the society.

22.4 Regulatory Issues

In Malaysia, there is no specific law that regulates regenerative medicine, either for clinical applications or research activities. Currently, the players in this area rely on the relevant regulations and guidelines. If stem cells are used in the production of bioprinted organs, it will fall under the purview of the guidelines on the use of stem cells for research and therapy, which were established in 2006 and amended in 2009 by the Ministry of Health (Malaysia, Guidelines for Stem Cell Research and Therapy 2009). Another relevant guiding rule would be the guiding notes for cell and gene therapy products manufacturing facilities established by the National Pharmaceutical Regulatory Agency (NPRA), which is the authority responsible in regulating pharmaceutical and cosmetic products in the country. The guiding notes require facilities to comply with the established good manufacturing practice (GMP). However, further examination is required to see how suitable these relevant guiding rules are with regard to the development of 3D bioprinted organs in Malaysia. The following sections touch upon a couple of issues relating to regulation that are worth noting in this chapter.

22.4.1 Risks Regulation and Responsibility

One of the key aspects of regulating the application of 3D bioprinting is determining the risks and responsibility for the end product. A question arises about who will be responsible for the safety and quality of the end product of this technology. This is because the development of a product will involve a few players, who play a key role in the production chain. This chain involves model designers, biologists, engineers, medical practitioners, and ethics committee members to name a few. Their input and role are vital to ensure that the end products are developed in a safe and responsible manner before they can be commercialised and introduced into a clinical setting.

Currently, in Malaysia, there is no comprehensive regulation for the 3D bioprinting process. Any new medical product has to go through the standard pathway. But it is not clear how this standard pathway is suitable for this technology. Will the product be subject to the existing risk-based assessment adopted by the NPRA to regulate cell-based products? In addition, there is also a need to determine who is responsible for the quality control. Will it be the responsibility of the regulators, 3D bioprinting providers or the medical entities that offer the service to their patients? Who should be responsible for the quality control, and who should be liable if there are any claims involving implanted bioprinted organs?

The transplantation of 3D bioprinted organs is not covered by the existing Human Tissues Act of 1974. The act only governs the "use of parts of human bodies of deceased persons for therapeutic purposes and for purposes of medical education and research" (Malaysia, Human Tissues Act 1974). It is important to determine how to regulate the implantation of 3D bioprinted organs because these organs are artificial organs and it is uncertain if it is appropriate to include them within the ambit of the existing regulation, which regulates natural tissues and organs. When human cells are involved, it may be relevant to also refer to the national organ, tissue and cell transplantation policy. This policy recognises that cell "transplantation holds the promise of treating many human diseases". (Malaysia, National Organ, Tissue and Cell Transplantation Policy 2007, para 9.1) Currently, the regulation of this area requires the involvement of the National Stem Cell Committee as one of the expert committees under the National Transplantation Technical Committee. (Malaysia, National Organ, Tissue and Cell Transplantation Policy 2007, para 9.3), However, it is uncertain whether 3D bioprinting is within the scope of the policy and the current committee members have adequate knowledge on 3D bioprinting.

Based on the discussion above, it shows that there is a need for Malaysian authorities, especially the Ministry of Health and relevant policymakers to look into the ambit of the existing regulation and the competence of relevant committees. It is worth highlighting that the application of stem cell therapy has existed in Malaysia for many years, but a comprehensive regulatory framework has not been developed to this date. With the emergence of other technologies, it would be even more difficult for Malaysia to catch up with the associated regulatory issues. In doing so, it is pivotal for the government to strike the balance in regulating regenerative medicine, which has been argued in a study focusing on stem cell research and therapy (Aziz et al. 2022). Regulation that is too strict may hinder progress, and if it is too lax, it may lead to unethical conduct and put public safety to risk. The same argument also applies for 3D bioprinting (Kirillova et al. 2020).

22.4.2 Commercialisation and Patenting

Another pertinent issue in regulating 3D bioprinting technology is the commercialisation and patentability of artificial human organs. The global approach to commercialisation and ownership of human organs and body parts has been widely discussed and regarded as ethically and morally unacceptable (Wilkinson 2004; Goold et al. 2014). Many countries across the globe have introduced law and guidelines to prohibit such an activity. The Malaysian national transplantation policy explicitly states that "the commercialisation of organ, tissue and cell transplantation and any act that may indirectly promote or lead to commercial transaction are prohibited". (Malaysia, National Organ, Tissue and Cell Transplantation Policy 2007, para 3.2). However, since 3D bioprinted organs are artificial organs and far from the actual human organs, one would argue that the position on commercialisation of 3D bioprinted organs should differ from the position on the actual human organs and it requires engagement among the relevant stakeholders in the country to decide a position that is apposite to the local context.

In addition, there is also an issue of commercialisation and ownership of digital models of the bioprinted organs. A question arises as to who will own the digital models? An example is the modern-day cardiopulmonary resuscitation (CPR) Annie. It is widely known that the origin of the model is from the body of an unknown person, a drowned woman who was pulled from the Seine in Paris. A death mask was made to identify the woman's origin. However, her image has since been used as a CPR model for many years. Using this as an example, one would argue whether it is ethical to use the face and the ownership of the face belongs to the manufacturer of the CPR dolls and the same question can be applied for the ownership of the digital models of bioprinted organs.

As for the issue of patenting or other intellectual property rights of bioprinted organs, it can be argued that these organs can be patented. This is because the 3D bioprinted organs are man-made, created with creative and highly skilled techniques and processes. This falls within the statutory definition of patentable inventions stated under section 11 of the Patent Act, which stipulates "an invention is patentable, if it is new, involves an inventive step and is industrially applicable" (Malaysia, Patents Act 1983). Also, they are created outside the human body. Therefore, it can be argued that intellectual property rights over such an invention and commercialisation of the product should be allowed to some extent. This is to encourage research and development in this area. However, section 13(1)(d) of the Patent Act stipulates "methods for the treatment of human or animal body by surgery or therapy, and diagnostic methods practised on the human or animal body" shall not be patentable (Malaysia, Patents Act 1983). Nonetheless, it is important to note that once the bioprinted organs are implanted, they need to be regarded as part of the human body and therefore the permissibility of their commercialisation shall cease.

It should not be ethical and legal for the recipient of the bioprinted organs to remove and sell the organs. However, such a concern is yet to be clearly addressed through law and regulations.

22.5 Conclusion

As has been discussed above, it is clear that Malaysia will benefit from developing 3D bioprinting particularly to address the issues of the increasing number of elderly patients with degenerative diseases, the well-being of people with disabilities, and the shortages of organ supply for transplantation. Notwithstanding the promising benefits, there are a number of ethical and regulatory issues that are yet to be clearly addressed in the Malaysian context. This includes the need to develop appropriate informed consent practice due to the unique nature of the technology that involves digital space; establish authority with members who are expert in this technology to develop ethical standards to ensure proper regulation of the use of different bioinks, clear distinction between the use of 3D bioprinting for 'therapy' and 'enhancement', ensuring the advancement of this technology takes into account safety and quality of the end products without jeopardising human rights and dignity while ensuring affordability and accessibility; as well as establish regulation that clearly addresses commercialisation, ownership, patentability, and the responsibility of relevant stakeholders taking into account human rights and dignity.

It is understandable that a low- and middle-income nation, such as Malaysia would have great challenges to translate this revolutionary technology from the laboratory to the market due to the ethical concerns and regulatory issues discussed above. While it may be difficult to establish specific regulation due to the fact that the technology develops rather rapidly and the development brings about new and emerging ethical and legal challenges and inevitably the associated regulatory costs, which might not be ideal for a LMIC nation, there is a great need to address both ethical and regulatory issues. Government needs to initiate regulatory interventions to address the lack of thereof. This can be done by improving the existing regulatory framework. In doing so, further investigation and engagement with relevant stakeholders are highly required. This is particularly vital to ensure that the future of this technology and the opportunity for Malaysia to benefit from this technology will not be disrupted by the lack of ethical standards and clarity in the regulation.

Conflict of Interests The author declares no conflict of interests.

References

Academy of Sciences Malaysia (2013) Advisory report on stem cells: ageing and regenerative medicine. Academy of Sciences Malaysia

ACFoAM Technologies, ACFoAMTSFo Terminology (2012) Standard terminology for additive manufacturing technologies. ASTM International

Aziz MFBA, Morrison M, Kaye J (2018) Regulating human stem cell research and therapy in low-and middle-income countries: Malaysian perspectives. New Genet Soc 37(1):2–20

Aziz MFBA, Morrison M, Kaye J (2022) Creative regulatory practices to develop stem-cell technology: the way forward for Malaysia. Regen Med 17(2):91–105

Bavanandan S et al (2015) The cost and utility of renal transplantation in Malaysia. Transplant Direct 1(10):e45

Berman B (2012) 3-D printing: the new industrial revolution. Bus Horiz 55(2):155–162

Code N (1949) The Nuremberg Code. Trials of war criminals before the Nuremberg military tribunals under control council law. 10(2):181–182

Crisostomo JLB, Dizon JRC (2021) 3D printing applications in agriculture, food processing, and environmental protection and monitoring. Adv Sustain Sci Engineer Technol (ASSET) 3(2): 0210201

Friesen P et al (2017) Rethinking the Belmont report? Am J Bioeth 17(7):15–21

Garcia J et al (2018) 3D printing materials and their use in medical education: a review of current technology and trends for the future. BMJ Simul Technol Enhanc Learn 4(1):27

Gilbert F et al (2018) Print me an organ? Ethical and regulatory issues emerging from 3D bioprinting in medicine. Sci Eng Ethics 24(1):73–91

Goold, I., et al., The human body as property? Possession, control and commodification. 2014, Inst Med Ethic 40(1):1–2

Guillemot F, Mironov V, Nakamura M (2010) Bioprinting is coming of age: report from the international conference on bioprinting and biofabrication in Bordeaux (3B'09). Biofabrication 2(1):010201

Hakam MS et al (2016) Evaluation of fibrin-gelatin hydrogel as biopaper for application in skin bioprinting: an in-vitro study. Biomed Mater Eng 27(6):669–682

Hunsberger J et al (2016) Bioengineering priorities on a path to ending organ shortage. Curr Stem Cell Rep 2(2):118–127

Islahudin F, Shahdan IA, Kua LM (2021) Kidney donation: bridging the gap in the shortage of kidney transplants in Malaysia. J Health Res 36(2):354–364

Kirillova A et al (2020) Bioethical and legal issues in 3D bioprinting. Int J Bioprint 6(3):272

Kumar A (2013) Malaysian biomedical centre's "major breakthrough" using stratasys 3D printing. CIO Asia, p 14

Lee M (2015) 3D printing may save your life if you need risky surgery. Star Online, p 25

Malaysia DoS (2021) Statistics on causes of death, Malaysia. 20 Sept 2022. Available from: https://www.dosm.gov.my/v1/index.php?r=column/cthemeByCat&cat=401&bul_id=R3VrRUhwSXZDN2k4SGN6akRhTStwQT09&menu_id=L0pheU43NWJwRWVSZklWdzQ4TlhUUT09

Malaysia, Guidelines for Stem Cell Research and Therapy (2009) Available at https://www.moh.gov.my/moh/resources/Penerbitan/Perkhidmatan%20OnG%20&%20Ped/STEM%20CELL/1._Guideline_For_Stem_Cell_Research_And_Therapy_.pdf

Malaysia, Human Tissues Act 1974 (Act 130) Available at https://www.moh.gov.my/index.php/database_stores/store_view/11?mid=31

Malaysia, National Organ, Tissue and Cell Transplantation Policy (2007) Available at https://www.moh.gov.my/moh/resources/auto%20download%20images/586f5007d800a.pdf

Malaysia, Patents Act 1983 (Act 291) Available at https://www.myipo.gov.my/wp-content/uploads/2016/08/Patents-Act-1983-Act-291.pdf

Mason J, Visintini S, Quay T (2019) An overview of clinical applications of 3-D printing and bioprinting. In: CADTH issues in emerging health technologies. Canadian Agency for Drugs and Technologies in Health, Ottawa, ON

Nations U (1948) Universal declaration of human rights

Otto I et al (2016) Ethical considerations in the translation of regenerative biofabrication technologies into clinic and society. Biofabrication 8(4):042001

Park R (2013) 3D printed surgical guides make their Malaysian debut. 3D Printing Industry
Pugliese L et al (2018) The clinical use of 3D printing in surgery. Updat Surg 70(3):381–388
Reverby SM (2012) Ethical failures and history lessons: the US Public Health Service research studies in Tuskegee and Guatemala. Public Health Rev 34(1):1–18
Saidi R, Kenari SH (2014) Challenges of organ shortage for transplantation: solutions and opportunities. Int J Organ Transplant Med 5(3):87
Sentral CV (2022) Heart disease in Malaysia. Available from: https://www.cvskl.com/heart-disease-in-malaysia/#:~:text=According%20to%20Department%20of%20Statistic,2020%20(Statistics%2C%202021
Supriyanto I (2018) Cardiovascular and biomedical research at Universiti Teknologi Malaysia. 20 Sept 2022. Available from: https://webarchive.nationalarchives.gov.uk/ukgwa/20170712142059/https://www.mrc.ac.uk/documents/pdf/universiti-teknologi-malaysia/
UNGA (2012) Report of the special rapporteur in the field of cultural rights, Farida Shaheed: the rights to enjoy the benefits of scientfic progress and its applications. New York
Vermeulen N et al (2017) 3D bioprint me: a socioethical view of bioprinting human organs and tissues. J Med Ethics 43(9):618–624
Vijayavenkataraman S, Lu W, Fuh JYH (2016) 3D bioprinting–an ethical, legal and social aspects (ELSA) framework. Bioprinting 1:11–21
Vollmann J, Winau R (1996) Informed consent in human experimentation before the Nuremberg code. BMJ 313(7070):1445–1447
Wilkinson S (2004) Bodies for sale: Ethics and exploitation in the human body trade. Routledge
World Medical Association (1964) Declaration of Helsinki - ethical principles for medical research involving human subjects, WMA General Assembly, Editor
Zakaria A (2018) Medical grade 3D print allows doctors simulate patients' surgery. 20 Sept 2022. Available from: https://rmc.upm.edu.my/news/medical_grade_3d_print_allows_doctors_simulate_patients_surgery-39913

Ethical and Regulatory Considerations for Sustainable Practices in Biomedical Applications

23

Yogeswaran Lokanathan and Angela Ng Min Hwei

Abstract

A wide range of biomaterials, both of natural and synthetic origin, have been studied for utilization in biomedical applications. The tremendous growth in science and technology of material science and biomedical engineering has enabled the influx of different kinds of materials, either obtained from natural sources or chemically synthesized, through various fabrications and bioprocessing techniques to be used in patient care. The progress in this field also raises many concerns and challenges from the ethical and legal aspects that need to be addressed to ensure its sustainability. This chapter discusses the ethical and regulatory factors including but not limited to the responsible and efficient use of resources, green synthesis, waste management, impact on the economy, society, and environment, and the legislation on a sustainable design approach that needs to be considered to ensure the sustainability of the materials and also its utilization in the biomedical field in the future.

Keywords

Sustainability · Regenerative medicine · Green synthesis · Circular economy · Waste reduction · Biomaterial · Biodegradable

Y. Lokanathan (✉) · A. N. M. Hwei
Centre for Tissue Engineering and Regenerative Medicine, Faculty of Medicine, Universiti Kebangsaan Malaysia, Cheras Kuala Lumpur, Malaysia
e-mail: lyoges@ppukm.ukm.edu.my

W. S. Wan Kamarul Zaman, A. A. Nurul (eds.), *Sustainable Material for Biomedical Engineering Application*, https://doi.org/10.1007/978-981-99-2267-3_23

Abbreviations

COVID-19	Coronavirus disease 2019
ECM	Extracellular matrix
EVs	Extracellular vesicles
FFP	Filtering facepiece
HLA	Human leukocyte antigen
MSCs	Mesenchymal stromal/stem cells
PPE	Personal protective equipment
SDG	Sustainable developmental goals
UN	United Nations

23.1 Introduction

Regenerative medicine is gaining attention in the medical fraternity as the future of medicine. Central to the concept of regenerative medicine is harnessing the body's natural regenerative or healing properties after cell/tissue/organ damage. In most cases, this can be achieved by enhancing the body's natural healing properties via nutrition, removing barrier such as scars that hampers the body's ability to heal or more commonly, introducing biologics such as cells or cell derivatives and tissue matrix to restore the original state of tissue or organ health. The practice of blood or bone marrow transfusion and tissue or organ grafting or transplantation can be seen as a pioneering success in regenerative medicine. Today, with the advancement in cell culturing techniques and stem cell science, allogenic cells can be cultured, propagated, and manipulated outside the body, and materials similar to components of extracellular matrix (ECM) can be synthesized and allogenic/xenogeneic materials can be decellularized. As the industry in this field gains greater momentum and the sustainability issues of human endeavors receive greater emphasis amid the climate change threat, the ethics of adopting or switching to such 'sustainable' practices will need to be discussed openly. In the following section, the efficient use of resources, green synthesis, responsible waste management, ethical R&D, ethical clinical translation and business practices, and impact on society and the environment will be presented. Finally, a new model of regenerative circular economy is proposed.

23.1.1 Overview of Sustainable Practices in Regenerative Medicine

Regenerative medicine has adopted sustainable practices in many ways. Tissue engineering, which is a subset of regenerative medicine started with a cyclic approach where tissue from the patient is manipulated in vitro and returned to the patient (Fig. 23.1). There are three main elements in tissue engineering, i.e., the

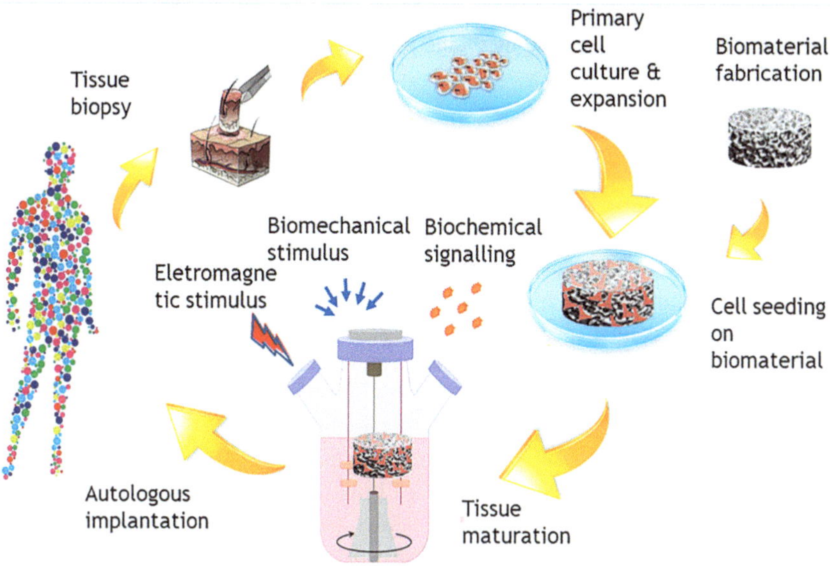

Fig. 23.1 The autologous approach to tissue engineering & regenerative medicine

living cells, biomaterials, and signaling molecules. Thus, sustainable approaches in regenerative medicine include the efficient use of these resources from nature, especially from waste materials, and also the adoption of green processes, if they are to be synthesized. Apart from this the efficient use of materials during the R&D phase, such as in vitro experimentation and preclinical testing especially animal testing should be considered, and these called for careful resource planning and the use of alternatives such as in vitro models or in silico modeling wherever possible. In the clinical translation of regenerative medicine products, reasonable scientific evidence of its efficacy should be obtained before the initiation of clinical applications. Understandably, the clinical outcome may only be appreciated in the longer term due to the required period of regeneration and inter-subject variation, and this should not be taken for granted with the excuse that they are safe anywhere. Hence, collaboration between the industry and academia is warranted to gain deeper mechanistic insight into such treatments and correlate efficacy with patient genetic and epigenetic factors. This is in line with the global initiatives toward precision medicine. A tool or assay may be developed to better predict the clinical outcome. While reimbursement of such treatment may be required even at the early phase of a trial to allow continued R&D in academia, commercialization of products at the early phase will be counterproductive. Often this results in dilution of the real effects and R&D investment. From the medical point of view, it is unethical to provide false hope or hinder patients from receiving more efficacious treatment. Because of this, regulatory control for the commercialization of regenerative medicine products seems inevitable. A sustainable business model will have to consider not only the monetary gains but also the science, ethics, and regulation of the production and administration of regenerative medicine products.

23.1.2 Sustainable Design

A sustainable design approach aims to reduce negative impacts on the ecosystem and environment while ensuring that superior-quality product is produced. Sustainable design approach has been advocated since the rapid industrialization, modernization, and overall improvement in lifestyle across the globe caused the production of industrial and consumer products to increase tremendously, and both raw material extraction and waste product disposal from and to nature, respectively, have been an increasingly critical problem to address. Exploring innovative materials, designs, and fabrication processes that will be sustainable in the long term and enable a circular economy is essential. In terms of sustainable or renewable materials, biomaterials have been at the forefront of biomedical applications as biomaterials are eco-friendly and biodegradable while having suitable properties to be manufactured into a variety of biomedical products for use in diagnostics (biosensors, in-vitro diagnostics, imaging, home care, and wearables), therapeutics (drug delivery vehicles, implants, tissue-engineered tissues or organs, surgical robotics), and assistive (prosthetics, rehabilitation systems applications, mobility aids) applications. Various biomaterials have been derived from natural resources and the long list of biomaterials includes metals, ceramics, starch, alginate, pectin, chitosan, carrageenan, chitin, cellulose, polysaccharides, oligosaccharides, lignin, silk, gelatin, fibrin, hyaluronic acid, collagen, and various bioplastics (Kumar Gupta et al. 2015). Besides, there are various biodegradable polymers such as Polycaprolactone (PCL), Polybutylene succinate (PBS), polylactic acid (PLA), and Polyglycolic acid (PGA), that can be chemically synthesized using monomers obtained from nature or synthetically obtained monomers (Samir et al. 2022). An example of sustainable biomaterial use in biomedical applications is the fabrication of biomaterial-derived hydrogel in wound healing applications. The global market for these products, mainly in the USA and European countries, is estimated to reach USD 16 billion by 2026 (Kumar Gupta et al. 2015). Biomaterials such as collagen and chitosan are suitable as wound dressing material due to their wound healing and antimicrobial properties and at the same time support sustainable production and disposal of the dressing material. As natural biomaterials may lack some properties, especially the low mechanical strength, quick degradation, and low heat stability, various sustainable design approaches can be incorporated into the fabrication of biomedical products that have similar or higher utility compared to the products derived from non-biodegradable materials. The modifications that can be applied include copolymerization, crosslinking, esterification, oxidization, enzymatic treatments, and free-radical polymerization. These modifications enable improved structure, biodegradability, mechanical strength, functionalities, stimuli-responsiveness, biocompatibility, and reduced cytotoxicity (Kumar Gupta et al. 2015).

In the following sections, an elaboration of the sustainable approaches and the associated ethical considerations are presented.

23.1.2.1 Sustainable Cell Sourcing and the Associated Ethical Concerns

A sustainable cell source has been derived from physiological waste such as urine, menstrual blood (Lv et al. 2018), hair, and deciduous teeth. A wide range of tissues can be tapped from discarded tissues from various surgical procedures. Typical abdominoplasty will result in abundant skin and fat tissues being discarded. Fetal tissues from aborted fetuses present an opportunity for deriving almost all cell types that can be used for regenerative medicine, but their supply is limited (Laurent et al. 2020). On the other hand, extra-embryonic tissues such as the placenta and umbilical cord, and amniotic fluid obtained during childbirth represent a source from the ongoing supply. Similar to fetal tissues, surplus fertilized eggs from in vitro fertilization procedures offer the opportunity for researchers to derive embryonic stem cells that can be further differentiated into almost all cell types.

Besides consideration of the sustainability and abundance of the cell sources, the potency of the cells that can be derived from these different sources differs in their potential for differentiation. Pluripotency is defined as the ability of cells to differentiate into all cell types except the germ cells while multipotency is defined as the ability of cells to differentiate into more than one cell type while unipotency is associated with the ability to differentiate into a single cell type. Embryonic stem cells are pluripotent while other adult stem cells are multipotent at best. Notwithstanding that, with the advancement of stem cell technology, cells can be manipulated to transdifferentiate into various cell lineages and to acquire a higher potency. Induced pluripotent stem cells are cells that are transduced with vectors to express pluripotent genes. This technology becomes the gateway to producing almost all cell types for regenerative medicine applications.

Another important point to note is the immunogenicity of the cells derived from the various cell sources, especially those of allogenic origin. Interestingly, mesenchymal stem cells (MSCs) are immune privileged due to their absence in HLA class I protein, low expression in HLA class II protein, and HLA-G and indoleamine 2,3-dioxygenase (IDO) expression. The low immunogenicity also applies to xenogeneic MSCs and fetal progenitor cells. With the advent of genetic engineering, animals such as pigs have been genetically modified to evade human immune detection and hence be farmed as a supply of cells and tissues for regenerative therapies in humans (Sykes and Sachs 2019).

23.1.2.2 Sustainable Material Sourcing and the Associated Ethical Concerns

Biomaterials are used as carriers for cells or drugs and as scaffolds to guide tissue regeneration. In certain instances, they can also serve as tissue spacers to inhibit scar tissue infiltration. Most of the sustainable materials are derived from discarded tissues from human surgery or livestock, food waste, and discarded portion of edible plants, crops, and seafood. Discarded human tissues are often cleaned, decellularized, and sometimes subjected to purification before they can be used as a biomaterial for regenerative medicine. Such a process aims to ensure that immunogens such as cells leave only the ECM. The ECM is the structural component of a tissue where cells reside and is composed of various macromolecules such as collagen, fibronectin, hyaluronic, elastin, laminin, proteoglycans, etc. Subsequent extraction of such ECM may be desirable to produce pure collagen, gelatin, hyaluronic acid, and glycosaminoglycan with consistent physicochemical characteristics (Fauzi et al. 2016; Selvarajah et al. 2020).

There is increasing interest in utilizing decellularized tissue-specific ECM as a ready-made scaffold for specific tissue regeneration. Tissue-specific ECM possesses tissue-specific biochemicals, biophysical, and mechanical properties that modulate cellular behavior, stem cell induction and activate highly regulated signaling pathways toward specific tissue development. The incorporation of micronized ECM into biomaterials has been shown to induce stem cell differentiation specific to the tissue where the ECM was derived (Shridhar et al. 2019).

A comprehensive review of the use of waste-derived materials for tissue engineering was conducted by Ahmad Zamri et al. (Zamri et al. 2021). It is nevertheless important to differentiate waste that has reached the end-of-life cycle from that which is being fed as raw material to another industry. The net value of waste utilization will need to be weighed when considering sustainability. The purposeful cultivation of live forms to substantiate its partial utilization only for one industry is to be avoided. An example of such practices is the harvesting of shark fins and not the utilization of the whole fish. Livestock wastes that are currently untapped or underutilized and can be used for producing biomaterials are feathers, animal hair, animal hooves, and animal blood. Examples of the discarded portion of edible crops or plants are palm kernels, rice husks, peels, and palm kernels can be processed into biomaterials (Shanmuganantha et al. 2022). Examples of food and seafood waste include eggshells, cockle shells, fish scales, fish bone, and the exoskeleton of crustaceans (prawn/shrimp/crab/lobster/snail), etc. (Leonga et al. 2015).

23.1.2.3 Sustainable Biochemical Factors and the Associated Ethical Concerns

Biochemical factors are used in tissue engineering and regenerative medicine for cell differentiation or conditioning and tissue maturation. The common biochemical factors used in tissue engineering are made up of growth factors, transcription factors, steroids, and small molecules. More recently, cytokines and chemokines that can mediate the immune cells are seen as biochemical factors that control the fate of tissue regeneration. These biochemical factors can be costly as some of them are currently manufactured using recombinant technology or chemically synthesized. However, the below have been explored as sustainable alternatives.

23.1.2.3.1 Secretomes

Co-culturing has long been practiced, as it is recognized that cell-cell communication can control cell fate. Secreted proteins or secretomes derived from the conditioned medium are usually discarded during the cell culturing process and can be used as a rich source of biochemical factors for controlling cell fate.

23.1.2.3.2 Extracellular vesicles

Extracellular vesicles (EVs) are lipid-membrane vehicles secreted by the cell that carry cargo such as proteins, peptides, and microRNAs for cell-cell communication. Exosomes are EVs in the 40–200 nm range and are found to be enriched with cargo that exerts significant therapeutic effects for regenerative medicine. These extracellular vesicles in micro and nano dimensions can be isolated from conditioned media and used for controlling cell fate. Nevertheless, the cost of isolating the EVs will have to be weighed against its effect on the crude conditioned media. Both secretomes and EVs have been explored as cell-free therapeutics in regenerative medicine and even in advanced cosmetics. The rampant use and availability of such secretomes and EVs as bi-products of cell culture are alarming. There is a possibility that animal products such as fetal bovine serum and growth factors are present in the conditioned medium. An important consideration is whether the benefit of using such cosmetic products is worth the risk and costs are borne by the user. As the regulation of cosmetic products is largely self-policing, a declaration of components in cosmetic products may not be accurate. Hence, the use of such biologics in cosmetics will need to be regulated separately as advanced cosmetic products.

23.1.2.3.3 Expired Human Platelet Lysate and Blood Components

Blood contains a variety of biochemical factors that support life. Blood has already been fractionated into various components, e.g., plasma, serum, and platelet lysate to support clinical treatment. A myriad of growth factors is present naturally in the plasma and serum. Fractionated blood such as human platelet lysate has a very short half-life. Upon expiry, this fractionated blood can be tapped for regenerative purposes. These blood components if procured via the National Blood Bank Programs would have undergone vigorous testing to minimize the risk of disease transmission and that consent has already been obtained from the donor for distribution of their blood products. The ethical concern in procuring expired blood components is the need for clear separation of entities for managing in-use blood from that for expired blood to mitigate the conflict of interests between the two entities. The quality system should be maintained by both entities to ensure product purity, integrity, and traceability.

23.1.2.3.4 Plant Extract

Plant extracts have been traditionally used as supplements for health. They often contain polyphenols and other biochemically active compounds that can stimulate cell growth and control cell fate. Although this represents a natural resource, the amount of plant material used to derive these biochemical factors and the cost of manufacturing will need to be considered. The ethical concerns for using the above natural resource are to ensure that food stock is prioritized before its use for other purposes.

23.1.2.3.5 Biophysical/Biomechanical and Electromagnetic Stimulation

Alternatively, cell conditioning and differentiation can be achieved via biomechanical or electromagnetic stimuli (Li et al. 2021; Safavi et al. 2022). Mechanical or physical stimulation, including surface topography, pore structure, dimension, hydrostatic pressure, etc., and electromagnetic stimulation can activate many intracellular signaling pathways and alter the intracellular microenvironment to affect cell migration, cell proliferation, and cell differentiation (Chen et al. 2019). The fine-tuning of such approaches will be required to control cell fate more accurately. This appears to be a sustainable solution to the high cost of using biochemical stimulation.

23.1.2.4 Sustainable Processing/Fabrication Methods and Its Associated Ethical Concerns

The natural materials sourced from waste would need to be cleaned, processed (decellularization or extraction), fabricated, and sterilized to be used as scaffolds for tissue engineering. The concept of sustainable processing or fabrication methods requires careful planning to reduce the amount of energy and toxic waste that poses environmental threats. In recent decades, green technologies have revolutionized many traditional manufacturing industries. The initiative to adopt green synthesis technologies in biomaterial fabrication is in its infancy. A review by Jahangirian et al. (2018), discusses the adoption of such methods in the manufacturing of scaffolds for tissue engineering intended for different tissues or organs. Transition to green synthesis primarily focuses on being energy efficient, clean waste, using natural starting material, and substitution of safer and recyclable solvents such as water, ethanol, ethyl acetate, etc. (Bedlovičová 2022).

Supercritical fluid technology is touted as a green technology that has transformed industries across various fields such as the oil and gas industry, food, pharmaceuticals, textile dyeing, etc. (de Aguiar et al. 2014). The use of supercritical carbon dioxide has substituted toxic solvents. Tissues can be delipidated and decellularized leaving the ECM as a scaffold for regenerative medicine (Wang et al. 2017). The ethical consideration for the adoption of such green technology is whether the talent and technology can be made available locally. Global collaboration for such technology to be deployed efficiently will be required.

23.1.3 Ethics of Translation and Commercialization of Regenerative Medicine Products

The clinical use of regenerative medicine products should be overseen by the Institutional Review Board for adherence to Good Clinical Practice Guidelines. The manufacturing and distribution of these products should be regulated by the National Regulatory Agency to ensure the safety, purity, and efficacy of the production and the adherence to good distribution practices across its supply chains. The ethical concerns revolving around the commercialization of regenerative medicine products make its manufacturing cost to be and making regenerative medicine products inaccessible to many. Such inequity has been criticized as most of these products had their early research and development either directly or indirectly funded by the public and carried out or benefited from research in academia and public research institutions.

23.1.3.1 Establishment of Cell and Tissue Banks and the Associated Ethical Concerns

Tissue banking has its initiatives traced back to the 1990s along with the successful launching of tissue and organ transplantation programs. Now there are over 100 tissue banks established across the globe. Tissue banks were initially operated as non-profit organizations with oversight by the Ministry of Health; however, due to the complexity of the supply chain, from tissue recovery to processing, storage, and distribution, many private entities have been engaged to partake in certain activities along the supply chain. Subsequently, private tissue banks such as cord blood banks have emerged especially in the USA including Malaysia. The common tissues that are banked to date are skin, bone, cornea, tendon, ligaments, nerves, vein, arteries, and heart valve due to the expansion of clinical use. Unlike organs that will expire within hours from their recovery from the donor's body, tissues can be processed and kept for months or even years.

From the nonexhaustive list of cell sources above, there have been public and private initiatives to establish cell establishments (cell banks) to facilitate the processing, storage, and distribution of these cells. The associated ethical concerns will need to be quickly addressed and guidelines lay before its global adoption. The Good Tissue Practice (GTP) guideline developed by the US Food & Drug Administration has been around since 2005 and has been adopted in many countries. GTP requirements include requirements for facilities, environmental control, equipment, supplies and reagents, recovery, processing, and process controls, labeling controls, storage, receipt and distribution, and donor eligibility determinations, donor screening, and donor testing (GTP Guideline, National Pharmaceutical Regulatory Agency 2015). The three most important criteria in the sourcing of such tissues and cells especially from humans are the robust review of the procedure by a well-trained ethics committee, written consent taking, and confidentiality on the part of the donor. While the cell banking facility should be managed, operated, maintained, and audited by an oversight body. In practice, the GTP guidelines may have adhered loosely as enforcement may not be in place in certain countries. For-profit-private cell banking is especially alarming and may not have been subjected to sufficient scrutiny in an environment where enforcement is weak.

Inevitably, for-profit cell banks will be answerable to their investors and will be pressured to maintain a revenue stream for the company. Hence, there is a conflict of interest on the part of the company when marketing such services to their clients. There is always an emotional undertone, giving a sense of urgency and exaggeration of its future potential against current scientific evidence.

On the other hand, public establishments in some countries may not have the resources to maintain such facilities. Hence, investors argued that there is a need to provide this option to the public in view of its promise and that the public should be given the right to try it. The truth is that many such regenerative therapies are in the early phase of research. More studies are required to calculate the real benefits of such therapy against its risk. Also, a more precise prediction of the clinical outcome will be required to justify the cost of such therapies and to delay access to other established therapies or to even put such resources (cost for therapy) to better use.

The Swiss model of Fetal Progenitor Cell Banking program has been purported as a holistic and sustainable approach for regenerative applications. The concept of qualifying one single fetal organ donation is sufficient to derive various cell types that can be cryopreserved, ultimately reducing multiple resource constraints, time, and costs. Such a model has also been successfully applied to ovine and equine fetal progenitor cells for veterinary use. The implementation of such a program around the globe will require engagement with the community, religious leaders, and regulators. As organ donation programs are slowly gaining acceptance in society, there is a need for the government and public academia to drive this awareness.

23.1.4 Ethical Flaws of Traditional Economic Model

The traditional economic model is linear based on Gross Domestic Product (GDP) growth is ethically flawed and has been viewed as a "take, make, use, dispose" or "from cradle to grave" approach. Such economic development is achieved at the expense of the environment and the exploitation of limited natural resources. As alternative, the circular economy development model establishes the principles of "Reduce, Reuse, and Recycle" as its core contents to achieve sustainable development, which has been viewed as a "from cradle to cradle" approach. Circular economy aims to foster a new ethical relationship, and proposes a new model of ethical production and consumption behavior (Qiao and Qiao 2013).

The concept of a circular economy was proposed by K. Boulding (1966), an American economist, in the wake of environmental degradation. The circular economy aims to transform the organization of the social system, natural resources and research and technology development, product manufacturing and consumption, and waste management, into an economy whose development depends on the recycling of ecological resources. Hence, the backbone of the circular economy is based on the "3R" principles, namely "reduction", "reuse", and "recycle". "Reduction" aims to reduce the amount of materials used for production and consumption; "Reuse" aims to extend the period of the product life cycle and maximize its value; "Recycle" aims to turn wastes into resources again (Qiao and Qiao 2013). In short, the circular economy organizes economic activities into a feedback process, namely "resources–products–renewable resources", so that resources can be reasonably recycled to protect the environment, reduce pollution, and achieve sustainable economic and social development. A life-cycle assessment of every product created will need to be carried out before any development endeavor.

Based on the circular economy concept, the below model (Fig. 23.2) is proposed for an ethical and sustainable approach to regenerative medicine. There should also be an emphasis on sourcing the required materials or resources from local sources and an interdependency and mutual benefit on a social obligation basis. In this model, the local industries will benefit from waste disposal costs which should be borne by the tissue processing and banking establishments. In organ donation, the donor will not benefit directly from this model but is motivated by conscience and a sense of social contribution. The circular economy promotes a holistic and

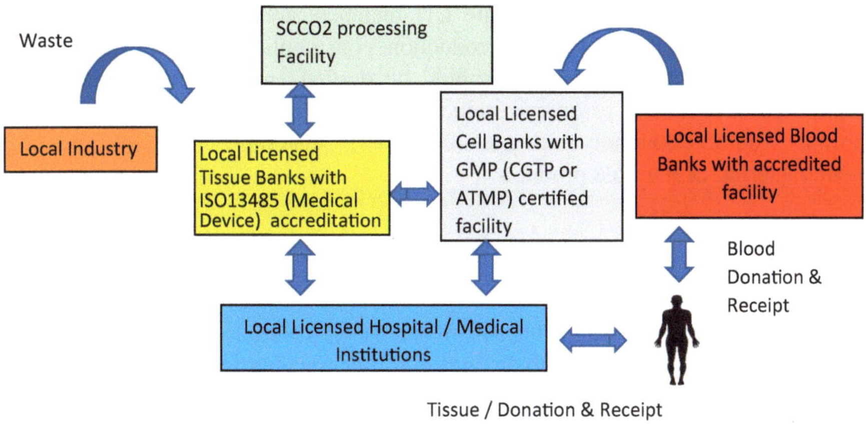

Fig. 23.2 A model for translation of regenerative medicine based on circular economic principles. *$SCCO_2$* supercritical carbon dioxide, *GMP* good manufacturing practice, *GTP* good tissue practice, *GDP* good distribution practice, *CGTP* cell and gene therapy products, *ATMP* advance therapy medicinal products

sustainable value principle, which can remedy the ethical flaws of the traditional linear model. However, such a model can only work where equity and fairness can be upheld not only by the legal framework and the national social system but in the conscience and morals of the people. Hence, an awareness program needs to be conducted and inequality and gaps in the social system need to be addressed.

23.2 Regulatory Considerations for Sustainable Materials in Biomedical Applications

Conventionally, the legislation, regulations, and formal guidelines to tackle the issue of sustainable materials in biomedical applications are mostly from biomedical waste or biohazardous waste management perspectives. This strategy mainly reduces or minimizes the environmental and sociological impact caused by biomedical waste. More recently, policies and regulations on the utilization of sustainable materials, which enable the manufacturing of a biomedical product that is manufactured without depleting non-renewable resources, produces less waste or zero waste, and is ecofriendly, is getting attention and gaining momentum.

In 2015, 193 members of the United Nations (UN) unanimously adopted the Sustainable Development Goals (SDGs), with the main agenda to ensure the sustainability of mankind by promoting sustainable development of the main three pillars of livelihood: economy, society, and environment. A minimum of 12 UN SDGs are directly or indirectly related to plastic and microplastic pollution and waste reduction (Walker 2021). It is of utmost importance to address issues that risk the environment, such as climate change, and also sustainability of natural resources.

With the adaptation of SDG, all the UN members are expected to reduce and eradicate unsustainable patterns of production, consumption, and waste generation. Guidelines, regulations, and laws should be drafted and adopted to reduce the consumption of natural resources, reduce the production of waste, especially hazardous and/or slow/non-biodegradable waste materials, and increase the manufacturing of reusable products.

23.2.1 The Circular Economy

The rise of issues of unsustainable design and use of materials and technologies that are energy intensive, produce toxic and/or non-degradable waste, and also release greenhouse gases has had a detrimental impact on the environment and also causes climate change. One such example is the production of plastic from fossil fuel, the manufacturing of a variety of industrial and consumer products at low cost using plastic, and the generation of an enormous amount of plastic waste. As single-use plastic products cannot be reused for an extended period and are affordable, they are often widely used as convenient and cheap products that often end up in landfills after a single use. This scenario is further worsened by the fact that plastic does not degrade quickly, end up piling up in landfills, and also filling up many rivers and ocean. The degradation products of various plastic materials are toxic to the ecosystem and environment.

Thus, the linear economy of the "buy, use, and throw" paradigm pattern needs to be changed to a circular economy that involves the design and manufacturing of products that last longer using sustainable materials and the products should be able to be reused, repaired, and remanufactured or recycled. By adopting sustainable design and a circular economy we will be able to reduce the consumption of raw materials, reduce energy usage, and also reduce waste generation. This, in turn, will minimize the detrimental effect on the environment and ecosystem while increasing the raw material and energy utilization efficiency.

Healthcare sectors produce a significant amount of waste, mainly due to single-use devices and products that are widely used in this sector to address patient safety and infection control. However, with the onset of COVID-19, healthcare waste has reached an all-time high due to the compulsory use of face masks, self-test kits, personal protection equipment (PPEs), and many other single-use products to avoid COVID-19 transmission, by patients and also the healthcare workers. It is reported that in March 2020, at the early onset of the COVID-19 pandemic, 1.6 million protective goggles, 76 million gloves, and 89 million face masks were used monthly worldwide. It was projected that gloves and facemasks amounting to 200 billion to be discarded every month during this period. This scenario can imply the burden posed to the environment due to healthcare waste produced during the COVID-19 pandemic alone, without taking other biomedical waste produced globally into consideration (Aragaw and Mekonnen 2021). Thus, it is of utmost importance to study, understand and move toward adopting sustainable designs and biomaterials in manufacturing healthcare devices and products that can contribute toward circular economy practices as much as possible.

A circular economy can be generated when the waste produced can be reduced and the waste itself can be recycled or repurposed into a product with economic value. A product becomes waste when it loses its value due to factors such as:

- Physical uselessness: the goods that have been damaged beyond economic restoration
- Functional uselessness: the goods are no longer required
- Technical uselessness: the goods are outpaced by the latest technology
- Economic uselessness: the goods are no longer economically beneficial
- Legal uselessness: the goods are made unlawful due to changes in law and regulations
- Desirability uselessness: the goods are no longer appealing to their users (Parnell n.d.)

To create and maintain a circular economy, the loss of value of the products must be minimized, while the obsolete products are recovered, restored, and reused for the same or different purpose as it is initially produced. It is a common perception that medical waste cannot be reused or repurposed due to safety reasons. However, according to the World Health Organization (WHO) report in 2017, up to 85% of the total amount of waste produced by healthcare providers was similar to household waste, and it can be classified as 'non-hazardous' or 'general healthcare waste'. Mostly administration departments, kitchens, and housekeeping tasks in healthcare facilities produce this kind of waste which includes packaging and construction waste too.

The European Commission (EC) drafted and adopted the Circular Economy Action Plan (CEAP) in 2015 to create more jobs, boost the economy, and importantly develop a carbon-neutral, resource-efficient, and competitive economy. Among key strategies to implement the circular economy plan was implementation of eco-design and energy labeling that includes regulations on materials efficiency such as readiness of spare parts, easiness of restoration, and enabling end-of-life treatment; empowering consumers by creating awareness on the environmental impact of products, organizations, and services throughout their lifespan, which enable consumers to make informed decision based on the eco-friendly nature of the products; revise waste legislation that includes increased recycling rate, reinforced rules and obligation on separate collection of waste types, extended producer responsibility (EPR) whereby the manufacturers will be accountable for the entire life span of the goods (from product entry into market till waste collection and recycling), and strengthening waste prevention and waste management procedures; introduction of Fertilizing Products rules that promote use of organic fertilizers generated using various production waste such as agricultural by-products and recovered bio-waste; and lastly regulations on disposable plastics items and fishing gear that ban the use of disposable plastic products made of petrochemicals and oxodegradable plastics, introduce steps to decrease food containers,

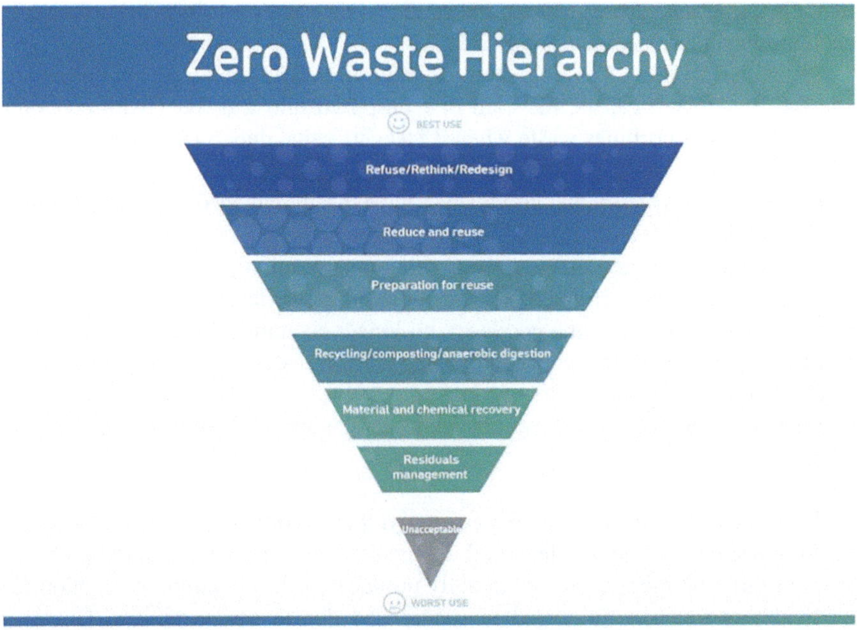

Fig. 23.3 Zero waste hierarchy-zero waste goal is a restorative and regenerative economy that keeps products, materials, and components at their highest utility and value at all times—a circular economy with no waste. (Reproduced with permission from Europe 2019)

incorporation of recycled plastic in beverage bottles and also reduce littering from ships and fishing gears. Zero Waste Europe, a network that connects and supports 35 local and national non-governmental organizations (NGOs) that work toward a sustainable environment and circular economy by adopting a zero waste policy, strongly advocates that all resources must be conserved and waste must be reduced to its core (Fig. 23.3).

23.2.2 Waste Management

Besides the COVID-19 pandemic, expanding healthcare sectors due to an aging society and also higher awareness of healthcare are also expected to contribute to the rise of healthcare waste production. Currently, the most common method to treat hazardous medical waste is by incineration and various undesirable compounds such as dioxins, furans, and heavy metals are released into the environment after this process (Kheirabadi and Sheikhi 2022). Another common method of hazardous medical waste treatment, autoclaving, poses the problem of the bulky appearance of the waste, which requires further treatment as disposing of autoclaved waste may need a bigger landfill area and may also release toxic compounds into the land and water. One sustainable alternative waste management will be reprocessing and reusing biomedical materials. One popular example will be the waste generated by the soaring use of single-use PPEs during the COVID-19 pandemic. A study by Jinia et al. found that ionizing radiation at doses less than 10 kGy sterilizes PPE for re-use at hospitals, including those handling patients with infectious diseases (Jinia et al. 2020). Another study found that both the gamma and electron beam ionizing radiations can sterilize FFP2 and FFP3 respirators, which are commonly used by healthcare workers in airborne infectious disease sectors, with condition that charge on the sterilized respirators is reapplied by conventional methods. This method can sterilize the respirators up to two times for safe reuse of the respirators (Pirker et al. 2021). The next option in medical waste management is recycling, recovering, and repurposing biomedical materials. As plastic is most commonly used in medical instruments and disposables, plastic-based medical waste recycling plays an important role in moving toward sustainability in biomedical products. Mechanical recycling of plastic-based biomedical waste is considered the more sustainable approach. Due to the tremendous increase in PPEs usage during the past few years, there has been more effort to recycle these items using mechanical and chemical methods to repurpose them in different industries. One such example is recycling PPEs into construction materials. A study by Mohan et al. (2022) showed that composite materials made of PPE waste and sand display enhanced performance in flexure, compression, and tension compared to other building materials like mud, bricks, and cement blocks. Furthermore, the composite material was also found to have minimal degradation and water absorption, which did not affect its strength after exposure to adverse conditions. These properties render them as suitable as construction material, especially as thermal insulation material in tropical countries. Another study by Koniorczyk et al. (2022), also reported a similar finding on the performance of concrete produced by mixing 2 g of fibers obtained from recycling face masks with 1 L of concrete and this proposed method found to have high recycling capacity. There are also reports on shredded face masks (SFM) that were added to hot mix asphalt (HMA) can increase the resistance to rutting and demonstrated exceptional permanent deformation resistance of the asphalt pavement, besides the SFM functioning as binder of the HMA aggregates.

23.3 Conclusion and Future Prospects

There is sufficient evidence that sustainable practices can be implemented in stages in the field of regenerative medicine ranging from sustainable material sourcing and green synthesis to zero-waste adoption. While the change may require time for the existing industrial player, a circular economic model for the regenerative medicine industry should be strategically planned during the research and product development phases through its implementation phase in the clinic, and then at the upscaling phase for commercial purposes. Funding agencies can set criteria for sponsoring projects that adhere to sustainable practices.

At the national level, an oversight committee to oversee the implementation of sustainable practices should be set up. Relevant government regulatory bodies such as the Ministry of Science and Environment, Ministry of Health, and Ministry of Trade and Consumer Affairs must convene to discuss the regulatory strategies and provision of incentives to promote the adoption of such circular economic models at all levels and to iron out barriers to their successful adoption. Guidelines and policies that provide an ethical framework for its implementation at academic institutions, government agencies, medical establishments, and industries should be made publicly available as soon as possible. An ongoing discourse with all stakeholders including civil society, on the social and economic impact of such practices will be imperative to address evolving and emerging issues from time to time.

Above all, such awareness of sustainability needs to be increased at all levels of society by creating knowledge on the effect of human activities that cause climate change, global warming, pollution, etc. Action to increase the ethical and legal responsibility of the citizens of the world should be carried out with a unified effort by all countries, maybe under the framework of the UN and SDG. Awareness of sustainability, the desire to lead a sustainable lifestyle or business practice, and a regulatory framework that limits harm to the environment and reduces the use of natural resources, will lead to a sustainable practice that becomes a norm, including in the healthcare and biomedical sector.

Acknowledgment This study was supported by the Fundamental Research Grant Scheme (FRGS/1/2020/SKK0/UKM/02/7) and the Grand Challenge Grant (DCP-2017-001/3).

Conflict of Interest Authors declare no conflict of interest.

References

de Aguiar AC, Martínez J, dos Santos P (2014) Supercritical fluid extraction as a green technology. In: Michalos AC (ed) Encyclopedia of quality of life and well-being research. Springer, Dordrecht, pp 6475–6478. https://doi.org/10.1007/978-94-007-0753-5_4218

Aragaw TA, Mekonnen BA (2021) Current plastics pollution threats due to COVID-19 and its possible mitigation techniques: a waste-to-energy conversion via pyrolysis. Environ Syst Res 10(1):1–11

Bedlovičová Z (2022) Green synthesis of silver nanoparticles using actinomycetes. In: Green synthesis of silver nanomaterials. Elsevier, Amsterdam, pp 547–569

Chen C, Bai X, Ding Y, Lee I-S (2019) Electrical stimulation as a novel tool for regulating cell behavior in tissue engineering. Biomater Res 23(1):1–12

Europe ZW (2019) Press release: a zero waste hierarchy for Europe. https://zerowasteeurope.eu/press-release/press-release-a-zero-waste-hierarchy-for-europe/

Fauzi M, Lokanathan Y, Aminuddin B, Ruszymah B, Chowdhury S (2016) Ovine tendon collagen: extraction, characterisation and fabrication of thin films for tissue engineering applications. Mater Sci Eng C 68:163–171

Jahangirian H, Lemraski EG, Rafiee-Moghaddam R, Webster TJ (2018) A review of using green chemistry methods for biomaterials in tissue engineering. Int J Nanomedicine 13:5953

Jinia AJ, Sunbul NB, Meert CA, Miller CA, Clarke SD, Kearfott KJ, Matuszak MM, Pozzi SA (2020) Review of sterilization techniques for medical and personal protective equipment contaminated with SARS-CoV-2. IEEE Access 8:111347–111354

Kheirabadi S, Sheikhi A (2022) Recent advances and challenges in recycling and reusing biomedical materials. Current Opin Green Sustain Chem 38:100695

Koniorczyk M, Bednarska D, Masek A, Cichosz S (2022) Performance of concrete containing recycled masks used for personal protection during coronavirus pandemic. Constr Build Mater 324:126712
Kumar Gupta G, De S, Franco A, Balu AM, Luque R (2015) Sustainable biomaterials: current trends, challenges and applications. Molecules 21(1):48
Laurent A, Hirt-Burri N, Scaletta C, Michetti M, de Buys Roessingh AS, Raffoul W, Applegate LA (2020) Holistic approach of Swiss fetal progenitor cell banking: optimizing safe and sustainable substrates for regenerative medicine and biotechnology. Front Bioeng Biotechnol 8:557758
Leonga LM, Sahalana AZ, Tana LH, Mustafaa NH (2015) Clarias batrachus collagen extract increases fibroblast cell adhesion, migration and proliferation. J Appl Pharmaceut Sci 5(3): 019–023
Li J, Liu Y, Zhang Y, Yao B, Li Z, Song W, Wang Y, Duan X, Yuan X, Fu X (2021) Biophysical and biochemical cues of biomaterials guide mesenchymal stem cell behaviors. Front Cell Dev Biol 9:640388
Lv H, Hu Y, Cui Z, Jia H (2018) Human menstrual blood: a renewable and sustainable source of stem cells for regenerative medicine. Stem Cell Res Ther 9(1):1–11
Mohan HT, Jayanarayanan K, Mini K (2022) A sustainable approach for the utilization of PPE biomedical waste in the construction sector. Eng Sci Technol Int J 32:101060
National Pharmaceutical Regulatory Agency (2015) Good tissue practice guideline. 2nd ed. Ministry of Health Malaysia. https://npra.gov.my/images/Guidelines_Central/Guidelines_on_Regulatory/GOOD_TISSUE_PRACTICE_MALAYSIA.pdf
Parnell H (n.d.) Using biomedical materials within a circular economy. https://www.ansys.com/blog/biomedical-materials-circular-economy. Accessed 18 Oct 2022
Pirker L, Krajnc AP, Malec J, Radulović V, Gradišek A, Jelen A, Remškar M, Mekjavić IB, Kovač J, Mozetič M (2021) Sterilization of polypropylene membranes of facepiece respirators by ionizing radiation. J Membr Sci 619:118756
Qiao F, Qiao N (2013) Circular economy: an ethicaland sustainable economic development model. Prakseologia 154:253–272
Safavi AS, Sendera A, Haghighipour N, Banas-Zabczyk A (2022) The role of low-frequency electromagnetic fields on mesenchymal stem cells differentiation: a systematic review. Tissue Eng Regen Med 19:1147–1160
Samir A, Ashour FH, Hakim A, Bassyouni M (2022) Recent advances in biodegradable polymers for sustainable applications. NPJ Mater Degrad 6(1):1–28
Selvarajah J, Mh Busra MF, Bin Saim A, Bt Hj Idrus R, Lokanathan Y (2020) Development and physicochemical analysis of genipin-crosslinked gelatine sponge as a potential resorbable nasal pack. J Biomater Sci Polym Ed 31(13):1722–1740
Shanmuganantha L, Khan MUA, Sulong AB, Ramli MI, Baharudin A, Ariffin HM, Abd Razak SI, Ng MH (2022) Characterization of titanium ceramic composite for bone implants applications. Ceram Int 48:22808
Shridhar A, Amsden BG, Gillies ER, Flynn LE (2019) Investigating the effects of tissue-specific extracellular matrix on the adipogenic and osteogenic differentiation of human adipose-derived stromal cells within composite hydrogel scaffolds. Front Bioeng Biotechnol 7:402
Sykes M, Sachs DH (2019) Transplanting organs from pigs to humans. Sci Immunol 4(41): eaau6298
Walker TR (2021) (Micro) plastics and the UN sustainable development goals. Current Opin Green Sustain Chem 30:100497
Wang JK, Luo B, Guneta V, Li L, Foo SEM, Dai Y, Tan TTY, Tan NS, Choong C, Wong MTC (2017) Supercritical carbon dioxide extracted extracellular matrix material from adipose tissue. Mater Sci Eng C 75:349–358
Zamri MFMA, Bahru R, Amin R, Khan MUA, Abd Razak SI, Hassan SA, Kadir MRA, Nayan NHM (2021) Waste to health: a review of waste derived materials for tissue engineering. J Clean Prod 290:125792

MIX
Papier aus verantwortungsvollen Quellen
Paper from responsible sources
FSC® C105338

If you have any concerns about our products,
you can contact us on
ProductSafety@springernature.com

In case Publisher is established outside the EU,
the EU authorized representative is:
Springer Nature Customer Service Center GmbH
Europaplatz 3, 69115 Heidelberg, Germany

Printed by Libri Plureos GmbH
in Hamburg, Germany